Advances in Preventing and Treating Violence and Aggression

Series Editor
Peter Sturmey, Queens College and The Graduate Center
City University of New York
Flushing, NY, USA

The series publishes books focused and developed across three domains. The first is understanding and explaining violence and aggression. Books in this domain address such subject matter as genetics, physiology, neurobiology, cultural evolution, biobehavioral, learning, cognitive, psychoanalytic, sociological and other explanations of violence. The second domain focuses on prevention and treatment for individuals and couples. Examples of books in this domain include cognitive behavioral, behavioral, counseling, psychopharmacological, psychosocial, couples, and family therapy approaches. They also explore extant treatment packages for individually focused treatments (e.g., mindfulness, cognitive analytic therapies). Within this domain, books focus on meeting the information needs of clinicians and professionals who work in youth facilities, emergency rooms, special education, criminal justice, and therapy settings. Finally, books in the third domain address prevention and treatment for groups and society, including topical focus on early intervention programs, school violence prevention programs, policing strategies, juvenile facility reform as well as socio-legal and ethical issues. Books in this series serve as must-have resources for researchers, academics, and upper-level undergraduate and graduate students in clinical child and school psychology, public health, criminology/criminal justice, developmental psychology, psychotherapy/counseling, psychiatry, social work, educational policy and politics, health psychology, nursing, and behavioral therapy/rehabilitation.

Peter Sturmey

Violence and Aggression

Integrating Theory, Research, and Practice

Springer

Peter Sturmey
City University of New York
Queens College City University of New York
Flushing, NY, USA

Advances in Preventing and Treating Violence and Aggression

ISBN 978-3-031-04385-7   ISBN 978-3-031-04386-4 (eBook)
https://doi.org/10.1007/978-3-031-04386-4

This Springer imprint is published by the registered company Springer Nature Switzerland AG
The registered company address is: Gewerbestrasse 11, 6330 Cham, Switzerland

To Dr. Michael "Chael" Scott Needle, Ph.D.

Preface

The aim of this volume is to provide an overview of the broad field of aggression and violence taking a functional approach that acknowledges the evolutionary, cultural, and operant causes of aggression and violence (Skinner, 1953). It will address the definition and nature of different forms of aggression and violence; epidemiology and risk factors; biological, cultural, and individual causes; and biological, cultural, and individual approaches to prevention and treatment. Finally, it will conclude that we can be cautiously optimistic regarding progress in reducing aggression and violence.

Aggression and violence are common in both nonhumans and humans, are costly in terms of time and energy, are risky to the aggressor, and involve lost opportunity to engage in other apparently more productive behavior, such as feeding, sexual behavior, and rearing offspring. Aggression and violence come at an enormous cost to humans in terms of the personal suffering of victims, perpetrators, relatives, and friends, and society generally. The economic costs are enormous and include prevention in families, schools and society generally; treatment and amelioration; and the justice, police, and incarceration systems, just to name a few. Even when we are not victims of violence directly, we all pay a price through increased taxes and insurance costs, and diversion of public and private money away from other programs and services. It also comes at huge costs in terms of lost opportunities: The time and energy invested in aggression and violence, and their prevention and treatment, could be invested in many other more productive activities. The time and money spent on policing, jails, and the justice system could be invested in building a better world by more investment in education, healthcare, and protecting the environment, for example.

A Functional Approach to Aggression and Violence

This volume takes a functional approach to explaining and changing aggression and violent behavior. This functional approach assumes that organisms engage in aggressive or violent behavior—even extreme and apparently pathological forms of aggression and violence—for a very good reason. These behaviors are in some sense useful, at least in some circumstances. Aggression and violence are caused by three selectionist processes: Biological, cultural, and operant selection. All three are conceptually parallel as they all involve variations in behavior and other features of the organism, such as morphology, physiology etc., and selection of some variations but not others by the environment (Skinner, 1953, 1990).

Preservation of the organism while under threat is an obvious function of aggression and violence. When the organism is under physical threat from extreme nonsocial or social stimulation not only does the organism withdraw from but it also lashes out to remove the threatening stimulus. This is true for both threatening physical and social stimuli. Thus, it is unsurprising that aggression and violence in both nonhumans and humans do indeed seem to have a substantial genetic component which is related to the unchanging aspects of the environment. Even in the most extreme examples, aggression and violence cannot be entirely biologically determined: The shape and texture of flesh and the victim's behavior all select, shape, and progressively refine the bite and attack of the aggressor. An evolutionarily well-designed organism will also benefit if it is well prepared to learn to avoid stimuli that predict threats to its integrity quickly and efficiently. Thus, organisms are prepared to quickly learn to avoid both the increasing heat as one approaches the hot pan on the stove before one is burned and also social stimuli such as the tiger rushing toward it before one may be eaten. When organisms are biologically prepared to learn these important behavioral repertoires faster and more efficiently, not only are these effective behaviors more likely to occur in the future within this organism, but also organisms and their genes will be more likely to survive for longer, will be able to contribute to the survival of their culture, and their genes will be more likely to survive and propagate into future generations.

Possibly the most important aspect of selectionist accounts of aggression and violence is selection of aggression and violence or their alternatives during the lifespan. One aspect of the environment which selects behavior during the lifespan is the physical environment, but much more important is the selecting social environment and the interlocking cultural contingencies that shape social behavior of many members of a culture. The cultural practices embedded in our families, schools, work, law enforcement, religion, and other social agencies take what our biology gives and then teach us to be nonviolent or violent during our lifespan.

A functional account of aggression and violence, especially in the area of operant learning, is challenged by the fact that most humans are quite familiar with their own and other people's behavior. They often feel or learn from others that one's mind or brain initiates action, and that the alleged, autonomous self choses to be violent or nonviolent. Recognizing environmental control of one's own behavior is often seen as demystifying or even dehumanizing, degrading, or disrespectful of the autonomous person. Much of this is embedded in our everyday language about behavior when we say "I felt angry, so I decided to hit her." Saying, "the probability of an aggressive response of waving a fist in the other person's face was increased by the establishing operation of sleep deprivation, the discriminative stimulus of an insulting personal slur and positively reinforced by observing the person wince, and cry, and negatively reinforced by the reduced probability of future insults" has less curb appeal than the language of the autonomous person. We might not say that in most circumstances any more than we might say "Pass the sodium chloride" to one's grandmother at Thanksgiving unless she is a chemist (Leigland, 1998), but to say "pass the salt" does nothing to deny or disconfirm the chemistry of the food we love to eat.

Skinner (1990) pointed out that the vernacular, everyday language of the household is full of words related to thinking and feeling, but that these words really refer to recent learning history and current reinforcers. When we say "I am hungry and I think I will get something to eat," we are really saying that "I have been deprived of food for a long time" and "If you give me some food, I will probably eat it" (Skinner, 1990). Such statements indicate recent histories of reinforcer deprivation and satiation and current reinforcers and punishers among other things. We can make the analogy with a statement such as "I feel angry, I could kill you right now." What we are really saying is something like "I have recently experienced aversive stimulation"—which could be aversive social interactions, high temperature, physical pain, goal frustration, etc.—and "If you or some other potential victim or a suitable object comes near, I will probably hit these people or objects to obtain certain consequences." Engaging in such aggressive or violent behavior often has the effect of removing the relevant aversive stimulation as when someone brings a cold drink, or turns on the air conditioning to cool down the aggressive person or the irritating person removes themselves. Thus, the vernacular, whether the language of everyday conversation, a therapy session, or of cognitive (neuro)science, describes recent histories and contingencies of reinforcement which renders reference to the initiating self unnecessary (Skinner, 1990) including when explaining aggression and violence.

Many people find such models of human behavior distasteful. This is often because they do not understand radical behaviorism and make false assumptions out of ignorance. Some critics may indeed understand radical behaviorism, but find it to be distasteful because it seems to take away our self-determination. In fact, in an important sense the very opposite is true. Understanding the environmental control of one's own behavior allows one to take action and determine how that happens and to then arrange the environment to make it easier to do the things one wants to do and to stop doing the things one does not want to do (Epstein, 1997). Behavioral approaches to self-management are directly related to each of us managing our own behavior on a day-to-day basis, for example, making it easier and more likely to control our emotional behavior, being more effective in solving life's problems, and avoiding temptations and the triggers for aggression and violence. This approach often involves conducting a functional assessment of one's own behavior and using that to develop a self-management plan which often involves goal setting and self-recording, making desirable behavior easier to engage in, and making undesirable more difficult, among other strategies. If we know we engage in angry behavior when our relatives irritate us during family occasions we can plan accordingly: We can cancel the trip, we can practice speaking up for ourselves, we can excuse ourselves from some parts of the event, we can look away from the annoying other person, we can ask the other person to behave differently, and we can reinforce their nonaversive behavior. We can control our own behavior by changing our environment to make prosocial, nonaggressive behavior more likely to happen. We can set a reminder on our watch or ask our significant other to make an excuse to leave early the irritating family occasion. Knowledge of the environmental control of our own behavior makes it more likely that we actively manage how we behave. It does not make us the passive recipient of stimuli. Rather it is the person without such a knowledge of the environmental control of one's own behavior and who lacks the relevant self-management skills who is a rudderless ship at sea without self-control (Epstein, 1996). Accounts of the self-managed lifestyle address many aspects of our own behavior that are very relevant to aggression and violence including managing social relationships, friendships, working in groups effectively and harmoniously, and managing mental health issues such as depression and stress (Williams & Long, 1979).

Behavioral self-management is also directly related to many interventions used in mental health interventions (Williams & Long, 1979). Three off the cuff examples include: (a) if someone with clinical anger learns to put a reminder out to use relaxation exercises (the controlling response) that may make it more likely that they engage in non-angry, prosocial behavior and less likely that they engage in angry behavior in the near future (the controlled response); (b) gradually exposing oneself to frustrations and provocations in a graded fashion that one has been avoiding for a long time (the controlling response) to reduce the probability of yelling (the controlled response) late on in the day when one's boss is being stupid; and (c) walking away from a crying child (the controlling response) to reduce the likelihood of yelling at and even striking the child (the controlled response) if the child continues to cry for 30 min. Although critics of radical behaviorism might view this model of environmental control of behavior to be distasteful, many might concur that the technology it yields is sometimes effective. Often when they see someone managing their own non-problematic and problematic behavior, it might appear to the critic that the person is behaving autonomously, even if the ultimate location of the cause of their behavior is in fact in the environment. Much of the literature on development of aggression and violence during the lifespan and many evidence-based therapies can be interpreted at the acquisition or failure to acquire a sufficient repertoire of effective self-management skills.

Why Did I Write This Book?

My background has primarily been in developmental disabilities including both children and adults with intellectual disabilities and/or autism spectrum disorders. I also trained as a clinical psychologist; hence, I have been long interested in the intersection between mental health and behaviorism (Sturmey, 1996, 2020). Although Applied Behavior Analysis (ABA) has been incredibly effective with these populations, like others, I have felt that ABA has become too focused on applied work with individuals with autism and intellectual disabilities, failed to continue its early interest in mental health as shown in behavioral work conducted in the 1960s and 1970s, and has lost its roots in basic research in learning, and thereby has failed to address many of humanity's important problems. Skinner's (1953) vision of behaviorism was just as concerned with nuclear war, pollution, and designing a better society after World War II and said little or nothing about autism. Thus, in 2017, I edited the *Wiley Handbook of Violence and Aggression* (Sturmey, 2017a, b, c) in part to expand my knowledge in an area of scholarship and interest that is of huge social significance. I am grateful to all the authors of that handbook for the education they gave me. Much of what I learned from them propelled and inspired me to write this book.

My values are that we should base our personal and public decisions on the best available evidence and explicit criteria for decision making. In this way we can base our decisions on truth and maximize the broad good. Thus, decisions should be made using science, well-designed experiments with replicable results, systematic reviews, and meta-analysis to guide public practice. This approach is imperfect. It goes wrong and sometimes it makes errors, but it eventually self corrects. We no longer believe in polywater, even though at some point it seemed to be a scientifically valid idea. Science sometimes helps us minimize harm, sometimes helps us do better at helping each other, and can be transparent and honest. Compared to the alternatives—appeals to authority and social convention, personal testimony, unquestioning faith of various kinds—it is better. Unfortunately, there are at least two types of forces opposing rational, explicit approaches to social problems. The first is that we live in a world with minimal scientific literacy. Many people still believe the earth is flat; that COVID can be cured with medications for animals, bleach, or worse; that dinosaurs roamed the earth 4000 years ago; that there are natural phenomena that physical science cannot explain; that some numbers seem to be lucky; that planets affect our behavior; that ESP, spirits, and magic are real; that UFOs are real and aliens visited ancient civilizations, etc. Further, these strange and superstitious, nonscientific beliefs are resistant or even impermeable to information and education (Impey et al., 2012) even when they lead to fatal errors on a large scale. The second is that vested interests often color public policy decisions. It is difficult for individuals who develop programs to evaluate their own work objectively, even when the conflicts of interests are relatively minimal and not explicit. Of course, when people develop proprietary methods, medications, programs, or clinics, the conflicts of interest are sometimes public and clear. Public figures and politicians too have vested and often short-term interests when evaluating programs. They may be more interested in personal notoriety, being seen to respond to the crisis of the day, and short-term goals like getting re-elected, rather than the public good. It is difficult for people making policy decisions to resist the temptations of self-aggrandizement, grabbing the microphone and the spotlight to get re-elected, or just taking a bribe and diverting money to themselves, friends, and family members. Practitioners too often have vested interests in their current practices even when ineffective, suboptimal, or even positively harmful. For many reasons they may be reluctant to abandon current practices, and learn and implement better practices. Hence, identifying, evaluating, and disseminating evidence-based practices for prevention and treatment of aggression and violence is very difficult. Nevertheless, despite these challenges society should heed science to maximize the public benefit.

Avoid Unnecessary Pathologizing of Aggression

Aggression and violence are not pathological, although it is comforting to think that this is so. Should we fear the stranger lurking in the bushes or the violent psychopath stalking us at night, or incidents of mass violence in malls? Mostly probably not. These fears are sometimes reasonable, but they are also strangely comforting in that they allow us to avoid a painful truth about most aggression and violence. If you want to see your future aggressor or murderer, just look across the breakfast table or around your place of work: Most aggression and violence, including sexual violence, takes place between people who know each other, often people who are members of one's own family, our friends and acquaintances, our neighbors, and our coworkers. Thus, common, everyday forms of aggression and violence include aggression and violence between family members, such as intimate partner violence, aggression and violence between caregivers toward children, and aggression and violence from other family members toward seniors; and abuse by paid caretakers of vulnerable populations, including individuals with intellectual and physical disabilities, seniors, youth in residential settings, and prisoners. Very common minor forms of aggression are often related to altercations between members of the public and others frustrated by delayed airline flights, someone stealing their parking space, other people cutting in line, being required to wear a mask in public, following instructions to do something one does not particularly want to do at that time, the dog, child, or significant other who will not stop whining, etc. Most people show some minor form of aggression in such situations.

This does not paint a very flattering portrait of human nature. It is consonant with the observation that in some situations many people commit acts of aggression and violence when the situation permits and encourages it. In some examples of group violence and genocide, it is the person who refuses to participate in aggression and violence who stands out as many ordinary people turn a blind eye or indeed participate. The Milgram experiments on obedience to authority to engage in aggressive behavior were done in part in response to the observation that many ordinary people committed terrible violence during World War II, and knowingly stood by and watched and many others participated while sometimes benefitting materially from that violence. Milgram's work was influenced by three factors. First, Asch's (1956) classic experiments on social influences on line judgment, where most people conform to the incorrect judgments of other people about how long a line is. These studies suggested that people are highly influenced to be compliant with the behavior of the group, even when plainly contrary facts stare them in the face. The second was his own family's Eastern European Jewish heritage, and the third was his personal ambition to be an influential academic and researcher. Milgram's experiments showed that the propensity to shock another person was a function of the situation, not the person (Blass, 2009). For example, participants were much more likely to shock the victim when the victim was not visible, was a greater distance away from them, and when the participant did not have to physically place the victim's hand onto a shock plate. Likewise, participants were more likely to shock the participants when the experimenter was visible rather than in another room, when the experimenter was physically close rather than distant, and when the shock was delivered indirectly by a machine without human contact with the victim. Finally, in experiments with groups of stooges, the participant was much more likely to follow the lead of the group whether the group decided to shock the participant or to defy the experimenter. Blass (2009) reported that these effects have been replicated in 11 countries with over 3000 participants. Milgram (1974) wrote that "Often, it is not so much the kind of person a man is as the kind of situation in which he finds himself that determines how he will act" (p. 205).

Modest Optimism

Despite commercial and media exploitation of violence and aggression, local flare ups in murder and other forms of violence and aggression, and recent trends in new forms of terrorism, we know that environmental factors, such as those identified in Milgram's experiments, cultural practices, and learning during the organism's lifespan all have strong influences on aggression and violence. Some may take Milgram's experiments as evidence that human nature is weak and too easily influenced by other people, but the glass may be seen as half full rather than half empty. If the environment, personal learning, and culture are strong determinants of aggression and violence, then we can change them to promote prosocial behavior and inhibit aggression and violence. Thus, evidence suggesting that historically human violence has reduced considerably the effectiveness of individual and other therapy, and progressive refinement of the effectiveness of cultural practices focused on the family, education, work, and evidence-based cultural design, to prevent or attenuate violence and aggression, give us reason for cautious optimism.

Flushing, NY, USA Peter Sturmey

References

Asch, S. E. (1956). Studies of independence and conformity: I. A minority of one against a unanimous majority. *Psychological Monographs: General and Applied, 70*(9), 1–70.
Blass, T. (2009). From New Haven to Santa Clara: A historical perspective on the Milgram obedience experiments. *American Psychologist, 64*(1), 37–45.
Epstein, R. (1996). *Self-help without the hype*. Performance Management Publications.
Epstein, R. (1997). Skinner as self-manager. *Journal of Applied Behavior Analysis, 30*(3), 545–568.
Impey, C., Buxner, S., & Antonellis, J. (2012). Non-scientific beliefs among undergraduate students. *Astronomy Education Review, 11*(1), 010111–010111. https://doi.org/10.3847/AER2012016
Leigland, S. (1998). Intentional explanations and radical behaviorism: A reply to Lacey. *Behavior and Philosophy, 26*(1/2), 45–61.
Milgram. (1974). *Obedience to authority*. Harper & Row.
Skinner, B. F. (1953). *Science and human behavior*. Norton.
Skinner, B. F. (1990). *Presidential address to the American Psychological Association*. Retrieved September 5, 2021, from https://www.youtube.com/watch?v=Bf-GKbcSFNo&t=632s
Sturmey, P. (1996). *Functional analysis in clinical psychology*. Wiley.
Sturmey, P. (Ed.). (2017a). *The Wiley handbook of violence and aggression. Volume 1. Conceptual issues*. Wiley.
Sturmey, P. (Ed.). (2017b). *The Wiley handbook of violence and aggression. Volume 2. Individual treatment*. Wiley.
Sturmey, P. (Ed.). (2017c). *The Wiley handbook of violence and aggression. Volume 3. Group interventions*. Wiley.
Sturmey, P. (Ed.). (2020). *Functional analysis in clinical treatment* (2nd ed.). Academic Press.
Williams, R. L., & Long, J. D. (1979). *Toward a self-managed lifestyle* (3rd ed.). Houghton Mifflin.

Contents

Part I
Introductory Concepts

The first part of this volume will deal with elementary concepts related to violence and aggression. These include definitions of violence, aggression, and related concepts. The second chapter will review ways to measure these concepts. The third chapter will selectively review the epidemiology of aggression and violence which identifies risk factors and so lays the foundation of what needs to be explained in theories of violence and aggression.

Chapter 1
Definitions of Violence and Aggression

Aggression and violence are all around us. Today's *New York Times* (July 7, 2020) includes stories that: (a) murder rates are sharply up across the USA, although violent crime and overall crime are down; (b) there have been several individual lethal shootings by the police which were decried by various mayors; and (c) reports of 64 shootings and 10 dead over the July 4, 2020 weekend. Even at the middle of the COVID pandemic in 2021, violence pushes itself to the front pages of the media. Many of us are very concerned for the safety of ourselves and loved ones including concerns over violence in families, schools, at work, and in public places. Many of us are aware, sometimes for many of us from personal experience, of the huge personal costs of aggression and violence and the terrible effects it can have on individuals, families, and communities. Many of us are also aware of the huge financial costs of aggression in violence. This can include loss of productivity such as time off from work or long-term disability following violence as well as loss of productivity in other family members when they must care for the victims of violence. Many of us are also aware of the huge financial costs of prevention programs, such as parent education, alcohol and drug education and programs, school- and community-based programs, and the huge costs of reactive programs, such as the police, legal system, probation and prison, mental health services, residential services for delinquents and offenders, and so on. In addition to the direct costs of these programs, is the indirect loss from the inability to invest the resources currently invested in dealing with aggression and violence that could be invested in healthcare, education and community services.

P. Sturmey, *Violence and Aggression*, Advances in Preventing and Treating Violence and Aggression, https://doi.org/10.1007/978-3-031-04386-4_1

1.1 What Are Aggression and Violence?

The terms "aggression" and "violence" are used very widely to refer to a broad array of behavior. The terms can refer to minor individual acts, such as a child pushing away another child in order to obtain a desired item such as a toy, or much more intense behavior, such as physical assaults that result in substantial physical injuries, psychological harm, and both. The terms can refer to both brief and extended individual acts, such as isolated violent outbursts, and to chronic violence within a relationship, organization, or society over years, decades, a lifespan or even to a pattern of behavior that endures within a culture for generations. For some, aggression and violence are a short-term, transitory personal experience. For others it is a long-term, daily personal experience or even life story, which has been part of the person's life since early childhood and continues throughout their lifespan either as perpetrators, victims or both. For some families and communities it is an experience that endures across generations.

Although aggression and violence necessarily involve at least two participants, aggression and violence can also be a group or society-wide phenomenon where groups of people engage in aggression and violence. Examples of aggression and violence in larger groups include bullying by groups of people; gang- and sports-related violence; mob violence, riots, lynchings; and society-wide aggression and violence such as religious, ethnic, and cultural pogroms, persecutions, and wars. A more recent form of group aggression and violence is online bullying and the so-called cyberviolence (Rodríguez-Domínguez et al., 2020). The family is one of the most common group settings for aggression and violence, especially against women, children, people with disabilities, and older family members. In educational settings various forms of aggression and violence are common. Common forms of aggression and violence in schools include relational violence, such as both physical and relational aggression (Swit & Slater, 2021). In making this distinction, Crick and Grotpeter (1995) wrote that "... we expected that girls would be most likely to harm peers through relational aggression (i.e., harming others through purposeful manipulation and damage of their peer relationships) whereas boys would be most likely to harm peers through overt aggression (i.e., harming others through physical aggression, verbal threats, instrumental intimidation)" (p. 711). Schools are also the location for aggression and violence during dating, and gang-related violence. In colleges and universities various forms of aggression and violence are also common. These include hazing rituals, which range from the offensive to lethal, and have a cultural history dating back hundreds of years (Fávero et al., 2020) and sexual violence, which is also common (Howard et al., 2019; Thomas, 2019).

Aggression and violence are also common forms of entertainment. Aggression and violence as entertainment ranges from verbal jokes, slapstick humor, gestural threats, pratfalls, and slipping on bananas. It can also include violent media, including TV, internet, video games and apps, and violent news. Art too can be violent, be it bloody paintings, the commercially highly successful violent movies of Quentin Tarantino among others, or art movies of indigenous people being shot dead from helicopters. Many people also enjoy and pay money to watch and participate in violent sports, such as martial arts, American football, soccer, rugby, and boxing and other ritualized aggression and violence as entertainment. Many sports fans also participate in sports-related violence such a violence between rival groups of supporters. Violence is also a part of regional conflicts, war, and genocide, which often includes sexual violence.

Cultures also vary considerably in the prevalence of aggression and violence. There are some cultures where aggression and violence are common, accepted, and encouraged as typical and valued behavior which are required as part of the culture, admired, and even glorified. Other cultures pursue and promote cooperation and pacifism which are valued and even designed explicitly into the culture.

Aggression is also common among nonhuman animals. This includes both individual acts of aggression, such as fights over resources, territories, and mates; protection of self, offspring and other relatives. Sometimes aggression in nonhumans is an apparently organized social behavior, such as the so-called "chimpanzee wars" involving cooperative aggression between group members, changes in group territories, and movement of fertile females between bands (Nishida et al., 1985).

Aggression and violence in both humans and nonhuman animals is at first paradoxical. It is expensive in that it involves expending time and resources in behavior that not only takes them away from other activities with more obvious immediate and long-term benefits, such as feeding, mating, and rearing offspring. It is also dangerous and sometimes lethal. In many but not all species there are obvious morphological and behavioral differences between males and females related to aggression and violence suggesting sexual selection for such differences. Animals also invest energy and physiological resources in weapons of aggression, such as claws and teeth, that could be invested in other organs and functions. Aggression and violence persists in nonhuman animals and humans. This presents a challenge for evolutionary scientists to explain how such behavior could be selected and where the evolutionary advantage of such costly behavior occurs.

We use the terms "aggression" and "violence" widely in everyday speech. We talk about an aggressive interviewer, a violent temper, and someone being pushy. Usually, this everyday use of these terms just means assertive or forceful. It does not necessarily imply intent to harm another, although this is sometimes true.

When considering aggression and violence, many authors usually exclude dangerous behavior that causes injury to another person but does not appear to be intentional. The car driver who swerves to avoid a child, but injures pedestrians at the bus stop would usually not be said to be aggressive or violent, even though she injured many people badly, because she never intended to harm them. The car driver, however, who planned ahead of time, waited for the time when most people could be hurt, and then deliberately swerved into the crowd in order to kill is said to be violent because of their apparent intention to harm others. Thus, when considering definitions of aggression and violence, many people consider intent to harm as central to the definition of these concepts (Baron & Richardson, 1994; Berkowitz, 1993; Green, 1990) which is reflected in the oft cited definition from Bushman and Anderson (2002) (see Table 1.1).

Table 1.1 Some definitions of aggression and violence

Tinbergen (1968) "aggression involves approaching an opponent, and, when within reach, pushing him away, inflicting damage of some kind, or at least forcing stimuli upon him that subdue him … such behavior tends to remove the opponent or at least to make him change his behavior in such a way that he no longer interferes with the attacker" (p. 1412)
Bushman and Anderson (2002) "… any behavior directed toward another individual that is carried out with the proximate (immediate) intent to cause harm. In addition, the perpetrator must believe that the behavior will harm the target and that the target is motivated to avoid the behavior" (p. 28)
Frederiksen and Peterson (1977) "attack behavior directed against either a live or inanimate target" (p. 60). "When inanimate targets have been used, aggression has been defined in terms of the topographical relevance of the response. responses that would be noxious if directed against another organism, e.g., pigeons pecking at a mirror image. Thus aggression has been defined as either (1) delivery of aversive stimuli to a live target or (2) the emission of a topographically relevant response directed toward an inanimate target" (p. 60)
Lee et al. (2016). Violence is "the intentional use of physical force or power, threatened or actual, against oneself, another person, or against a group or community, that either results in or has a high likelihood of resulting in injury, death, psychological harm, maldevelopment, or deprivation"
Salimi et al. (2019). "The definition of aggression falls into two categories as instrumental aggression and hostile aggression. While instrumental aggression is used to achieve an external goal, hostile aggression intends to harm others and includes two forms of explicit aggression (physical and verbal aggression) and social aggression. Physical aggression refers to purposefully harming someone to cause pain such as beating, pushing, throwing objects, etc. Application of the aggressive language such as yelling and screaming that causes hurting emotions and credit of a person and lowers a person's value is called verbal aggression. Social aggression refers to actions to harm others' social statuses and friendships. Heavy silence and negligence are some examples of social aggression" (p. 159)

Many of the everyday examples of aggression and violence are less dramatic and intense. They are so routine that, in the whirlwind of related media events, we may not pay attention to or remember many of them and may often find them amusing. Consider the following everyday examples of aggression and violence.

- A cohabiting couple engage in pushing, shoving, and maybe sometimes punching perhaps involving alcohol or illegal drugs. Sometimes they argue over something perhaps trivial such as why someone did not put gas in the car again, or perhaps highly personally significant such as how to raise children or family finances.
- A cohabiting couple rarely engage in physical fighting. One partner, however, is constantly threatening, demeaning, and controlling the other, asking where they are going, when they are coming back home, restricting access to money, friends and family, and frequently implying the worthlessness of the other partner. They often check the other partner's cell phone for signs of infidelity or lying. The other partner is fearful and vigilant for any indication of displeasure in the other and engages in secretive and covert behavior.
- The angry customer in the store, frustrated over waiting too long, not getting a product, poor service, or being disrespected by staff or other customers, yells,

and threatens staff who cannot resolve the issue quickly. Eventually they storm off after pushing the staff and throwing a few objects around the store.

- The frustrated and angry airline passenger at the ticket desk who had missed their flight or who had the wrong ticket and is not going to make their meeting, get home, and may have to sleep on the airport floor or pay extra money for new tickets becomes aggressive. They threaten the desk staff, pound the desk, and shout for the supervisor.
- The belligerent and drunken airline passenger threatens and perhaps actually fights with airline staff and refuses to sit down when instructed to do so. When the police arrive to remove them from the plane they alternately apologize and again threaten as the police restrain and remove them from the plane as other passengers cheer and jeer them.
- The member of the public is stopped by the police for a traffic offense. The driver becomes angry and or entitled. The incident escalates from refusal to follow police instructions to yelling, and when the police attempt to handcuff them, they struggle, yell, curse, threaten the police, and drop to the floor. Sometimes the police remain calm and controlled and other times they too escalate and threaten.
- Some parents and other family members get angry at school, frustrated by apparent mistreatment of their children at school, such as getting a bad grade, safety issues, such as not being picked up by the school bus, or during sports, and apparently unfair or inaccurate referee calls. The parents yell, scream, and threaten school staff or referees in order to remedy the situation and prevent it from happening again.
- A person in charge of a program—perhaps a teacher, residential staff, prison officer or teacher—or a parent asks and then requires compliance from another person such as an individual student, resident, prisoner or child who then refuses to follow the instruction. As the supervisor or parent intensifies the interaction through repeated and perhaps louder requests, getting closer and making gestures, the subordinate's behavior also escalates and may eventually result in aggression such as threats and physical aggression to the supervisor. The supervisor or parent too may reciprocate with physical aggression. These incidents may result in occasional compliance with requests or mutual avoidance of requests except in particular circumstances when the supervisor or parents are especially motivated to obtain compliance. The constant air of coercion occasionally boils over into violence and is followed by apologies. For some couples it is followed by, sex, and promises of love and fidelity, and so the cycle begins again.
- At lunchtime on the first day of a new high school semester, students are milling around the school exit with teachers telling them they cannot not leave campus. The students and teachers all escalate to speaking more and more loudly. Suddenly, one of the biggest students pushes past a teacher blocking their egress from school and a crowd of students pour out of the school cheering. When the teachers complain to the principal, she says she thinks it is completely unacceptable and expresses the hope that those students will not be back in school the next day. She takes no further action.

- At school or outside schools, groups of high school students trade escalating insults and threats, perhaps relating to sports, race, gender, gang membership or other forms of group identity. After a period of escalating tensions perhaps involving escalating threats on line which may last hours, days or weeks, groups of students assemble and individual altercations break out and group fights break out. The situation continues and eventually breaks up as security or police arrive or after extreme and sometimes lethal violence.
- Two groups of soccer (football) fans congregate around an intersection a couple of hours before a game, each group of fans occupying different pubs. There is much jeering and threatening gestures from the fans conducted almost ritualistically and from a distance. The crowd cheers and laughs at first, egging the would-be combatants on. A few fans then separate from the rest and ritualistic fighting at a distance escalates into contact with chairs, pint glasses, and fists. Later, everyone disperses and their fellow supporters restrain them as the police arrive. Later video clips are posted on social media. Some comments include "Beautiful kick by that lad in the beginning" and "Haha good fight young boys."

Most of these examples involve predominantly verbal aggression, such as threats, hitting objects, such as banging on desks and throwing things, and sometimes less intense forms of physical aggression, such as shoving and pushing. They all have the potential to escalate into more severe forms of physical aggression, such as fist fights or worse. Many of these examples appear to be goal-directed behavior, such as getting onto an airline flight, escaping school, or resolving some interpersonal issue and also involve frustration at not being able to obtain these ends quickly or at all. Many of them can also be characterized by both: (a) the absence or ineffectiveness of prosocial behavior to obtain these valued outcomes, such as problem solving and assertiveness; and (b) few, uncertain, and likely delayed negative consequences for engaging in aggression and violence.

These individual incidents, however, do not capture the chronic and dynamic nature of aggression in some relationships or over the lifetime; they do not describe what happened in between each incident of aggression or violence. In many such relationships, there are ongoing problems in between individual incidents of aggression and violence. Within individual relationships examples include mutual avoidance, avoidance of certain triggering activities, minor escalations of threatening or disrespectful behavior that do not always escalate into aggression or violence, etc. Within lifetime development examples include the dynamics and developmental pathways of early precursor behavior that later develops into aggression or violence, changes in adaptive behavior such as language, tolerance of delay to access reinforcers, problem solving skills, etc. and periodic increases and decreases in aggression and violence across the lifespan are also not captured.

When discussing defining aggression and violence, it is often noted that some harmful behavior can even be prosocial and nonproblematic because ultimately its intent appears to be benevolent. Painful medical, dental, and physical therapy procedures may involve hurting others, but this hurtful behavior is usually seen as nonproblematic when the person consents to the procedure and there is some long-term beneficial outcome for the person who is hurt. Similarly, injuring someone when preventing greater harm may be seen as legal, ethical, or even required, for example, when someone attempts to prevent someone falling but injures the person while stopping the fall.

Harming oneself or others through exercise and sports is also often seen as non-problematic, even when one deliberately exposes oneself or others to aggression and violence from others, again because of the apparently voluntary and consensual nature of the behavior and because of the entertainment and enjoyable nature of the activity and potential long-term benefits. Thus, exposing oneself to being beaten badly during boxing matches may be seen as gruesome yet entertaining to many. Since the participants apparently volunteer, there are some limits on the violence, and there are prizes of various kinds, the activity is not seen as problematic violence by some. Interestingly, historically, boxing has become less violent with the prohibition of punching below the belt, prohibition of spiked shoes, required boxing gloves and helmets and limiting the number of bouts (Leclerc & Herrera, 2000). Others note the long-term effects include irreversible brain damage and risk of dementia and argue that this form of violence is unethical as it exploits the boxers for public entertainment and financial gain. Recent images of professional boxers such as Mohammed Ali with Parkinson's disease warn us of the long-term effects of this form of ritualized violence.

Finally, consensual behavior that involves pain, such as sadomasochistic sexual behavior, is also not typically seen as instances of aggression and violence, as the participants consent and get some pleasure from the activities. Some, however, have argued that sadomasochism is a dimensional construct and that more extreme forms of sadomasochism are problematic and involve aggression and violence (Longpré et al., 2018).

These examples show that there are several examples of violence, including some forms which result in severe injury that are not seen as problematic. This is because either there is no intent to harm or if there is intent, there is some apparent consent, there are known rewards, and/or the harms are delayed.

The terms "violence" and "aggression" are widely used by authors from many disciplines—psychologists, psychiatrists, other mental health practitioners, sociologists, criminologists, legal scholars, educators, politicians, evolutionary, and other biological scientists, etc.—and over a wide span of time. Thus, one should be sensitive to the different ways in which different authors use these terms. Sometimes these terms are used interchangeably and synonymously. Sometimes but not always an author specifies specific meanings to each term. Sometimes authors use the terms with different connotations which may be implicit or explicit. Let the reader beware and examine carefully the manner in which each author uses these terms!

A sample of some definitions of violence and aggression are summarized in Table 1.1 (see above). A number of observations can be made on these definitions. First, some authors use the distinction between aggression and violence to refer to differences in intensity of behavior, that is, aggression refers to less severe or intense behavior than violence. Others put it this way, that violence is a subset of aggression, that is, violence necessarily involves aggression, but not vice versa (Bartol &

Bartol, 1989). Sometimes the term aggression is used to refer broadly to behavior which is harmful to others, but which may not involve any physical contact, and violence is reserved for behavior involving physical contact (Bartol & Bartol, 1989). This distinction is theory-neutral. It just describes the topography of the behavior but it does not say anything about why any such differences if any between aggression and violence exist. Salimi et al. (2019) definition is more wide-ranging than the other definitions and specifies and defines different types of aggression. Indeed, although it includes physical aggression like most others, it includes behavior such as a "heavy silence" as a form of aggression even though the harm may not be as explicit, direct, or immediate as physical aggression. Some definitions include distinctions within the definition, such as the distinction between instrumental and hostile aggression (Salimi et al., 2019). In contrast, Frederiksen and Peterson (1977) provided an interesting analysis of definitions of aggression from a behavior analytic perspective (see Table 1.1). Many early behavioral studies of aggression defined aggression in pigeons as either pecking at a ball in the Skinner box or pecking at a live conspecific, often physically restrained in the Skinner box next to the potential aggressor (Azrin et al., 1966). Here aggression was measured as the rate of operationally defined aggressive behavior, such as the number of pecks per minute against the restrained conspecific or the ball.

In behavior analytic studies of aggression in humans the issue of defining intent—which is central to the definition of aggression for other academic authors and lawyers and the law—is avoided because "intent" cannot be defined or measured operationally. The resulting strategy is to defined aggression in one of three ways: (1) delivery or actual or ostensibly aversive stimuli to another organism, e.g., pushing a button to deliver a possible aversive stimulus to another organism; (2) topographically relevant behavior that is likely to result in physical harm and likely to be labeled aggressive, e.g., punching; and (3) symbolic measures of aggression, such as hitting a doll. The first two definitions are least problematic because they involve observable and measurable behavior without the need to define "intent" or "symbolic" whereas the third definition is problematic because the term "symbolic" is subject to wide interpretation.

1.2 Aggression or Violence?

Others make the distinction between aggression and violence implying that the difference is a meaningful one. Table 1.2 summarizes these alleged differences and Table 1.3 illustrates these differences with two contrasting examples. The differences between aggression and violence, as shown in the two examples in Table 1.3, may seem obvious. Indeed, it probably would not be difficult to train raters to categorize such very different examples reliably. A more important question would be whether these reliable differences have any meaning in terms of the nature, causes, and treatment of aggression and violence. An alternative view is that these incidents lie along the extremes of a continuum of severity of aggression-violence, but do not

Table 1.2 Some dimensions that may distinguish aggression and violence

Variable	Aggression	Violence
Cause	Predominantly environmental	Predominantly biological
Normality	Normal/typical	Abnormal/pathological
Understandability	Understandable	Inexplicable
Personal control	Person can control aggression	Person cannot control violence
Responsibility	Person is responsible for aggression	Person is not responsible for violence
Treatment responsiveness	Aggression is treatable and relatively easy to treat	Violence is untreatable or difficult to treat
Type of treatment	Psychosocial/environmental treatments	Biological treatments
Blameworthiness	Blameworthy since the person could/should have controlled their behavior	Not blameworthy because violence could not have controlled their behavior

reflect any fundamental differences between them. An example of a dimensional approach to measuring and thinking about aggression and violence comes from an observational study of aggression in bars. Graham et al. (2006) distinguished three levels of intention of aggressive acts: (1) *Defensive intent*, referred to using minimal force necessary to protect oneself, such as intervening to prevent greater harm, and heated verbal aggression due to the urgency of the situation; (2) *Probable intent* referred to more intense aggressive behavior for the context, but without apparent obvious intent; and finally, (3) *Definite intent* referred to aggression with clear anger with obvious potential harm to the victim and behavior that clearly exceed self-defense (p. 286). Notice that in this dimensional approach to aggression and violence the authors rated only degrees of intent and not other potential dimensions of aggression and violence, such as intensity, duration, resultant harm, and degree of planning, etc.

1.2.1 Less Immediately Physically Harmful Forms of Aggression

Over time, the definitions of aggression have expanded to include less physically directly harmful forms of behavior. Five examples are bullying and more recently cyber bullying, microaggression, indirect aggression, symbolic violence, and economic aggression. Although these forms of behavior are less immediately physically harmful than physical aggression and violence, some claim that they are indeed harmful, and perhaps seriously so, in other ways because they may result in suffering and even suicide.

Bullying is often characterized by intentions to harm, asymmetrical power (with stronger, more powerful bullies and weaker, vulnerable victims), and, persistence over time rather than a single or few acts of bullying, although children who are

Table 1.3 Some examples of aggression and violence that illustrate the commonly assumed differences between them

Variable	Aggression	Violence
Examples	A 4-year-old child smacked their sibling on the arm and grabbed a toy their sibling was playing with	A person who has never before committed such a violent act breaks into the house of an elderly couple whom he does not know, tortures them, sexually assaults the women and murders them. When he left he took an Ipad and $200 cash
Cause	The cause is apparently clear: The child wanted the sibling's toy and was frustrated that could not get it quickly through other means	If the violent person has never engaged in such behavior before then the violence behavior seems inexplicable: How could they have possibly learned to commit such a heinous act? Such extreme behavior seems unprovoked: The victims did not even know the perpetrator. The value of the items seems trivial in relation to the behavior. So to say "He murdered the elderly couple it to steal the ipad and $200" seems an inadequate explanation
Normality	This is normal behavior as many 4-year old children do such things	This seems abnormal. Very few people ever do this except under circumstances that might be more understandable, such as self-defense or war. The lack of any obvious explanation seems to render this behavior abnormal
Understandability	It seems understandable, if undesirable. Many children want to play with toys and it is frustrating if they cannot get them quickly	This behavior is very hard to understand. How did the perpetrator commit such an extreme act out of the blue and apparently for no reason?
Personal control	The child could and perhaps should control this behavior. They could and should learn to ask their peer directly or ask an adult for help. Caretakers could teach this child to learn to ask and wait	Many people might think that the perpetrator might be able to control such extreme behavior, but many people might assume the person must have been "out of control," have extreme anger, have a brain dysfunction or be mentally ill to do such a thing and hence be unable to control their behavior
Responsibility	The child and/or parents are responsible for this because of the child's failure to be patient or the parent's failure to teach their child patience	If the person has limited or no control for their act it is hard to hold them accountable. We might lock them up for protection of society, maybe for rehabilitation if that were possible, but not for punishment of an act for which they could not control

(continued)

Table 1.3 (continued)

Variable	Aggression	Violence
Treatment responsiveness	This seems an easy problem to treat. Some caretakers would readily smack the child's hand and take the toy away as punishment to stop aggression happening again. Other caretakers would try to teach the child to ask nicely, wait patiently or ask for help, perhaps by showing them how to do this, pointing out peer models, or practicing the skills directly and systematically. The child is young and has not had a lot of time to learn this behavior yet and there is lots of time left to learn to do something else	We would probably be pessimistic that we could rehabilitate such an offender. The act happened only once and its causes are unknown. How could such behavior be treated if we cannot allow it to occur again because it is too dangerous and because we do not know its causes? This person is an adult and this problem might reflect a lifetime of bad experiences and bad learning. It may well be an ingrained pattern of behavior. An adult is unlikely to unlearn such a pattern of behavior that might reflect a lifelong history of related problems
Type of treatment	Teaching communication, self-control (learning to wait) and social skills may both well work if done competently	Perhaps there is some biological treatment, such as psychotropic medication for the violence, some antilibidinal drug or castration for the sexual violence. Maybe there is no effective treatment available
Blameworthiness	We could assign blame to the child for being impatient or to the caretakers for poor child-rearing practices or some balance of both	Some people might assume that this person knew what they were doing and should be blamed for their act, but others might think this is so unusual and abnormal that they should not be blamed for their behavior

bullied themselves emphasize physical aggression, harassment, and verbal aggression (Vaillancourt et al., 2008). Bullying can occur in many contexts including school and work. More recently, research has identified cyberbullying as a relatively new phenomenon. Smith et al. (2008) defined cyberbullying as "an aggressive, intentional act carried out by a group or individual, using electronic forms of contact, repeatedly and over time against a victim who cannot easily defend him or herself" (p. 376). Cyberbullying can include online flaming, tricking, outing, and online stalking among other behaviors (Menesini & Nocentini, 2009) and may result in a wide range of harm to some individuals including suicidal thoughts and attempts (Hinduja & Patchin, 2010).

Microaggression is "… subtle verbal and nonverbal insults directed toward non-Whites based on one's race, gender class, sexuality, gender immigration status, phenotype, accent or surname" (Solórzano et al., 2002, p. 17). These authors give examples such as being ignored for service, being stared at because of one's color, body language, and use of terms such as "you people" and "highly qualified Black." A broader definition comes from Ogunyemi et al. (2020) who defined it as "everyday brief, low-intensity events that convey negative messages toward marginalized groups" (p. 97). These definitions of microaggression do not refer to physical aggression and violence or direct threats, but rather to more subtle and indirect behavior that is viewed as demeaning, insulting, or invalidating.

Some authors have distinguished indirect aggression from direct or physical aggression. Vaillancourt et al. (2008) defined indirect aggression as "a set of behaviors that are socially manipulative and circuitous in nature. For example, indirect aggression entails acts such as encouraging others to dislike others, befriending another as a form of revenge, telling a person's secrets to another, saying bad things about another behind a person's back and telling others to avoid a person" (p. 314). This category of aggression appears "soft" in the sense that, although it can be very hurtful to others emotionally, it produces less immediate direct physical harm than many forms of physical aggression and it overlaps with other categories of aggression. Similar categories and terms include "relational aggression" and "social aggression." Archer and Coyne (2005) reviewed the formal and operational definitions of these three terms and the developmental course of such forms of behavior. They concluded that these terms referred to essentially the same thing. In early childhood, indirect aggression begins with threatening to end friendships if peers do not comply with requests. In middle childhood and preadolescence, it includes gossip and practical jokes. In adulthood it includes various forms of bullying and exclusion at work and in the context of intimate relationships includes making others feel guilty or manipulation through feigning emotional injury, flirting, and infidelity.

Another indirect form of aggression is *sabotage*. This word is derived from the practice of nineteenth century French workers throwing their wooden shoes (*sabots*) into their employers' machines to harm their employers economically. The word now has extended uses to imply deliberately doing something to harm another person, as in deliberately playing music badly to frustrate other people.

In a volume on masculine domination, Bourdieu (2001) introduced the idea of *symbolic violence*. He defined this notion as "the dominated apply categories constructed from the point of view of the dominant to the relations of domination, thus making them appear natural" (p. 35). This notion has been used by Ericsson and Ellingsen (2005) to describe the harm done to children of Nazi soldiers and Norwegian women in occupied Norway when they were defined by their governments and follow citizens by terms such as "German Brat" and subsequently mistreated in numerous ways, both directly and indirectly. Again, terms such as "symbolic violence" are slippery and difficult to operationalize and measure objectively, although they may provide a useful label for discussion of how people are mistreated and sometimes subjected to physical aggression and violence.

Sometimes the act of witnessing violence is also referred to as an example of violence itself, even when the verbal or physical aggression or violence is directed at someone else. In certain contexts, such as children witnessing aggression or violence is indeed associated with harm to the witness. For example, young children witnessing intimate partner violence in their home may be at a greater risk for a range of child problems (Kitzmann et al., 2003) such as later aggression and violence themselves and mental health issues such as substance abuse, especially if the exposure to intimate partner violence involves multiple exposures on a prolonged basis and exposure to other risk factors such as direct violence (Kulkarni et al., 2011). Thus, many authors might agree that the parents' failure to act, that is failing to prevent exposure to witnessing violence, is negligent and harmful, but not all would define this as an example of violence per se.

Finally, several terms related to aggression are used in sociology, history, politics and economics. These definitions are often very broad and not clearly operationalized. For example, the term "economic aggression" is used in contrast to armed aggression to refer to "Any unilateral act by a state directed against another state or group of states which is intended to and which does in fact deprive such state or states of essential resources or which seriously impairs its normal economic activities, thereby threatening the state's sovereign right to make its own political decisions, is an act of economic aggression" (Domb, 1978, p. 99). An example would be one state deliberately cutting off the water supply to a neighboring state in order to harm the citizens of the second state. The term "economic" or "slow aggression" is also used in economics and politics and refers to achieving political objectives through economic actions. For example, during the Cold War the US took economic action to impoverish the USSR by reducing their income from energy exports, increasing the costs of the cold war, and fomenting unrest in the USSR and its allies thereby hastening the end of the Cold War over many years (Conroy, 2019.) Others have used the term *structural violence* to refer to "violence that results in harm but is not caused by a clearly identifiable actor" (Vorobej, 2008, p. 84) and Galtung (1969) wrote that "Violence is present when human beings are being influenced so that their actual somatic and mental realizations are below their potential" (p. 168).

A related term is "structural violence" is used to refer to a wider set of cultural practices resulting in violence, often against devalued groups. One example of this comes from a review by Monchalin et al. (2019) who analyzed structural violence in North America against Indigenous peoples, especially indigenous women. Whereas other attempts to understand aggression and violence focus on immediate and proximal factors, such as alcohol, interpersonal disputes, etc., structural violence focuses on the broader and social picture. Monachlin et al., for example, point out that high rates of violence against indigenous Canadians are part of a broader picture of very high rates of poverty, poor health, education services, and even lack of access to such basics, such as clean water in Indigenous communities. These problems have been recognized for many years, but when they are made public they are typically met with apologies, promises of change, slow, and usually ineffective governmental responses. Monachin et al. frame these problems as part of broader societal practices rooted in settler colonialism, oppression of native peoples through police, residential schools, and devaluation of indigenous women that permit their abduction and murder without meaningful consequences. Thus, in contrast to other definitions of violence, structural violence emphasizes macro variables including historical and cultural practices that encourage violence against members of devalued groups within society whose effects are similar to genocide.

The broadening of the terms aggression and violence raises some interesting challenges. The first is that these less obvious forms of behavior must be measured reliably. One might ask, for example, how one measures ephemeral concepts such as human potential and its absence. The second, which is more substantive, is that the validity of grouping these rather different behaviors under the term "aggression" may be questionable to some. The issue of validity has been addressed by some psychometric studies of measures of measures of microaggression (e.g., Nadal, 2011; Wright & Wegner, 2012), but suffice it to say that problems in the definition of aggression continue to plague the field (Edens & Douglas, 2006).

1.3 Types of Aggression and Violence

Many authors have attempted to categorize aggression and violence into different types. This may be done for rather different purposes. Classification of different types of crimes may be done to allocate different degrees of culpability which are associated with different kinds of punishment. Another approach to classifying aggression and violence is to do so based on different reasons for the development or maintenance of aggressive behavior. This answers a different question from the first example, namely, it attempts to classify aggression and violence in order to understand their development. This section reviews several attempts to do classify different kinds of aggression and violence.

1.3.1 Hostile and Instrumental Aggression and Violence

A commonly made distinction—which dates back thousands of years—is between hostile and instrumental aggression. This distinction is sometimes referred to as "hot" and "cold" aggression (Bushman & Anderson, 2001). That is, some forms of aggression appear to be responses to antecedent stimuli, such as provocations like insults, frustrations, or pain, etc. This form of aggression is sometimes referred to as "affective," "hot," "angry," "impulsive," or "retaliatory" aggression. A prototypical example would be the partner who returns home unexpectedly to find their significant other in bed with a third person and, in a fit of rage, kills them both. Other forms of aggression appear to be primarily influenced by its consequences. Such instrumental aggression may appear premeditated and involve extensive preplanning. A prototypical example might be the person who plans the staged car accident in order to inherit money from the victim.

Behavior analysts might note that the distinction between hostile and instrumental aggression appears to be one in which the two types of aggression are distinguished primarily based on whether it is controlled by its antecedents or its consequences. Additionally, the hot blooded variety of aggression occurs quickly and in the moment. There seems to be little or no planning and if there are weapons involved, the perpetrator just grabs the closest available object. In contrast, with cold blooded variety of aggression consists of a long chain of behavior distributed over time and weapons may be carefully researched, planned, and stored before the act. From a cognitive perspective, the distinction is made in terms of "automatic controlled," hostile aggression which involves little conscious or effortful thinking that the perpetrator can label and is aware of versus "information processed," instrumental aggression in which the perpetrator is very aware of their own effortful thinking and planning before the aggressive or violence act (Bushman & Anderson, 2001). Such a binary distinction is, indeed, quite old and goes back to the legal idea of "hot blooded," impulsive murder, for which the perpetrator has less responsibility, versus "cold blooded," premeditated murder, for which the perpetrator is fully responsible or even held to be especially culpable for the act.

Although many authors make the distinction between hostile and instrumental aggression, Bushman and Anderson (2001) have questioned its usefulness. Using a cognitive-structural approach, their reasons for doubting this distinction's utility are as follows. First, they claim that reactive aggression is aggression in which the individual acts quickly and efficiently with little or no self-conscious or cognitively-controlled thinking, whereas the individual who engages in instrumental aggression acts with deliberate self-controlled and effortful thinking toward an articulated goal. Cognitively-controlled thinking is characterized by "(a) conscious intention of what the control will accomplish, (b) a sense or feeling of control, (c) an expenditure of effort in the control action, and (d) a (closed-loop) monitoring of the control output" (p. 275). Automatic cognitive processing involves none or few of these features. Hence, they claim that the two types of aggression are confounded with two types of information processing.

Second, Bushman and Anderson (2001) claimed that different types of aggression and motives are confused and that there may be both hostile and instrumental aspects to a single act of aggression. For example, if a child pushes another child off their own bike one might ask, "Is this hostile aggression with the intent to harm the other child because the first child's goal-directed behavior was frustrated, or was it instrumental aggression to obtain the bike and remove the child?" In a second example, a person might engage in violence a long time after an initial insult. They might be angry at the time of the insult, calm, and purposeful in between while planning the violence, and engage in a mix of both "hot" and "cold" violence during the final violent rampage.

Therefore, when Bushman and Anderson posed the question "Is this reactive or instrumental violence?", and they suggested there is no clear distinction to be made here. Rather, the better approach is to consider the types of cognitive processes that underlie aggressive behavior.

1.3.2 Structural and Functional Categories

It is also common to categorize aggression and violence during literature reviews using everyday language and familiar concepts that seem to make sense. One such example can be found in Table 1.4 based on the work of Alvarez and Bachman

Table 1.4 Types of aggression and violence using Alvarez and Bachman's (2019) taxonomy

Type of aggression and violence	Definition
Animal aggression	Includes predatory aggression to kill prey; male-on-male aggression to achieve "supremacy"; aggression in response to a perceived threat where there is no escape; maternal protection of young; irritable aggression in response to pain, frustration, deprivation; sex-based aggression for mating purposes; instrumental aggression in experiments; territorial aggression to defend land
Collective violence	Violence committed in groups such as gangs
Assault	There is a continuum of violence from assault to murder in which assault covers a wide range of behavior from threats to near fatal attacks. Aggravated assaults include the use of a weapon
Homicide and murder	Homicide refers to any killing and murder refers to illegal killing. Homicide may be justifiable, e. g., some police killings and some killings while serving in the military. Homicide may be justifiable, such as killing in defense of oneself or others. Criminal homicides (murders) are often classified in terms of degree of premeditation and planning, whether or not the person acted with malice and whether it was committed during another felony
Violence at home	Violence between various family members, other intimate partners and former partners, such as spouses, ex-spouses, partners, ex-partners, including stalking; parents and other caregivers and children and other young family members; sibling-on-sibling violence; and abuse of seniors within the family
Rape/sexual assault	These are usually legal terms that evolve over time and differ from place to place, but currently recognize both women and men as potential victims and perpetrators. Perpetrators are often known to victims. Rape and sexual assault occur commonly with female victims and occurs commonly on university campuses, military services, and during war and genocide
Stranger violence	Includes robberies for money, goods or thrills and is less common than violence between individuals who know each other
Group violence	Group violence which includes crowds and mobs. Crowds that engage in violence may have collective purposes, such as legal and social protests, race riots and may be organized. Mobs may include violent riots to protest political situations, drafts into the armed forces, or be violent against specific ethnic or religious groups. Crowd violence may also include lynchings and vigilantism, where crowds take on quasi legal action in the absence of an effective police force or legal system
Terrorism	Involves a wide range of aggression from kidnapping, bombing, theft, hijacking and mass murder which strikes fear in society. It is usually done to obtain political goals and may result in eventual incorporation of terrorist groups into conventional political life
Genocide	Involves large scale mass killings of specific tribes, ethnic, religious or other culturally identifiable goups. This may be done directly through mass killings or through actions, such as mass starvations, relocations, prevention of births or forcibly transferring children to other cultural groups. It may result in material and other benefits for aggressors

(2019). Such taxonomies are useful in organizing information around ideas that are familiar to the reader, such as perpetrator and victim demographics (gender, age, and social relationships, etc.); the severity of injury (stalking, assault, aggravated assault, murder); and various legal and pseudolegal ideas that have crept into everyday language (homicide, murder, etc.). These taxonomies are also useful because they also reflect the way many human services are delivered and funded, such as prevention of gang violence, services for battered women, training to residential service staff to prevent elder abuse etc. There are many such attempts to classify aggression and violence using everyday language and concepts. For example, Rodríguez-Domínguez et al. (2020) conducted a systematic review of cyberdating aggression and violence. They categorized the articles in terms of two super-ordinate dimensions such as nonsexual and sexual violence. They further categorized nonsexual violence as verbal and emotional, such as threats and insults, relational violence, deliberately provoking jealousy, and controlling and monitoring the other person. They also categorized sexual cyberviolence as sexual cyber abuse, digital sexual coercion, and coercive sexting. Notice here that these authors use the term "violence" very broadly to encompass behavior such as written threats that are not directly physically harmful or, indeed, do not involve any physical contact. In this example, various forms of cyber abuse are categorized intuitively using everyday language which can be a useful tool for initial communication among researchers and practitioners, but which may include categories that are invalid, may split categories unnecessarily or omit important functional distinctions.

A second attempt to categorize aggression and violence comes from ethology, which often takes a very functional approach to defining different subtypes of aggression and violence. Tinbergen (1968), for example, distinguished interspecific and intraspecific fighting and offensive and defensive forms of aggression. This distinction was made not only to distinguish the type of victim of aggression, but also to try to understand the reasons behind or functional significance of different forms of aggression. Moyer (1972), again coming at this problem from an ethological perspective, made a more elaborate typology of aggression which distinguished seven different types of non-human animal aggression that occur in natural habitats and can be observed in unmodified, naturalistic rather than contrived, experimental conditions. It is interesting to contrast the manner in which these two approaches categorize aggression and violence. Table 1.4 uses a predominantly structural approach to categorizing violence, that is, it focuses on the topography of behavior describing people's observable behavior as if similar instances of a particular behavior are the same in some important way. This is done using everyday, less scientific language. In contrast, the approach illustrated in Table 1.5 is a much more functional approach, that is it describes categories of aggression in terms of their apparent purposes, rather than topography. This approach usually includes some description of the environmental triggers and consequences of aggression and perhaps also states or implies their possible evolutionary functions.

One functional approach to classification is based on or influenced by learning theory. This approach has been applied to a wide range of problematic behavior (Sturmey, 2020) including aggression. An example of this approach to the

Table 1.5 Moyer (1972) taxonomy of aggression in nonhuman animals

Type of aggression	Description
Predatory	Aims to kill and perhaps eat prey of another species; aggression is elicited by hunger and the sight of the prey; inhibited by fear, no strong sexual differences; typically conducted quietly without displays
Intermale	Directed against males who are usually conspecifics; may occur before maturity; often for access to female conspecifics; may sometimes result in death of other male; may be ritualized and without serious injuries; submissive behavior in other may diffuse aggression and establish dominance hierarchies within colonies or groups of animals
Fear-induced	Shown by defensive animals unable to escape a threat; may be preceded by freezing or submissive posturing and attempts to escape
Maternal	Directed by mothers against intruders threatening offspring, often during lactation
Sex-related	Males attack a female in order to gain access for sex, such as aggressive courtship
Instrumental	Aggression elicited by experimenters' manipulations to reinforce aggression with consequences
Territorial	Territorial animals spend most time in their territories patrolling and sometimes defending their territory; may involve scent marking or calls to defend but avoid actual aggression; seen in many primates

classification of aggression and violence comes from Dixon and Wride (2021) who began by noting the limitations of classification of types of aggression and violence to predict treatment by structural features, such as severity and form of behavior, etc. They note that any classification system should be completely explicit as to the purpose of the classification system. Thus, in this case they proposed a classification system whose aim was to predict the best treatment. To do this they proposed eleven types of motivations or "systems" for aggression and violence based on Ward and Carter's (2019) functional offending behavioral classification framework. For each system there are nonoffending and offending examples of problematic behavior that may occur if prosocial or adaptive ways of meeting these needs are blocked or are ineffective. For example, the "bodily regulation system" regulates basic needs such as "warmth, fluids, nutrition, sleep, sex, and to regulate temperature." If adaptive ways of meeting these needs are blocked or ineffective, then nonoffending, problematic ways of meeting these needs include problems with health care, housing, and nutrition and offending, problematic ways of meeting these needs include theft to gain resources and sexual offenses. The provisional list of typologies of aggression and violence systems include: (1) body regulation; (2) aggression (i.e., responding with aggression); (3) fear and security; (4) disgust (i.e., avoidance of harmful materials); (5) status (i.e., competition for high social status); (6) mating and pair bonding; (7) attachment and caregiving; (8) affiliation with nonfamily members of the group; (9) reciprocity and acquisition (i.e., fairness of interactions and protection of personal resources; (10) play systems (i.e., acquisition of various skills in a nonthreatening context); and (11) curiosity (i.e., acquisition of knowledge including knowledge of other people.) These typologies of motivations for aggression and violence appear to specify potential types of reinforcers. This may be quite useful in that it can guide clinical case formulation, and perhaps predict maximally effective and iatrogenic interventions, although data on this possibility has yet to be collected.

1.3.3 Categorization by Severity of Aggression and Violence

As shown in Tables 1.2 and 1.3, many authors distinguish aggression and violence at least in part in terms of severity of behavior. In other contexts, researchers have also distinguished and sometimes operationalized different degrees of severity of aggression.

When ethologists, such as Goodall (1986) and Wrangham et al. (2006), for example, observed aggression in chimpanzees (*Pan troglodytes*), they distinguished three levels of aggression as follows: "'attack" was defined as contact aggression, including hits, kicks, or slaps delivered in passing (level 1), more extended episodes of pounding, dragging, and biting (level 2), and similar incidents lasting more than 30 s or leading to serious injury (level 3)" (p. 15). Such definitions may be useful in distinguishing different intensities of aggression and perhaps observers can make reliable observations using such codings. Such definitions, however, mix up topographies of aggressive behavior, context ("in passing"), and outcomes ("serious injuries"). Outcomes, of course, may have more to do with (bad) luck than the intensity of the aggressive behavior itself.

There are also many distinctions in law relating to the severity of aggression and violence. See the section below on legal definitions of aggression and violence.

1.3.4 Policy-Driven Definitions

Different types of aggression and violence may also be proposed to frame policy and organizational initiatives on aggression and violence. For example, the World Health Organization (2016) *Violence Prevention Alliance* (Lee et al., 2016) proposed dividing violence into family/partner violence (specifying child, partner, and elder victim) and community violence (specifying acquaintance and stranger), and whether the mode of violence consisted of physical, sexual, psychological, or deprivation/neglect. Thus, this five-by-four typology resulted in 20 different types of violence. Finally, the WHO typology distinguished types of violence based on the victim-perpetrator relationship by distinguishing self-directed, interpersonal, and collective violence.

A second example of this approach comes from Crawshaw's (2009) attempt to develop a taxonomy of workplace bullying. Crawshaw noted the plethora of terms that had been used in this context (abuse, emotional harassment, mobbing, workplace psychological violence, etc. to name a few) and the confusion that this has engendered. As a solution, Crawshaw proposed a systematic nomenclature in which the term "workplace abuse" refers to a broad range of abuse, harassment, and workplace violence and the term "workplace psychological harassment" is used as a synonym for workplace bullying. The proposed taxonomy then specifies two dimensions of the incident: the perpetrator(s) and target(s) (superior, peer, subordinate, etc.) Thus, an incident might be classified as "workplace psychological harassment, individual to group (including superiors, peers, and subordinates), manifested in verbal abuse" (p. 266).

These policy-driven approaches to classifying aggression and violence often have aims such as apparently neutral description and promoting uniformity of reporting across time, agencies, and countries.

1.3.5 Theoretically-Driven Approaches

An alternate way to divide up different types of aggression and violence is to go back and forth between theory and data and continuously refine behavior categories in this manner. For example, Blanchard and Blanchard (1989, 2003) provided a different, more systematic, and intellectually satisfying approach to defining types of aggression in the laboratory and wild rat (*R. norvegicus* and *R. rattus*). Rats typically live in colonies and have strong male dominance hierarchies. Blanchard and Blanchard distinguished between two forms of aggression—offensive and defensive—defined by both topography and function. Defensive aggression included attacking conspecifics and dangerous objects, such as shock prods. In response to a predator approaching, the rat will flee if possible, but if the predator gets closer, the rat will vocalize, jump attack, and bite the predator's head. If escape is impossible, the rat may freeze while oriented toward the predator, increase muscle tension, startle more readily and then vocalize, jump attack, and bite the predator's head. In contrast, during offensive aggression an alpha male approaches an intruder, piloerets, and initiates aggression to the other rat, terminating in biting the back of the other rat's back. There is a characteristic chain of behavior including chasing, lateral attack, and standing on top of the other rat in order to gain access to biting the intruder's back.

There is substantial experimental evidence of the construct validity of the distinction between defensive and offensive aggression as shown by physiological correlates and differential effects of different pharmacological agents on offensive and defensive aggression. First, brain sites thought to mediate irritability, rage, and aggression, such as ventromedial hypothalamus, affect defensive aggression but either do not affect offensive aggression. There are also specific effects of pharmacological agents. For example, low dose ethanol increases offensive aggression but not defensive aggression (Blanchard & Blanchard, 1984).

Provocatively, Blanchard and Blanchard (1984) also speculated about potential parallels between rodent and human aggression. Namely, they proposed that angry aggression often relates to disputes over property and privilege, inviting the parallel to territorial defensive aggression. In contrast, human angry aggression being more likely to involve slaps and closed fist blows, whereas human fearful aggression being more likely to involve screaming, biting and scratching.

1.3.6 Measurement Systems that Define Aggression and Violence

Definitions of aggression and violence are also buried within or implied by measurement systems, even if the definitions are not made explicit. When we use the cut off score on a psychometric measure or count the number of forms of aggressive or violent behavior when making a psychiatric diagnosis related to aggression or violence, we are using an implicit definition of aggression or violence. The decision to use a particular cut off score to define aggression or "clinically significant" aggression is one way to define aggression.

An example of this approach comes from a study by Vitaro et al. (1998) who measured proactive and reactive aggression using questionnaire measures. The authors administered a broad battery of psychometric measures to 742, low-income boys. To measure proactive and reactive aggression, they selected items used in previous research to measure these two forms of aggression. For example, reactive aggression was measured using items such as "When this child gets teased or threatened, he gets angry easily and strikes back," whereas proactive aggression was measured by items such as "This child gets others children to gang upon a peer that he does not like". Although ratings of both types of aggression were quite closely correlated ($r = 0.71$, $p < 0.01$), there was evidence of construct validity in measuring these two different kinds of aggression. That is at age 12 years, proactive but not reactive aggression predicted delinquency at 15 years. Studies such as this both define and validate psychometric approaches to measure aggression and its different types.

1.3.7 Behavioral Approaches

Finally, one specific example of defining aggression and violence comes from basic and applied behavior analysis. These definitions are made in the context of both basic behavioral research often with nonhuman animals and in applied contexts with humans and sometimes nonhuman animals. In basic behavior analytic research aggression is often measured by specific responses such as the number or rate of pecking at a ball or a conspecific, for example, in response to an environmental event such as a painful stimulus or denial of previous reinforcement. In applied behavior analytic research the definitions are usually developed in the context of understanding the environmental control of aggression and using this pretreatment information to reduce aggression and increase prosocial behavior such as communication. Thus, applied behavior analytic studies involve taking observational data on aggressive responses or other relevant behavior in real world environments.

The first step in measuring aggressive behavior in behavioral studies is to develop a measurement system that includes an operationalized highly reliable definition of the target and other relevant behaviors. Some examples of such definitions of aggression include: "firmly grabbing and pulling on the experimenter or cursing" (Allen et al., 2013; p. 686); "scratching … pinching … hitting … and kicking (Saini et al., 2015; p. 318); and "hitting, kicking, biting, hair pulling and grabbing" (Dupuis et al., 2015, p. 670). It is worth noting that these definitions of aggression differ from one another as they are each defined specifically for each participant and context. Behavioral approaches focus on the observable behavior and its context, but eschew attempting to infer covert behavior, such as intent or ability to control aggressive behavior. A second measurement decision is to identify the specific dimension of aggressive behavior that is being measured (See Table 1.6). The researcher or practitioner must decide which dimension(s) are most relevant, meaningful, and sensitive for the study or intervention being conducted. In addition to measuring the behavior, behavior analytic studies of aggression also involve clear definitions of environmental variables. For example, Dupuis et al. describe the types of noises they exposed the participant to as follows "Sounds included a therapist singing "wheels on the bus," a novel person singing other nursery rhymes (Nursery Rhymes 1), the sound of students in circle time, and three adults talking. Nursery Rhymes 1 songs varied within the session and did not include "Wheels on the Bus." During the second 30-dB condition, the therapist sang the same songs as the novel person to evaluate generalization (Nursery Rhymes 2) …. Brandon did not have

Table 1.6 Some dimensions of aggressive behavior

Dimension	Examples
Frequency	Natasia hit her sister 8 times
Rate	Natasia hit her sister 8 time in 20 min for a rate of 24/h or 0.4/min
Duration	Natasia ran after her sister for 3.5 min
Percentage of time	Natasia ran after her sister for 3.5 min during 20 min of observation for 17.5% of the time
Latency	Natasia hit her sister 3 s after her sister said "No, you cannot play with me"
Intensity	When Natasia hit her sister at 11:20 am it did not leave a red mark on her sister's skin. When Natasia hit her sister at 11:27 am it left a 3cm red mark on her arm
Variability	Between 11:20 and 11:40 on Monday Natasia emitted 1 aggressive response (Calling names) seven times. On Tuesday between 3:20 and 3:40 pm Natasia emitted three different aggressive responses: (1) called names (7 times), (2) threatened (2 times), and (3) physical aggression (3 times)
Opportunity-based measures	Following eight instances of aggression, Natasia's parent put her in time out on 2/8 (25%) of opportunities and during 3, 5 min periods with no aggression Natasia's parent gave her a point as part of her behavior plan on 1 of 3 (33%) of occasions

access to leisure items." (p. 670). Finally, in applied behavior analytic studies, treatments are both based on the results of a pretreatment assessment of the environmental control of aggression, use behavior analytic concepts and are clearly operationalized so they can be replicated.

1.3.8 Summary

Aggression and violence refer to a very wide range of nonhuman and human behavior. It includes behavior that varies greatly in intensity and duration from brief episodic, low-intensity behavior to long-term severe forms of violence; short-term aggression between dyads; aggression in response to pain; developmentally typical behavior, such as children pushing other children to obtain toys; long-term abuse in intimate partner relationships; aggression and violence between family members other caregivers and vulnerable family members, such as children or older adult; complex social phenomena such as institutional abuse, violence related to gangs, sports, and, religious and other cultural organizations; and organized ethnic, political, and religious violence, purges, wars. Hence, these terms refer to a wide range of different behavioral phenomena, which perhaps require different kinds of definitions and measurement strategies appropriate for different contexts.

1.4 Legal Definitions

In current US law, homicide is "the killing of a person by another" (Charette & Hickey, 2017). Homicide may be legal, for example, when defending oneself, state-sanctioned homicide, such as fighting for one's country, and, conducting a legal execution. Murder is an illegal homicide. First degree murder is an illegal form of homicide in which the killer had *malice aforethought*, that is, they intended to kill the other person as shown by their actions prior to the killing, such as lying in wait, purchasing poisons, or buying armor-piercing bullets. Second degree murder involves extreme recklessness including committing a felony, but does not involve malice aforethought. For example, if someone killed another person during an armed robbery without malice aforethought, this would be a second degree murder because the armed robbery was extremely reckless and likely to put people at risk for injury and death. Some state laws distinguish other different degrees of murder and, generally, US states have developed progressively more and more granular taxonomies of manslaughter and homicide (Charette & Hickey, 2017). In some states these different degrees of murder determine the eligibility for the death penalty.

Violence and aggression have been defined by legal scholars and authorities over millennia. Some four thousand years ago, the first extant legal code of Ur-Nammu (2100–2050 BCE) distinguished many modern-sounding violent crimes including murder and rape. Medieval English law distinguished three types of criminal aggression based on the degree of harm done. First, *assault* was another person's action that would make a reasonable person jump from the french *saltare*. Second, *battery*, another person's action involving touching from the French *battre*. Finally, if the battery rendered the person unable to participate in battle, say due to their arm used in archery being severed so they could no longer use a bow and arrow, then it was the crime of *mayhem* from the French to maim. Hence, assault and battery often go together, because a person may threaten and then touch the other person. Such distinctions have been incorporated into contemporary British, American, and other law (Charette & Hickey, 2017). Strangely, European historical records rarely mention infanticide as a crime. Rather infanticide was regarded more as a sin than a crime. Later, legally recognized crimes against minors increased substantially to include many nonlethal forms of harm that were not recognized in historically earlier times (Eisner, 2014).

In most legal definitions of crimes involving aggression and violence, issues of whether the person intended to harm another person, the degree of planning before the crime, whether the person was legally competent at the time of the crime, all affect the type and severity of the crime and related punishment. Legal systems often distinguish different degrees of a crime. For example, some laws distinguish rape from aggravated rape. If the rape took place in the context of another crime and/or with weapons, or if the act was especially heinous because the victim was particularly vulnerable, for example, if the victim was a minor, a person with a disability, a relative, or an older person, then the crime may be classified as aggravated rape in some jurisdictions carrying heavier penalties than rape. Similarly, crimes which are partly or wholly motivated by personal characteristics, such as disability, gender, nationality, race or ethnicity, religion, sexual orientation, etc., today are often classified as hate crimes which are deemed to be more heinous than crimes without such motivation (today, in some states the police and other law enforcement and military personnel are also included in protected groups). Hate crimes also usually result in enhanced penalties.

Legal definitions of aggression and violence are tricky as the behavior itself is insufficient to make the act a crime. Without intent there may be no crime, but the greater the intent the more serious the crime. This aspect of legal definitions of murder raises the difficult issue of judging people's intentions at the time of the crime, which may be many years prior to a trial. Thus, judging intent and degree of intent is not simple and may often be inaccurate. Legal definitions of violence and aggression often bleed over into the research and practice literature. This may not necessarily be a bad thing, since conviction of a crime or risk factors for a crime may be necessary for access to certain treatment programs or may set the context for a research study. Thus, the law is yet another source of definitions of violence and aggression with its own framework and concerns, and hence may be another source of confusion.

1.5 Normal Versus Abnormal Violence and Aggression

1.5.1 *Statistical Approaches*

There are several approaches to defining normal and abnormal aggression and violence. Statistical approaches take normative data and use some criteria to identify individuals with unusually high or low levels of aggression and violence. For example, most children show some aggression between the ages of 2 and 5 years and most appear to "grow out of it" quickly without any obvious or formal intervention. Normative data on aggression are available for children between the ages of 2 and 11 years in Canada from between 1994 and 2000 (Nagin & Tremblay, 1999; Nantel-Vivier et al., 2014). These data allow identification of children within each age band who show rather more or rather less aggression than their peers. Likewise, children who persist in aggression after age 5 years are also statistically unusual as most children appear to "grow out" of aggression by the age of 5 years or so. Furthermore, children who persist in aggression throughout childhood and adolescence and adolescents who become aggressive and violent as teenagers and persist in aggression and violence are statistically unusual as most children do not do this. Similarly, onset of aggression and violence in adulthood or late life is also relatively unusual (see Chap. 8 for a more detailed account of the development of human aggression here).

An example of this approach to defining statistical unusual aggression comes from a longitudinal study by Girard et al. (2019) who modeled the developmental trajectories of physical, indirect, proactive, and reactive forms of aggression in a large sample of children aged 6–13 years based on data collected between 1997 and 1998 in Quebec, Canada. Using multivariate analyses, they grouped the children into five groups. There were two groups with low levels of aggression, two with initially high levels of aggression that abated over time, and one group that showed high levels of all four forms of aggression persistently. This latter group accounted for only 4.9% of the sample. Thus, this latter group could be considered to show abnormal aggression because (a) of the very high levels of aggression compared to other children, and (b) the unusual persistence of aggression, which is not typical of most children.

In addition to charting the differences in level and persistence of aggression, Girard et al. then went on to compare the five groups on other variables not used to classify the children into the five groups as a way to validate this classification. Using multivariate analyses, they found that the high and persistently aggressive group of children were distinguished from other children in terms of maternal education, estimated maternal IQ, coercive parenting, never breastfed, and being male. The authors suggested that these observations were useful in designing early intervention for aggression in that the high-risk children and mothers can be readily identified at age 6 years or earlier. They also suggested that the constellation of risk factors within the high-risk group was highly idiosyncratic, that is each high-risk child-mother might have a combination of some, but not all, of the identified risk factors. Hence, it might be possible to develop early intervention for high-risk children-mother dyads and tailor the early intervention for each child based on these and other relevant factors.

A statistical approach to defining abnormal aggression has some limitations. Obviously, the norms used have to be accurate and relevant to the population that they are applied to. The cut off scores also have to be valid for the purpose used. Furthermore, these norms and other statistical procedures have to be robust across different populations. That is, the results derived from the original sample have to be largely applied to the sample we are interested in today. Note that the data from the two studies discussed in this section were both collected in relatively urban areas of Canada some 20–30 years ago. Such studies are expensive, effortful to conduct, and take time to analyze and interpret. Hence, obtaining contemporary data for specific geographical regions and cultures is very difficult. Finally, although high rates of aggression and violence might be statistically unusual, they may or may not be socially significant and may not represent an important problem for the individual or society. Thus, high rates of aggression may be relatively unproblematic in one context, for example, where aggressive behavior is the norm and valued, whereas in another context much lower rates of less intense aggression may still result in socially significant negative outcomes, such as expulsion from school because of the interaction of specific behavior and context.

1.5.2 Association with Psychopathology

A second approach to defining abnormal aggression is by its possible association with psychopathology. An interesting line of research that attempts to address this question comes from work by Haller and colleagues who attempted to develop animal models of abnormal aggression in order to develop models of aggression which could be useful in both understanding abnormal aggression and developing novel psychopharmacological treatments for abnormal aggression (Haller, 2017, 2020). Haller and Kruk (2006) noted that aggression is associated with many forms of psychopathology, specifically, they noted that risks of aggression, violence, and homicide are elevated for disorders related to brain damage, such as traumatic brain injury and Alzheimer disease and diverse DSM diagnoses, including schizophrenia, alcohol abuse, Attention Deficit Hyperactivity Disorder, Conduct Disorder, and Antisocial Personality Disorder, etc.

Haller and Kruk (2006) proposed that there were three possible approaches to model abnormal aggression in rats. The first model was called *hypoarousal-driven aggression*. They defined normal aggression as the behavioral pattern typically observed, namely, rats typically bite nonvulnerable body parts where there are no vital organs and where the body is covered with muscle and bone. Attacks to vulnerable body parts occur in only 3–7% of attacks and hence are statistically unusual and more dangerous to the victim. Hence, typical or apparently normal aggression is not as dangerous as it might be to either parties. These authors argued that abnormal aggression was indicated by a higher proportion of attacks to vulnerable body parts in two situations. The first situation was when aggression was elicited by electrical stimulation of the hypothalamus. The second situation was when rats were glucocorticoid deficient. (Glucocorticoids are steroids secreted from the adrenal gland's cortex and are involved in a wide range of physiological functions, including regulation of stress, immune response, and inflammation. These effects are familiar to us as steroid treatment for allergic reactions, asthma, etc.) Thus, glucocorticoid deficiency is associated with hypoarousal.

Haller's group induced glucocorticoid deficiency experimentally by removing the adrenal gland and using subcutaneous pellets to maintain low levels of corticoids. Aggression in these two preparations were said to be abnormal because it occurred: (a) more frequently than in control animals; and (b) in unusual contexts, such as within the animal's' own territories against smaller conspecifics, rather than extraterritorial aggression against larger intruders. Glucocorticoid deficient rats also showed decreased heart rate and reduced social interaction, but no changes in exploratory behavior suggesting that the effects were not general suppression of behavior, but relatively specific to social interactions. The authors drew comparisons to abnormal human aggression associated with hypoarousal, such as antisocial personality disorder.

The second form of abnormal aggression was labeled *hyperarousal-drive aggression*. Hyperaroused aggression occurred: (a) during frustration, that is when previously reinforced behavior is no longer reinforcer (i.e. operant extinction), and (b) in response to the sight or odor of an unfamiliar conspecific. Aggression that occurred in these contexts was said to be abnormal because it occurred more frequently than usual. Haller speculated that this form of aggression may model aggression associated with chronic fatigue and posttraumatic stress disorder.

The third form of aggression was called *attacks out of context*. This form of aggression typically occurred as part of a fairly predictable behavior chain involving signaling to a conspecific, ritualistic behavior, and submissive behavior. In glucocorticoid deficient rats, aggression not only occurred more frequently, but also with fewer precursor behavior. In hypothalamic stimulated aggressive rats, aggression occurred without provocation from conspecifics and may even have involved attacks on dead rats. Haller and Kruk (2006) speculated that this form of aggression might be analogous to impulsive, uncontrollable, genetically-driven human aggression.

1.6 Summary

We often talk about aggression and violence without thinking carefully about the exact meaning of these and related terms. Because these terms are used in everyday life, by researchers and practitioners, by the legal system in many different contexts, and with both nonhumans and humans, they are used in many different ways resulting in considerable imprecision and confusion. Common distinctions include distinctions between less severe (aggression) and more severe (violent) behavior, but not all authors agree that this distinction is a useful or fundamental one. It is also common in everyday life, legal, research, and applied contexts to distinguish "hot blooded," impulsive, reactive aggression, and violence that the person cannot control at the time and is not responsible for and "cold blooded," planned, instrumental aggression, and violence which the is culpable for, although again not all agree that this distinction is a valid one. In addition, various authorities have categorized different types of aggression and violence in many different ways reflecting the different purposes of classification. Such attempts vary considerably in their reliability, accuracy, and intellectual cohesion. Sometimes they fail because of being unclear as the purpose and guiding principles behind each attempt at classification. One such approach has been to distinguish normal from abnormal aggression and violence which uses either statistical approaches or association with psychopathology to make such distinctions. Each of these approaches has their own limitations. Finally, measurement of aggression and violence is closely tied up with conceptual, measurement, and ethical questions which are often muddied if the purpose of assessment is not explicit.

References

Allen, M. B., Baker, J. C., Nuernberger, J. E., & Vargo, K. K. (2013). Precursor manic behavior in the assessment and treatment of episodic problem behavior for a woman with a dual diagnosis. *Journal of Applied Behavior Analysis, 46*(3), 685–688.

Alvarez, A., & Bachman, R. D. (2019). *Violence: The enduring problem* (4th ed.). SAGE.

Archer, J., & Coyne, S. M. (2005). An integrated review of indirect, relational, and social aggression. *Personality and Social Psychology Review, 9*(3), 212–230.

Azrin, N. H., Hutchinson, R. R., & Hake, D. F. (1966). Extinction-induced aggression. *Journal of the Experimental Analysis of Behavior, 9*(3), 191–204.

Baron, R. A., & Richardson, D. R. (1994). *Human aggression*. Springer.

Bartol, C. R., & Bartol, A. M. (1989). *Juvenile delinquency: A systems approach*. Prentice Hall.

Berkowitz, L. (1993). *Aggression: Its causes, consequences, and control*. McGraw-Hill.

Blanchard, D. C., & Blanchard, R. J. (1984). Inadequacy of pain–aggression hypothesis revealed in naturalistic settings. *Aggressive Behavior, 10*(1), 33–46.

Blanchard, D. C., & Blanchard, R. J. (2003). What can animal aggression research tell us about human aggression? *Hormones and Behavior, 44*(3), 171–177.

Blanchard, R. J., & Blanchard, D. C. (1989). Antipredator defensive behaviors in a visible burrow system. *Journal of Comparative Psychology, 103*(1), 70–82.

Bourdieu, P. (2001). *Masculine domination*. Stanford University Press.

Bushman, B. J., & Anderson, C. A. (2001). Is it time to pull the plug on hostile versus instrumental aggression dichotomy? *Psychological Review, 108*(1), 273–279.

Bushman, B. J., & Anderson, C. A. (2002). Violent video games and hostile expectations: A test of the general aggression model. *Personality and Social Psychology Bulletin, 28*(12), 1679–1686.

Charette, C. N., & Hickey, E. W. (2017). Legal definitions of violence and aggression. In P. Sturmey (Ed.), *The Wiley handbook of violence and aggression* (pp. 1–14). Wiley.

Conroy, M. E. (2019). Economic aggression as an instrument of low-intensity warfare. In M. E. Conroy (Ed.), *Reagan versus the Sandinistas* (pp. 57–79). Routledge.

Crawshaw, L. (2009). Workplace bullying? Mobbing? Harassment? Distraction by a thousand definitions. *Consulting Psychology Journal: Practice and Research, 61*(3), 263.

Crick, N. R., & Grotpeter, J. K. (1995). Relational aggression, gender, and social-psychological adjustment. *Child Development, 66*(3), 710–722.

Dixon, L., & Wride, A. (2021). Classification of intimate partner aggression. *Aggression and Violent Behavior, 59*, 101437.

Domb, M. (1978). Defining economic aggression in international law: The possibility of regional action by the Organization of American States. *Cornell International Law Journal, 11*(1), 4.

Dupuis, D. L., Lerman, D. C., Tsami, L., & Shireman, M. L. (2015). Reduction of aggression evoked by sounds using noncontingent reinforcement and time-out. *Journal of Applied Behavior Analysis, 48*(3), 669–674.

Edens, J. F., & Douglas, K. S. (2006). Assessment of interpersonal aggression and violence: Introduction to the special issue. *Assessment, 13*(3), 221–226.

Eisner, M. (2014). From swords to words: Does macro-level change in self-control predict long-term variation in levels of homicide? *Crime and Justice, 43*(1), 65–134.

Ericsson, K., & Ellingsen, D. (2005). Life stories of Norwegian war children. In K. Ericsson & E. Simonsen (Eds.), *Children of World War II: The hidden enemy legacy* (pp. 93–112). Bloomsbury Publishing.

Fávero, M., Pinto, S. C., Del Campo, A., Moreira, D., & Sousa-Gomes, V. (2020). Power dressed in black: A comprehensive review on academic hazing. *Aggression and Violent Behavior, 2020*, 101462.

Frederiksen, L. W., & Peterson, G. L. (1977). Schedule-induced aggression in humans and animals: A comparative parametric review. *Aggressive Behavior, 3*(1), 57–75.

Galtung, J. (1969). Violence, peace, and peace research. *Journal of Peace Research, 6*(3), 167–191.

Geen, R. G. (1990). *Human aggression*. Open University Press.

Girard, L., Tremblay, R. E., Nagin, D., & Cote, S. M. (2019). Development of aggression subtypes from childhood to adolescence: A group-based multi-trajectory modelling perspective. *Journal of Abnormal Child Psychology, 47*, 825–838. https://doi.org/10.1007/s10802-018-0488-5

Goodall, J. (1986). Social rejection, exclusion, and shunning among the Gombe chimpanzees. *Ethology and Sociobiology, 7*(3-4), 227–236.

Graham, K., Osgood, D. W., Wells, S., & Stockwell, T. (2006). To what extent is intoxication associated with aggression in bars? A multilevel analysis. *Journal of Studies on Alcohol, 67*(3), 382–390.

Haller, J. (2017). Studies into abnormal aggression in humans and rodents: Methodological and translational aspects. *Neuroscience & Biobehavioral Reviews, 76*, 77–86.

Haller, J. (2020). *Neurobiopsychosocial perspectives on aggression and violence: From biology to law enforcement*. Springer.

Haller, J., & Kruk, M. R. (2006). Normal and abnormal aggression: human disorders and novel laboratory models. *Neuroscience & Biobehavioral Reviews, 30*(3), 292–303.

Hinduja, S., & Patchin, J. W. (2010). Bullying, cyberbullying, and suicide. *Archives of Suicide Research, 14*(3), 206–221.

Howard, R. M., Potter, S. J., Guedj, C. E., & Moynihan, M. M. (2019). Sexual violence victimization among community college students. *Journal of American College Health, 67*(7), 674–687. https://doi.org/10.1080/07448481.2018.1500474

Kitzmann, K. M., Gaylord, N. K., Holt, A. R., & Kenny, E. D. (2003). Child witnesses to domestic violence: A meta-analytic review. *Journal of Consulting and Clinical Psychology, 71*(2), 339–352. https://doi.org/10.1037/0022-006X.71.2.339

Kulkarni, M. R., Graham-Bermann, S., Rauch, S. A. M., & Seng, J. (2011). Witnessing versus experiencing direct violence in childhood as correlates of adulthood PTSD. *Journal of Interpersonal Violence, 26*, 1264–1281. https://doi.org/10.1177/0886260510368159

Leclerc, S., & Herrera, C. D. (2000). Sport medicine and the ethics of boxing. *Western Journal of Medicine, 172*(6), 396.

Lee, B. X., Donnelly, P. D., Cohen, L., & Garg, S. (2016). Violence, health, and the 2030 agenda: Merging evidence and implementation. *Journal of Public Health Policy, 37*, 1–12. https://doi.org/10.1057/s41271-016-0011-6

Longpré, N., Guay, J. P., Knight, R. A., & Benbouriche, M. (2018). Sadistic offender or sexual sadism? Taxometric evidence for a dimensional structure of sexual sadism. *Archives of Sexual Behavior, 47*(2), 403–416.

Menesini, E., & Nocentini, A. (2009). Cyberbullying definition and measurement: Some critical considerations. *Journal of Psychology, 217*(4), 230–232.

Monchalin, L., Marques, O., Reasons, C., & Arora, P. (2019). Homicide and Indigenous peoples in North America: A structural analysis. *Aggression and Violent Behavior, 46*, 212–218.

Moyer, K. E. (1972). *Territorial aggression: Is this concept useful? Totus Homo, 4*, 104–117.

Nadal, K. L. (2011). The racial and ethnic microaggressions scale (REMS): Construction, reliability, and validity. *Journal of Counseling Psychology, 58*(4), 470–480. https://doi.org/10.1037/a0025193

Nagin, D., & Tremblay, R. E. (1999). Trajectories of boys' physical aggression, opposition, and hyperactivity on the path to physically violent and nonviolent juvenile delinquency. *Child Development, 70*(5), 1181–1196.

Nantel-Vivier, A., Pihl, R. O., Côté, S., & Tremblay, R. E. (2014). Developmental association of prosocial behaviour with aggression, anxiety and depression from infancy to preadolescence. *Journal of Child Psychology and Psychiatry, 55*(10), 1135–1144.

Nishida, T., Hiraiwa-Hasegawa, M., Hasegawa, T., & Takahata, Y. (1985). Group extinction and female transfer in wild chimpanzees in the Mahale National Park, Tanzania. *Zeitschrift für Tierpsychologie, 67*(1-4), 284–301.

Ogunyemi, D., Clare, C., Astudillo, Y. M., Marseille, M., Manu, E., & Kim, S. (2020). Microaggressions in the learning environment: A systematic review. *Journal of Diversity in Higher Education, 13*(2), 97–119. https://doi.org/10.1037/dhe0000107

Rodríguez-Domínguez, C., Pérez-Moreno, P. J., & Durán, M. (2020). Cyber dating violence: A review of its research methodology. *Anales de Psicología, 36*(2), 200–209.

Saini, V., Greer, B. D., & Fisher, W. W. (2015). Clarifying inconclusive functional analysis results: Assessment and treatment of automatically reinforced aggression. *Journal of Applied Behavior Analysis, 48*(2), 315–330.

Salimi, N., Karimi-Shahanjarini, A., Rezapur-Shahkolai, F., Hamzeh, B., Roshanaei, G., & Babamiri, M. (2019). Aggression and its predictors among elementary students. *Journal of Injury & Violence Research, 11*(2), 159–170. https://doi.org/10.5249/jivr.v11i2.1102

Smith, P. K., Mahdavi, J., Carvalho, M., Fisher, S., Russell, S., & Tippett, N. (2008). Cyberbullying: Its nature and impact in secondary school pupils. *Journal of Child Psychology and Psychiatry, 49*(4), 376–385.

Solórzano, D., Allen, W. R., & Carroll, G. (2002). Keeping race in place: Racial microaggressions and campus racial climate at the University of California, Berkeley. *Chicano-Latino Law, 23*, 15–112.

Sturmey, P. (2020). *Functional analysis in clinical treatment*. Academic Press.

Swit, C. S., & Slater, N. M. (2021). Relational aggression during early childhood: a systematic review. *Aggression and Violent Behavior, 2021*, 101556.

Thomas, R. (2019). College student peer aggression: A review with applications for colleges and universities. *Aggression and Violent Behavior, 48*, 218–229.

Tinbergen, N. (1968). On war and peace in animals and man. *Science, 160*(3835), 1411–1418.

Vaillancourt, T., McDougall, P., Hymel, S., Krygsman, A., Miller, J., Stiver, K., & Davis, C. (2008). Bullying: Are researchers and children/youth talking about the same thing? *International Journal of Behavioral Development, 32*(6), 486–495.

Vitaro, F., Gendreau, P. L., Tremblay, R. E., & Oligny, P. (1998). Reactive and proactive aggression differentially predict later conduct problems. *The Journal of Child Psychology and Psychiatry and Allied Disciplines, 39*(3), 377–385.

Vorobej, M. (2008). Structural violence. *Peace Research, 2008*, 84–98.

Ward, T., & Carter, E. (2019). The classification of offending and crime related problems: A functional perspective. *Psychology, Crime & Law, 25*(6), 542–560.

Wegner, D. M., & Bargh, J. A. (1998). Control and automaticity in social life. In D. T. Gilbert, S. T. Fiske, & G. Lindzey (Eds.), *The handbook of social psychology* (pp. 446–496). McGraw-Hill.

World Health Organization. (2016). *Ethical and safety recommendations for intervention research on violence against women: Building on lessons from the WHO publication putting women first: Ethical and safety recommendations for research on domestic violence against women*. WHO.

Wrangham, R. W., Wilson, M. L., & Muller, M. N. (2006). Comparative rates of violence in chimpanzees and humans. *Primates, 47*(1), 14–26.

Wright, A. J., & Wegner, R. T. (2012). Homonegative microaggressions and their impact on LGB individuals: A measure validity study. *Journal of LGBT Issues in Counseling, 6*(1), 34–54.

Chapter 2
Measurement of Violence and Aggression

Measurement is "the process of quantifying an attribute of a dimension or individual or behavior" (Fisher et al., 2017, p. 632). Definitions, conceptualization, and measurement of aggression and violence are closely linked to one another. Only when one knows what one is attempting to measure can one be sure that one has chosen an adequate measure of the problem; that is, the measurement system must be valid for the specific question asked. Since measurement questions about aggression and violence are so varied and take place in so many different contexts and with many different populations, one assessment method may be adequate for one question and inadequate for another. Researchers and all practitioners (clinicians, attorneys, advocates, etc.) should be aware of this issue and reflect when measuring aggression and violence as to whether their measurements are the right ones for the questions asked. Assessment of violence and aggression is an extensive topic addressed in many different contexts, for example, Browne et al. (2017) devoted an entire book to assessment of violence in forensic settings (Chap. 10 of this volume, provides a more detailed review of assessment).

Any measurement system must have certain desirable properties for it to be credible, including reliability and validity (Fisher et al., 2017). Although discussion of these issues usually is associated with psychological measures, such as questionnaires and observational measures, they do in fact apply to any measurement system. Thus, issues of reliability and validity apply just as much to legal distinctions of severity and judgements of culpability of offenses as they do to psychometric measures of callous and indifferent traits.

Fisher et al. (2017) proposed a number of psychometric principles that underlie all measurement systems. The first is that psychometric evidence is conditional. That is measures vary in their relevance and technical properties across contexts. Thus, a predictor of violence may be reliable and valid under one condition and invalid in another, for example, across members of different cultural groups. Second, psychometric properties are unstable; they may change over time. For example, a measure of bullying might be valid in the 1960s but be woefully incomplete in the

P. Sturmey, *Violence and Aggression*, Advances in Preventing and Treating Violence and Aggression, https://doi.org/10.1007/978-3-031-04386-4_2

2020s. Third, adequacy of psychometric measures might vary across dimensions of individuals. For example, a good predictor of violence in children may be an inadequate predictor of violence in adolescents. Fourth, the content validity of a measure may not be equivalent across different individuals or groups. For example, a measure that might capture relational conflict in adolescents might be relevant or incomplete or contain irrelevant items when measuring relationship conflict in older adults. The stability of psychometric measures can also be affected by the context of measurement. This lead Fisher et al. to state that "Psychometric evidence for a measure is conditional and cannot be presumed to be stable (i.e. equivalent) across dimensions of individual difference, assessment goals, sources of data, aspects of the targeted construct, contexts or time" (p. 636). Thus, attempts to demonstrate the absolute reliability and validity of a measure of aggression and violence are futile; thus statements such as "Questionnaire X is a reliable measure of intimate partner violence" should be regarded with suspicion and may be meaningless in some instances. Rather, researchers and practitioners should be content to demonstrate that the measure is reliable, valid, and appropriate for the purpose and context in which it was used.

There are a variety of methods that have been used to measure aggression and violence including databases and archives, interviews, self- and other-report questionnaires, observational, and physiological measures. Measurement systems may address aggression and violence directly or may measure related and overlapping concepts such as anger, or cognitions, attitudes, and beliefs related to aggression and violence and related concepts. The following sections will review these different types of measurement.

2.1 Archival Measures

We often depend on archival sources of data for state, national, and global statistics on aggression and violence, especially those related to crime. Examples include police statistics on violence and violent crimes, social services data on child and elder abuse, etc. Such data are potentially useful to determine the epidemiology of some forms of aggression and violence, to identify high risk areas or populations, and to evaluate social programs, such as crime prevention programs. Thus, when we read or hear in the media that "violent crime is up," we should ask ourselves, where these data came from and exactly what such statements might mean.

There are two main archival sources of US crime statistics: (a) Police reporting of crimes to states; and (b) national databases and surveys of the general public to obtain their reports of victimization (Alvarez & Bachman, 2019). Police recording of crimes into state or national databases is a potentially useful source of data to measure changes in society-wide rates of aggression and violence both over time and geographically. For example, Alvarez and Bachman (2019) chapter on assault and murder showed that contrary to commonly held beliefs, US violent crime has generally *decreased substantially* from about 1970 to 2014; this was true for various forms of assault and murder. US murder rates in 2014 also varied from state to state by a factor of about 6. These trends are part of a much broader trend of very large reductions in violent crime over the last few hundred years or so (see Chap. 7, for a more detailed discussion of this interesting phenomenon).

Data from recording of crimes by the police—even violent crimes—may not be as accurate as we might hope. Especially in the case of less severe forms of aggression and violence. Such data may be inaccurate because victims fail to report the crime, police encourage victims to withdraw allegations, the reporting process is arduous or embarrassing, crimes are recorded as being less serious that they actually were, etc. Even actual murders may not be recorded as such. Furthermore, in cases of missing persons it may be difficult to determine if they have purposefully disappeared, have been abducted or murdered and their bodies have not been found. In certain cases, such as the disappearance of indigenous women in the US and Canada, it may be that the government does not try too hard to determine what happened. In addition, governments might be motivated to misrepresent such statistics, for example, to exaggerate the effectiveness of public programs, minimize public concern when violent crime does in fact increase, and cover up institutional abuse and violence, including homicides and murders, when programs are run by the government itself or contracted organizations, such as religious organizations running residential programs.

Many governments also conduct surveys of potential victims by interviewing members of households, teenagers who might commit aggressive or violent crime, and individuals in relationships who might be victims of intimate partner violence. These can include phone surveys of random samples of the general population. Such surveys of self-reports may be useful, especially to evaluate violent crime which is not reported to the police. (The majority of crime, including violent crime, is not reported). Again, the accuracy and completeness of these data is questionable as potential victims might misunderstand questions or exaggerate or misreport. Making questions as explicit as possible and adding timelines may help. Questions such as "Have you been a victim of domestic violence?" may elicit a few accurate, positive answers resulting in an undercount of aggression and violence. On the other hand, questions that start with explicit timelines, such as "In the past 12 months have you ….", and making language explicit, such as "receive anal sex … perform oral sex … slapped you ... burn you on purpose ..." etc., may result in a more complete and accurate report of the phenomenon.

An example of the use of archival data on aggression and violence comes from Yampolskaya and Banks (2006) who attempted to identify predictors of recurrence of maltreatment of children. They analyzed archival data on child maltreatment in nearly half a million children in Florida who had been maltreated between 1996 and 2003. The authors used extensive state databases to code features during initial reports of aggression and violence to the children and used them to predict recurrence of aggression and violence. The databases were administrative databases and were not designed for research and so the researchers had to recode data and make assumptions about the meaning of these data. For example, Florida had two state databases related to child abuse which have been combined into one database; data were generally quite complete but missing data had to be imputed for about 0.1–1% of some variables, including race. The authors excluded children who died or who aged out of the system (i.e., were older than 18 years). They coded the severity of aggression and violence on a rating scale going from unconfirmed allegations of absence of a caregiver to abuse and other forms of maltreatment.

The authors found that there was a gradual reduction in the likelihood of recurrence of aggression and violence over a 7-year period. The authors evaluated predictors of recurrence of aggression and violence at 12, 24, and 36 months after an initial report. The main predictors of recurrence of aggression and violence to the children were neglect at initial report, alcohol and drug abuse, prior referral, and multiple maltreatments.

Such archival data are useful in evaluating the extent of aggression and violence (some 15% of children in Florida were in the database) and the effectiveness of public programs in preventing and treating aggression and violence. Furthermore, the identification of predictors of recurrence of aggression and violence has important implications for identifying: (a) the most at risk children; and (b) choosing optimal interventions, such as parent training, monitoring to ensure parents are physically present in the home, and/or programs to reduce alcohol and drug use and abuse.

2.1.1 Archival Data on War Crimes, Genocide, Ethnic Cleansing

If accurate state and national data on violent crime is difficult to obtain, obtaining accurate data on state and supranational mass violence is especially difficult to obtain. It may be genuinely difficult with the best will in the world to measure violence and murder that has occurred on a large scale that takes place over an extended period of time and/periodically. Bodies may be hidden or destroyed. People may be unable or unwilling to report what happened. A related problem is that some deaths might be directly related to war, such as deaths that occur in a battle, but other deaths may be less clearly related to a war such as civilian deaths due to firebombing of urban areas, deliberate starvation, or scorched earth policies imposed directly or indirectly on a population sometimes with the explicit intent to kill. There is also the issue of how to count deaths that have occurred in internment or labor camps, and enforced labor and enslavement during wartime. Furthermore, governments are clearly motivated to deny, lie, dissimulate, and prevaricate as to what occurred even in the face of clear facts to the contrary. Thus, counting how many people died during wars and related conflicts is genuinely difficult. A further problem is that the classification of such large-scale violence into certain categories, such as "pogrom," "ethnic cleansing," etc. is problematic. For example, the definition of "genocide" requires an intent on the part of a government to wipe out an ethnic group. Thus, a government might claim that, although they did indeed murder large numbers of members of an ethnic group, their intention was only to reclaim land or protect their own population, but not to eradicate the people who lived there and that the eradication of a particular cultural group was never the government's intension (Alvarez & Bachman, 2019).

In order to measure large-scale violence using archival data, it is necessary for scholars and international organizations such as WHO and NATO to construct databases using the best historical information available. One example comes from Leitenberg (2006a, 2006b) among others who have attempted to construct databases measuring worldwide deaths during the twentieth century. This was done by setting limits on the dates studied—1900–2000 in one case and 1945–2000 in another—and listing conflicts and wars that had occurred. Then, for each conflict or war estimates were made for the number of military and civilian deaths that had occurred. Leitenberg assembled and evaluated the best evidence available for each conflict or war. The resulting data often resulted in widely varying estimates that differed perhaps by factors of five to ten depending on the dates set and definitions of conflict-related deaths in each case. For example, some reports might focus narrowly only on military deaths during a conflict whereas others would include post conflict deaths in refugee camps if the deaths were deliberately engineered, for example, by victors running refugee camps for displaced persons from the losing side of the war. In addition, Leitenberg estimated deaths in gulags in the USSR, China, and North Korea. Table 2.1 summarizes these data that indicated approximately 140 million deaths due to war and a further 43 million deaths in Russian, Chinese, and North Korean gulags and enforced labor camps. In addition there were other large-scale deaths from deliberate starvation and forced relocations of people, for example, in Russia between 1930 and 1950 under Stalin (6–9 million deaths), in China under Mao Zedong (15–55 million deaths) during the Great Leap Forward, and Cambodia under the Pol Pot (1.5–2 million deaths).

Table 2.1 A summary of twentieth century deaths related to wars, gulags, work camps and related events

War/conflict	Number dead
World war I	13–15M
Russian civil war and Russian–Polish conflict (1918–1922)	12.5M in Russia plus others
Mexican revolution (1909–1916)	1M
Spanish civil war (1936–1939)	0.6M
World war II	65–75 M
Wars and conflicts 1945–2000	41M
Total from wars 1900–2000	136.5–148.5
Russian gulags	35M
Chinese gulags	6M
North Korean gulags	0.6M
Armenian genocide	1M

Establishing reliable and accurate databases to document aggression and violence is indeed difficult. The above example appears especially difficult as there is no single responsible body to assemble such data and considerable motivations to under-report data. There are, of course, but similar problems exist in maintaining agency, state, and national databases related to aggression and violence as often there are insufficient resources to maintain the database accurately and individuals who enter and analyze the data may have considerable latitude in how to code events.

2.2 Interview Measures

Interviews are commonly used by both researchers and practitioners to assess aggression and violence and related behavior in many contexts and for different purposes. For example, someone might be interviewed concerning their personal experience during their lifetime or a specified time period such as adolescence, or more recent experience of aggression and violence. They might be interviewed concerning their experiences as either victim or perpetrator or related to a specific event, such as a crime. These interviews might take place in the context of an applied problem, such as ensuring the immediate safety of victims and family members, or as part of a research project, for example, to determine the prevalence of intimate partner violence, its risk factors, and/or related service planning needs. Interviews produce data based on interviewee verbal or written reports rather than direct measures or permanent product of behavior. Hence, they are subject to many potential problems in accuracy of those reports. If this is true for relatively neutral topics, this may be even more true for emotionally-charged topics, such as personal experience of aggression and violence or safety of oneself or family members.

A good example of this approach comes from an analysis of three surveys which produced substantially different estimates of the prevalence of domestic violence in Nicaragua. Ellsberg et al. (2001) noted that three surveys estimated that the prevalence of domestic violence was either 28%, 52% or 69%! The first survey was a large-scale national survey and the last two surveys were smaller scale regional surveys. Ellsberg et al. compared the interview procedures used in terms of the types of questions posed, the strategy used to present questions, the presence of safety procedures for the interviewee, the presence or absence of other people during the interview, the interviewer general behavior and training and support for the interviewer.

Ellsberg et al. (2001) framed the problem in terms of the under-reporting of violence during interviews. That is, many victims of domestic violence might not report domestic violence because the questions are unclear or that the interviewees were reluctant to report violence because of shame, embarrassment, lack of clarity of the definition of violence, or presence of other family members, and other safety issues during the interview. One important factor is the comfort of the interviewee with the interviewer. These authors addressed this in part by having female rather than male interviewers. They also had the interview open with framing the issue of domestic violence as a common occurrence for many. Another relevant factor is the question format. If a question is posed rather broadly (e.g., “Are you the victim of domestic violence”) interviewees are less likely to report violence than if the questions is more specific (e.g., “Has your husband/domestic partner threatened you with a knife or other weapon over the last 12 months?). The authors also noted that, since these interviews were conducted in people’s family homes, other relevant people, such as the perpetrator or his immediate family, might be present or nearby resulting in fear of reprisal for reporting. In the larger national survey, the authors reported that there was time pressure to complete interviews quickly resulting in rushed, incomplete, and superficial assessment. Finally, Ellsberg et al. reported that interviewers also needed support, as in when they learned about domestic violence and believed that the woman remained at risk, but the interviewers were unable to offer the support needed to keep the woman safe. These considerations led the authors to conclude that the large-scale national survey was inaccurate due mostly to under-reporting of violence due to the additive influence of many of these factors. Thus, the much higher prevalence figures in the smaller regional surveys appeared to be more accurate.

As a consequence of these problems and as a part of their Violence Against Women initiative, the World Health Organization (2016) developed guidelines for how to conduct interviews in a safe and ethical manner. These guidelines attempted to both keep women safe during and after interviews, to ensure that the results of the interviews address the problem of under-reporting and are complete and accurate, and that staff conducting the interviews are also safe and experience minimal trauma. The eight recommendations were as follows. First, the top priority is to ensure the safety of interviewees and that should guide all project decisions. Second, prevalence studies should be well designed in order to ensure that under-reporting of violence does not occur. Third, researchers should protect interviewees’ confidentiality to ensure interviewee safety and data integrity. Fourth, research team members and interviewers should be selected and trained carefully. They should also receive ongoing support. Fifth, the research protocol should include steps to minimize distress to interviewees. Sixth, there should be mechanisms to refer women who are at risk of harm to local services and/or create short-term support mechanisms where there are few local resources. Seventh, researchers and donors should ensure that the results are disseminated accurately and used to advance policy and services. Finally, when questions about violence are inserted into existing interview and other protocols, the researchers should ensure that this is done in an ethical and methodologically sound manner.

Interviews are one of the most commonly used and easy to administer ways to assess aggression and violence. Although commonly used, interviews come with numerous limitations. Some of these problems can be minimized by changing the interview format in various ways or supplementing the results of interview data with other data sources to validate the interview data and hopefully arrive at some valid conclusion.

2.3 Psychometric Measures

Psychometric measures of aggression and violence are commonly used by psychologists, mental health professionals, and other social scientists. Psychometric measures are used for multiple purposes. They are often used to screen for the presence of aggression and violence at intake or annual or other periodic evaluation. They are sometimes used to measure differences between individuals, and also to measure changes in behavior over time, such as age-related changes or changes in behavior during treatment trials. A key feature of psychometric measures is that their format and administration is typically standardized. This is done to reduce extraneous and irrelevant variability, for example, by limiting the questions asked, phrasing scoring, and other aspects of administration and interpretation. For example, psychometric instruments standardize both the questions and the manner in which respondents answer them. For example, a person might be asked to rate the frequency of their aggressive behavior on a seven-point Likert scale. Psychometric measures might be completed by the person themself or by relevant others, such as staff or family members. Additionally, some psychometric instruments do not focus only on aggression, but may measure a wider range of related maladaptive behavior, psychiatric symptoms, and adaptive behavior. (See below for an example.)

Psychometric measures of aggression and violence and related behavior may include ratings of the frequency, intensity, and duration of behavior. There are very many psychometric measures of aggressive behavior for use with many different populations, contexts, and purposes. Here we will consider these three examples of many such questionnaires

One example of a psychometric measure is the *Overt Aggression Scale* (OAS) (Yudofsky et al., 1986) which aims to measure verbal and physical aggression, aggression against oneself and others. Each item consists of a scale which is weighted based on intensity. A total score is generated based on the sum of the weighted items. In the original study, Yudofsky et al. evaluated the OAS with 16 children with 54 episodes of aggression, 21 adults with 35 episodes of aggression, and eight adults with 70 episodes of aggression. They showed that the OAS was easy to complete and reliable. The OAS has been widely used in practice and research. Applications have included evaluation of changes in aggression in response to clinical drug trials (Silver & Yudofsky, 1991), and for a variety of populations such as psychiatric in-patients (Sorgi et al., 1991), individuals with autism (Hellings et al., 2005), and intellectual disabilities (Oliver et al., 2007).

The second example is the 52-item *Buss-Perry Aggression Questionnaire* to measure aggression (Buss & Perry, 1992). They developed this measure because they judged that previous psychometric measures were outdated and perhaps technically inadequate and so they generated items from previous psychometric measures and eliminated or revised items to clarify wordings, etc. They administered the measure to 1,253 undergraduates enrolled in introductory psychology classes alongside various personality and other measures. Statistical analysis (factor analysis) revealed four dimensions of aggression labeled physical aggression, verbal aggression, anger, and hostility. They found that men reported being more aggressive than women, especially on physical aggression; there were no sex differences on anger; and reports of aggression were correlated with personality traits such as assertiveness, competitiveness, impulsiveness, and emotionality. This measure focused on behavioral topographies, such as fighting, and cognitions and emotions, such as feeling suspicious of others, but did make distinctions between functional types of aggression, such as respondent-operant reactive-proactive types of aggression (Raine et al., 2006).

A third study illustrates a different application of psychometric measures of aggression, namely to evaluate the effects of psychotropic medication on aggressive behavior in children with autism, that is, the measure was designed to measure change in behavior in response to psychotropic medication for autism and perhaps aggression. McDougle et al. (1998) conducted a randomized controlled trial to evaluate the effect of a then new psychotropic medication, risperidone, on a variety of behavior in adults with autism, including aggression. The authors measured aggression using the *Self-Injurious Behavior Questionnaire* (SIB-Q), a 25-item, unpublished psychometric instrument that measured a wide range of externalizing problem behaviors, including aggression, self-injury, property destruction, and other maladaptive behavior. They found that scores on the SIB-Q reduced much more in the risperidone group than the placebo group at 4, 8, and 12 weeks of the trial and that other adults with autism who were given an open-label trial of risperidone showed similar changes. Thus, in this approach to measuring aggression the SIB-Q was sensitive to changes in response to psychotropic medication over time, but perhaps muddied the water by mixing up items on aggression with other behavioral items.

There are very many psychometric instruments to measure aggression in many different contexts for many different purposes. Readers should be familiar with the details of each instrument cited within each study.

2.4 Physiological Measures

Although physiological measures do not measure aggressive and violent behavior directly, they are often used as a research tool and sometimes to evaluate treatment in order to understand the nature of aggression and violence and to evaluate theories of aggression and violence. For instance, researchers have used related measures of physiological reactivity test theories of psychopathy and offending and the development of psychopathy.

Patrick and Drislane (2015) distinguished three broad classes of psychophysiological measures. First were autonomic measures, such as heart rate (HR) and skin conductance (SC). HR may be measured with electrodes on each side of the body often placed on the arms or wrists which detect ventricular contractions. HR measures are often taken at rest and may include simple beats per minute and HR variability. SC is typically measured by placing sensors on the palm or fingers to measure the activity of sweat glands as a measure of sympathetic nervous activity which controls many body functions related to threat and survival, reaction to stress, and fight or flight, such as arousal, vasoconstriction, pupil dilation, and gut mobility. Both HR and SC measures may be taken at rest (tonic) or after some stimulus is presented (phasic). As new technologies continue to develop there are more and more readily available and cheap sensors that can be worn continuously that may measure autonomic function.

The second class of physiological measures are electrocortical measures that detect changes in the electrical activity on the surface of the brain. These measures may be taken at rest with eyes closed or open or after some stimulus is presented. The measures derived from EEGs are coded in terms of the time from the onset of stimulation and the direction (positive or negative) of the polarization.

Finally, neuroimaging measures assess either brain structure or brain processes. Magnetic resonance imaging (MRI) allows visualization of soft tissues such as the brain. Functional MRIs visualize blood flow and oxygenation. Hence functional MRIs can index the processes going on in the brain rather than the brain's structures as indexed by changes in metabolism reflected in blood flow and oxygenation.

Researchers use electrophysiological measures to understand aggression and violence better in a wide range of populations. For example, electrophysiological measures have indicated that in children who are aggressive and violent show lower resting HR and SC. They are more reactive to threatening stimuli than nonaggressive children; although children who show proactive aggression show reduced responsivity to threatening stimuli. Such observations are useful in building theoretical models as to the causes and processes that may underlie aggression and violence (Patrick & Drislane, 2015).

An example of this use of physiological measures comes from White et al. (2016) who studied physiological correlates of aggression in 56 youth. Thirty of the youth showed Disruptive Behavior Disorder (DBD) (such as Conduct and Oppositional Defiant Disorders). The researchers split the DBD group into those with high and low callous-unemotional traits. (Callous and unemotional traits are reflected in lack of empathy for others, lack of guilt, and lack of emotional expressivity). There was also a group of 26 youth with no DBD. Participants played a "fairness game" while undergoing MRIs in which money could be divided evenly or unevenly and in which the participant could accept and offer or punish the game partner's unfair offers with fines of different magnitudes. The tendency to retaliate was measured as the average fine imposed on the hypothetical game partner. All groups accepted fair offers and punished unfair offers similarly, but individuals in the DBD group were more likely to punish mildly unfair offers than the control group. Neurological correlates of these behavioral findings indicated that individuals with DBD and low callous-unemotional traits demonstrated exaggerated basic threat circuitry (amygdala and periaqueductal gray). This study illustrates how physiological measures can be used to describe the relationship between behavior and brain function and obtain a better understanding of the biology of aggression and violence.

2.5 Experimental-Analog Measure

Experimental psychologists have studied aggression and violence in analog settings. Such studies use tasks where participants can engage in various forms of retribution, punishment, or symbolic aggression and violence, with real or contrived other people who act as potential victims of aggression and violence. These studies measure aggression as the quantity of an aversive substance, such as delivering unpleasant hot sauce, pleasant or nasty tasting drinks, blasts of air to the facial area, electric shocks (real or imagined), and monetary or points fines to a hypothetical victim. The main advantage of such measures is that they provide quantitative measures of aggression in terms of the frequency or stimulus magnitude, in the context of an experiment in which the experimenter can control the situation systematically in order to have strong internal validity and greater confidence that the independent variable caused a change in aggression and violence.

An example of this approach comes from Verona et al. (2009) who compared the effects of impersonal with interpersonal provocations on aggression and underlying neural functioning. The experimenters told participants in the impersonal provocation condition that the magnitude of air blasts was delivered by a computer at random whereas they told participants in the interpersonal condition that the magnitude of the air blasts was determined and delivered by another person based on the quality of an essay they had written. In fact, participants in both conditions received the same air blasts. There was also a control group that did not receive any air blasts. Individuals who received air blasts showed more aggression and more left frontal cortical activity as measured by EEG recordings than the control group, but the authors did not find differences between the two groups who had received air blasts. Such studies contribute both to understanding of aggressive behavior, such as the validity of the distinction between impersonal and interpersonal antecedents to aggression and to the potential neurobiological processes underlying aggression and violence.

Most of these studies are firmly within the tradition of experimental psychology and often involve contrived tasks and settings performed by psychology undergraduates working for course credit and other volunteers. Thus, this approach to measuring and understanding aggression and violence trades internal validity for potential problems with external validity. That is, it is unclear if these results apply to other contexts, populations and measures of actual aggression and violence. Sometimes the situations are so contrived and the participants are so psychologically-minded that debriefing of participants shows that a significant proportion do not believe the experimenters' cover story. For example, in Verona et al.'s (2009) study just under 10% of the participants had to be dropped because they stated that they did not believe the experimenters' cover story; of course, it is unknowable what percentage of those participants who stated they believed the cover story actually did not believe it!

2.6 Observational Measures

Observational measures of aggression and violence are typically associated with ethology and Applied Behavior Analysis and will be illustrated elsewhere in this volume. Typically, this involves systematically training observers to criterion to take reliable data either in the natural or contrived environments or using videotapes or audiotapes. Typically, collecting observational data on aggression begins with pilot data and observer training to clarify response definitions, identify areas of disagreement, refining response definitions, and training observers until they have adequate agreement. Additionally, many such observational studies also involve developing categories of relevant environmental events, such as crowding, temperature, presence of conspecifics and nonconspecifics, interactions, and types of interactions from other people and the presence or absence of certain tasks.

In developing an observational measurement system of aggressive behavior, one would have to decide first what the applied or research question was and develop a measurement system that reflected the question at hand. For example, if one were interested in the effects of competition for mates on male aggression in primates one would have to develop a series of behavior codes reflecting the relevant environmental features, such as number of females, their estrus status, and age, and the number of other males and their position in a dominance hierarchy, etc. One would also have to develop measures of relevant behavior, such as aggression, dominant, and submissive social behavior, and measures of their relevant dimensions. For example, the rate of aggressive behavior might be only partially useful, whereas the latency to aggression given a threat from another male given the presence or absence of an available mate might be a much more meaningful measure.

Observational measures of aggression and violence can measure specific dimensions of aggressive and violent behavior including the frequency, rate, duration, proportion of time, latency, behavioral variability, and behavior product, such as environmental damage or injuries to others. For example, if one was to measure the frequency of aggressive acts, one would have to define both the behavior and what constitutes one act of aggression. Only if the observational periods were of the same duration would frequency be a meaningful measure. More typically, observational periods in real world settings differ in duration and so the number of acts per unit time expressed as rate of probability might be more meaningful. Duration of behavior requires observers to agree as to the onset and offset of the behavior or to use time sampling to estimate duration. Again, since observation periods might commonly differ in duration, this measure might be more meaningful expressed as the percentage of time observed. Latency is the time between the onset of a stimulus and the behavior. For example, an Applied Behavior Analysis study might measure the number of seconds between a parent asking their child to do something and the first instance of the child hitting their parent. Variability refers to differences in some relevant aspect of behavior, such as variability in intensity or sequencing of behavior (Johnston & Pennypacker, 2010). Finally, behavioral product can be measured through environmental damages such as the number and / or value of broken items (See Table 1.6 in Chap. 1 for further details and examples of this approach).

To consider this approach to observational measurement of human aggression, consider this interesting naturalistic observational study from Graham et al. (2006), discussed briefly in Chap. 1, who observed aggression and violence in Canadian bars. Using trained observers, they used a manualized approach to code incidents of aggression during 1334 nights in 118 bars and clubs. They observed 1052 incidents involving 1754 patrons. About two-thirds of incidents involved men and about 94% of incidents involved minor or moderate levels of harm. Forty-two percent of incidents were defensive and 33% of incidents were definitely intentional incidents of aggression. There were gender differences in aggressive behavior. Women exhibited more incidents involving defensive intent, less intense aggression, and less anger than men; and more abusing, swearing, insulting and demeaning forms of aggression than men. Thus, this study used naturalistic observation to provide a simple description of the types of aggression humans displayed in a specific context and gender differences in the frequency and types of aggressive behavior.

2.7 Economic Evaluation

As noted in Chap. 1, the economic cost of aggression and violence is very great and one of the reasons for intervention. Hence, many program evaluations report not only reductions in aggression and violence, injuries, and crime, but also the economic benefits, if any, of the program. Program costs include the initial costs of developing the program, designing and piloting materials and procedures, and initial training of intervention agents, and ongoing costs include the costs of delivering the service, such as disseminating materials, training local change agents, recruiting participants and evaluation. Costs can include the cost of staff and consultant time, materials, and overheads, such as the costs of renting and using office space and related costs such as utilities. Often staffing costs are the largest parts of such costs. Savings can accrue from the avoidance of immediate costs of violence such as medical treatment and avoidance of lost productivity. Savings can also accrue from avoidance of days lost from work, and long-term and lifelong costs, such as ongoing medical and psychological treatment, disability costs, such as costs of surgery and rehabilitation, and special education and residential costs in some physically abused children, etc. The economic evaluations of violence programs might also be different depending upon the context. Thus, primary prevention programs may have to treat many individuals to avoid the costs of one incident of violence, where are programs for existing violence may sometimes have greater initial costs because of the more severe nature of the violence, but also greater opportunity for savings as the lifelong costs of the untreated effects of violence may be very large. Costs and benefits can also be calculated from the perspectives of different people. For example, the costs and benefits of intervention for a victim of intimate partner violence might be different from the costs and benefits to society.

There are three main approaches to measuring the costs and benefits of social programs. Cost–benefit analysis (CBA) compares money spent with the money saved and often expresses the outcome as a ratio such as "for every $1 spent, society saved $10." Cost effectiveness analysis (CEA) compared the costs per unit outcome, such as the costs of each future violence incident that was avoided. For example, "It costs $10,000 to avoid one future incident of a child with brain damage due to violence from a family member." Finally, cost-utility analysis (CUA) presents the costs of some utility or outcome, such as the costs of an additional year of a disability-free life (DALY). For example, "In this program the cost of one additional DALY was $15,000."

Peterson and Kearns (2021) conducted a systematic review of economic evaluations of violence reduction programs published between 2000 and 2019. They identified 28 studies of which 12 addressed community violence including sexual assault (3 studies) and firearms assaults (1 study), bullying (2 studies), elder abuse (1 study.) Most studies addressed both primary and secondary prevention. The most common strategies were education (11 studies), home visiting (5 studies), hospital-centered violence prevention (4 studies), multisystemic therapy (2 studies), social service (2 studies), violence hotspot coordination (2 studies), and alcohol policy change, built environment change, and counseling (1 study each.) Most studies indicated that violence prevention programs were economically effective, i.e., resulted in savings both in studies using short- and long-term windows. The studies reported widely diverging values of savings, which sometimes varied by a factor of ten.

Measurement of the economic benefits of violence prevention programs is based on a modest number of studies. This literature has produced consistent evidence of the economic benefits of violence prevention programs regardless of the type of program and population in which the study was conducted, although the size of the savings varied widely.

2.8 Risk Assessment

In many situations related to aggression and violence people want to predict what the risks are of future aggression and violence to determine the answers to important questions such as the necessary type of supervision, staffing or other actions to prevent future violence. Some examples of these questions include legal disposal, such as prison, placement in psychiatric, locked or unlocked facilities, or outpatient and other community services; protection of potential future victims; and release from prison and other restrictive settings and if so what conditions should apply, etc. Risk assessment has become a common practice to show that a service took reasonable action to prevent future aggression or violence. For example, a school might assess the risks after an incident of aggression, or prior to readmitting a suspended student to school, etc. Risk assessment is particularly important for assessment of aggression and violence as it may help understand aggression and violence, direct applied work to the most important cases, aid in clinical decision making, such as decisions for admission and discharge, and may be integrated into forensic case formulation (Hart et al., 2011; Sturmey & McMurran, 2011) (see Chap. 10). These concerns are heightened by the responsibility of clinicians and administrators to both provide effective treatment and to keep the public and others safe and the move away from long-term institutional services to shorter placements and community settings.

Within specific populations who are at risk for violence, such as repeat offenders with histories of violence or psychiatric in-patients, predicting *who* will be violent is relatively easy as a small minority of individuals often account for a large proportion of violent incidents. For example, people with repeated histories of arrests for violence and psychiatric disorders who are inpatients in acute admission locked units are more likely to be violence than individuals without such histories (Douglas & Skeem, 2005.) This is also reflected in restraint data which is similarly skewed toward a small minority of individuals and situations (Sturmey, 2015). However, predicting when and under what circumstances this may occur is more difficult as information on client history may be incomplete or missing and/or it may be difficult to anticipate accurately what future circumstances might occur that might precipitate an episode of aggression or violence.

When predicting future violence, it has been recognized that it is useful to distinguish between factors that cannot be changed readily—static variables—such as age, gender, and history, and factors that are mutable—dynamic variables—such as social skills, degree of intoxication, and some environmental provocations. In a review of dynamic risk factors for violence, Douglas and Skeem (2005) identified the following mutable rick factors: impulsiveness; negative affectivity, including anger and negative mood; psychosis; antisocial attitudes; substance use and related problems; interpersonal relationships; treatment alliance; and adherence for both medication and psychosocial treatments. Dynamic risk factors also vary along a dimension reflecting how stable they are. For example, some dynamic risk factors may be more trait-like and less likely or just more difficult to change than other risk factors that are more situational and liable to acute change such as drug and alcohol use. This framework has been used to develop inventories of static and dynamic risk factors related to prediction and prevention of violence in many studies.

Douglas and Skeem (2005) argued convincingly that clinicians and services must do something more difficult than predict who will be violence, rather they must predict the changing probability of future violent behavior, take action to prevent future violence for a specific person, and allocate resources to do so in a responsible and effective fashion. An assessment of risk at a single point in time—especially for a high risk individual—is insufficient as it will not assess the changes in dynamic risk factors over time in order to adjust the prediction of future aggression and violence. Thus, to predict future violence well, practitioners and services must assess changes in dynamic risk factors frequently for high risk individuals. Furthermore, to discharge their ethical duty to the individual and to keep others safe, they must also take effective action to prevent that violence, in part through modifying those dynamic risk factors. For example, if a high risk forensic outpatient begins carry weapons with him to work and social events, the clinician might identify the dynamic factors that have contributed to this change in behavior, such as increased paranoid thoughts, medication noncompliance, and loss of work, and then develop a plan to increase medication compliance, use cognitive behavior therapy to challenge paranoid beliefs and work with him to get a job or equivalent daytime activity. Note that this example illustrates the intimate relationship between risk assessment and forensic case formulation in that many dynamic risk factors are incorporated into forensic case formulation.

Douglas and Skeem (2005) argued that an important part of risk assessment and forensic case formulation is to identify the *causal* predictors of future violence rather than *risk markers*. They argued that the modification of the causal predictors is more likely to produce change than correlated. They give the example of an increase in violence associated with alcohol consumption. If alcohol consumption is a dynamic risk factor which is a causal variable, then direct modification of alcohol consumption is a likely effective risk management strategy. If, however, alcohol consumption is merely a correlate of a true causal dynamic risk factor, such as involvement with an antisocial and violent peer group, then modification of alcohol consumption will have missed the mark. This may, by focussing on a dynamic noncausal risk factor, result in suboptimal intervention that may not reduce the risk of future violence. A final important implication of this approach to assessing and managing dynamic risk factors is that the frequency of assessment of dynamic risk factors should be tailored to the risk for each individual. For example, an individual with low static risk factors generally should receive lower frequency assessments of risk, unless there is some clear sign of increases in dynamic risk factors; whereas an individual with high static risk factors might receive much more frequent assessments of risk than a low risk individual. This more frequent risk assessment can be useful in allowing the clinician and service to identify which risk factors they should address, whether or not their interventions to reduce dynamic risk factors has been sufficiently effective and whether further action should be taken.

Over the last 15 years or so, this approach has been adopted widely in services in many countries, but has this approach delivered better, safer interventions for violent individuals? A recent systematic review by Viljoen et al. (2018) was cautious about the answer to this question concluding that practice is far ahead of research. Their systematic review of risk assessment tools and reduction of violence identified 73 studies with 31,551 offenders or patients and 10,002 professionals. There were 34 risk assessment tools and studies used a range of outcomes including supervision, dispositions and violence. Of these, only 12 studies (7350 patients/offenders) evaluated the effects on reducing violence or offending and included three randomized controlled trials. Two of the three randomized controlled trials reported reduction in violence and one did not. Sometimes these studies included risk assessment as part of a larger battery of assessments and sometimes staff advice and training on how to reduce violence. Six nonexperimental studies found no effect on reduction in violence. Additionally, eight studies (2617 offenders or patients and 102 professionals) evaluated potential strategies to improve the effectiveness of risk assessment to reduce violence. These studies compared risk assessment versus risk assessment plus training, guidelines or policies. Outcomes were mixed. One study found clear improvements that were associated with staff acquiring intervention skills resulting in reduced recidivism and staff engagement with support activities for clients. Although training, guidelines, and policies improved some processes related to risk assessment, risk assessment was more likely to reduce violence when associated with staff training rather than guidelines, suggestions and policies alone.

A similar note of caution was sounded regarding measurement of risk by Heffernan et al. (2019). They noted that measurement of risk requires reliable and accurate measurement of changes in both the dynamic factors and changes in offending or offending-related behavior. In addition, researchers and practitioners should distinguish between predicting reoffending and predicting violent reoffending or some other more socially significant form of aggression and violence. Measurement of dynamic risk factors is also challenging because changes in dynamic risk factors today may not be reflected in changes in aggressive and violent reoffending until some unknown point in the future.

Today, risk assessment is widely used by many practitioners and different kinds of services related to violence and aggression. When done well and combined with other strategies, such as appropriate individual treatment, perhaps forensic case formulation (see below), and staff training and ongoing support, it has the potential to reduce violence in high risk populations.

2.9 Summary

Measurement of aggression and violence is fundamental to both research and practice. Since people come to this topic from many different intellectual and practice fields, and the reasons for measuring aggression and violence vary considerably from one study to another, it is unsurprising that approaches to measurement of aggression and violence vary so much. When we attempt to measure aggression and violence it sometimes forces us to think carefully about what we mean by these terms and the purpose for which we are measuring aggression and violence and related concepts. In the next chapter, *Epidemiology of Violence and Aggression*, we go on to explore the question of how common aggression and violence. In so doing we will explore risk factors and their implications for understanding, preventing, and treating aggression and violence.

References

Alvarez, A., & Bachman, R. D. (2019). *Violence: The enduring problem*. Sage Publications, Incorporated.

Browne, K. D., Beech, A. R., Craig, L. A., & Chou, S. (Eds.). (2017). *Assessments in forensic practice: A handbook*. John Wiley & Sons.

Buss, A. H., & Perry, M. (1992). The aggression questionnaire. *Journal of Personality and Social Psychology, 63*(3), 452.

Douglas, K. S., & Skeem, J. L. (2005). Violence risk assessment: Getting specific about being dynamic. *Psychology, Public Policy, and Law, 11*(3), 347–383.

Ellsberg, M., Heise, L., Pena, R., Agurto, S., & Winkvist, A. (2001). Researching domestic violence against women: Methodological and ethical considerations. *Studies in Family Planning, 32*(1), 1–16.

Fisher, J. E., O'Donohue, W., & Haynes, S. N. (2017). Principles and foundations of psychological assessment. In P. Sturmey (Ed.), *The Wiley handbook of violence and aggression* (pp. 1–12). Wiley UK.

Graham, K., Osgood, D. W., Wells, S., & Stockwell, T. (2006). To what extent is intoxication associated with aggression in bars? A multilevel analysis. *Journal of Studies on Alcohol, 67*(3), 382–390.

Hart, S., Sturmey, P., Logan, C., & McMurran, M. (2011). Forensic case formulation. *International Journal of Forensic Mental Health, 10*(2), 118–126.

Heffernan, R., Wegerhoff, D., & Ward, T. (2019). Dynamic risk factors: Conceptualization, measurement, and evidence. *Aggression and Violent Behavior, 48*, 6–16.

Hellings, J. A., Nickel, E. J., Weckbaugh, M., McCarter, K., Mosier, M., & Schroeder, S. R. (2005). The overt aggression scale for rating aggression in outpatient youth with autistic disorder: Preliminary findings. *The Journal of Neuropsychiatry and Clinical Neurosciences, 17*(1), 29–35.

Johnston, J. M., & Pennypacker, H. S. (2010). *Strategies and tactics of behavioral research*. Routledge.

Leitenberg, M. (2006a). *Deaths in wars and conflicts in the 20th century*. Cornell University, Peace Studies Program.

Leitenberg, M. (2006b). *Deaths in wars and conflicts between 1945 and 2000* (3rd ed.) Peace Studies Paper #29. Retrieved 30 July, 2021, from https://drum.lib.umd.edu/bitstream/handle/1903/7964/deathswarsconflictsjune52006.pdf?sequence=1&isAllowed=y

McDougle, C. J., Holmes, J. P., Carlson, D. C., Pelton, G. H., Cohen, D. J., & Price, L. H. (1998). A double-blind, placebo-controlled study of risperidone in adults with autistic disorder and other pervasive developmental disorders. *Archives of General Psychiatry, 55*(7), 633–641.

Oliver, P. C., Crawford, M. J., Rao, B., Reece, B., & Tyrer, P. (2007). Modified overt aggression scale (MOAS) for people with intellectual disability and aggressive challenging behaviour: A reliability study. *Journal of Applied Research in Intellectual Disabilities, 20*(4), 368–372.

Patrick, C. J., & Drislane, L. E. (2015). Triarchic model of psychopathy: Origins, operationalizations, and observed linkages with personality and general psychopathology. *Journal of Personality, 83*(6), 627–643.

Peterson, C., & Kearns, M. C. (2021). Systematic review of violence prevention economic evaluations, 2000–2019. *American Journal of Preventive Medicine, 60*(4), 552–562. https://doi.org/10.1016/j.ampere.2020.11.002

Raine, A., Dodge, K., Loeber, R., Gatzke-Kopp, L., Lynam, D., Reynolds, C., & Liu, J. (2006). The reactive–proactive aggression questionnaire: Differential correlates of reactive and proactive aggression in adolescent boys. *Aggressive Behavior: Official Journal of the International Society for Research on Aggression, 32*(2), 159–171.

Silver, J. M., & Yudofsky, S. C. (1991). The overt aggression scale: Overview and guiding principles. *The Journal of Neuropsychiatry and Clinical Neurosciences, 3*(2), S22–S29.

Sorgi, P., Ratey, J. J., Knoedler, D. W., Markert, R. J., & Reichman, M. (1991). Rating aggression in the clinical setting: A retrospective adaptation of the overt aggression scale: Preliminary results. *The Journal of Neuropsychiatry and Clinical Neurosciences, 3*(2), S52–S56.

Sturmey, P. (2015). *Reducing restraint and restrictive behavior management practices*. Springer.

Sturmey, P., & McMurran, M. (Eds.). (2011). *Forensic case formulation* (Vol. 49). John Wiley & Sons.

Verona, E., Sadeh, N., & Curtin, J. J. (2009). Stress-induced symmetric frontal brain activity and aggression risk. *Journal of Abnormal Psychology, 118*(1), 131.

Viljoen, J. L., Cochrane, D. M., & Jonnson, M. R. (2018). Do risk assessment tools help manage and reduce risk of violence and reoffending? A systematic review. *Law and Human Behavior, 42*(3), 181–214. https://doi.org/10.1037/lhb0000280

White, S. F., VanTieghem, M., Brislin, S. J., Sypher, I., Sinclair, S., Pine, D. S., & Blair, R. J. R. (2016). Neural correlates of the propensity for retaliatory behavior in youths with disruptive behavior disorders. *The American Journal of Psychiatry, 173*(3), 282–290. https://doi.org/10.1176/appi.ajp.2015.15020250

World Health Organization. (2016). *Ethical and safety recommendations for intervention research on violence against women: building on lessons from the WHO publication putting women first: ethical and safety recommendations for research on domestic violence against women*. WHO.

Yampolskaya, S., & Banks, S. M. (2006). An assessment of the extent of child maltreatment using administrative databases. *Assessment, 13*(3), 342–355.

Yudofsky, S. C., Silver, J. M., Jackson, W., Endicott, J., & Williams, D. (1986). The overt aggression scale for the objective rating of verbal and physical aggression. *The American Journal of Psychiatry, 143*(1), 35–39. https://doi.org/10.1176/ajp.143.1.35

Chapter 3
Epidemiology of Violence and Aggression

Understanding the extent of aggression and violence is important for many reasons. First, data on the extent of aggression and violence are useful descriptions of the extent and distribution of the problem globally, nationally, regionally, locally, and over time. Second, these data can be useful in identifying high- and low-risk populations and contexts and may thereby inform what theories of aggression and violence have to explain. Third, these data may be useful in testing the predictions from theories of aggression and violence. Finally, these data can be useful in evaluating how effective public policy efforts—individual treatments, policy changes, environmental design, etc.—are in reducing aggression and violence on a large scale.

This chapter will begin by describing some epidemiological concepts and then move on to consider the epidemiology of aggression and violence in humans and nonhumans. In the section on human violence we will begin looking broadly at global and regional epidemiology of aggression and violence and then focus down on specific populations, contexts, and types of aggression and violence of concern including an extensive list of risk factors. We will also illustrate the epidemiology of aggression and violence in animals with the examples of aggression in domestic dogs. Aggression and violence takes place in many different contexts and so this chapter can only sample this literature unsystematically to illustrate the broad trends in this field.

3.1 Epidemiological Concepts

Epidemiology is the scientific study of disease, disability, and other social problems at the level of a population for the purposes of prevention, control, and treatment (Rothman, 2012). Epidemiology often observes the temporal and geographical patterns of disease, disability, and social problems by time, place, and person and by events that are potentially related to the onset, maintenance, or cause of the problem

P. Sturmey, *Violence and Aggression*, Advances in Preventing and Treating Violence and Aggression, https://doi.org/10.1007/978-3-031-04386-4_3

and protective factors. Although originating in the study of infectious diseases in the eighteenth and nineteenth century, epidemiology now studies disease, disability, and social problems defined very broadly, including social problems such as aggression and violence. For example, today, social problems such as intimate partner violence (IPV) and police violence are often described as "epidemics" and "health care crises" in part because of the injuries and disability that arises from them.

The aims of epidemiology are both to understand, prevent, and treat these problems but also to promote good health and wellbeing. Thus, epidemiology is often closely involved in identifying the distribution and magnitude of the problem, risk factors, and causes, and generating data for the planning and evaluation of treatment and prevention, since understanding the causes and distribution of the problem is often useful in directing prevention and treatment.

In order to conduct epidemiology, one has to define a case of the condition of interest reliably and accurately. Second, one has to find all the cases completely or estimate the number of cases based on samples from the population. Finally, one has to define the population to be studied and measure or estimate the number of persons in the population. Once these things have been done, the epidemiologist can begin to look for geographical and temporal patterns of distribution of the condition of interest and events that might be related to risk or cause in order to generate and test hypotheses. If one wishes to study changes in the distribution of the problem over time, one has to use the same methodology with the same population over repeated observations. One change between two observations could simply reflect the problem increasing and decreasing from observation to observation. Thus, at least three observations are needed to detect any consistent trend to rule out the possibility that the data are merely going up and down rather than consistently increasing or decreasing. Furthermore, the observation must be scheduled sufficiently apart in time to detect any true trend. If there is a very steep trend, fewer observations spaced closely together may detect the trend, but if the trend is not so steep then more observations spaced out over a longer period of time may be required to detect the trend.

Epidemiology uses distinction between the incidence (number of new cases) and the prevalence (total number of cases) of a condition. These data are often expressed in terms of ratios of number of cases per base population and with some specification of the time period studied such as a specific point in time, which is referred to as the "point prevalence," or a during a specified longer period of time, which is referred to as the "period prevalence." For example, an epidemiological study of aggression and violence might measure the period prevalence as the number of cases today, during the last month or each person's lifetime. Thus, the number of new cases of a condition might be 10 during the month of June 2018 in a population of 100. On a particular day there might be five cases for a point prevalence of 5% but for the month of June 2018 there might be ten cases for a period prevalence 10%. If we looked at the lifetime prevalence of aggression, we might calculate that there were already 20 other people who had experienced aggression and violence before June and so, adding in the ten new cases, we might calculate a lifetime prevalence of 30%.

The observed prevalence is the outcome of several factors. These factors include the cumulative number of new cases, the degree to which case finding is complete and accurate, mortality, and the duration of the condition. The observed prevalence will also reflect any changes in the size of the base population and any pattern of migration in and out of the population studied. Thus, the observed prevalence of a particular condition over time might vary substantially due to changes in factors related to the condition studies such as the number of observed cases, mortality, and recovery from the problem. The prevalence may also change due to changes in the size and composition of the population. For example, if low-risk people leave a geographical area the prevalence of aggression and violence might increase, not because of new cases, but because the base population has reduced in size and composition. Similarly, if high risk people move into a geographical area—perhaps for economic reasons or to access better services—the prevalence might increase, again, not because of new cases but because of mobility in the base population.

Epidemiological studies that attempt to identify potential causes can do so using odds ratios (OR) which express the degree of association between a risk factor and an outcome. The OR expresses the odds of the problem occurring given the exposure to the risk factor compared to the odds of the outcome occurring given the absence of the risk factor. For example, an OR of 8 means that individuals exposed to this risk factor are 8 times more likely than the rest of the population to show the problem and an OR of 0.3 means that individuals exposed to this risk factor are 70% less likely than the general population to show the problem.

To illustrate these epidemiological concepts consider a study of epidemiology of disability-related sexual violence conducted by Basile et al. (2016) who used data from a national US telephone survey of penetrative and nonpenetrative sexual violence analyzed by gender and disability status. The sample consisted of 9,086 women and 7,421 men who responded to the survey. Approximately, 1.8% of women with a disability and 0.9% of women without a disability reported being raped in the previous year and the reported prevalence of rape among men was too small to estimate reliably but was very low. Thus, the OR for women with a disability was 3.3 indicating that women with a disability were more than three times at risk of rape than the overall sample studied. The reported prevalences of other sexual violence, such as unwanted touch, kissing, other forms of unwanted sexual contact, and being forced to penetrate another person, were 6.9% and 5.2% of women with and without a disability respectively. It was 6.3% and 5.1% of men with and without a disability respectively. The ORs associated with having a disability were 2.2 for women with a disability and 1.9 for men with a disability indicating that disability was associated with double the risk of this form of unwanted sexual contact than the general sample studied. The OR data from this study is useful because it shows that women with and without a disability are at much greater risk of rape than men and that women with a disability are at greater risk that women without a disability.

Additionally, for other forms of unwanted sexual behavior, some of which were violent, the most consistent differences were to do with disability status rather than gender. These data have implications for what theories of aggression and violence must explain and which groups of people to target for prevention of treatment of rape and other sexual violence.

The application of epidemiological concepts to aggression and violence, like many other social problems, is tricky. This is because researchers must be clear about the question asked, the operational definition of a case of aggression and violence must be good, all the cases must be found completely, the population must be defined and sampled accurately, and the risk factors must also be defined and measured well. In addition, epidemiological research must be free from political interference with and from government agencies. This does not happen all the time as government agencies often fund such research and may have motivations to minimize or maximize the problem at hand or are influenced by political lobbying from many groups.

A well publicized example comes from a study by Kellermann et al. (1993) who found that gun ownership was associated with *increased* rather than decreased violence in the home. The study was funded by the Center for Disease Control and Prevention. Consequently, the National Rifle Association successfully lobbied congress to ban funding on gun violence through the 1996 Dickey amendment, named for the republican representative who added this provision into legislation: "None of the funds made available for injury prevention and control at the Centers for Disease Control and Prevention may be used to advocate or promote gun control." This resulted in defunding of gun violence by the Center for Disease Control between 1996 and 2019. US Federal research into gun violence was only reinstated after decades of advocacy by researchers and family members of victims of gun violence.

3.2 Human Aggression and Violence

3.2.1 Global Population Studies

Numerous studies have attempted to measure the prevalence of different types of violence and aggression in a wide variety of human populations, but recently the World Health Organization has taken the lead in estimating global rates of murder, and associated burdens and costs. They have also investigated and developed public policies, both generally and for specific populations, such as women, children, etc. Murder is only one form of aggression and violence, so, although it does not index all aspects of aggression and violence, it does indicate the widespread nature of even the most dangerous forms of aggression and violence.

3.2.1.1 Global Murder Rates

In 2014, Mikton et al. (2016) surveyed 190 World Health Organization member states to obtain information on national homicide rates; national action plans related to murder; prevention policies, programs, and laws; and, various services for victims of violence. One hundred-thirty-three countries with a population of 6.1 billion (representing 88% of the world's population) participated. The authors estimated that there were approximately 475,000 murders for a rate of 6.7/100,000. These estimates were broadly similar to estimates from two other World Health Organization sources. Subsequently, Butchart et al. (2014) noted that there was much that could be done to prevent aggression and violence that was evidence-based, but implementation and enforcement was highly varied from country to country, in part because of the resources that required to address such an enormous problem.

These data are interesting but are just the tip of the iceberg in terms of epidemiology of aggression and violence. They exclude many other forms of aggression and violence and may contain incomplete datasets, such as murder victims being reported as "missing" or disappeared" rather than murdered, and they also exclude people who were violently assaulted and nearly died. They do show clearly, however, that even the most severe and serious form of violence—murder—is a widespread and global phenomenon today.

3.2.1.2 Women

In 2013, The World Health Organization published a report on global statistics on violence against women on the context of IPV and non-IPV. The report focused on physical abuse and excluded emotional abuse, such as humiliation and controlling behavior, such as preventing family contact, movement, and access to money (World Health Organization, 2013). The authors compiled data from online searches of the literature in multiple languages and estimated rates of violence using statistical methods to infer missing values.

The authors estimated that approximately 30.0% of women worldwide who had ever been partnered had experienced IPV with wide variations between regions and countries. A striking finding was the IPV starts early with about a third of women aged 15–19 years experiencing IPV. Estimates of non-IPV were around 7.2%. Additionally, around 36.5% of women having experienced either forms of violence during their lifetime.

Effects on the health of women and their children were pervasive and often severe. They included physical injuries, including death; psychological disorders, such as anxiety, suicide, depression; and physical health problems, such as physical injuries, and STDs, such as HIV. Approximately, 38% of all murdered women were murdered by their intimate partners. In contrast, only 6% of men who were murdered were murdered by intimate partners. Violence against women was also associated with negative effects on their children, including low birth weight and prematurity.

Alhabib et al. (2010) conducted a systematic review of this literature and found broadly similar results. They identified 134 studies published in six databases between 1995 and 2006 of women aged 18–65 which excluded women with specific disabilities and diseases. The median prevalences of physical, sexual, and emotional abuse were of the order of 20–30% with wide ranges across countries in high quality studies. Prevalences were much higher in psychiatric and medical settings, such as obstetrics and gynecology settings. Thus, these studies agree that around a third of women experience various forms of violence at some point during their lifetime.

Older women too are likely to be victims of aggression and violence. Cook et al. (2011) conducted a systematic review of aggression and violence in the lives of older women. They reviewed papers published between 1980 and 2009 that reported data on aggression and violence in women aged over 65 in community settings. They found that studies reported a lifetime prevalence of around 6–8% for physical violence and 8% for sexual violence and a 1-year prevalence of around 1%. A minority of affected women reported multiple incidents sometimes over many years.

The data from these studies are a good example of the provocative nature of epidemiological data: it prompts us to ask why are women but not men so likely to be killed by their intimate partners? It also prompts us to ask who else, if not their intimate partners, are killing men? What would explain these large differences in patterns of violence? Given these large differences, perhaps aggression and violence in the lives of women are quite different from aggression and violence in the lives of men.

3.2.1.3 Children as Victims

Children are especially vulnerable to many forms of aggression and violence from family caretakers, including siblings, and other family members; in schools; in non-family residential settings, such as prisons and the justice and mental health systems; and, in their places of work. Unfortunately, very many children worldwide also observe and experience aggression and violence indirectly.

Hillis et al. (2016) conducted a systematic review of 38 reports from 96 countries which reported past year violence against children defined broadly to include physical, emotional, and sexual forms of violence. They found that globally the minimum prevalence of last year violence against children was over 50% with more than one billion children aged 2–17 years being affected. Prevalence estimates varied widely between regions of the world from 12% in Europe to 64% in Asia for children aged 2–17 years. As part of this global effort to understand and prevent aggression and violence against children, the CDC has undertaken ten studies in a variety of countries.

Children with various kinds of physical and mental disabilities are at particular risk of aggression and violence. Jones et al. (2012) conducted a systematic review of such studies from 12 electronic databases for publications between January 1,

1990 and August 17, 2010 which estimated prevalences of violence against children under 18 years old with disabilities. They retained 16 studies and calculated a pooled prevalence of 26.7%, 20.4%, and 13.7% for combined violence measures, physical, and sexual violence, respectively. Children with disabilities were at approximately three times greater risk for violence than other children.

3.2.1.4 Parents as Victims

When one thinks of family violence one immediately thinks of violence between the partners and from adults to infants and children, so unsurprisingly, violence by children to other family members has received less attention. Yet, there is evidence that violence does occur from adolescents to parents. A series of studies by Miles and Condry (2015) reported the results of a survey of aggression and violence from children to parents. The study was based on police files and interviews with relevant parties including family members and police based in 33 London boroughs. The average age of the perpetrators was just over 16 years. Most injuries to parents were none, threat of injury or minor, but around 5% were rated to be of moderate to severe injuries. Behind the injuries themselves, Miles and Condry (2015) also uncovered patterns of coercive and controlling behavior from some adolescents toward their parents which they described as paralleling intimate partner abuse, that is, threats or actual violence from adolescents to parents was the tip of the iceberg. Abusive adolescents also threatened and coerced parents into doing unwanted things, controlled or attempted to control their parents' social life and manner of dress, and were often generally disrespectful and demeaning to them.

Since this problem was relatively hidden until this research, there was little in terms of policy to address it. Additionally, when various authorities, such as the justice system, youth service and the police were involved, they sometimes took little or no action, some of their actions may have been dismissive or even made things worse (Miles & Condry, 2015).

3.2.2 Violence in the USA

The USA—a very wealthy country with a relatively effective justice and policing system—is often portrayed as an unusually violent and murderous country that has not yet outgrown its "wild wild west" history. Indeed, compared to other wealthy industrialized countries, the USA has an unusually high murder rates: If you want to get safe in the USA, just head north and cross the border to peaceful Canada—a country which is economically quite similar—and your risk of murder will drop substantially. The USA also exports its violence on a worldwide scale. It produces violent entertainment exported on a global scale and it exports violent gang members from California to Central and Latin American countries which has also caused large increases in violence there and eventually further increases in violence back in the US when some gang members return.

Media headlines on spikes in murder rates are common. The media report on this especially in inner city areas; mass shootings at public events, movie theaters, malls, and nightclubs; riots on the streets; the violent response of the state to protesters; and dramatic incidents of people driving cars into protesters or other groups of people all get everyone's attention. The media also report on spikes in murder rate over holiday weekends such as July 4. Many have commented on the fact that murder rates in the USA are higher than other ostensibly similar countries. What do the data show?

3.2.2.1 Murder Rates

Compared to many other countries, murder rates in the USA are relatively *low* – many other countries have substantially higher murder rates. Over time murder rates have continued to decline for decades and including recent decades. Dwyer-Lindgren et al. (2018) reported data on mortality due to interpersonal violence in the USA between 1980 and 2014. They found that there were 760,749 deaths due to interpersonal violence during that time period for an average murder rate of 5.7 per 100,000. They analyzed these rates by county and found they ranged 1.0 to 35.7 per 100,000 with the Mississippi, Alabama, Alaska, Montana, North Dakota, South Dakota, Kansas, Michigan, Maryland, Virginia, and North Carolina with unusually high rates of deaths due to interpersonal violence. Interestingly, rates of deaths due to interpersonal violence fell by 44.9%—nearly half!—during this time period. In fact, worldwide approximately 95% of counties experienced such a decrease. These temporal patterns are broadly similar of violent crime in the USA generally, including various kinds of assault and violent sexual crimes as well as broader reductions in murder rates over historical periods of time. (See Chap. 7, on cultural evolution and human cultures for a more detailed discussion of historical trends in murder and violence and evidence for very large historical reductions in murder rate worldwide over the last 800 years.)

Youth violence is a particular concern in the US as rates of violence and murder rates are especially high among teenage and young adult men in their twenties. David-Ferdon et al. (2015) reported national data on youth violence in the USA. In a sample of individuals aged 10–24 in 2013, fully 4481 youths were homicide victims for a murder rate of 6.9 per 100,000. Homicide was the third leading cause of death in this age group.

3.2.2.2 Other American Aggression and Violence

Violence in American families is very widespread, leading some to nominate the family as "the cradle of violence"—the place where we first experience and first learn violence. A classic series of studies comes from Murray Straus and colleagues who published two large scale surveys of violence in American families (Straus et al., 1990; Straus & Gelles, 1990). They interviewed nationally representative samples of 2,143 families in 1975 and 3,520 families in 1985 by telephone to measure overall violence, child abuse, spouse violence, and wife beating using the *Conflict Tactics Scale* (Straus, 1979). To be included a family had to have a male and female adult with at least one aged over 18 years. Straus and Gelles (1990)selected adults and children at random to select the adult's respondent and referent child.

In 1975, Straus and Gelles (1990) found overall violence occurred in 630/1000 children (i.e. two-thirds of all children). There were 140/1000 incidents of severe violence (i.e., 14%) and 36/1000 (3.6%) incidents of very severe incidents of violence. The most common forms of violence to children were minor and included slapping and spanking (582/1000) and pushing, grabbing, and shoving (318/1000). Common severe violent acts included hitting (134/1000) and kicking, biting, and hitting with fist (32/1000). The overall pattern in 1985 was very similar, except that there were large reductions in minor violence. Reporting of items such as "threw something" fell from 54 to 27/1000); severe violence (kicking, biting and hitting with fist) fell from 32 to 13/1000); severe violence fell from 140 to 107/1000; and very severe violence fell from 36 to 19/1000). In terms of violence between spouses, overall violence occurred in around 11–12% of couples and there was a nonsignificant 6.6% decline between these two time periods. So, there was an overall and consistent reduction in all forms of violence within the family between these two time periods.

Since this classic study there have been many other studies published on this topic. Desmarais et al. (2012a, 2012b), for example, reported two related systematic reviews of victims and perpetrators of physical violence in intimate relationships of research published between 2000 and 2010. They reported that 22.4% of persons in intimate relationships and about one in four women and one in five men experienced physical violence in an intimate relationship. In terms of perpetrating violence, about 25% of partners reported perpetrating violence with slightly more women (28.3%) than men (21.6%) reporting perpetrating violence against their partner.

A more recent telephone survey of last year and lifetime prevalence of IPV (Black et al. 2011) found that 36% of women and 29% of men had experienced rape, physical violence or stalking by an intimate partner. In addition, about 24% of women and 14% of men had experienced severe physical violence from an intimate partner (such as hit with a fist or something hard, beaten or slammed against something) during their lifetime. Most violence sexual assaults are committed by spouses, other family members, friends, and acquaintances, such as coworkers and neighbors (Basile et al., 2016).

These studies converge on the conclusion that that family is a major source of aggression and violence between adults and children. Although there is evidence that this has reduced, the American family remains a violent one.

3.3 Risk Factors in Humans

Identifying risk factors for aggression and violence is useful for many reasons. First, it describes what needs to be explained. For example, one might ask, why are teenage boys and young men so likely to be murdered and kill other teenage boys and young men? Second, risk factors can give us hints as to explanations for increased risks of protections against aggression and violence. Third, risk factors for aggression and violence can help us direct resources to prevention and treatment of aggression and violence and guide us as to where to look to evaluate these social programs. Finally, risk factors for aggression and violence can help us to identify goals for social justice related to aggression and violence and to evaluate whether efforts toward social justice have been effective.

When considering risk factors for aggression and violence, one should be mindful of which risk factors are modifiable and which are not. For example, male gender and aged 15–30 years are not variables we can change. So, although they may be useful in directing our attention to at-risk populations, they are not helpful in changing the problems. In contrast, poverty; alcohol and drug use; social, problem-solving, and self-management skills; and attitudes and beliefs are all variables that are potentially modifiable. These variables are useful in directing our attention to variables that can potentially be changed and hence reduce aggression and violence.

3.3.1 Interpersonal Disputes

Arguments, disagreements, and minor fights that escalate are a very common antecedent to violence. For example, Lyons et al. (2016), in a report on violence deaths in 17 states in the USA in 2013, reported that the overall rate of homicides was 4.9/10,000 population. Men were about 3.5 times more likely to be killed than females: The homicide rates were 7.7 and 2.2 per 100,000 population for males and females, respectively. Homicide rates were highest among people in their twenties (about 10–12/10,000 population). Infants were at much greater risk than children aged 1–4 years (7.6 versus 2.1 per 10,000 population). The most at risk group of men were aged 20–24 years (20.4/100,000).

Lyons et al. (2016) reported that violent deaths were most likely between acquaintances, friends, and intimate partners. The most common reported events 2 weeks prior to the homicide were family arguments (22.5%) and jealousy (16.3%). In addition, two of the three main precipitants for violent deaths of females intimate partnerships were arguments and interpersonal conflicts. One-third of homicides took place in the context of another crime, such as a robbery or burglary. Firearms were used in two-thirds of homicides. Of those who were tested for alcohol and drug use, alcohol was present in about 40% of cases and of those about two-thirds were probably drunk (Blood Alcohol Concentration >0.08 g/dL, i.e. illegal in all states). Other drugs were also present including marijuana (42.5%), opiates (18.3%), and cocaine (15.7%). These data paint a picture of homicide that primarily involves interpersonal disputes between family members, intimate partners, friends, and acquaintances which is more likely after a recent period of poor relationships in situations involving alcohol and drugs with firearms present.

There were also 222 deaths involving law enforcement or other legal personnel. Most involved males (93.7%), aged 30–34 years; involved firearms (92.3%); in the context of another crime (97.6%); and mostly occurred in people's residences (41.4%), public places, such as a street or highway (25.2%), or a vehicle (8.6%). Three-quarters of incidents involved alcohol and drug use. In terms of ethnicity, the rate of homicides was much higher for Black males (1.2/100,000 population) than non-Hispanic White males (0.3/100,000 population), although non-Hispanic White men accounted for the largest percentage of deaths. These data paint a picture of deaths involving law enforcement mostly in the context of ongoing crime in people's residences and in the streets or highways with individuals, many of whom are drunk or high.

Lyons et al.'s (2016) data are a good example of the potential usefulness of epidemiological causes because these data identify at-risk populations, and at-risk social and physical contexts. These data can be used to evaluate public health prevention programs. In addition, they can be used to guide what those programs should address These include: (a) building better more respectful relationships between family, friends, and acquaintances, and interpersonal conflict resolution, such as social skills training, problem-solving, and anger management; (b) removal and reduction in access to firearms or perhaps safer use of firearms, such as keeping them locked away and difficult to access readily; and (c) reduction or safer use of alcohol and drugs.

3.3.2 Alcohol and Illicit Drugs

Alcohol consumption has been known for thousands of years (Brown & Leonard, 2017) and as noted above is a common factor correlated with homicide. Acquisitive violent crime such as robbery and burglary are often done to support drug addiction. Furthermore, violent crime by drug-related gangs is well known and correlates with international patterns of murder rates across the world. These facts are old and well known. So, what, if anything new, does contemporary behavioral science have to offer?

Contemporary behavioral science confirms the very robust association between alcohol and aggression and violence: This has been confirmed across at least 32 meta-analyses and one meta-meta-analysis. Duke et al. (2018) analyzed the results of 32 meta-analyses and found a median effect size of $d = 0.45$, which is a medium effect size, which was highly robust across populations, research methods (such as lab-based vs. nonlab-based studies), and types of dependent variables (i.e., different types of measures and forms of aggression and violence). There was also a significant and independent effect of illicit substances on aggression and violence.

The effect size for illicit drugs alone was $d = 0.42$, i.e. similar to the overall effect of alcohol alone and the effect of combined alcohol and illicit drugs was somewhat larger ($d = 0.55$). There was also a similar effect size for the relationship between alcohol and aggression and violence for victims ($d = 0.42$). Duke et al. concluded that the relationship between alcohol and illicit drugs with aggression and violence was highly robust. They also concluded that treatment of aggression and violence should address alcohol and illicit drug use and that similarly treatment of alcohol and illicit drug use should also address aggression and violence.

Brown and Leonard's (2017) review of alcohol and aggression and violence noted that the effects of alcohol on aggression and violence is often mediated by other variables. For example, some studies have found a dose–response relationship between alcohol and aggression and violence. A very active area of research is the role of mediating behavior and events after or during alcohol consumption which may potentiate or mitigate the effects on aggression and violence. Examples include: The effects of expectations of aggression and violence after alcohol consumption; the normalization of aggression and violence after alcohol is consumed, in part by the contexts in which this occurs; priming (or inhibiting) effects of environmental events, such as hearing certain words or experiencing other cues.

3.3.3 Crowding

Crowding is often mentioned as a cause of aggression and violence in many different contexts including public spaces, such as sports events, and congregate residential settings including prisons, psychiatric facilities, among others. Crowding is also mentioned as a factor that may promote aggression and violence in the context of differences of density in different residences, neighborhoods, or other geographical units. Measuring crowding is complex. First, changes in number of people per square unit of space may covary with other relevant measures such as population size, staff to client ratios in residential settings, changes in the behavior of people, changes in management and supervision practices, and other environmental variables. Second, crowding can be measured in different units. For example, when considering prisons, crowding can be measured as square feet of space per inmate, inmate population divided by facility capacity, and percentage of prison cells that are overcrowded (Franklin et al., 2006). Thus, observing that crowding is related to increased aggression and violence is only the first step in understanding the mechanisms of how crowding may increase aggression and violence.

It is commonly believed that crowding in prisons causes or increases the risk of aggression and violence. What do the data say? There have been several meta-analyses on this topic and all concur that crowding has *little* direct effect on aggression or related behavior by itself! (Franklin et al., 2006; Gendreau et al., 1997; Gonçalves et al., 2014). Franklin et al. (2006), for example, conducted a meta-analysis of 16 studies and found that effect sizes were only very slightly larger than 0. The unweighted effect size was a mere 0.066, i.e., not much different than 0! The only significant relationship was that the relationship between crowding and aggression and violence was mediated by offender age. The effect size for offenders aged 18–25 was 0.34 ($p < 0.05$) but was again essentially 0 for all other age groups.

Several studies have investigated the effects of crowding in psychiatric and other mental health facilities and have found evidence for a positive association (Ng et al., 2001; Palmstierna et al., 1991) but not all studies have found this (Hardie, 1999). Intervention studies which have redesigned psychiatric units that involved multiple changes including reduced crowding may result in reduced prn ("as needed" or emergency) medication for aggression and restraint (Ulrich et al., 2018). Rago et al. (1978), for example, found that increasing available space in the residence of 14 individuals with profound intellectual disabilities was associated with decreased aggression. This increase in space, however, may also have been associated with many other environmental and social improvements and changes in social behavior.

These studies suggest that, contrary to popular belief, crowding by itself has a *weak* effect on aggression and violence. Crowding effects on aggression and violence, when they occur, tend to be small and mediated by other variables which hitherto have not been clearly identified. Perhaps a clearer relationship might emerge with more extreme degrees of crowding or with combinations of crowding with one or more other variables.

3.3.4 Temperature

Many people have commented that hot weather is associated with aggression and violence and cold climates with the lack of aggression and violence. We often use temperature metaphors for aggression and violence, such as "being hot under the collar" or "boiling with rage" and taking time to "cool off" or "chill out." Social geographers have often commented that southern Europeans appear more emotional than people from northern Europe, not only because of their alleged "hot" and "cold" blood, but also because of differences in temperature. Others have noted that most of the countries with high homicide rates are located around the equator (Coccia, 2017). American newspapers also report outbreaks of violence during heat waves in American summers. How much truth is there to this stereotypical view of temperature and violence?

This topic has been investigated extensively over the years using multiple methods and paradigms (Anderson et al., 2000.) These include at least three approaches. First, some studies have compared rates of aggression and violence between locations that vary by temperature. Second, some studies have compared rates of aggression and violence across times when temperature varies such as seasonal, monthly, daily variations in temperature. Some studies have looked at the possible effects of global warming on aggression and violence (Plante & Anderson, 2017). Finally, some experiments have systematically manipulated the temperature and the experimenters have then observed its effects on some measure of aggression and violence in laboratory studies. The first two methods are potentially problematic in that changes in temperature can be confounded with many other variables. The latter method is potentially problematic because the results, though reliable, might not be generalizable to the natural world.

There are several reviews and many empirical studies published over many years and triangulation and integration of these data, especially by Anderson and colleagues. These studies have resulted in the conclusion, albeit a nuanced one, that the effects of temperature on aggression and violence are relatively robust (Anderson, 1989; Anderson et al., 2000). For example, when cities are matched by other potentially relevant factors, such as size and poverty, temperature alone independently predicts aggression and violence: The hotter the city gets, the more likely we are to be aggressive and violent. This is true across different counties and whether we measure drivers honking horns, baseball players throwing the ball directly at the pitcher, and trainee police officers drawing weapons during training (Vrij et al., 1997).

The explanations for these observations are more problematic. For example, some have suggested that changes in temperature are confounded with other changes that might explain the increase in aggression, such as there being less structured activities, for example, less school for teenagers, more people being out on the street and/more alcohol consumption (Cohen & Felson, 1979). Although the relationship between temperature and aggression and violence is relatively robust in naturalistic human studies, results from experimental studies have produced mixed results. This may perhaps reflect artifacts of the experimental methods, the range of values of variables studied, and differences in measures of aggression and violence.

It is unclear if the relationship between temperature and aggression and violence is linear, curvilinear, or tails off at very high temperature. Furthermore, this relationship might hold between lower and moderate temperatures, up to 85 °F, but might level off thereafter. (Perhaps at some point it gets so hot we cannot do anything, even be aggressive). The effects of typical, local temperatures might also be relevant: A temperature of 85 °F might be a pleasant morning in Phoenix, AZ, but an insufferable summer day in Montreal. Finally, researchers have paid much less attention to the effects of cold on aggression and violence.

Unlike crowding, the effects of temperature on aggression and violence are relatively robust, leading Anderson (1989) to describe the effects as "ubiquitous." These effects have interesting implications for environmental design related to aggression and violence. Today, air conditioning and heating allow us to both control temperature and enable escape from unpleasant heat and cold. Such technology gives us the possibility of interventions for aggression and violence based on temperature manipulations and/or control. Global warming, however, presents particular challenges to aggression and violence, as global warming comes with numerous and significant frustrations, such as threatening access to food and water, environmental destruction, and national and civil war. These environmental challenges may present potential triggers for the development of aggression and violence in young, developing children (Plante & Anderson, 2017).

3.3.5 *Access to Weapons*

There have been nearly a million firearm-related deaths in the USA between 1933 and 1982. For example, tragically, 88 children aged 0–14 years were unintentionally shot and killed by themselves or peers in California between 1977 and 1983 (Wintemute et al., 1987). Thus, changes in firearm-related deaths is the major driver in changes in homicide rates in the US (Wintemute, 1987). Purchase of firearms is also closely associated with subsequent suicide by firearms, especially among women (Studdert et al., 2020). Access to weapons may be a risk factor for aggression and violence both at the level of individuals and communities and is often part of individual risk assessment in settings for violent or potentially violent offenders. Indeed, there is robust evidence that the prevalence of gun ownership in the USA is consistently associated with a range of violent gun-related domestic crimes (see the discussion of Kellermann et al.'s (1993) classic study above).

Such firearms-related injuries and deaths are pervasive and extensive in the USA: Geier et al. (2017) estimated that there were more than 31,000 gun-related deaths and 74,000 nonfatal gun-related injuries annually in the USA. Using US national data on firearm-related deaths (FRDs) from 1999 to 2014, Geier et al., found that men were 5.4 times more likely than women to be victims of firearm-related deaths and 24 times more likely to be victims of law enforcement incidents involving a firearm.

Geier et al. (2017) also reported that there were also very large differences in FRDs by geographical region at the level of census division, state, and region. For example, large metropolitan areas had much higher rates of gun-related violence than the overall US. About half (52%) of all US households reported posessing firearms varying from 35% in New England and 68% in East South-Central USA. Others have reported similar associations between access to guns and gun-related violent crime (Siegel & Rothman, 2016). In addition, firearm ownership is associated with other risky behavior such as heavy alcohol consumption, leaving guns out, and driving while intoxicated (Wintemute, 2011). These combined factors may well place individuals at greater risk of gun-related aggression and violence and homicide.

This strong association between firearms and violence and homicide is the basis for many different interventions to reduce firearm related violence and homicide. Examples of such interventions include reducing accessibility of firearms to risky individuals (Wintemute et al., 2001) and teaching children how to behave safely when they find a gun in their home (Miltenberger et al., 2004, 2020).

3.3.6 *Poverty and Income Inequality*

The association between poverty and crime, including violent crime including homicide is robust and is found in comparisons across countries (Hsieh & Pugh, 1993; Pridemore, 2011). For example, currently murder rates vary from over

50/100,000 per year in low income countries such as Honduras and Venezuela and are around 0.5/100,00/year in high income countries Japan and Iceland (Ouimet et al., 2018). Poor people are more likely than rich people to both commit and be the victims of violent crimes. This association has been noted for thousands of years. It is the stuff of the yellow press and populist presidents and other politicians. It has led some politicians to demonize poor people and their apparently notoriously violent neighborhoods and also to lead others to promise to greatly reduce violent crime by reducing poverty and inequality.

But although the association between poverty and violence is robust, poverty is confounded with many other variables that could account for this association. A commonly identified alternative explanation is income inequality rather than poverty per se. Poverty refers to absolute *income* or deprivation whereas *inequality* refers to one's relative income or level of deprivation compared to the rest of the population. Pridemore (2011) noted that many cross-national studies have found a strong and robust relationship between poverty and homicide, although studies in the US have made inconsistent findings; however, almost inevitably poverty was confounded with inequality.

Pridemore (2011) attempted to unconfound poverty and inequality by replicating previous studies, which had shown a close association between inequality and homicide, and then examining the effects of poverty independently of inequality. Using data from a sample of 46 countries, Pridemore confirmed earlier research on the association between income inequality and homicide. In a subsequent analysis of the same data, he added a measure of absolute poverty independent of income inequality and found that the association between income inequality and homicide disappeared in two of three analyses and absolute levels of poverty independently predicted homicide rate both overall and separately for men and women, suggesting that absolute levels of poverty rather than income inequality was the more powerful predictor of homicide.

While the association between poverty and violent crime is robust, explaining this association is more difficult and unclear as there are many potential causes confounded with one another. An interesting empirical study which attempted to untangle some of these variables comes from Ouimet et al. (2018). This study was part of a larger study to understand global patterns of homicide. This study developed a two-variable model to explain global patterns of homicide and then evaluated it empirically. The first variable was called *Adverse Social Conditions* and was measured by indices of collective and individual poverty, inequality, and a young population structure. The second variable, which was of most interest, was called *Formal Social Control*. These authors noted that violence often escalates when police disappear; during blackouts; and, periods of societal chaos resulting in widespread increases in theft, sexual assault, and murder. They measured formal social control in each country by getting local and national experts and scholars to rate countries that they were familiar with on fundamental rights, corruption, order and security, and constraints of government power and corruption. By way of validation, the authors showed that these expert ratings correlated quite well with other similar independent measures of similar constructs. A notable positive feature of this study was that it included data from a large sample of 145 countries with a very wide range of measures of homicide, adverse social conditions and formal social control.

Adverse social conditions and adverse social control were very strongly negatively correlated ($r = -0.86$), that is countries with high poverty and youthful populations were very likely to have weak rule of law, high levels of corruption, and weak criminal justice systems. Formal social control predicted homicide somewhat better ($r = -0.41$) than adverse social conditions ($r = 0.33$), but both predicted national homicide rates independently of one another. The authors concluded that in countries with high adversity and ineffective formal social control, citizens can engage in violent behavior with impunity and sometimes may do so as a form of citizen-initiated justice for wrongs that formal mechanisms of social control are ineffective.

When one looks at the national differences in homicide and the correlated cultural practices, it is interesting to ask what people actually do that promote homicide or peaceful behavior? What cultural practices in Japan result in Japan having a homicide rate that is 10% of economically similar countries (such as the USA) and what maintains these cultural practices?

Finally, hidden among these national differences in homicide is large within-country variations. These include: (a) differences in distribution across time and place; and (b) different types of homicide (i.e., differences in individual reactive and acquisitive types of murder, gang-related murder, political murders, multiple murders organized by the government or other political entities). These differences also require explanation beyond or within the framework of the explanations of these national differences.

3.3.7 At Risk Settings

Aggression and violence do not occur at random across time and space: There are safe and dangerous times and places. Analysis of the temporal and spatial distribution of crime is sometimes called "hot spotting." A "hot spot" is a geographical area or time that has higher rates of crime than surrounding areas. Hotspotting may be useful in understanding why aggression and violence occurs and in developing and evaluating interventions to reduce aggression and violence. Ratcliffe (2010) classified hotspots in a three-by-three matrix resulting in nine types depending on their spatial and temporal patterns. Spatial distribution could be diffuse, relatively clustered around a particular area or concentrated completely in one location. Temporal distribution could be diffused, focused, or acute depending on the degree of concentration around a particular time.

One good example comes from spatial and temporal patterns of aggression and violence related to football (soccer). Kurland et al. (2017) provided a case study of the spatial and temporal distribution of aggression and violence around football matches at Sheffield Wednesday's football ground. They mapped 1,388 violent incidents between 2004 and 2005 and 2009 and 2010 of which 742 occurred on days when there was a football match scheduled and 646 on other, comparison days. Spatial distribution of incidents of aggression and violence were broadly similar on match and nonmatch days suggesting common environmental features related to crime generally, not just specifically to soccer. There were also a few differences between match and nonmatch days in that more aggression and violence occurred in only about eight particular hotspots during match days than control days. Analysis of the temporal distribution of aggression and violence on match days indicated that more aggression and violence occurred during and 8 h running from 4 h before to 4 h after a match which also included the time the match was being played than at other times of the day. Depending upon the specific soccer match, violence between rival teams supporters may be especially likely at pubs, transport locations, and intersections where rival fans may face off against each other.

These data are broadly similar to much data on hotspotting and the physical and temporal distribution of crime. Some have proposed a 20–80 rule where 20% of the physical and temporal locations account for 80% of the crime (Groff & McCord, 2012).

3.3.8 What Makes a Hot Spot?

If crime is not randomly distributed, what environmental features are associated with high and low rates of crime? An example of such an analysis comes from Groff and McCord (2012) who analyzed the distribution of various crimes, including violent crimes, across parks in Philadelphia and the environmental correlates of these crime locations. The authors described parks as "contested spaces": They are publicly owned spaces in that they belong to everyone and nobody at the same time. Thus, it is ambiguous who is supposed to surveille these spaces and maintain order. Park environments vary in terms of many features including physical size; openness; shrubbery; lighting; and the so-called activity generators, such as sports fields and recreational centers; and their environs, such as public transport and local neighborhood features. Parks also vary socially in terms of the number and types of people present and their different social roles.

In their empirical study, Groff and McCord (2012), gathered data on environmental features of 249 smaller parks in Philadelphia. Most parks were about 10 acres in size. They surveyed the parks and coded types of recreational activities (such as children's playgrounds), indications of public presence and involvement, (such as bulletin boards), and security features, (such as lighting and posted rules). They also collected data on violent-, property-, and disorder-types of crime throughout the city, in parks, and within 400, 800, and 1200 feet of the park perimeter. The dependent variable was a location quotient (LQ) which the authors defined as the ratio of crime rates locally to the overall crime rates throughout the city.

They first showed that parks were associated with all three forms of crime with LQs being around 2.0 to 2.5. Furthermore, crime LQs were very unevenly distributed across the parks. Fifteen percent of parks accounted for around 50% of park-related crimes yet around 10–15% of parks were crime-free. In terms of violent crime, they found large effects of sports facilities, such as basketball courts, baseball fields, tennis courts, soccer fields, and fieldlighting that were associated with an approximately 50% reduction in violent crime. (Data were similar for the two other types of crimes). Larger parks with more activity generators were also modestly associated with reductions in all three types of crime. Crime was lowest immediately surrounding parks, increased at 400–800 feet and fell again beyond 800 feet from the park perimeter.

Groff and McCord (2012) speculated that these physical park features generated activities associated with an increase in the number of legitimate users with clear social roles, such as referee, coach, player, spectator, etc., that included informal rule enforcement and place managers. The authors concluded that aggression and violence in parks could be reduced by problem-oriented policing focused on hot spots and high-risk parks; should involve citizen groups and other citizen agencies; and, by physical design of parks adding features associated with lower crime rates.

3.3.9 Ethnicity and Cultural Groups

The rates of homicide and other forms of aggression and violence differ substantially by ethnicity and cultural groups in studies conducted both in the US and elsewhere, both in terms of victims and perpetrators. In a systematic review of a wide range of risk factors for IPV Capaldi et al. (2012), in reviewing studies which reported ethnic differences in IPV, reported that in studies conducted in the US " being a member of a minority group is a risk factor for IPV, with findings of greater risk being most consistent for African Americans" (p. 8). Furthermore, in some studies, this was true when controlling for multiple confounding factors, with African American women being twice as likely as other women to be victims of IPV and African American men being approximately twice as likely as other men to be perpetrators. These findings were inconsistent across studies and at least one study which controlled for age, income, marital status, etc., found no effect of ethnicity (Vest et al., 2002). Other studies conducted outside the US have also found higher rates of IPV in non-White groups. For example, in a New Zealand study, Marie et al. (2008) found that after controlling for potentially confounding factors, Moari women and men were more likely to report being perpetrators and victims of IPV.

In terms of gun-related violence in the US, Geier et al.'s (2017) study (see above) reported that there were large differences between ethnic groups with African Americans accounting for 54% of FRDs although they are only 13% of the US population; with Whites and Asians/Pacific Islanders accounting for many fewer FRDs given their percentage of the population. Hispanics / Latinos were at slightly higher risks of FRDs than other groups given their proportion of the population.

3.3.10 Medical Conditions

A number of medical conditions may be associated with aggression and violence. Examples of relevant medical conditions include hormonal conditions, such as hyperthyroidism; seizure disorders, and Traumatic Brain Injury (TBI), although there are many others.

Hyperthyroidism is a condition in which the thyroid gland produces excessive amounts of thyroxin, a hormone that influences digestion, heart and muscle function, metabolic rate, and a number of other functions. Individuals with hyperthyroidism may feel hyperaroused and driven, and experience heart racing, sleep disturbances, and emotional problems which sometimes include irritability, anger, and aggression. Effective treatment of hyperthyroidism through medication and sometimes surgery removing part of the thyroid gland may reduce aggression in some individuals (Mandelbrote & Wittkower, 1955).

A seizure is a sudden onset in change in brain activity that results in a change in consciousness. There are many different types of seizures and if they reoccur the person is diagnosed with epilepsy. Although rare, seizure disorders may sometimes be associated with aggression. In a retrospective review of over 1,000 clinic-referred patients with epilepsy, Tassinari et al. (2005) identified only 11 individuals with seizure related aggression. The aggression was often preceded by an aura, described as a strange feeling; intense fear or anger; and, an inability to control the aggressive behavior. The aggressive behavior was often described as motorically simple, sometimes involving biting, and automaton-like, and was often evoked by the nearby presence of other people. Biological treatment of this aggression is sometimes attempted using treatments for epilepsy, such as seizure medications, and rarely even psychosurgery (Faria, 2013) (see Chaps. 11 and 12, for reviews of biological treatment). Sometimes, seizure disorders are even used as a legal defense for violence due to the apparent lack of intent of any aggressive or violence act that was indeed committed.

TBI is brain damage resulting from physical insult to the brain, including blows to the head, such as those from personal assault or car wrecks, etc., or brain damage due to jolting the head, for example, due to shaking infants or whiplash during a car wreck or similar event. TBI occurs to people from birth to late adulthood and is due to a wide range of events and hence the outcomes are quite varied depending upon the age of the person and the types and severity of injury. TBI is commonly associated with a wide range of conditions including suicide, substance abuse, post-traumatic stress disorder, sleep disorder, and impulsivity. Aggressive and violent behavior can occur immediately after the injury, for example, when the person is disoriented and confused. For some they can occur as a long-term consequence of TBI, due to long-term behavioral inhibition, which may result in long-term care including psychiatric and residential care in some cases.

Based on a large national community sample, Vaughn et al. (2019) found that TBI over the preceding year was quite rare—only about 0.5% of the entire sample. They found that criminal behavior was more likely than those with TBI, including both violence crime (Adjusted OR [AOR] = 1.65, 95% CI = 1.03–2.65) and nonviolent crime (AOR = 1.84, 95% CI = 1.25–2.70). Using a ten-item index of antisocial behavior, they also found that the likelihood of TBI increased monotonically from a low of 0.38% among those reporting 0 items of antisocial behavior to 4.0% among those reporting 7 or more items of antisocial behavior (i.e., an approximately tenfold increased risk). Thus, TBI was associated with an increase of both violent crime and general increase in aggression and broadly defined antisocial behavior (see Chap. 19, for a more detailed review of this issue).

3.3.11 COVID and Aggression and Violence

Apparent increases in aggression and violence, including IPV and other forms of family violence, have received some attention in the media. The possibility that COVID increased violence within the family is plausible as COVID restrictions increased risk factors for aggression and violence such as crowding, loss of income, social isolation, individuals with poor social relationships being forced to spend time together when in the past they have successfully avoided one another, and increased experience of frustration, such as being unable to leave the home when something important is needed and cannot be readily accessed. Indeed there is a small literature from previous quarantines suggesting that quarantine may be associated with increased aggression and violence especially IPV (Brooks et al., 2020). Plausible as it is that aggression and violence increased in families during COVID one should ask what is the evidence that this happened?

A study by Gosangi et al. (2021) studied the rate and type of injuries seen during the first 3 months of the COVID lockdown with injuries in an emergency room during the corresponding seven-week window during the preceding three years. Data included both self-reports of IPV intimate partner violence and data on types and severity of injuries from radiological findings. They found that rates of injuries to women from probable IPV was nearly twice as high and that the risk of severe injuries was five times higher during the pandemic lockdown. The overall number of IPV victims was reduced during the COVID lockdown, suggesting that victims of IPV were waiting longer and perhaps experiencing more IPV before presenting at the emergency room. Other studies have also found evidence of increased IPV during the COVID pandemic lockdown with increasing rates of IPV being mediated by COVID-related anxiety, depression, and stress (Cannon et al., 2021).

A review of COVID-related violence to children revealed a somewhat similar pattern. Cappa and Jijon (2021) reviewed 48 recent empirical papers on this topic, almost all of which were related to family violence such as physical maltreatment and neglect and emotional and psychological violence. They found four main results. First, referrals to police and child protection services may have reduced during the pandemic. Second, there was no systematic change in data on calls to various kinds of helplines. Third, there was an increase in child injuries seen in emergency rooms in studies from the US and Britain. Finally, based on multiple studies from around the world, surveys often reported increases in family violence. Some studies found decreases in certain kinds of violence, such as school-based gun violence. Thus, the data on child-related violence during times of COVID parallel the above data on IPV.

This limited sample of studies suggests that there is evidence that COVID-related quarantine has increased the rates of IPV and violence toward children. This situation is, of course, highly varied, and fluid. It takes place in many different cultural contexts and different stages of the epidemic which is still unfolding. Thus, it is premature to make any broad conclusions at this time as these results may change in the future.

3.3.12 Summary

There are many risk factors for human aggression and violence and this section has reviewed only a sample of the more commonly researched ones. When examining these risk factors, it is always important to consider which ones can be altered readily and which ones have a large effect. Some static risk factors, such as age, gender, ethnicity, cannot be altered and are not particularly useful for intervention. Other factors could be altered, but it would be difficult to do so quickly: Poverty and income inequality are examples although over time public policy could change these variables which might result in reduction in aggression and violence among other outcomes. Other variables might be changeable but only in certain circumstances and perhaps have only modest and unreliable effects under certain circumstances. Temperature and crowding might be examples of these kinds of variables. Some variables are potentially readily alterable and have potentially large effects, perhaps both for certain individuals and for people more generally. The most obvious examples here include our own social behavior related to interpersonal disputes, alcohol, and drug use, presence of weapons, and, alterable features of hot spots. If people were more prosocial with each other, especially in problem situations; if people refrained from alcohol and drug use or used them in a safer manner; if people refrained from having access to or using guns; and, if we modified the physical and social aspects of hot spots; then perhaps there would be less aggression and violence in the world.

3.4 At Risk Populations

There are a number of populations that are specifically at risk for aggression and violence. Information on this topic is important as it identifies something to be explained. It is also important for developing social interventions to prevent and treat aggression and violence in at-risk populations where the pay back for intervention might be large.

3.4.1 Psychiatric Disorders

It is a part of popular culture that if someone commits a heinous act of violence "they must be mad." Indeed, there is a popular stereotype that people with schizophrenia are particularly violent. Furthermore, there are often reports in-patient psychiatric and mental health facilities are dangerous to residents and staff alike. What data do we have on these issues?

Iozzino et al. (2015) reported a systematic review of empirical papers published between 1995 and 2014 on the prevalence and risk factors for violence in psychiatric in-patient units. They located 35 studies with 23,972. They reported that on average 17% (CI 14–20%) of participants had engaged in at least one act of violence and that there was a very wide range of figures between studies. Lower quality studies reported higher percentages, but this did not affect the overall conclusion that about one in five in patients showed at least one incidence of violence. The mediators with the largest impact on the likelihood of violence were the proportion of males, lifetime history of violence, history of alcohol abuse (all percentages of variance around 40%), and, proportion of involuntary admission. A diagnosis of schizophrenia had more moderate effects (all percentages of variance around 20%). Two multivariate analyses indicated that involuntary admission, schizophrenia, alcohol abuse, and male gender all independently predicted the proportion of individuals who were violent.

More recently, Farrington et al. (2017) reviewed risk factors for broadly defined violence, offending, and delinquency and located two systematic reviews published between 2000 and 2016. The authors commented on the narrow range of predictors included in the meta-analyses: many of the included variables, such as history of violence, were not amenable to change. These authors highlighted a meta-analysis of risk factors by Derzon (2010) on family factors identified in prospective studies. This meta-analysis found that low parental education, poor parental supervision, poor child rearing skills, parental discord, and large family size all had small effect sizes (*r*'s = about 0.2–0.3); notably, four of these five predictors are potentially modifiable and, thus, hint at avenues for prevention and treatment.

At least one systematic review has specifically focused on aggression and violence in schizophrenia. Large and Nielssen (2011) conducted a systematic review and meta-analysis of violence in first-episode psychosis. The authors reported that the prevalence of any, severe, and serious violence was 35%, 17%, and 1%, respectively. Risk factors included involuntary treatment (OR = 3.2), hostile affect (IOR = 3.5), manic symptoms (OR = 2.9), illicit substance use (OR = 2.3), less education (OR = 2.0), younger age (OR = 1.9), male (OR = 1.6), duration of untreated psychosis (OR = 2.7), and total symptoms (OR = 2.1). Violence after initiation of treatment was predicted strongly by involuntary status (OR = 5.7). The authors concluded that less severe forms of violence were common but the most severe forms were rare.

These studies suggest that aggression and violence are common in individuals with psychiatric disorders—at least those individuals who are in in-patient units. Of course, aggression and violence is likely to be one of the main reasons for admission to such units. In addition, this literature is useful in that it identifies predictors that are modifiable and hence a potential part of prevention and treatment of aggression and violence.

3.4.2 Individuals with Developmental Disabilities

About 10–15% of individuals with Intellectual Disability show some form of challenging behavior including aggression (Einfeld et al., 2011) which is often rated as among the most severe problem behavior (Jacobson, 1982b). In these studies, aggression and violence are often subsumed under the broad label of challenging behavior. For example, Crocker et al. (2006) reported a large-scale epidemiological study of aggression among 3,165 adults with Intellectual Disability. They found a 1-year prevalence of aggression of 52%; this figure subsumed 1-year prevalence of 24% for property damage, 38% for verbal aggression, 24% for self-injury, 24% for physical aggression, and 10% for sexually aggressive behavior. Despite these high figures only 5% of individuals were sufficiently aggressive to result in injuries to others. Cooper et al. (2009) reported a point prevalence of aggression of 10% and Tyrer et al. (2006) reported that 14% of adults with Intellectual Disability showed aggression. Therefore, these studies do agree that this population is at much higher risk of aggression and violence than the general population and the prevalence figure lies somewhere around 10–20%.

Risk factors for aggression among individuals with intellectual disability include greater degree of intellectual disability; comorbid autism spectrum disorder; male gender; deficits in adaptive behavior, especially communication skills; and living in more restrictive residential settings, such as group homes and institutions. Diagnoses also are correlated with rates of aggression. For example, individuals with Down syndrome tend to show less aggression than other individuals with intellectual disabilities (Jacobson, 1982a, 1982b; Tyrer et al., 2006).

Individuals with Autism Spectrum Disorders have a greater risk of challenging behavior generally, including aggression and other externalizing behavior problems, than individuals with Intellectual Disability when matched for gender and degree of disability (Tsakanikos et al., 2006). For example, Kanne and Mazurek (2011) reported that 68% of 1,038 children and adolescents with Autism Spectrum Disorders had aggressed to a caregiver and 49% to noncaregivers. Risk factors identified in the literature include greater degree of intellectual disability, communication deficits, and repetitive stereotyped behavior (Dominick et al., 2007).

Finally, there are also differences in the rates of violence and aggression among different genetic syndromes associated with intellectual disabilities. For example,

Powis and Oliver (2014), reported a systematic review of empirical studies of the prevalence of aggression in seven genetic syndromes associated with intellectual disabilities (Cris du Chat, Smith Magenis, Fragile-X, Angelman, Cornelia de Lange, Prader-Willi, Williams and Down syndromes.) They found an overall prevalence of around 50%, which is very high and rather higher in studies of individuals with intellectual disabilities generally. The prevalence of aggression was markedly lower than the other groups—around 15%—in William and Down syndromes.

These studies show that individuals with a variety of developmental disorders are at special risk for aggression. Additionally, certain genetic syndromes are associated with both higher and, in some cases, a markedly lower prevalence of aggression.

3.4.3 *Youth in Residential Settings*

Violence in some residential facilities is sometimes common, and includes violence between youth and between youth and staff. Some youth facilities, such as Rikers Island, New York City, have become notorious for such problems, including staff on youth violence.

In a review of youth facilities and violence, Egozy and Cox (2017) estimated that about a third of youth in residential facilities had been beaten up or threatened while in residential care. A quarter of these incidents involved staff and the remainder involved other residents. Four percent of youth had been coerced into sexual activity of which about half involved staff and half involved other youth. Some states report about a third of youth had been restrained each month on a monthly basis with some instances of isolation lasting over 8 h.

3.4.4 *Older Adults in Residential Care*

Aggression and violence in nursing homes is a relatively common problem. For example, Leonard et al. (2006) analyzed data sets from five states containing information on aggression in over 100,000 individuals aged over 60 years in nursing homes with dementia. They found that about 6.9% had been aggressive in the preceding week. Risk factors included psychiatric symptoms, such as hallucinations and delusions, and some physical symptoms including constipation.

Much aggression takes place between staff and residents, for example, during care routines, but aggression between residents also occurs and may be potentially fatal, but is under-studied. Ferrah et al. (2015) reported a systematic review which identified six empirical studies of resident-to-resident aggression. Unfortunately, it was unable to estimate the prevalence of resident-to-resident aggression with confidence due to widely different reporting practices in the articles, but did note that around one in five residents had been injured by other residents. One study reported that most injuries were due to pushing, grabbing, and pinching. The most common triggers were communication difficulties, for example due to weak verbal skills, hearing problems, etc., and invasion of personal space.

3.4.5 Comment: Some Common Factors?

The preceding sections review aggression and violence in rather different populations, but they do share some things in common. Namely, many individuals in these populations are at risk for residential care either to protect the public, the individual, or provide intervention that community services do not offer. This is of course a chicken and egg situation: It is difficult to disentangle preadmission aggression from the potential effects of residential care in causing and maintaining aggression and violence. Nevertheless, placement in residential treatment programs is an index of the severity of the problem and the inability or unwillingness of communities to prevent, manage, and treat these forms of aggression and violence in community settings.

3.5 Nonhuman Animals

Nonhumans are both aggressors and victims of aggression and violence. Aggression is common in many nonhuman animals (Lorenz, 2005). Thus, there is an existing literature on aggression and violence in nonhuman animals which is often interested in differences in the epidemiology of aggression and violence between different species and breeds of domestic animals. It also examines predictors and risk factors for aggression and violence, and nonhuman animals as victims of human aggression and violence to nonhumans and aggression toward humans. (See Chap. 4, which deals with evolution of aggression and violence in nonhuman animals in more detail than here).

Humans benefit tremendously from nonhuman animals as sources of food, labor, entertainment, clothing, affection, in research, and as fashion items. Indeed, humanity has spent thousands of years of efforts in domesticating and breeding nonhuman animals for these very purposes. Today there are veritable industries dedicated to breeding and raising nonhuman animals for people's benefits and people pay large amounts of money for nonhuman animals as sources of energy, food, and companionship. We also breed animals for aggressive and nonaggressive behavior: We sometimes breed dogs to fight and sometimes to be passive and solicitous nonaggressive companions.

Although we often appreciate our nonhuman animal companions, people are sometimes aggressive, violent, and cruel to nonhumans, including their pets. Sometimes we breed and raise nonhuman animals specifically to be aggressive and to take part in human cultural rituals, such as bull fighting, bear baiting, dog fighting, and cock fighting. These activities are often associated with betting, monetary gain, and careers for some of the active human participants and entertainment for the audience. Such behavior is sometimes labeled "animal cruelty" is may be seen as a sign of human psychopathology and criminality, however, there may also sometimes be more mundane explanations for some of these forms of aggression and violence. Beating and whipping the beast of burden will get your cartload delivered faster and may help you win a horse race and gain a monetary prize. Yelling, hitting you pet dog with a newspaper and pulling on its choke chains may result in termination of your pet dog's sometimes aversive behavior, perhaps making it less likely that it will snarl at you in the future, get it to move out of your way or begin to walk, and perhaps get it to comply with instructions faster.

3.5.1 Aggression in Dogs

Violence and aggression among canids are well known and it is likely that early ancestors of domestic dogs including wolves and hyenas indeed attacked and killed humans for food (Lockwood, 2016). Today, such behavior is uncommon in domestic dogs in most situations, although society remains concerned about dogs biting humans. For example, a Dutch study by Cornelissen and Hopster (2010) using three surveys of 200,000 people found that there were around 136,000 dog bites in The Netherlands for a rate of 8.3 per 1000 people one in five of which resulted in skin punctures. (Bites are probably under-reported, especially by dog owners who sometimes relabel bites as "nips" or "playing"). Sheep and cattle dogs, pinchers, schnauzers, and terriers were most likely to bite people whereas Spaniels, retrievers, Maltese, and Yorkshire terriers were least likely to bite people. Children were most likely bitten at home when unsupervised with their pets while playing with them. These data suggest that prospective owners can choose their breed of pet accordingly and parents should supervise and teach their children how to play with their pet dogs properly in order to prevent dog bites to their children.

As noted earlier, people are sometimes cruel to nonhuman animals. A study by Baglivio and Epps (2016) found that among a very large sample of juvenile offenders that self-reported animal cruelty was rare; only 0.6% of 292,649 juvenile offenders reported engaging in animal cruelty. Other reviews have found widely differing prevalences from 5 to 70% depending on the sample and methodology (Connor et al., 2018). Thus, it is difficult to determine the true prevalence of human aggression and violence to animals as much of it occurs privately and data depends on accuracy of self-reports.

There is an extensive literature on behavior management in dogs including aggression. These include papers which include behavior analytic approaches including true experimental functional analyses, function-based treatment of aggression in dogs (Dorey et al., 2012; Winslow et al., 2018), and training dog owners and improving relationships between dog owners and their dogs (Feuerbacher & Wynne, 2016).

3.6 Summary

Human aggression and violence are common and distributed very unequally. When it comes to homicide, young men are often the perpetrators and victims of homicide and many women experience violence in the context of intimate personal relationships. When it comes to aggression and violence, the family is the main setting for varying intensities of aggression and violence between intimate partners, caregivers, their children and other family members. Aggression and violence toward children in many forms is very common worldwide, although somewhat less common in wealthy than developing countries. Aggression and violence are also closely related to many environmental factors that have potential for prevention and treatment. These include alcohol and illicit drugs, crowding, temperature, access to weapons, poverty and inequality, physical location and time of day and year, and medical conditions. The next section of this book goes on to explore the various causes of aggression and violence and these theories must explain the observations made in this chapter.

References

Alhabib, S., Nur, U., & Jones, R. (2010). Domestic violence against women: Systematic review of prevalence studies. *Journal of Family Violence, 25*(4), 369–382.

Anderson, C. A. (1989). Temperature and aggression: Ubiquitous effects of heat on occurrence of human violence. *Psychological Bulletin, 106*(1), 74.

Anderson, C. A., Anderson, K. B., Dorr, N., DeNeve, K. M., & Flanagan, M. (2000). Temperature and aggression. *Advances in Experimental Social Psychology, 32*, 63–129.

Baglivio, M. T., & Epps, N. (2016). The interrelatedness of adverse childhood experiences among high-risk juvenile offenders. *Youth Violence and Juvenile Justice, 14*(3), 179–198.

Basile, K. C., Breiding, M. J., & Smith, S. (2016). Disability and risk of recent sexual violence in the United States. *American Journal of Public Health, 106*(5), 928–993. https://doi.org/10.2105/AJPH.2015.303004

Black, M. C. (2011). Intimate partner violence and adverse health consequences: implications for clinicians. *American Journal of Lifestyle Medicine, 5*(5), 428–439.

Brooks, S. K., Webster, R. K., Smith, L. E., Woodland, L., Wessely, S., Greenberg, N., & Rubin, G. J. (2020). The psychological impact of quarantine and how to reduce it: Rapid review of the evidence. *The Lancet, 395*(10227), 912–920.

Brown, W., & Leonard, K. E. (2017). Does alcohol cause violence and aggression? In P. Sturmey (Ed.), *The Wiley handbook of violence and aggression* (pp. 1–13). Wiley.

Butchart, A., Mikton, C., & Krug, E. (2014). Governments must do more to address interpersonal violence. *The Lancet, 384*, 2183–2185. https://doi.org/10.1016/S0140-6736(14)62124-3

Cannon, C. E. B., Ferreira, R., Buttell, F., & First, J. (2021). COVID-19, intimate partner violence, and communication ecologies. *American Behavioral Scientist, 65*(7), 992–1013. https://doi.org/10.1177/0002764221992826

Capaldi, D. M., Knoble, N. B., Shortt, J. W., & Kim, H. K. (2012). A systematic review of risk factors for intimate partner violence. *Partner Abuse, 3*(2), 231–280. https://doi.org/10.1891/1946-6560.3.2.231

Cappa, C., & Jijon, I. (2021). COVID-19 and violence against children: A review of early studies. *Child Abuse & Neglect, 116*, 105053.

Coccia, M. (2017). A Theory of general causes of violent crime: Homicides, income inequality and deficiencies of the heat hypothesis and of the model of CLASH. *Aggression and Violent Behavior, 37*, 190–200.

Connor, M., Currie, C., & Lawrence, A. B. (2018). Factors influencing the prevalence of animal cruelty during adolescence. *Journal of Interpersonal Violence, 36*, 3017–3040.

Cohen, L. E., & Felson, M. (1979). Social change and crime rate trends: A routine activity approach. *American Sociological Review, 44*, 588–608.

Cook, J. M., Dinnen, S., & O'Donnell, C. (2011). Older women survivors of physical and sexual violence: A systematic review of the quantitative literature. *Journal of Women's Health, 20*(7), 1075–1081.

Cooper, S. A., Smiley, E., Jackson, A., Finlayson, J., Allan, L., Mantry, D., & Morrison, J. (2009). Adults with intellectual disabilities: Prevalence, incidence and remission of aggressive behaviour and related factors. *Journal of Intellectual Disability Research, 53*(3), 217–232.

Cornelissen, J. M., & Hopster, H. (2010). Dog bites in The Netherlands: A study of victims, injuries, circumstances and aggressors to support evaluation of breed specific legislation. *The Veterinary Journal, 186*(3), 292–298.

Crocker, A. G., Mercier, C., Lachapelle, Y., Brunet, A., Morin, D., & Roy, M. E. (2006). Prevalence and types of aggressive behaviour among adults with intellectual disabilities. *Journal of Intellectual Disability Research, 50*(9), 652–661.

David-Ferdon, C., Simon, T. R., Spivak, H., Gorman-Smith, D., Savannah, S. B., Listenbee, R. L., & Iskander, J. (2015). CDC grand rounds: Preventing youth violence. *MMWR. Morbidity and Mortality Weekly Report, 64*(7), 171.

Derzon, J. H. (2010). The correspondence of family features with problem, aggressive, criminal, and violent behavior: A meta-analysis. *Journal of Experimental Criminology, 6*(3), 263–292.

Desmarais, S. L., Reeves, K. A., Nicholls, T. L., Telford, R. P., & Fiebert, M. S. (2012a). Prevalence of physical violence in intimate relationships, Part 1: Rates of male and female victimization. *Partner Abuse, 3*(2), 140–169.

Desmarais, S. L., Reeves, K. A., Nicholls, T. L., Telford, R. P., & Fiebert, M. S. (2012b). Prevalence of physical violence in intimate relationships, Part 2: Rates of male and female perpetration. *Partner Abuse, 3*(2), 170–198.

Dominick, K. C., Davis, N. O., Lainhart, J., Tager-Flusberg, H., & Folstein, S. (2007). Atypical behaviors in children with autism and children with a history of language impairment. *Research in Developmental Disabilities, 28*(2), 145–162.

Dorey, N. R., Tobias, J. S., Udell, M. A., & Wynne, C. D. (2012). Decreasing dog problem behavior with functional analysis: Linking diagnoses to treatment. *Journal of Veterinary Behavior, 7*(5), 276–282.

Duke, A. A., Smith, K. M., Oberleitner, L., Westphal, A., & McKee, S. A. (2018). Alcohol, drugs, and violence: A meta-meta-analysis. *Psychology of Violence, 8*(2), 238–249.

Dwyer-Lindgren, L., Bertozzi-Villa, A., Stubbs, R. W., Morozoff, C., Shirude, S., Unützer, J., & Murray, C. J. (2018). Trends and patterns of geographic variation in mortality from substance use disorders and intentional injuries among US counties, 1980-2014. *Journal of the American Medical Association, 319*(10), 1013–1023.

Egozy, O., & Cox, A. (2017). Youth facilities and violence: An overview. In P. Sturmey (Ed.), *The Wiley handbook of violence and aggression* (pp. 1–11). Wiley UK.

Einfeld, S. L., Ellis, L. A., & Emerson, E. (2011). Comorbidity of intellectual disability and mental disorder in children and adolescents: A systematic review. *Journal of Intellectual and Developmental Disability, 36*(2), 137–143.

Faria, M. A. (2013). Violence, mental illness, and the brain–A brief history of psychosurgery: Part 3–From deep brain stimulation to amygdalotomy for violent behavior, seizures, and pathological aggression in humans. *Surgical Neurology International, 4, 91.*

Farrington, D. P., Gaffney, H., & Ttofi, M. M. (2017). Systematic reviews of explanatory risk factors for violence, offending, and delinquency. *Aggression and Violent Behavior, 33*, 24–36.

Ferrah, N., Murphy, B. J., Ibrahim, J. E., Bugeja, L. C., Winbolt, M., LoGiudice, D., & Ranson, D. L. (2015). Resident-to-resident physical aggression leading to injury in nursing homes: A systematic review. *Age and Ageing, 44*(3), 356–364.

Feuerbacher, E. N., & Wynne, C. D. (2016). Application of functional analysis methods to assess human–dog interactions. *Journal of Applied Behavior Analysis, 49*(4), 970–974.

Franklin, T. W., Franklin, C. A., & Pratt, T. C. (2006). Examining the empirical relationship between prison crowding and inmate misconduct: A meta-analysis of conflicting research results. *Journal of Criminal Justice, 34*(4), 401–412.

Geier, D. A., Kern, J. K., & Geier, M. R. (2017). A longitudinal ecological study of household firearm ownership and firearm-related deaths in the United States from 1999 through 2014: A specific focus on gender, race, and geographic variables. *Preventive Medicine Reports, 6*, 329–335. https://doi.org/10.1016/j.pmedr.2017.04.004

Gendreau, P., Goggin, C. E., & Law, M. A. (1997). Predicting prison misconducts. *Criminal Justice and Behavior, 24*(4), 414–431.

Gonçalves, L. C., Gonçalves, R. A., Martins, C., & Dirkzwager, A. J. (2014). Predicting infractions and health care utilization in prison: A meta-analysis. *Criminal Justice and Behavior, 41*(8), 921–942.

Gosangi, B., Park, H., Thomas, R., Gujrathi, R., Bay, C. P., Raja, A. S., & Khurana, B. (2021). Exacerbation of physical intimate partner violence during COVID-19 pandemic. *Radiology, 298*(1), E38–E45.

Groff, E., & McCord, E. S. (2012). The role of neighborhood parks as crime generators. *Security Journal, 25*(1), 1–24.

Hardie, T. J. (1999). Crowding and violent behaviour: The influence of patient density on violent and self-harming behaviour at a medium secure unit. *Medicine, Science, and the Law, 39*(2), 161–166.

Hillis, S., Mercy, J., Amobi, A., & Kress, H. (2016). Global prevalence of past-year violence against children: A systematic review and minimum estimates. *Pediatrics, 13, 320–327.* https://doi.org/10.1542/peds.113.2.320

Hsieh, C. C., & Pugh, M. D. (1993). Poverty, income inequality, and violent crime: A meta-analysis of recent aggregate data studies. *Criminal Justice Review, 18*(2), 182–202.

Iozzino, L., Ferrari, C., Large, M., Nielssen, O., & De Girolamo, G. (2015). Prevalence and risk factors of violence by psychiatric acute inpatients: A systematic review and meta-analysis. *PLoS One, 10*(6), e0128536.

Jacobson, J. W. (1982a). Problem behavior and psychiatric impairment within a developmentally disabled population I: Behavior frequency. *Applied Research in Mental Retardation, 3*(2), 121–139.

Jacobson, J. W. (1982b). Problem behavior and psychiatric impairment within a developmentally disabled population II: Behavior severity. *Applied Research in Mental Retardation, 3*(4), 369–381.

Jones, L., Bellis, M. A., Wood, S., Hughes, K., McCoy, E., Eckley, L., & Officer, A. (2012). Prevalence and risk of violence against children with disabilities: A systematic review and meta-analysis of observational studies. *The Lancet, 380*(9845), 899–907.

Kanne, S. M., & Mazurek, M. O. (2011). Aggression in children and adolescents with ASD: Prevalence and risk factors. *Journal of Autism and Developmental Disorders, 41*(7), 926–937.

Kellermann, A. L., Rivara, F. P., Rushforth, N. B., Banton, J. G., Raey, D. T., Francisco, J. T., & Some, G. (1993). Gun ownership as a risk factor for homicide in the home. *New England Journal of Medicine, 329*(15), 1084–1091. https://doi.org/10.1056/NEJM199310073291506

Kurland, J., Johnson, S. D., & Tilley, N. (2017). Hotspotting and football violence: Current statistics and implications for prevention. In P. Sturmey (Ed.), *The Wiley handbook of violence and aggression* (pp. 1–15). Wiley UK.

Large, M. M., & Nielssen, O. (2011). Violence in first-episode psychosis: a systematic review and meta-analysis. *Schizophrenia Research, 125*(2-3), 209–220. https://doi.org/10.1016/j.schres.2010.11.026

Leonard, R., Tinetti, M. E., Allore, H. G., & Drickamer, M. A. (2006). Potentially modifiable resident characteristics that are associated with physical or verbal aggression among nursing home residents with dementia. *Archives of Internal Medicine, 166*(12), 1295–1300.

Lockwood, R. (2016). Ethology, ecology and epidemiology of canine aggression. In J. Serpell (Ed.), *The domestic dog: Its evolution, behaviour & interactions with people* (pp. 160–181). Cambridge University Press.

Lorenz, K. (2005). *On aggression*. Routledge.

Lyons, B. H., Fowler, K. A., Jack, S. P., Betz, C. J., & Blair, J. M. (2016). Surveillance for violent deaths—national violent death reporting system, 17 states, 2013. *Morbidity and Mortality Weekly Report. Surveillance Summaries, 65*(10), 1–42.

Mandelbrote, B. M., & Wittkower, E. D. (1955). Emotional factors in Graves' disease. *Psychosomatic Medicine, 17*(2), 109–123.

Marie, D., Fergusson, D. M., & Boden, J. M. (2008). Ethnic identity and intimate partner violence in a New Zealand birth cohort. *Social Policy Journal of New Zealand, 33*, 126–145.

Mikton, C. R., Butchart, A., Dahlberg, L. L., & Krug, E. G. (2016). Global status report on violence prevention 2014. *American Journal of Preventive Medicine, 50*(5), 652–659.

Miles, C., & Condry, R. (2015). Responding to adolescent to parent violence: Challenges for policy and practice. *British Journal of Criminology, 55*(6), 1076–1095.

Miltenberger, R. G., Flessner, C., Gatheridge, B., Johnson, B., Satterlund, M., & Egemo, K. (2004). Evaluation of behavioral skills training to prevent gun play in children. *Journal of Applied Behavior Analysis, 37*(4), 513–516.

Miltenberger, R. G., Sanchez, S., & Valbuena, D. (2020). Pediatric prevention: Teaching safety skills. *Pediatric Clinics, 67*(3), 573–584.

Ng, B., Kumar, S., Ranclaud, M., & Robinson, E. (2001). Ward crowding and incidents of violence on an acute psychiatric inpatient unit. *Psychiatric Services, 52*(4), 521–525.

Ouimet, M., Langlade, A., & Chabot, C. (2018). The dynamic theory of homicide: Adverse social conditions and formal social control as factors explaining the variations of the homicide rate in 145 Countries. *Canadian Journal of Criminology and Criminal Justice, 60*(2), 241–265.

Palmstierna, T., Huitfeldt, B., & Wistedt, B. (1991). The relationship of crowding and aggressive behavior on a psychiatric intensive care unit. *Psychiatric Services, 42*(12), 1237–1240.

Plante, C., & Anderson, C. A. (2017). Media, violence, aggression, and antisocial behavior: Is the link causal? In P. Sturmey (Ed.), *The Wiley handbook of violence and aggression* (pp. 1–12). Wiley UK.

Powis, L., & Oliver, C. (2014). The prevalence of aggression in genetic syndromes: A review. *Research in Developmental Disabilities, 35*(5), 1051–1071.

Pridemore, W. A. (2011). Poverty matters: A reassessment of the inequality–homicide relationship in cross-national studies. *The British Journal of Criminology, 51*(5), 739–772.

Rago, W. V., Parker, R. M., & Cleland, C. C. (1978). Effect of increased space on the social behavior of institutionalized profoundly retarded male adults. *American Journal of Mental Deficiency, 82(6)*, 554–558.

Ratcliffe, J. (2010). Crime mapping: Spatial and temporal challenges. In A. R. Picquero & D. L. Weisburd (Eds.), *Handbook of quantitative criminology* (pp. 5–24). Springer.

Rothman, K. J. (2012). *Epidemiology: An introduction*. Oxford University Press.

Siegel, M., & Rothman, E. F. (2016). Firearm ownership and suicide rates among US men and women, 1981–2013. *American Journal of Public Health, 106*(7), 1316–1322.

Straus, M. A. (1979). Measuring intrafamily conflict and violence. The conflict tactic (CT) scales. *Journal of Marriage and the Family, 41*, 75–88.

Straus, M. A., & Gelles, R. J. (1990). How violent are American families? Estimates from the national family violence resurvey and other studies. In M. Straus & R. Gelles (Eds.), *Physical violence in American families: Risk factors and adaptations to violence in* (pp. 297–298). Routledge.

Straus, M. A., Gelles, R. J., & Smith, C. (1990). Physical violence in American families: Risk actors and adaptations to violence in 8,145 families. In M. Straus & R. Gelles (Eds.), *Physical violence in American families: Risk factors and adaptations to violence* (pp. 95–112). Routledge.

Studdert, D. M., Zhang, Y., Swanson, S. A., Prince, L., Rodden, J. A., Holsinger, E. E., & Miller, M. (2020). Handgun ownership and suicide in California. *New England Journal of Medicine, 382*(23), 2220–2229.

Tassinari, A. C., Tassi, L., Calandra-Buonaura, G., Stanzani-Maserati, M., Fini, N., Pizza, F., & Meletti, S. (2005). Biting behavior, aggression, and seizures. *Epilepsia, 46*(5), 654–663.

Tsakanikos, E., Costello, H., Holt, G., Bouras, N., Sturmey, P., & Newton, T. (2006). Psychopathology in adults with autism and intellectual disability. *Journal of Autism and Developmental Disorders, 36*(8), 1123–1129.

Tyrer, F., McGrother, C. W., Thorp, C. F., Donaldson, M., Bhaumik, S., Watson, J. M., & Hollin, C. (2006). Physical aggression towards others in adults with learning disabilities: Prevalence and associated factors. *Journal of Intellectual Disability Research, 50*, 295–304. https://doi.org/10.1111/j.1365-2788.2005.00774.x

Ulrich, R. S., Bogren, L., Gardiner, S. K., & Lundin, S. (2018). Psychiatric ward design can reduce aggressive behavior. *Journal of Environmental Psychology, 57*, 53–66.

Vaughn, M. G., Salas-Wright, C. P., John, R., Holzer, K. J., Qian, Z., & Veeh, C. (2019). Traumatic brain injury and psychiatric co-morbidity in the United States. *Psychiatric Quarterly, 90*(1), 151–158.

Vest, J. R., Catlin, T. K., Chen, J. J., & Brownson, R. C. (2002). Multistate analysis of factors associated with intimate partner violence. *American Journal of Preventive Medicine, 22*(3), 156–164.

Vrij, A., van der Steen, J., & Koppelaar, L. (1997). The effects of distraction on police officer shooting. In S. Redondo, V. Garrido, J. Pérez, & R. Barberet (Eds.), *Advances in psychology and law* (p. 155). De Gruyter. https://doi.org/10.1515/9783110801163

Winslow, T., Payne, S. W., & Massoudi, K. A. (2018). Functional analysis and treatment of problem behavior in 3 animal shelter dogs. *Journal of Veterinary Behavior, 26*, 27–37.

Wintemute, G. J. (1987). Firearms as a cause of death in the United States, 1920–1982. *Journal of Trauma and Acute Care Surgery, 27*(5), 532–536.

Wintemute, G. J. (2011). Association between firearm ownership, firearm-related risk and risk reduction behaviours and alcohol-related risk behaviours. *Injury Prevention, 17*(6), 422–427.

Wintemute, G. J., Teret, S. P., Kraus, J. F., Wright, M. A., & Bradfield, G. (1987). When children shoot children: 88 unintended deaths in California. *Journal of the American Medical Association, 257*(22), 3107–3109. https://doi.org/10.1001/jama.1987.03390220105030

Wintemute, G. J., Wright, M. A., Drake, C. M., & Beaumont, J. J. (2001). Subsequent criminal activity among violent misdemeanants who seek to purchase handguns: Risk factors and effectiveness of denying handgun purchase. *Journal of the American Medical Association, 285*(8), 1019–1026.

World Health Organization. (2013). *Global and regional estimates of violence against women: prevalence and health effects of intimate partner violence and non-partner sexual violence*. World Health Organization.

Part II
Causes

Having established some basic facts concerning violence and aggression in Part I of this volume, this Part goes on to consider the three principal causes of violence and aggression using a Skinneran selectionist framework, namely biological, cultural, and operant learning. Chapters 4 and 5 consider biological causes such as evolution, genetics and specific biological mechanisms such as brain structures, neurotransmission, and hormones. Chapters 6 and 7 consider cultural evolution as a cause of violence and aggression and, Chap. 8 consider learning during the lifespan as a cause of violence and aggression, and finally, Chap. 9 considers some theories of violence and aggression.

Chapter 4
Biological Evolution of Violence and Aggression. I. Evolution and Genetics

4.1 What Is a Cause?

The next few chapters examine different theories of causation of aggression and violence including biological, cultural, and individual learning as causes of aggression and violence. Before we consider the specific theories, we should be clear about what is meant by a "cause." For one variable to be said to cause a change in another variable in the simplest forms of causality, the cause always precedes the effect, the effect always occurs in the presence of the cause (Brown & Leonard, 2017), the cause must be temporally and/or physically proximate to the effect, and, the relationship between the cause and the effect must be reliable. Such causes might be considered "simple causes." One might think that simple causes of aggression and violence are rare, but they do exist. For example, highly reliable relationships between environmental variables and some forms of aggression and violence and have been demonstrated extensively in the field of applied behavior analysis. A second example might be the reliable relationship between electrical stimulation of certain brain regions that is often followed by observable aggressive behavior (see Chap. 5).

More typically, however, there may be multiple antecedent conditions or chains of causality necessary for causality to be demonstrated. For example, it may be that only when the temperature and crowding increase over a certain value that an individual responds to a provocation with aggressive behavior. A second example of long complex chains of causation might be the developmental processes or lifelong learning that result in some terminal aggression or violence. For example, a clinical case formulation might chart a person's lifelong history of failure to learn important social, communicative, and problem-solving skills, acquisition of unusual reinforcers, such as seeing people suffering, and acquisition of mild forms of aggressive behavior resulting in aggression and eventually violence (Gresswell & Hollin,

P. Sturmey, *Violence and Aggression*, Advances in Preventing and Treating Violence and Aggression, https://doi.org/10.1007/978-3-031-04386-4_4

1992). A final example of long chains of causation might be the gradual biological evolution of aggression over millions of years in which particular forms of behavior that eventually result in current aggression are selected over time.

One relevant approach to thinking about causality and aggression and violence comes from Tinbergen (1963) who wrote a position paper on the then emerging field of ethology which he defined as "the biological study of behavior" (p. 411). His work was strongly influenced by the work of Konrad Lorenz (1971), who provided a large body of observational research and theory on animal behavior. Lorenz's work was based on direct observation of animals in their natural environments and in semi-natural environments and his personal observations of them when these animals lived with Lorenz. Ethologists, at least at that time, eschewed and were skeptical concerning experimental psychology. Tinbergen write that "… this 'return to nature' was a reaction against the tendency prevalent at that time in psychology to concentrate on a few phenomena observed in a handful of species which were kept in impoverished environments to formulate theories claimed to be general, and to proceed deductively to test these theories experimentally" (p. 411). Tinbergen was concerned that observing behavior out of context—in experimental settings or zoos—might result in behavior being inexplicable and that careful and patient observation in the natural world might result in a better understanding of the behavior. Thus, one of his concerns was the balance between inductive observation in the natural environment and experimental deductive science, sometimes carried out in the laboratory and sometimes carried out in modified natural settings.

In addition to emphasizing direct observation in the natural environment, Tinbergen (1963) emphasized the importance of function of not only morphology, but also behavior. He characterized the work of early ethologists as trying to answer the question "Why do these animals behave as they do?" (p. 411). He regarded behavior as a kind of organ-system which, like physical organs, served some purpose and contributed toward the survival of the animal and its genes. Thus, aggression including defenses against predation might usefully contribute to the survival of the organism and its genes just as much as a claw or sharp beak might.

In order to clarify the question of causality, he famously proposed four kinds of questions. The first two questions are related to function of behavior and to more distant kinds of causation. They were: (1) What is the adaptive value of the behavior to the organism? and (2) How did the behavior evolve (phylogeny, i.e. biological evolution)? The two remaining questions regarding the proximate causes of behavior. They were: (3) What is the immediate cause or mechanism of behavior; and (4) How did the behavior develop during ontogeny, (i.e. from embryo to adult form of the organism). These four questions can be posed specifically regarding aggression. First, what is the adaptive advantage to the organism to engage in aggressive behavior? Second, how did aggression evolve in this species and perhaps but not other organisms? Third, what was the immediate cause of that incident of aggressive behavior—what happened immediately before and after aggression that might be the cause of the aggressive behavior? (This is parallel to behavior analytic questions of causation.) Finally, how did aggressive behavior develop from fetal development, infancy, childhood through adulthood in that individual? (This is parallel to developmental studies of aggression across the lifespan). Some of these questions are parallel in various chapters of this book.

Tinbergen (1963) noted that certain forms of animal behavior are highly stereotypical or invariant in form and sequence, species-specific, apparently relatively uninfluenced by experience. He called these apparently invariant patterns of behavior "fixed action patterns." They appear to be "released" by specific environmental stimuli. He termed these "innate releasing mechanisms" with the implication that, since they were innate, learning and experience were relatively unimportant in determining the form and environmental control of the behavior. Fixed action patterns are in some ways similar to unconditioned reflexes in that they circumvent inefficient learning during the lifespan and contribute to the survival of the organism and its genes, although they are often more motorically complex than most reflexes.

In earlier work, Lorenz had shown that specific stimuli could affect animal behavior. For example, the placement of pine cones on a beach could affect whether or not bee wolves returned with their prey to their nest sites on sandy beaches accurately; changing the location of the pinecones changed where the beewolves returned to after flying around. (See the discussion below for several similar examples from the work or Lorenz). This work illustrated a proximate for of causation.

The combination of the two ideas of specific environmental stimuli affecting behavior and behavior being apparently released by those stimuli with relatively little learning converged in his classic studies related to aggression is his studies of aggression in the three-spined sticklebacks (*Gasterosteus aculeatus*). In nature, three-spined sticklebacks attack other sticklebacks by charging at them if the other stickleback has a red underbelly and in nature the intruder would often back off. Using wooden mockups of sticklebacks, Tinbergen showed that only models with red underbellies provoked attacks, whereas the size, general shape, and movement of the model did not influence attack behavior (Tinbergen, 1952). Thus, specific environmental stimuli seemed to "release" sometimes complex, evolutionarily important animal behavior, including aggression. Tinbergen (1963) also mentioned aggression and related behavioral phenomena, such as defense against predators, in several places as other examples of the evolution of aggression-related behavior. In other studies, Tinbergen also showed that the spines were quite useful to the stickleback; if the spines were removed, they were more likely to be eaten by other predator fish than if they kept their spines (Morris et al., 1956). Thus, it appeared that biological evolution has endowed the stickleback with both behavioral and morphological features serving similar functions.

The next section addresses the first of Tinbergen's four questions. (The reader is also directed to parts of Chaps. 13, 14, and 15, that addresses selection of cultural practices related to pacific or violent cultures (Glenn, 1988, 2004).

4.2 Biological Causes

Many authors have proposed that aggression—and especially its more severe forms—is caused by biological factors. These biological causes might include biological evolution and specific genetic mechanisms, abnormal brain structures and functioning, neurotransmitters, and hormones. Others have suggested mechanisms by which biological factors might interact with environmental factors in various ways. In this chapter we review some of these ideas. It is sometimes tempting to think that one day science may find a single cause of aggression and violence, but this is a mistake. Sometimes scientists have indeed been able to show simple causality of aggression and violence in a few highly specific cases. Examples include electrical stimulation of the brain of certain animals that sometimes elicits attack. For example, electrical stimulation of the periaqueductal gray matter, locus coeruleus, substantia innominata, and central nucleus of the amygdala elicits defensive reactions in the cat reliably (Miczek et al., 1994). Aggressive behavior can be turned on and off reliably in some applied behavior analytic studies, typically in children with disabilities, but even these analyses do not preclude other kinds of causes and thus represent an incomplete explanation of aggression. In more typical situations, variables combine and interact with each other, with the individual's existing biology, culture, and learning history to render qualified or probabilistic relationships among potential causes. Hence, hoping to find the single biological causes of aggression such as neurotransmitter, brain site or gene to explain aggression and violence is a foolish pursuit.

A good example of these interactions between genes and environment comes from Caspi et al. (2002), one of the most widely cited empirical research studies on aggression and violence. Caspi et al. investigated the interactive effects of childhood maltreatment and the neurotransmitter-metabolizing enzyme monoamine oxidase (MAOA) the participants were the male members of the Dunedin Multidisciplinary Health and Development study a birth cohort of 1037 New Zealand children assessment 10 times between ages 3 and 26 years with 96% of the original sample being retained at age 26 years. Childhood maltreatment was rated as severe, probable, and no maltreatment in 8%, 28%, and 64% of the sample, respectively. Using psychometric instruments, aggressive behavior was measured as a composite index of adolescent conduct disorder, convictions of violent crimes, a psychological assessment of personality dispositions toward violence and antisocial personality. The MAOA gene is located on the X-chromosome and metabolizes Norepinephrine 5-Hydroxytryptamine and dopamine rendering them inactive; hence, MAOA deficiencies results in an excess of these neurotransmitters which is correlated to aggression in humans and nonhumans. Due to a rare genetic mutation, some men have only a single copy of the MAOA gene and hence are functionally "a human knock out" for this gene and do not metabolize 5-hydroxytryptamine.

Caspi et al. (2002) found a clear interaction between childhood maltreatment and MAOA status. That is, individuals who were knock outs for MAOA were much more sensitive to the effects of childhood maltreatment than those who had the

MAOA genes. Namely, individuals with no or probable maltreatment showed relatively low aggression, but individuals with low MAOA *and* severe maltreatment showed about four times as much aggression than individuals with high MAOA activity. Of course, even in the highest risk group (low MOAO/severe maltreatment) there remained considerable variability including individuals who were not aggressive; and within the low risk groups there were still individuals who were very aggressive indicating there is still much else to be explained.

Subsequently, Kim-Cohen et al. (2006) replicated these observations on a separate sample and found similar effects on measures of other childhood disorders such as ADHD and emotional problems. They also conducted a meta-analysis of five studies (including two of their own) and found evidence from pooled results that this interaction between genotype and childhood maltreatment on aggression was robust. These studies illustrate the importance of interaction of variables causing aggression and violence.

4.3 Biological Evolution of Aggression

Darwin's *Origin of Species and Expression of Emotions in Animals and Man* (Darwin & Prodger, 1998) was interested in the reasons why animals engage in aggression, why males often engage in more aggression than females, and why in other species these differences between males and females are not apparent. Darwin was also well acquainted with animal breeders and must have been aware that animal breeders often selected animals for many aspects of temperament including aggression. Darwin noted that sexual dimorphism was often (but not always) characterized by larger males with weaponry, such as horns, claws, and teeth, which were usually used for male-on-male aggression rather than for defense from predators. This suggested that evolution had selected certain morphologies and behavior to enable males of many species to be more aggressive to other males in order to succeed in obtaining mates. He termed this particular form of natural selection sexual selection.

Sexual selection occurs through two mechanisms. In intrasexual selection members of one sex compete with each other for access to the other sex. In intrasexual selection in which members of one sex choose certain members of the opposite sex. Competition between males for access to females is the most common form of intersexual selection. Female choice of mate characteristics indicating genetic fitness, such as peahens' selection of peacocks with large tail feathers, is an example of intra-sexual selection. These two processes may not be independent of one another as when females chose the most aggressive males for mating, who have both the most effective forms of weaponry and other signs of genetic fitness not related to aggression. Finally, sexual dimorphism is common, occurring in about half of all species, but it is by no means ubiquitous. Indeed, it is very difficult for humans to discriminate between males and females in many species of birds and even some mammals such as dolphins. So, one should ask why it is selected for in some but not other species (Lindenfors & Tullberg, 2011).

4.3.1 The Contribution of Konrad Lorenz

Lorenz's (1971) classic text, *On Aggression*, provides a vivid and personal description and evolutionarily informed discussion of aggression and violence in nonhuman animals. Lorenz was familiar with many species of fish, birds, and mammals, both through personal experience of them as members of his household and through careful observation of their behavior in the natural environment. Lorenz repeatedly asked the evolutionary question of what aggression was good for, given its costly nature. Since it persisted from generation to generation in many animals, there must be some evolutionary advantage to its inheritance—there must be some benefit to the aggressor, despite its costs.

Lorenz distinguished the functions of inter- and intraspecies aggression. Interspecies aggression appeared to primarily be about defense of organisms from attacks from members of other species and was relatively uncommon compared to intraspecies aggression. For example, he noted that highly territorial and aggressive species of fish coexist with few problems between members of different species, perhaps because they exploit unrelated ecological niches and, hence, although physically close to one another, do not compete with each other for resources since they exploit different ecological niches. Once conspecifics intrude into a territory, however, many species of fish engage in displays of bright colors and dramatic behavior visible at long distances which have the effect of warding off the intruder, often before engaging in mutually dangerous aggression is necessary. Some birds and mammals also engage in warnings of potential aggression at a distance through characteristic calls and behavioral displays that keep conspecific intruders at bay.

In his work on geese, Lorenz noted that specific stimuli appeared to release aggressive behavior. That is, when adult geese vigorously and repeatedly extended their necks, their offspring would chase away other goslings belonging to other parents. Indeed, if Lorenz or his students made similar movements with their arms to ducklings that they had imprinted on Lorenz and his students, the goslings would also aggress against other unrelated goslings. In this manner, adult geese established a social hierarchy among themselves and also transmitted this pattern of social behavior to their offspring. Lorenz speculated that this social behavior was highly useful in that it minimized the amount of time and energy spent in aggression to all members of the flock thereby making time available for other activities such as feeding and mating that contributed to survival.

Lorenz noted that males are often (but not always) more aggressive than females. He speculated that male-to-male aggression mostly functioned to obtain mates directly, through access to and control of females for reproduction, or indirectly through acquisition of territory or resources, such as food and water, wherein females may select male mates with the best resources to increase the inclusive fitness of the females. Both males and females, depending upon the species considered, may also engage in aggression relating to defense of offspring. Male-to-female aggression was much less common and, Lorenz thought, mostly functioned to control the fertility of females and to prevent mating with other males.

For Lorenz, the expression of aggression was the result of competing instincts or internal drives. The drive to engage in aggression is like a pressure that builds up and will eventually be discharged, and hence is a hydraulic or "toilet" model of aggression somewhat similar to psychoanalytic models of aggression. If, however, a drive for a competing behavior, such as fear or sexual behavior is greater, then aggressive behavior may not be expressed. This model is difficult to test as it readily becomes circular as it is difficult to measure aggressive drive independently of aggressive behavior.

The work of Lorenz and other ethologists aroused considerable interest and controversy. As this work entered popular culture through books and television shows in the 1960s with book titles such as *The naked ape* (Morris, 1967) such ideas became part of popular culture. These ideas also attracted much academic criticism with charges of superficiality, false analogy, errors of fact, and just plain bad science (Montague, 1976a, 1976b, 1976c) (see Sect. 4.6 for a more detailed discussion of these critiques of Lorenz's and others' work).

4.3.2 Sexual Selection and Aggression

The work on evolution of aggression has come a long way since the 1940s and 1950s and had resulted in more nuanced, and sophisticated models of the causes of aggression that allows predictions to be made and different explanatory models to be evaluated.

4.3.2.1 The Selfish Gene

An intuitive understanding of biological evolution might suggest that parents should raise as many healthy children as possible to maturity in order to increase the likelihood of the survival of the species. But there seems to be more to evolution than that. A simple observation of the behavior of many organisms reveals some apparently paradoxical observations including (a) organisms vary tremendously in the number of their offspring; (b) parents sometimes kill their offspring as do siblings; (c) organisms sometimes kill other relatives; (d) organisms engage in apparently selfless acts that seem to involve self-harm, in that they expend time an energy without any obvious benefits to themselves; and (e) some organisms are sterile, choose not to have children, or are functionally sterile for a good part of their life, as in infancy and old age. How could evolution explain such apparently harmful behavior?

An approach to address some of these apparent problems comes from work such as Trivers (1972) who analyzed the different interests of parents and their offspring. He analyzed the parent-offspring conflict regarding how long parents should invest in their children and when they should stop doing so: It is in the children's interest to maximize parental investment, but parents may have other current and future offspring to invest in instead of current children. Thus, there may be selection pressure for offspring to be solicitous of parental care and also selection pressure for parents to stop investment in the current offspring under certain circumstances. Trivers noted that during early rearing parents may solicit their infants to feed, in the middle period there is less parental solicitation, and closer to weaning offspring are solicitous of parents who become progressively less available and sometimes even aggressive to their offspring.

Trivers (1972) noted that parent-offspring conflict is just one special case of conflicts of interest between relatives. That is there may be benefits for relatives to invest in one another, such as investing in siblings and raising nephews and nieces with whom they share many genes, but, as with the parent-offspring conflicts, depending on the circumstances, at some point the investment in the genes of one's siblings, nieces, and nephews may not pay off.

Interestingly for this chapter, Trivers also discusses the selection pressure for altruism, behavior that does not directly benefit one's own genes and also "nasty acts" which might include aggression. Trivers suggested that there may be selection pressure for altruism, including altruism toward unrelated conspecifics when that altruism results in fitness benefits to offspring or pay back either to oneself or relatives. Likewise, there could be selection pressure for nasty/aggressive acts when there are inclusive fitness benefits for aggression.

4.3.2.2 Sexual Selection and Aggression

Generally, but not always, males in many species are more aggressive than females, reflecting, according to Archer's (2009), the lower parental investment in offspring by fathers compared to mothers. In many species, females invest more heavily in offspring through production and maintenance of fewer, larger, more "expensive" gametes than males' more numerous "cheaper" gametes; greater time spent in mate selection than males; pregnancy; and, feeding, guarding, and defending their young. Thus, maternal abandonment of their offspring comes at a greater cost to females than males. In this circumstance, females become a limiting resource that males will compete for access to fertilize females and related resources. These resources can include territories, nesting sites, and food accessed through inter-male aggression, and strategies, such as sperm competition, and morphological and behavioral displays that aid access to fertilize females.

This inter-male competition for females results in much greater variability in male fertility due to differential outcomes of inter-male competition. Inter-male competition results in some males having access to many females and many offspring and many males having no offspring at all despite considerable investment in attempts to do so. Inter-male competition also results in less variation in females' fertility as most females will have some offspring. In species where both sexes are required to raise offspring, there is less sexual dimorphism and less male aggression.

In some circumstances, sexual selection results in the opposite—females who are consistently more aggressive and larger than males of the same species, resulting in morphological and behavioral changes apparently similar to those that result from intrasexual selection in males such as female ornamentation, vocalization, and weaponry (Rosvall, 2011). Rosvall (2008), for example, demonstrated that female aggressiveness predicted acquisition of nest sites when nest sites were experimentally reduced in number. This observation suggested that intrasexual, female–female competition for limited resources directly related to reproduction can be selected for.

4.3.2.3 Selection of Human Aggression?

Archer (2009) went on to examine the potential application of these ideas to evaluation of human aggression. Archer argued that humans tend to be either monogamous or polygamous and even monogamous males tend to be polygamous under certain circumstances. Archer considered seven lines of evidence to support the notion of sexual selection of aggression in human males.

First, men kill other men much more often than women and, second, men kill much more than women do. This is especially true for young men with fewer resources than other young men. Consequently, men have evolved to engage in more risky behavior of all kinds whereas women tend to be risk averse. Thus, sex differences in human aggression are largest for the most dangerous and risky forms of aggression, such as physical fighting, but are less marked for less risky forms of aggression such as verbal and indirect forms of aggression. Archer cited several meta-analyses of large groups of studies of sex differences in different forms of aggression in humans to support this conclusion.

Second, sex differences in aggression start very early on—before the age of 2 years—suggesting to Archer that experience and learning are relatively unimportant and that genetic factors prevail. Some studies, for example, have found that these sex differences are stable between ages 17 and 29 months (Baillargeon et al., 2007). Archer suggested that this is contradictory to an environmental explanation of aggression, which would predict that aggression should increase rather than remain stable. Although human aggression reduces between ages 2 and 6 years, this is primarily due to children learning not to aggress.

Third, factors that mediate aggression seem different for the two sexes. Men seem to be insensitive to risks, anticipate less upset or harm, and even enjoy being aggressive whereas women are more fearful of harmful outcomes of aggression and enjoy aggression less. Some studies have found this in very young children, again suggesting the importance of genetic predisposition rather than experience.

Fourth, there is a large degree of variability in male aggression reflecting several factors, including testosterone levels, antisocial behavior and traits, dominance, and aggressiveness. Again these differences are present before puberty.

Fifth, violence among me including homicide is related to several relevant environmental conditions. These include married men and men with more resources are less violent than single men and men with fewer resources. Men are more violent when the ratio of men to available fertile women is higher indicating greater competition for fertile females. Some studies have reported that when there are more women than men aggression in *women* increases. Societies with stronger rule of law have reduced violence due to increased costs of violence, whereas when societies have weak rule of law the cost of male aggression is reduced and thus male aggression is more likely.

Sixth, differences in aggression are part of a broader behavioral adaptive complex associated with polygynous mating systems. (a) Human polygyny was common historically, and today, Archer argued, remains in other forms such as high-status men having multiple sexual partners, and remarrying younger women. (b) The behavioral adaptive complex includes differences in visual and vocal signals, such as male facial hair, brow ridges, squared jaws, broad necks, deep voices, and lowered larynxes. These stereotypical masculine features predict lifting strength (which might be a proxy for genetic fitness) in some studies. (c) On multiple measures and in many societies, men generally have greater size, strength, musculature, larger hearts and lungs, and blood oxygen capacity than women. (d) There are longer maturation rates of men than women. This perhaps enables men to become more competitive with older men when younger men are finally sexually mature, thereby reducing the costs of male-on-male aggression. (e) Mortality is much higher in men than women from conception and throughout life, but peaks at ages 20–24 years when three men die for each woman who dies. This is associated with more male fetuses at conception perhaps due to selection based on greater male than female mortality.

Seventh, and finally, Archer reviewed evidence concerning physical aggression between opposite sex partners and, based on several meta-analyses, he found *little* evidence of differences in rates of aggressive behavior including physical aggression; an exception was high rates of injuries to women from men. There were, however, quite large variations in rates of aggression between same sex partners between one society and another. Specifically, aggression from men to women partners was more likely when female alliances were weak, when there was a lack of support or availability of natal kin, when there were stronger male alliances, when there women was economically dependent, and, when aggression was a private rather than public matter; some women were more physically violent toward male partners when retaliation was unlikely. In trying to explain this Archer suggested that the likelihood of aggression between opposite sex partners reflected a calculus of the likely benefits and costs of engaging in aggressive behavior.

Archer (2009) noted that this biological approach based on sexual selection does not exclude a role for other factors. Namely, he acknowledged a role for learning and changes in societal practices, including changing sex roles norms, but considered these to be imposed upon a more basic biological process driven by evolution and sexual selection of aggression in humans.

Archer concluded that there were large differences in aggression between men and women with men generally being more aggressive and especially more physically aggressive than women. He concluded that this reflected evolutionary processes, namely sexual selection of male aggression in order to access the largest number of fertile females by a relatively small number of men in a usually polygamous species. This was seen as part of a broader pattern of behavioral and physical adaptation which included but was not limited to aggressive behavior. Archer also concluded that sexual selection was a better explanation than other social psychology accounts of sex differences in aggression such as Social Role Theory. (See Chap. 9, for a more detailed discussion of Social Role Theory).

4.3.2.4 Killing Your Family

The male lion that takes over a pride of lions and kills the offspring of the previous unrelated male seems to be carrying rational genes. Killing the unrelated baby lions reassigns resources from the stranger's baby lions that are unrelated to the new male lion to the new lion's offspring. In addition, the new male lion killing the babies is associated with the females in the pride coming into estrus relatively quickly, thereby making the likelihood of the new male lion passing his genes to the next generation faster and even more likely. These mechanisms seem to make it much more likely that the genes from the new male lion will survive in the next generation, hence, killing the stranger's babies seems to make evolutionary sense (Trivers, 1972).

Such behavior is not uncommon in many animals including humans. Yet, human parents often kill their own offspring and family members often kill one another. This seems strange as at first sight killing one's relatives removes copies of some of one's own genes. In fact, human parents killing their own children, infanticide, and other examples of killing relatives are common under some circumstances in non-human animals. For example, infanticide by mothers accounts for at around 10% of infant deaths in Belding's ground squirrels (*Urocitellus beldingi*) (Morton & Sherman, 1978). Infanticide occurs in many species, including a variety of birds, sea lions, and many primate species, so it should come as no surprise that humans kill their offspring too. Whereas killing unrelated organisms makes intuitive sense, killing your relatives at first seems to make no evolutionary sense at all, especially your siblings, parents, and your own offspring, who carry half of your genes. It seems to make even less sense to kill yourself when you are 100% related to yourself! or remain functionally sterile (i.e., have no surviving children) and have no copies of one's own genes present in one's potential offspring in the next generation. Yet, all these phenomena are widely observed in many human and nonhuman species under certain circumstances. How could selfish genes behave in such an apparently foolish way? How could such behavior be selected?

In a recent review of fratricide in nonhuman animals, Morandini and Ferrer (2015) found an extensive literature on siblicide, especially in birds. Siblicide can occur through direct action of a parent or sibling killing another sibling, for example, when older kittiwakes push their younger, smaller, and weaker siblings out of the nest to their deaths when there is a shortage of food (Dickins, 2021; Dickins & Clark, 1987), or indirectly by starving the weakling in the nest. If such behavior is selected for by evolution, then there must be some evolutionary advantage for it to be inherited.

Hrdy (1979) noted five possible types and functions of infanticide. First, is the direct exploitation of the infant as a food source or other resource as in cannibalism.

Infants can also be exploited in other ways. For example, they may be used as "buffers" in fights when parents may sacrifice their children to protect themselves. For example, some kangaroos jettison their joeys when attacked by predators. Another example of exploitation of infants by their parents is others using them as "props" or objects in "play mothering" by a female who is not the infant's mother who "kidnaps" another mother's infant. In some cases, this leads to the death by starvation of the kidnapped infant. Second, infanticide decreases competition for limited resources from a conspecific thereby increasing the resources for all conspecifics. Resources can include food, maternal milk in lactating mothers, or increased access to nest sites. Infants in species who do not protect their young, and infants of subdominant females are vulnerable to this form of infanticide. Reports of killing infants of other species are relatively rare, perhaps again reflecting the lack of competition between species if they do not compete for common resources and hence there is a lack of benefits of killing unrelated infants. Rather, infanticide is common within a family, suggesting that the evolutionary benefits of infanticide accrue to relatives of the same species rather than strangers of the same species and to members of other species. Third, infanticide may benefit parents, for example, by avoiding the costs of continuing investment in offspring who are unlikely to survive, which is also a relatively common practice in humans. In dangerous environments, mothers may absorb their fetuses, naturally abort, or cannibalize their fertilized eggs. This form of infanticide is often related to the availability of resources, that is infanticide is more likely when resources are limited. Fourth, sexual selection refers to competition between members of the same sex for access to the other sex for reproduction; hence, destroying the offspring of another member of the same sex is advantageous to the killer as it may make the fertile female available for reproduction.

Most examples of infanticide concern males, but there are examples in some species of females killing each other's children to compete for male investment. In a gruesome experiment, Mallory and Brooks assigned three groups of female collared lemmings (*Dicrostonyx*) with litters to either a strange male or the father; there was also a control group with no intervention. The stranger male killed nearly half of the offspring, but the father killed none. Furthermore, the females in the group exposed to the strange male had their next litter after 21 days compared to after 28 days for the females exposed to the offspring's father, thus, they were in fact available for reproduction more quickly than mothers in the other group. A similar phenomenon is the "Bruce effect" in which pregnant females absorb fetuses when exposed to the odor of an unfamiliar male. The final type of infanticide is called "social pathology," which Hrdy (1979) defined as infanticide with no overall fitness benefits or decrease in fitness benefits to the parent. This is hard to identify because the failure to identify the fitness benefits may indicate a lack of imagination or data on the part of the scientist. For example, if a caged animal kills her offspring when disturbed by her human keepers, is this a pathological behavior with no fitness benefits to the mother or is it a rational choice to defer investing in offspring when threats have passed? Many writers have evoked pathological explanations of infanticide especially in primates including humans, but perhaps this is more sentimentality than science.

4.3.2.5 Human Infanticide and Biological Evolution

Parallels between infanticide in nonhuman animals with human aggression and violence are controversial, but have been explored. Daly and Wilson (1988), for example, noted that fratricide and parricide are commonly components of myths and stereotypes in many cultures. They quoted Gelles and Straus (1985) who wrote that "the family is perhaps the most violent social group, and the home the most violent social setting …. A person is more likely to be hit or killed in his or her home by another family member than anywhere else or by anyone else" (p. 88); indeed, the statistics on prevalence of aggression and violence continue to endorse this impression of the violent family. (See Chap. 3). Do evolutionary perspectives on aggression and violence in nonhuman animals shed light on human aggression and violence within the family?

In considering this possibility, Daly and Wilson (1988) began by supposing that evolution has selected human behavior that is nepotistic, that is invest predominantly in close relatives in order to promote survival of the genes each organism is carrying. Yet, humans commonly kill family members of all kinds. For example, nonrelated family members such as step-parents and nonrelated household members are much more likely to kill other family members, especially unrelated children aged 0–2 years thereby reducing investment in unrelated offspring and potentially releasing resources for investment in their own offspring.

Evolutionary accounts have emphasized the asymmetrical role of fathers versus mothers in investing in children perhaps reflecting the different evolutionary interests of fathers and mothers. One aspect of this is that fathers should ensure that the offspring that they raise are their own. In contrast, mothers should select fathers who are the most genetically fit who may or may not be their mate. Evolutionary biologists have hypothesized that these two different genetic interests may have evolved distinct evolutionary strategies. That is, male should be suspicious and vigilant for other males impregnating their mates. Thus, evolution of degrees of male coercion of females may pay males off in that it may be more likely to ensure that the males raise their own children.

Likewise, collaborative killers of family members are quite likely to be related with each other, but not with their victims. The stereotype of cruel step-parents is common in many cultures, perhaps reflecting the fact that step-parents are not genetically close to their step-children. This may be the reason why step-parents are often over-represented in child abuse and why step-parents are much more likely than natural parents to kill their step-children.

In some societies cultural practices, such as cousins marrying, grandparents raising orphaned grandchildren, and widows marrying their husband's brother, might be seen as cultural practices that reflect inclusive fitness. Parents are (generally) not related to one another, but share genes with their common children, hence, usually both should be interested in cooperating in the survival of their shared offspring. Two factors mitigate against this perfect harmony: Extra-pair reproduction and parents' nepotistic interests in other relatives. Spousal conflicts can arise when there are children from former relationships and are less likely when there are shared children. Thus, men, according to this view, often take a proprietary view of women and guard them through various practices such as claustration, controlling their access to other men, and having adultery laws that punish men more harshly than women. Sexual jealousy is often the cause of many forms of violence and murder by men in heterosexual relationships. For example, about half of women killed by their husbands are likely to be involved in separation or divorce, whereas almost no men who are killed by their wives are involved in separation or divorce, again suggesting that men's attempts to control women's sexual behavior in heterosexual marriage is an important correlate of aggression and violence in intimate partner violence (Daly & Wilson, 1988).

Daly and Wilson identified three situations when parents might be reluctant to invest in their own children: (a) Doubt that the offspring is actually related to the parents; (b) poor offspring quality; and (c) extrinsic circumstances, such as food shortages, a lack of social support, and demands from other older siblings. Rates of infanticide are a function of maternal age: older mothers are less likely to kill their infants, perhaps because the older mothers have fewer chances of raising other children to maturity than younger mothers. Both mothers and fathers are more likely to kill their younger rather than their older children thereby cutting their losses when investment is smaller and when opportunities to invest in other children still exist; unrelated strangers tend to kill adolescents rather than younger family members.

Offspring also kill their parents and this is related to the fitness value of the parents. That is, offspring are more likely to kill older parents who are less likely than younger parents to produce new offspring who will share genes with existing offspring. Interestingly, evolution produces some novel ideas concerning father-offspring conflicts regarding the arrival of new siblings. The interests of the father's genes is to have new children quickly whereas the interests of the children's genes are to delay the arrival of siblings to maximize parental investment in themselves rather than their future siblings, which seems to be an evolutionary explanation of sibling rivalry as a form of offspring control of parental investment in offspring. Daly and Wilson (1988) concluded that family violence is readily amenable to evolutionary explanations, although translation of these ideas from nonhumans to humans is not always simple.

To test this possibility, a disturbing study of familial filicide by Harris et al. (2007) compared the rates and methods of infanticide by five groups of people within the family: Genetic mothers, genetic fathers, adoptive mothers, adoptive fathers, and non-kin household members. To do this they analyzed a Canadian police database containing details of 378 cases of filicide and compared the rates, methods, and other characteristics of these homicides. As predicted, filicides were disproportionately more common among step-parents and were more likely than other filicides to involve abuse and death by beating. Mothers who committed filicides, especially if their children were older, were more likely to have mental illness than other mothers. In contrast, filicides by genetic fathers were more likely to be accompanied by marital discord, uxoricide, and paternal suicide than other murderers. The only examples of familicide (killing of both parents and offspring) involved genetic fathers. Filicide by all groups was more likely if the children were of poor health. The authors concluded that these results "yielded a pattern uniquely consistent with selectionist accounts based mainly on parental investment theory" (p. 2007).

4.3.2.6 Human Cultural Practices

There may be at least two parallels to human cultural practices. The first is the exposure of infants as a mechanism to limit family size, which kills less viable infants, and / or free up resources for investment in other more viable offspring. Examples of infanticide as a cultural practice are widespread in human cultures. Examples include classical Greek (Patterson, 1985), Roman, and Muslim cultures (Harris, 1982).

In the middle ages, exposure of infants was widespread throughout European cultures until quite recently (Boswell, 1988). Children were commonly abandoned to religious institutions in order to prevent infanticide, due to the cost of dowries, and embarrassment over illegitimate births. For example, at Brunelleschi's Ospedale Degli Innocenti in Florence, Italy, one can see the foundling wheel where parents could anonymously leave their babies and the wheel turned to deliver the offspring to the care of the institution's staff who were not genetically related to the family.

The second parallel is discussion which continues today in discussions of whether or not fetuses with disabilities should be aborted or allowed to be born and if born what treatment infants with very severe disabilities should receive and when they should be allowed to die. The widespread abortion of fetuses with Down syndrome and other disabilities in countries where screening of fetuses is available is one aspect of this practice (e.g., de Groot-van der Mooren et al., 2018). In other cases, the failure to offer medical care to children with Down syndrome, such as heart and other forms of potentially lifesaving surgery, which would be provided to infants without Down syndrome, might be another such example. This demonstrates that this form of killing of family members continues to today.

4.4 Genetics

Evidence that aggression and violence is caused at least in part by genetics comes from three types of research. First, there are differences in aggression between different species. Second, humans have bred behavioral traits in domestic animals, including differing degrees of aggression or nonaggression. Finally, there are human genetic studies of aggression, including twin studies in the general population and genetic studies of forms of psychopathology potentially related to aggression and violence including the genetics of psychopathy; and, human genetic syndromes related to high or low levels of aggression and violence.

4.4.1 Interspecies Differences in Aggression

Casual observation indicates that there appear to be differences between species in their temperament generally, including their readiness to aggress, suggesting a possible genetic basis for aggression. Some of our evolutionary near relatives seem to differ greatly from each other in aggressive behavior. For example, Bonobos (*Pan paniscus*) appear to be docile and very peaceful whereas chimpanzees, who are closely related to Bonobos, appear to be highly aggressive under some circumstances and engage in organized group aggression. As Diamond (1997) noted in *Guns, Germs, and Steel*, interspecies differences in temperament including propensities to aggression are quite large. He speculated that these differences, among other factors, have affected which animals humans have domesticated. (See the next section for further discussion of Diamond (1997) commentary on domesticability of animals and heritable temperament.)

A more systematic study of inter-species differences in aggression comes from a study by Clarke and Mason (1988) who compared the reaction of rhesus, bonnet, and crab eating macaque monkeys. Anecdotally, staff who handled the monkeys described bonnet macaques as docile and friendly, rhesus monkeys as frenetic, belligerent, and difficult to handle, and crab eating macaques as fearful. To test the veracity of these impressions of the different species, Clarke and Mason used observation and time sampling with eight categories of behavior reflecting responsiveness, hostility, and fear. The monkeys were captive-born experimentally naive females aged 25–38 months. The experimenters observed them in three situations including in the presence of a passive human observer. The objective observational data confirmed the impressions of large interspecies differences with rhesus monkeys being more responsive to the passive human observer and more hostile generally. Similar interspecies differences in degree of aggressiveness have also been reported in studies of other primates (Sussman et al., 2013.)

4.4.2 Breeding Domestic and Laboratory Animals

Humans have bred animals as food, labor, and companions: Many of us like to eat animals, induce them to give us rides or pull vehicles or plows, and many of us have them as objects of our affection. Such animal breeding historically was often done informally for physical, physiological, and behavioral characteristics, such as size, morphology, efficient food production, and temperament. There are many social institutions, from pigeon fanciers, dog and cat shows, pet stores to county shows, which breed animals to satisfy many human desires. Either directly or indirectly, human selection of domestic animals often involves selection of behavioral traits, including aggression or opposite behavioral traits, such as passivity or social deference.

4.4.2.1 Domestication of Animals

Domestication of animals and plants refers to the practice of taking naturally occurring wild organisms into the house (or *domus*.) Hence, humans have chosen certain kinds of wild animals and plants and selected variations that are useful and / or pleasing. Examples include taking in animals suitable for food, labor or companionship and entertainment and taking in plants that are useful as food, pleasing items or for other purposes. In the process of choosing some but not other forms of wild forms of organisms and subsequently selecting and discarding or killing some but not other forms, humankind has greatly modified naturally occurring organisms in many ways. Domestication should be distinguished from taming animals. "Taming" means to subdue, subjugate, or indeed restrain. Taming occurs during the animals's lifetime as when a circus captures or buys a lion and subdues it and renders it tractable. Some animals that are very difficult to domesticate can be tamed but their wild nature is not changed but only suppressed. This often seems to occur using positive punishment as with the lion tamer's whip, and the bear's shackle.

Diamond (1997) noted that humans in fact have domesticated relatively few animals, although they also have tamed others. Hence, humans may capture and tame lions and zebras and even induce them to pull carts if they train them thoroughly and carefully (often with painful positive punishment) but humans have not succeeded in breeding either and hence have been unable to select more desirable behavioral traits readily. Lions tend to eat their owners before their owners can breed them for better manners. Zebras are bad tempered, biting and kicking anyone who is foolish enough to try to ride them. Humans have not succeeded in domesticating zebras and, if they must have an animal to ride or pull a cart, can use more pliable alternatives such as horses and donkeys, that humans can breed relatively quickly. Zebras and lions can of course be tamed to do such tasks as pulling carts, but it takes considerable persistence and often misdirected bravery to do so.

Some of the best examples of domestication and genetics of aggression comes from canids, including wolves, foxes, and perhaps jackals. It is hard to imagine that our pet lap dogs ancestors were wild wolves perhaps at least only 30,000 years ago, but it does seem so. Indeed, many of the wide variety of breeds of dogs have only been selected in the last 200 years, suggesting that breeding for behavioral and other characteristics can occur quite quickly (Wang et al., 2013).

Human selection of Huskies for long distance sled racing provides an interesting example of how quickly humans can influence the genetics of dogs for both physique, physiology, and temperament. Prior to sled racing, Siberian Huskies and other related breeds of Arctic dogs were used by native peoples to pull sleds in cold weather as work dogs. With mining in Alaska, Siberian Huskies were important to pull sleds. Once sled racing began as a sport, humans organized betting on sled racing in the bars of Alaska around 1908. Thus, humans took existing Siberian huskies and other similar breeds, bred them with other dogs, and selected the offspring that were of similar size, had considerable endurance and cooperated with each other when pulling sleds over great distances quickly (Skoglund et al., 2015.) Note that by selecting cooperative dogs humans may be selected out aggressive traits. Hence, within only a few years Siberian huskies and other dogs were selectively bred to become modern Huskies. Social institutions, such as kennel clubs, now apply contingencies to humans in the form of prizes and recognition among other contingencies, to breed huskies to conform to predetermined social standards of appearance and perhaps temperament in a consistent manner.

Humans have bred dogs for both high and low aggression in different circumstances. We have bred the dogs of war and other breeds of dogs for fighting, such as pit bulls. We have also selected nonaggressive traits, such as when dogs are bred to find and gently return the game, rather than eat it or eat the people. A specific example of rapid selection of behavioral characteristics of canids comes from a project in which Dmitri K. Belyaev, a Russian genetics researcher, deliberately bred foxes into aggressive and nonaggressive strains in a remarkably short time. He did this by selectively breeding foxes that behaved in a nonaggressive manner, i.e. did not bite his gloved hand when he put it in the foxes' cages. The two strains of foxes were maintained for over 50 years (Goldman, 2010). Kukekova et al. (2008) reported that there were very large behavioral differences between the two strains of foxes. These authors used systematic testing and behavioral observation to compare the two strains of foxes. They found that the aggressive strain readily attacked humans. When people approached individual dogs from the aggressive strain and offered their gloved hand, the foxes adopted alert postures to get ready to attack, were alert, pinned their ears back, and made aggressive sounds. In contrast, foxes in the nonaggressive strain wagged their tails, tolerated human's approaches, and licked (rather than bit) humans' hands when people offered their hands. They behaved like many pet dogs. Interestingly, cross-breeding between the aggressive and nonaggressive strains produced mixes of behavioral traits of the aggressive and nonaggressive strains. Importantly, although bred for behavioral traits, the foxes also changed their physical appearance becoming more juvenile and "cute" in physical appearance. Thus, Kukekova et al. demonstrated that behavioral traits of aggressiveness or nonaggressiveness could be bred very quickly in canids such as foxes.

Another species that has been selected for aggression as well as for a wide variety of beautiful and elaborate appearances by humans is the fighting fish (*Betta splendens*). Fighting fish, especially the males, are naturally aggressive to one another, but human cultural practices selected them for greater aggression. In Thailand and Malaysia the nineteenth century or earlier, people caught wild fighting fish and won prizes for the fish that persisted in fighting for the longest period of time, in other words, humans selected those fish that persisted in fighting for the longest duration. In the 1840s the King of Thailand licensed fighting and collected the fish. Subsequently, they were imported to Europe and selected for both appearance and aggression (Goldstein, 2015). Several early basic behavior analytic studies of aggression used fighting fish as subjects, perhaps in order to make a point that even in organisms that are highly selected for aggressive behavior, that environmental control of aggressive behavior was possible, even in organisms highly selected for aggressive behavior (Thompson & Sturm, 1965). For example, Thompson and Sturm (1965) showed that fighting fish would emit chains of operant behavior to access preferred images of other aggressive male fighting fish, suggesting that images of other aggressive male fighting fish were positive reinforcers for their behavior. This example illustrates the interaction of evolutionary, cultural, and learning processes in the evolution of aggressive behavior.

4.4.2.2 Laboratory Animals

Laboratory animals are often bred for differing degrees of aggression both indirectly and directly. Thus, commercially available strains of rats are often less aggressive than wild strains. Perhaps this occurred inadvertently by selection of individuals who were easier to handle and sell (de Boer et al., 2003).

Although researchers have studied the genetics of aggression in a wide range of animals, one of the most commonly studied examples is the domestic mouse (*Mus musculus*). Researchers have perhaps chosen the mouse as a research model of aggression because it is cheap, small, breeds very quickly, and lives in relatively large colonies of individuals whose environment can be manipulated readily in a number of ways. The domestic mouse produces relatively large litters and is not so aggressive that the researchers and lab staff get harmed badly other than a few occasional small bites.

In nature, mice live in extended breeding units called "demes" consisting of parental generation, prepubertal juveniles, and pups. A few breeding males, who account for more than 90% of all offspring, defend their territories to exclude or dominate other males who, hence, rarely have the opportunity to breed. Dominant females defend food and nesting sites against invading males and females, especially postpartum, when mothers defend their offspring and related resources such as nest sites.

Researchers usually study aggression by inserting an unfamiliar intruder mouse into the other mouse's cage and measure the latency to the first incident of aggression. There are numerous variations in this procedure. Other strategies have included pain-induced aggression by pinching or shocking the mouse's tails.

Researchers have bred mice aggression in a number of ways resulting in several strains of aggressive and nonaggressive mice (Miczek et al., 2001). For example, researchers bred the Turku Aggressive and Turku Nonaggressive strains of mice by taking mice from an existing strain of albino mice, placing them in a standard situation with a male opponent, rating the severity of aggression on a seven-point scale, and then mating high- and low-aggressive mice with the sisters of other high- or low-aggressive mice. Differences in aggressivity were apparent after only two generations. These differences in aggression were also observed when offspring were raised by nonrelated parents, suggesting a genetic basis for these behavioral differences.

Despite the relative ease of breeding for high and low aggression, no single gene for aggression has been identified suggesting that aggression is determined by the effects of multiple rather than single genes. Additionally, although inbreeding for high and low aggression is straightforward, Miczek et al. (2001) emphasized that aggression is strongly affected by other environmental and learning history factors.

Some examples of these environmental factors included: (a) what kind of environment including conspecifics the home mouse is exposed to prior to the arrival of the intruder; (b) degree of environmental isolation prior to exposure; and (c) the intruder mouse can vary in terms of degree of aggressiveness/dominance status, genetic modification etc. Thus, aggression/nonaggression is readily bred in mice, but even in the aggressive strains, environment remains a salient cause of aggression.

4.4.2.3 Genetics, Culture, and Learning Interact

The preceding examples include all three forms of evolution of behavior. Whether huskies, fighting fish, mice or rats, breeding starts with some initial genetic propensity for aggression. Human breeding of animals based on desirable forms of animal behavior selects the genes associated with those traits. This happens surprisingly quickly, sometimes in only a few generations. Humans often do this in response to cultural contingencies, such as winning races, maintaining one's dogs as good examples of a breed, or having laboratory animals that are easy to handle and can be sold to laboratories readily. In addition, operant contingencies on human behavior, such as having working dogs that cooperate and pull sledges efficiently and easily, and our pets' cute and nonaggressive behavior when we approach them, mutually reinforce the behavior of domestic animals and their owners.

4.4.3 Human Genetic Studies

There is an extensive literature on genetics of human aggression. These studies include twin studies, single gene studies, studies of the genetics of psychopathology, and rare genetic conditions such as XYY syndrome, that appear to be related to either high or low levels of aggression and violence.

The relative influence of genetics is usually expressed as a statistic called heritability (h) which varies from 0 to 1 where 0 indicates no genetic influence and 1indicates complete genetic influence. The values of h may vary not only by genetic influences, but also by measurement error and differing degrees of environmental influence that may operate in different populations, times, and settings. It is important to note certain characteristics of heritability. First, it is a characteristic of a *population* not an individual. Thus, values of heritability in a certain population may say little about the roles of genetics and environment in particular individuals. Second, heritability may vary substantially between populations and within populations over time: A value of h in one context may or may not be applicable to another context. Third, heritability might also vary at different ages, since the relative contributions of genes and environment vary at different points in the lifespan. Finally, heritability may be substantially influenced by measurement issues. That is, if the measurement of a trait has high reliability, then heritability may appear high but if measurement of the trait is done with low reliability, then heritability may be low within the same population at the same time due to measurement error influencing the value of *h*.

4.4.3.1 Human Twin Studies

There have been many studies of heritability using human twins on the genetic influences on aggression and aggression-related traits, such as anger. For example, Rhee and Waldman (2002) conducted a systematic review and meta-analysis of 51 twin studies of broadly defined antisocial behavior. They found that heritability estimates varied widely from 0.00 to 0.72! (k = 100+ studies). They estimated that the proportion of variance due to different sources were as follows: (a) Additive genetic influences—0.32; (b) nonadditive genetic influences—0.09; (c) shared environmental influences, such as parental practices, home environment, etc.—0.16; and (d) nonshared environmental influences, such as different schools, chance events, error measurement—0.46. These moderate heritability estimates allow for large environmental influences on human aggression and violence, especially for specific individuals and settings.

More recently Porsch et al. (2016), in an analysis of two cohorts of twins aged 6–14 years from The Netherlands and United Kingdom reported heritability of approximately 50–80% in the Dutch cohort and around 60% in the British cohort. Furthermore, heritability was fairly stable across this age range in both cohorts. Thus, twin studies suggest moderate to high heritability of aggression but with a large range of values observed in different studies.

4.4.3.2 Single Human Genes

Recently, there has been much excitement over the possibility of identifying the relationship between single genes and behavior and medical issues and this is true for the possible relationship between single genes and aggression and violence. Based on the serotonin hypothesis of aggression, namely that serotonin plays an important part in regulating aggression (Ficks & Waldman, 2014) many studies have investigated the relationship between genetic polymorphisms and aggression, including genes that may regulate serotonin. (See Chap. 5, for a further discussion of serotonin.) This issue is beginning to have social importance as some attorneys are using "genes for aggression" as a defense in criminal cases which is sometimes used to reduce sentencing of violent criminals (Feresin, 2009). Crassly, this has been referred to as the "psycho gene" (Hunter, 2010).

Ficks and Waldman (2014) conducted a meta-analysis of a large sample of studies of the relationships between two candidate genes related to serotonin and Monoamine Oxidase and aggressive and antisocial behavior. Although they found some relationship the effects were quite small, suggesting that individual genes had only modest effects on aggression. In the same year, Vassos et al. (2014) also reported a meta-analysis of genetic association studies of violence and aggression based on 185 studies of binary or quantitative associations between 31 candidate genes and aggression and violence. Using relaxed criteria for association, they found no association between any genetic polymorphism and aggression and violence both generally and within specific subgroups. They concluded that to date there is little evidence for such an association and that specific genes are of little value in predicting future aggression and violence in individuals.

These results do not square with other human and nonhuman genetic studies indicating relatively high heritability of aggression and violence, including psychopathy. Vassos et al. (2014) speculated that this negative result might reflect the polygenic nature of the genetics of aggression and violence, small sample sizes in individual reports and a failure to distinguish reactive-impulsive and controlled-instrumental aggression and violence.

4.4.3.3 Genetics of Psychopathy

As might be expected, several authors have investigated the genetics of psychopathy, expecting higher heritability of aggression and violence than in the general population as one might believe this to be a condition where there was a more biologically determined form of aggression and violence than other forms of aggression and violence. Some studies do indeed report similar or sometimes higher heritability of psychopathy compared to everyday aggression (Tuvblad et al., 2014). Furthermore, psychopathic traits, which include lack of empathy, anxiety, and remorse, and difficulties maintaining close relationships, are present in less than 1% of population, but are present in 15–20% of prison populations, again suggesting an association between psychopathic traits that might have a strong genetic basis and more severe forms of violent criminal behavior.

In order to evaluate the heritability of psychopathic traits, Waldman and Rhee (2006) conducted a meta-analysis of ten independent samples published before 2003. They found that approximately 49% of the variance in psychopathic personality was due to genetic factors; approximately 51% was attributable to the nonshared environment. Note that this estimate of heritability is indeed somewhat higher than that reported by Rhee and Waldman (2002) for aggression. Dhanani et al. (2018) refined these analyses by examining the heritability of Hare's two factor psychopathic traits. Using a total of 24 studies they found that psychopathy had relatively high heritability, but the highest heritability was found for callous-unemotional traits. These studies suggest that psychopathic traits are more heritable than run of the mill aggression and violence.

4.4.3.4 The XYY Syndrome

In 1961, while karyotyping the father of an apparently typical man who was the father of a daughter with Down syndrome, a group of geneticists serendipitously discovered that the father—a six-foot-tall man of average intelligence—had an extra Y chromosome. Since the Y chromosome is associated with male gender and men are often more aggressive than women, then one might suspect that males with XYY syndrome might be hypermasculine and especially aggressive, thereby providing further evidence of the genetic basis of aggression and violence. Thus, the XYY syndrome was sometimes referred to as "double male syndrome" (Very rarely there are cases of men with up to four Y chromosomes which tends to be associated with more pronounced intellectual delay and features of autism, but not violence (Demily et al., 2017)). Subsequently, the prevalence of the XYY syndrome is approximately 1/1000. In the 1960s and 1970s the observation that the XYY genotype was more common than expected in high security special hospitals in Britain, France, and the USA—around 2% or so—raised the salacious possibility that the "male chromosome" was a gene for violent criminality (Re et al., 2015).

Although men with an extra Y chromosome are somewhat different from other men, they are not *very* unusual physically or psychologically. On average, they are about 3 inches taller than other men, do not have very obvious physical stigmata, and have approximately average lifespan. They may have a higher than expected rate of some medical problems, but do not have elevated testosterone. They tend to have low-average intelligence, with IQ about 10–15 points below average, have delays in speech and language, and may have some features of autism. Thus, many individuals with XYY syndrome are not recognized as such.

Despite the initial lurid reports, subsequent empirical studies found little evidence linking XYY syndrome to violence and many violent criminals who did not have XYY syndrome. This led Re et al. (2015) to conclude that the only robust finding in this literature is that individuals with XYY syndrome are a bit taller than average. Despite this, having the XYY syndrome is sometimes raised as a defense to reduce guilt in some criminal cases.

4.4.3.5 Disability-related Genetic Syndromes

As noted in this chapter, various genetic syndromes related to intellectual and developmental disabilities and autism are correlated with increased (autism spectrum disorders) or decreased (Down syndrome) rates of aggression. These observations also suggest a genetic basis for aggression.

4.5 Epigenetics and Aggression

In the past it was assumed that there was a linear relationship of genes causing behavior and any suggestion of environment changing genes and those genetic changes being passed on to future generation was dismissed as Lamarckian heresy. More recently the field of epigenetics has expanded and demonstrated the two-way interaction of how not only do genes affect behavior, but environmental factors modify the functioning of genes which may also affect behavior. Thus, epigenetics has been defined as "a functional modification to the DNA that does not involve an alteration of sequence" (Meaney, 2010, p. 57). This can involve environmental functional modifications to the DNA and variation in gene expression and function that may occur from conception and throughout life and which may also be short or long term and may be reversible. Epigenetic mechanisms, for example, may account for phenotypic variation, including phenotypic variation in genetically homogenous or identical organisms. For example, epigenetic research has addressed how variation in maternal behavior influences gene expression in children and subsequent behavioral changes in the children.

Meaney (2010) pointed out that genetics is often misrepresented in popular media and sometimes by academics too. Statistical associations between genes and behavior fail to elucidate the precise mechanism(s) by which genes influence phenotype. In a similar manner, "discovery" of single genes for aggression, such as the so-called "warrior" or "psycho" genes fail to describe how such genes result in aggressive behavior.

4.5.1 Cichlid Epigenetic Model

One epigenetic model of behavior that in part addresses aggression is that of Cichlid fish (Alvarado, 2020). There are about 2000 species of Cichlids mostly in the great lakes of east Africa that date back a mere 10 million years. Cichlids are highly diverse in morphology, especially color, and behavior. Sexual selection of different color phenotypes might be important in turbid river waters as color intensity may be a male fitness proxy that is the basis of female selection of mates. Despite considerable phenotypic variation, Cichlids species are genetically quite homogenous. Thus, one might ask, if not genes, then what accounts for this high degree of variation in physiology, morphology, and behavior?

There are robust relationships between different phenotypes related to environmental factors such as crowding, status, and reproductive potential. For example, in *Astatotilapia burtoni* pigmentation can change dramatically from yellow to blue and relatively quickly—in 28 days of less—in response to background sand color. These changes in appearance are reflected in changes in the number and color of the pigmentation cells in the skin. Color induction results in pronounced behavioral manifestations in which fish of one color are socially dominant and fish of another color are recessive socially. When paired with one another, each phenotype shows a typical pattern of social behavior. Markov chain analysis revealed that the socially dominant phenotype shows increased aggressive and sexual behavior to its conspecifics whereas the second socially subdominant phenotype spend more time in their home territories but displays little aggression and sexual behavior (Dijkstra et al., 2017; Korzan et al., 2008).

What is important for the genetics of aggression is that color changes and associated epigenetic mechanisms can be induced in a number of environmental and other mechanisms. For example, color changes can be induced by pharmacological manipulation, environmental manipulation of the background color of the sand in the fish tanks, crowding, and other social cues. These effects reflect changes in the functioning of the DNA in which methylation of the DNA apparently "turns off" functioning genes, a process which can be reversed again through environmental and other manipulations.

This nonhuman epigenetic model of social behavior in humans has implications for understanding, prevention, and treatment of human aggression. For example, environmental risk factors for aggression, such as parent behavior, poverty, etc., and evidence of gene x environment interactions can be reinterpreted more broadly in terms of possible epigenetic effects. DNA methylation in humans, for example, can be used as a marker for environmental events that affect aggression, but unlike Cichlids however, specifying the precise epigenetic mechanisms that might underlie this in humans is at present obscure (Tremblay, 2015). Indeed, Tremblay (2015) concluded that these epigenetic effects are most likely to occur during pregnancy and the first year of life. Hence, the implications for prevention seem to be those made by Lucien Bovet in 1951, who also concluded that the best approach to prevention of aggression was to support future mothers who already have a history of aggression and other behavior problems themselves and who will become the mothers of future delinquents if society does not deliver effective primary prevention.

4.6 Critiques of Evolutionary and Genetic Models

This chapter has outlined several evolutionary and genetic explanations for aggression and violence. Yet, biological explanations of aggression and violence are controversial, especially when findings from nonhumans are extended to humans. Thus, these findings received attention from many critics early on. Criticisms include that biological explanations are too simple, that the results from nonhumans do not apply to humans, the scholarship is bad, that the policy implications are false and possibly harmful, that they involve numerous ethical problems including giving violent perpetrators excuses, and sexism.

Several commentators have suggested that biological explanations are too simple to apply to human aggression and violence. Montague (1976a, 1976b, 1976c) made a systematic critique of popular biological science such as Ardrey's (1961) *African genesis*, *The territorial imperative …* (Ardrey, 1969), Morris (1967) *The naked ape*, and Lorenz's work. Not holding anything back, Montagu described their work variously as "offered pretentiously, with at least the appearance of support from apparatus of scientific learning, observation, discovery, facts, and authorities … " (p. viii); "the experts have found the theories of Ardrey and Lorenz unsound " (p. viii); and "they [Ardrey and Lorenz] overlook or evade bodies of fact that do not support their theories" (p. ix).

Montague (1976b) makes three specific critical points regarding the idea of an instinct for aggression being an explanation for human aggression. First, he suggested that Lorenz's conclusion that instinct is an explanation for aggression and territoriality in lower animals is incorrect. He noted that, although Lorenz admits that the location and boundary of territories may be learned, the basis of territoriality and aggression to defend the territory must be instinctive and unlearned. According to Montagu, in so doing Lorenz made the error of excluding the role of learning and experience in acquiring such a complex social behavior. Second, in attempting to extend these ideas to human behavior, Montagu argued that these authors ignore or grossly underplay a fundamental aspect of human behavior. Namely, learning and experience is characteristic of human behavior and its flexibility, even if some aspects of human behavior have significant genetic influences. A third argument is that even if aggression in lower animals is instinctual, these authors do not give an adequate explanation of how this might explain human aggression, that is, the explanation and connection between the two was unconvincing. The fourth critique is the accusation of bad scholarship. One aspect of this is, according to Montagu, errors of fact. For example, according to Montague, these authors portray early humans and their close relatives as aggressive, carnivores, and cannibals, when in fact both early humans and their close relatives were predominantly vegetarian with little consumption of meat. Furthermore, Montagu claimed that, although aggression occurs in close human relatives, most primates are generally peaceful and cooperative within one another. In a similar vein, Montague argued that early humans were actually peaceful and had limited contact with other groups as the population density was low. Another error, according to Montagu, is that Lorenz characterizes the Ute American Indians as aggressive, whereas the original anthropologists made opposite conclusions (Beatty, 1976). Montagu also claimed that this scholarship is biased. For example, he accuses Ardrey of consistently interpreting tools as weapons, burned humanoid bodies as evidence of aggression, when other explanations are possible and should be considered: Tools such as scrapers might just be tools for scraping and not weapons, and burned early humanoid bodies might have been burned after they were killed by other nonaggressive means. Thus, Montagu's volume represents a strong attack on the validity of this early work on the biological explanation of aggressive behavior as inadequate and poor scholarship.

Berkowitz (1969) provided another critique of this same literature that overlapped in part with Montagu's book and which provided further detailed critiques. One aspect of Berkowitz's critique is the contrast between Lorenz's drive model and a learning model of aggression. In the drive model, unobserved aggressive energy builds up in the nervous system and, if not, discharged in some other manner, explodes into spontaneous and unprecipitated aggression. A suggested course of action to prevent aggression is to engage in other activities that discharge the energy before it explodes through acceptable activities, such as sports or through breaking objects, rather than people as vigorously as possible. It is a cathartic model of aggression.

Berkowitz pointed out that Lorenz failed to address or greatly minimized the role of environmental factors affecting aggression and violence and that Lorenz and other ethologists make two errors in assuming that the environment does not affect aggression. Specifically, Berkwitz suggested that they fail to discuss the frustration-aggression hypothesis and the large amount of evidence that goal frustration or denial of access to reinforcers are often antecedents to aggressive behavior, even though such evidence was available at that time. Similarly, according to Berkowitx, Lorenz failed to address other environmental control of aggression though learning mechanisms, such as classical conditioning to withdraw from noxious stimuli (Azrin et al., 1964), modeling (Bandura, 1973) and reinforcement of aggressive behavior (Cherek et al., 1973); again, both of these phenomena were well documented at that time and hence deserved some comment or rebuttal from Lorenz. The main thrust of Berkowitz's first critique is that Lorenz's analysis of aggression was incomplete.

A second critique from Berkowitz (1969) is that the policy implications of their position are misguided because they are not based on well-founded empirical studies to back them up and, indeed, the implied policies may even increase aggression and be harmful. For example, if Lorenz's position is correct, engaging in nonaggressive behavior, such as aggressive fantasy, violent sports, and aggression toward objects should *reduce* future aggression through catharsis of the aggressive energy. Subsequent empirical papers in fact suggested that just the opposite is the truth. Children observing aggressive behavior are more rather than less likely to be aggressive (Bandura et al., 1963) leading some to conclude that treatments based on this approach such as catharsis are in fact both harmful and unethical.

Berkowitz's (1969) third critique is that the terms such as "instinct" is never clearly defined and hence can be used to explain almost anything by invalid analogies. Thus, a term that starts as an "instinct for aggression" is rapidly expanded to "instincts for territoriality," etc. so, if the term is not defined clearly it is impossible to measure and study it. Thus, a friendly, social smile when two people greet might superficially look like other primates aggressively baring their teeth at one another and hence both can be explained by a sublimated aggressive drive in the first case and a regular aggressive drive in the second. Berkowitz asked, but how can one be sure of this? How can one test such an imprecise idea scientifically? Berkowitz went on to note that although human and nonaggressive behavior might appear similar, it does not necessarily follow that the causes of these apparently similar behavior are the same: Similar behavior might be acquired by different mechanisms and science should investigate this. Bekowitz also pointed out that Lorenz and others use superficial analogies from language to draw parallels with nonhuman aggression. Thus, although we may say that we *attack* a problem and *get our teeth* into it, according to Berkowitz, it is questionable if this has anything to do with aggression other than a superficial resemblance based on vigorous activity. To claim that it does requires proof on the part of those who make such a claim.

A third group of critiques of biological explanations of aggression is that this approach has numerous ethical problems including giving violent perpetrators excuses for their aggressive and violent behavior, and sexism. There are numerous examples here, but some include perpetrators of violent crimes being given reduced sentences because of the presence of certain genes or that higher rates of aggression and violence in men is due to their aggressive nature or instinct and hence they are less responsible for their actions including violence against women. Underlying this are two opposing views of the nature of humans. The dark, conservative view is that human nature is sinful, evil, and unredeemable and that society's evils are due to humanity's dark and unchanging nature. The other liberal, utopian, "American optimistic" view is that humanity is perfectible and that society's problems are due to environmental flaws that can be identified and corrected (Berkowitz, 1969; Montague, 1976a, 1976b, 1976c).

4.6.1 Critiques from Feminist Theories

The word "feminism" refers to many different things. Everyday use of the work may connote having a goal of equal rights and opportunities for women, which is sometimes framed in terms of equal rights for any discriminated group, including discrimination due to race, skin color, language, culture, sexual orientation, political beliefs, class, nationality, and wealth status. Its everyday use is often concerned with gender stereotypes, work opportunities, and payment for women's work, and parenting and child rearing practices. Feminist theories are attempts to analyze and explain issues such as sex and gender differences and inequalities, gender stereotypes, patriarchy, pornography, and violence against women, etc. An important distinction is made between the term "sex," which connotes biological differences between two genders which are relatively innate, static, and essential, and "gender" which refers to learned and culturally determined behavior which is fluid and nonessential, and lies along one or more continua. Feminist perspectives have been around for several hundred years in literature and elsewhere, but in recent history feminism has been especially active. It now infuses many disciplines, including biology, psychology, and behavior analysis, and, this, has produced multiple theories and perspectives. Sociobiology, which explicitly addresses sex and gender differences and their causes, has attracted considerable attention from feminist theories.

Some feminist critiques of sociobiology include the following. Hubbard (1990) stated that production of knowledge is a social, nonobjective process influenced by the perspective of those producing the knowledge. In the case of academic sociobiology, knowledge is produced by privileged, predominantly White, middle class, highly educated men and is done so with pretended objectivity. Thus, Hubbard claims that sociobiology makes claims for universal truths about humans based on European sexual and gender differences that are androcentric and fail to take into account or minimize the perspectives of other groups and broader society. Hubbard also claims that the belief from biology of sex differences—men are producers in the outside world and women are nurturers who stay at home with children—is used as a justification from discriminatory practices such as exclusion from better paying jobs and underpayment of women.

There have been several responses to these critiques including a movement within sociobiology by feminists to address these issues. Fehr (2012), for example, while agreeing that sociobiology did indeed produce sexist stereotypes in some circles of sociobiology, noted that some critiques misrepresented and mischaractered sociobiology and missed the variety of views within sociobiology, including those from feminist evolutionary scientists, that actually exist. Furthermore, she noted that some Darwinian feminist scholars have taken up the challenge and investigated the robustness of sex differences in a wider range of species including primates and found that, in fact, these differences were not as robust as first thought. Fehr advocated for action both in terms of better public discourse including with the public and media, as well as improving evolutionary science. This could include better nonconfrontational and constructive engagement by sociobiologists and evolutionary psychologists with feminist critics and vice versa which could result in better, feminist-informed evolutionary psychology.

4.7 Summary

These different sources of evidence show that evolutionary and genetic influences on aggression are important, at least in some contexts and perhaps more important for more severe forms of aggression. Genetic factors, however, do not act in an isolated fashion, but rather interact with a wide range of historical and contemporary environmental variables. Hence, the magnitude of these genetic effects varies considerably depending upon the context, populations studied, different types of aggression and context. But even studies with evidence of high genetic influences in certain species, populations and genetic syndromes do not preclude important environmental influences on aggressive behavior.

References

Alvarado, S. G. (2020). Molecular plasticity in animal pigmentation: Emerging processes underlying color changes. *Integrative and Comparative Biology, 60*(6), 1531–1543.

Archer, J. (2009). Does sexual selection explain human sex differences in aggression? *Behavioral and Brain Sciences, 32*(3-4), 249–266.

Ardrey, R. (1961). *African genesis*. Collins.

Ardrey, R. (1969). *The territorial imperative: A personal inquiry into the animal origins of property and nations*. Dell.

Azrin, N. H., Ulrich, R. E., Hutchinson, R. R., & Norman, D. G. (1964). Effect of shock duration on shock-induced fighting. *Journal of the Experimental Analysis of Behavior, 7*(1), 9–11.

Baillargeon, R. H., Zoccolill, M., Keenan, K., Cote, S., Perusse, D., Wu, H. X., & Tremblay, R. E. (2007). Gender differences in physical aggression: A prospective population-based survey of children before and after 2 years of age. *Developmental Psychology, 41*(1), 13. https://doi.org/10.1037/0012-1649.43.1.13

Bandura, A. (1973). *Aggression: A social learning analysis*. Prentice-Hall.

Bandura, A., Ross, D., & Ross, S. A. (1963). Vicarious reinforcement and imitative learning. *The Journal of Abnormal and Social Psychology, 67*(6), 601–607.

Beatty, J. (1976). Taking issue with Lorenz on the Ute In: The nature of human aggression. In A. Montagu (Ed.), *The nature of human aggression* (pp. 111–115). Oxford University Press.

Berkowitz, L. (1969). Perspectives: Simple views of aggression: An essay review. *American Scientist, 57*, 372–383.
Boswell, J. (1988). *The kindness of strangers: The abandonment of children in western Europe from late antiquity to the renaissance*. Pantheon Books.
Brown, W., & Leonard, K. E. (2017). Does alcohol cause violence and aggression? In P. Sturmey (Ed.), *The Wiley handbook of violence and aggression* (pp. 1–15). Wiley.
Caspi, A., McClay, J., Moffitt, T. E., Mill, J., Martin, J., Craig, I. W., & Poulton, R. (2002). Role of genotype in the cycle of violence in maltreated children. *Science, 297*(5582), 851–854.
Cherek, D. R., Thompson, T., & Heistad, G. T. (1973). Responding maintained by the opportunity to attack during an interval food reinforcement schedule. *Journal of the Experimental Analysis of Behavior, 19*(1), 113–123.
Clarke, A. S., & Mason, W. A. (1988). Differences among three macaque species in responsiveness to an observer. *International Journal of Primatology, 9*(4), 347–364.
Daly, M., & Wilson, M. (1988). Evolutionary social psychology and family homicide. *Science, 242*(4878), 519–524.
Darwin, C., & Prodger, P. (1998). *The expression of the emotions in man and animals*. University Press.
de Boer, S. F., van der Vegt, B. J., & Koolhaas, J. M. (2003). Individual variation in aggression of feral rodent strains: A standard for the genetics of aggression and violence? *Behavior Genetics, 33*(5), 485–501.
de Groot-van der Mooren, M. D., Tamminga, S., Oepkes, D., Weijerman, M. E., & Cornel, M. C. (2018). Older mothers and increased impact of prenatal screening: Stable livebirth prevalence of trisomy 21 in the for the period 2000–2013. *European Journal of Human Genetics, 26*, 157–165. https://doi.org/10.1038/s41431-017-0075-1
Demily, C., Poisson, A., Peyroux, E., Gatellier, V., Nicolas, A., Rigard, C., & Rossi, M. (2017). Autism spectrum disorder associated with 49, XYYYY: Case report and review of the literature. *BMC Medical Genetics, 18*(1), 1–8.
Dhanani, S., Kumari, V., Puri, B. K., Treasaden, I., Young, S., & Sen, P. (2018). A systematic review of the heritability of specific psychopathic traits using Hare's two-factor model of psychopathy. *CNS Spectrums, 23*(1), 29–38.
Diamond, J. (1997). *Germs and steel: The fates of human societies*. Norton.
Dickins, D. W. (2021). "Punishing" pecks and "siblicidal" pecks in kittiwake (Rissa tridactyla) chicks. *Learning and Motivation, 73*, 101695. https://doi.org/10.1016/j.lmot.2020.101695
Dickins, D. W., & Clark, R. A. (1987). Games theory and siblicide in the kittiwake gull, Rissa tridactyla. *Journal of Theoretical Biology, 125*(3), 301–305.
Dijkstra, P. D., Maguire, S. M., Harris, R. M., Rodriguez, A. A., DeAngelis, R. S., Flores, S. A., & Hofmann, H. A. (2017). The melanocortin system regulates body pigmentation and social behaviour in a colour polymorphic cichlid fish. *Proceedings of the Royal Society B: Biological Sciences, 284*(1851), 20162838.
Fehr, C. (2012). Feminist engagement with evolutionary psychology. *Hypatia, 27*(1), 50–72.
Feresin, E. (2009). Lighter sentence for murderer with 'bad genes'. *Nature*. https://doi.org/10.1038/news.2009.1050
Ficks, C. A., & Waldman, I. D. (2014). Candidate genes for aggression and antisocial behavior: A meta-analysis of association studies of the *5HTTLPR* and *MAOA-uVNTR*. *Behavior Genetics, 44*, 427–444. https://doi.org/10.1007/s10519-014-9661-y
Gelles, R. J., & Straus, M. A. (1985). Violence in the American family. In A. J. Lincoln & M. A. Straus (Eds.), *Crime and the family* (pp. 88–110). Thomas.
Glenn, S. S. (1988). Contingencies and metacontingencies: Toward a synthesis of behavior analysis and cultural materialism. *The Behavior Analyst, 11*(2), 161–179.
Glenn, S. S. (2004). Individual behavior, culture, and social change. *The Behavior Analyst, 27*(2), 133–151.
Goldman, J. G. (2010). Mans's new best friend? A forgotten Russian experiment in fox domestication. *Scientific American Guest Blog*. https://blogs.scientificamerican.com/guest-blog/mans-new

Goldstein, R. J. (2015). *The betta handbook (B.E.S. pet handbook)*. Barron's educational Series.

Gresswell, D. M., & Hollin, C. R. (1992). Towards a new methodology for making sense of case material: An illustrative case involving attempted multiple murder. *Criminal Behaviour and Mental Health, 2*(4), 329–341.

Harris, G. T., Hilton, N. Z., Rice, M. E., & Eke, A. W. (2007). Children killed by genetic parents versus stepparents. *Evolution and Human Behavior, 28*(2), 85–95.

Harris, W. V. (1982). The theoretical possibility of extensive infanticide in the Graeco-Roman world. *The Classical Quarterly, 32*, 114–116.

Hrdy, S. B. (1979). Infanticide among animals: A review, classification, and examination of the implications for the reproductive strategies of females. *Ethology and Sociobiology, 1*(1), 13–40.

Hubbard, R. (1990). *The politics of women's biology*. Rutgers University Press.

Hunter, P. (2010). The psycho gene: While the idea of a 'criminal gene' is nonsense, there is growing evidence that some psychopathic behaviour might indeed be grounded in genes. *EMBO Reports, 11*(9), 667–669. https://doi.org/10.1038/embor.2010.122

Kim-Cohen, J., Caspi, A., Taylor, A., Williams, B., Newcombe, R., Craig, I. W., & Moffitt, T. E. (2006). MAOA, maltreatment, and gene–environment interaction predicting children's mental health: New evidence and a meta-analysis. *Molecular Psychiatry, 11*(10), 903–913.

Korzan, W. J., Robison, R. R., Zhao, S., & Fernald, R. D. (2008). Color change as a potential behavioral strategy. *Hormones and Behavior, 54*(3), 463–470.

Kukekova, A. V., Trut, L. N., Chase, K., Shepeleva, D. V., Vladimirova, A. V., Kharlamova, A. V., & Acland, G. M. (2008). Measurement of segregating behaviors in experimental silver fox pedigrees. *Behavior Genetics, 38*(2), 185–194.

Lindenfors, P., & Tullberg, B. S. (2011). Evolutionary aspects of aggression: The importance of sexual selection. *Advances in Genetics, 75*, 7–22. https://doi.org/10.1016/B978-0-12-380858-5.00009-5

Lorenz, K. (1971). *On aggression*. Routledge.

Meaney, M. J. (2010). Epigenetics and the biological definition of gene× environment interactions. *Child Development, 81*(1), 41–79.

Miczek, K. A., Maxson, S. C., Fish, E. W., & Faccidomo, S. (2001). Aggressive behavioral phenotypes in mice. *Behavioural Brain Research, 125*(1-2), 167–181.

Miczek, K. A., Weerts, E., Haney, M., & Tidey, J. (1994). Neurobiological mechanisms controlling aggression: Preclinical developments for pharmacotherapeutic interventions. *Neuroscience & Biobehavioral Reviews, 18*(1), 97–110.

Montague, A. (1976a). *The nature of human aggression*. Oxford University Press.

Montague, A. (1976b). Introduction. In A. Montagu (Ed.), *The nature of human aggression* (pp. vii–xiv). Oxford University Press.

Montague, A. (1976c). The new litancy of "innate depravity," or original sin revisited. In A. Montagu (Ed.), *The nature of human aggression* (pp. 3–17). Oxford University Press.

Morandini, V., & Ferrer, M. (2015). Sibling aggression and brood reduction: A review. *Ethology Ecology and Evolution, 27*(1), 2–16.

Morris, D. (1967). *The naked ape: A zoologist's study of the human animal*. Random House.

Morris, D., Tinbergen, N., & Hoogland, R. (1956). The spines of sticklebacks (Gasterosteus and Pygosteus) as means of defence against predators (Perca and Esox). *Behaviour, 10*(1), 205–223.

Morton, M. L., & Sherman, P. W. (1978). Effects of a spring snowstorm on behavior, reproduction, and survival of Belding's ground squirrels. *Canadian Journal of Zoology, 56*(12), 2578–2590.

Patterson, C. (1985). "Not worth the rearing": The causes of infant exposure in ancient Greece. *Transactions of the American Philological Association, 115*, 103–123.

Porsch, R. M., Middeldorp, C. M., Cherny, S. S., Krapohl, E., Van Beijsterveldt, C. E., Loukola, A., & Bartels, M. (2016). Longitudinal heritability of childhood aggression. *American Journal of Medical Genetics Part B: Neuropsychiatric Genetics, 171*(5), 697–707.

Re, L., Birkhoff, J. M., Sozzi, M., Andrello, L., & Osculati, A. M. (2015). The choking game: A deadly game. Analysis of two cases of "self-strangulation" in young boys and review of the literature. *Journal of Forensic and Legal Medicine, 30*, 29–33.

Rhee, S. H., & Waldman, I. D. (2002). Genetic and environmental influences on antisocial behavior: A meta-analysis of twin and adoption studies. *Psychological Bulletin, 128*(3), 490.

Rosvall, K. A. (2008). Sexual selection on aggressiveness in females: Evidence from an experimental test with tree swallows. *Animal Behaviour, 75*(5), 1603–1610.

Rosvall, K. A. (2011). Intrasexual competition in females: Evidence for sexual selection? *Behavioral Ecology, 22*(6), 1131–1140.

Skoglund, P., Ersmark, E., Palkopoulou, E., & Dalén, L. (2015). Ancient wolf genome reveals an early divergence of domestic dog ancestors and admixture into high-latitude breeds. *Current Biology, 25*(11), 1515–1519.

Sussman, A. F., Ha, J. C., Bentson, K. L., & Crockett, C. M. (2013). Temperament in rhesus, long-tailed, and pigtailed macaques varies by species and sex. *American Journal of Primatology, 75*(4), 303–313.

Thompson, T., & Sturm, T. (1965). Classical conditioning of aggressive display in Siamese fighting fish. *Journal of the Experimental Analysis of Behavior, 8*(6), 397–403.

Tinbergen, N. (1952). The curious behavior of the stickleback. *Scientific American, 187*(6), 22–27.

Tinbergen, N. (1963). On aims and methods of ethology. *Zietschrift fur Tierpsychologie, 20*, 410–433. https://doi.org/10.1111/j.1439-0310.1963.tb01161.x

Tremblay, R. E. (2015). Developmental origins of chronic physical aggression: An international perspective on using singletons, twins and epigenetics. *European Journal of Criminology, 12*(5), 551–561.

Trivers, R. L. (1972). Mother-offspring conflict. *American Zoologist, 12*, 648–648.

Tuvblad, C., Bezdjian, S., Raine, A., & Baker, L. A. (2014). The heritability of psychopathic personality in 14-to 15-year-old twins: A multirater, multimeasure approach. *Psychological Assessment, 26*(3), 704.

Vassos, E., Collier, D. A., & Fazel, S. (2014). Systematic meta-analyses and field synopsis of genetic association studies of violence and aggression. *Molecular Psychiatry, 19*(4), 471–477.

Waldman I. D., & Rhee, S. (2006). Genetic and environmental influences on psychopathy and antisocial behavior. In C. J. Patrick (Ed.), *Handbook of psychopathy* (pp. 205–228). Guilford.

Wang, G. D., Zhai, W., Yang, H. C., Fan, R. X., Cao, X., Zhong, L., & Zhang, Y. P. (2013). The genomics of selection in dogs and the parallel evolution between dogs and humans. *Nature Communications, 4*(1), 1–9.

Chapter 5
Biological Evolution of Violence and Aggression. II: Brains, Neurotransmitters, and Hormones

This chapter continues to review the role of biological causes of aggression and violence. Specifically, it addresses neurological structures, neurotransmitters, and hormones. It is important to note that although the information in this chapter is presented initially in a linear fashion, however, none of these individual factors act independently, either from one another or independently of the person's genetic makeup, learning history, and the current environment including their culture. Rather, biological causes form a reverberating, dynamic system that is influenced by the environment in numerous ways. Both environmental influences during development and in the present environment produce aggressive and violent or peaceful, adaptive behavior.

Many authors do not consider aggression and violence to be a unitary biological phenomenon and, hence, different biological mechanisms may underlie different forms of aggression and violence. For example, Haller (2020) distinguished at least three biologically distinct forms of aggression that may have different associated biological mechanisms. These were: (1) elicited aggression or "defensive rage" that seems similar to reactive, "hot," aggression that is based on classical conditioning (i.e. elicited by unconditioned and conditioned stimuli); (2) predatory aggression that is related to obtaining food and is one form of operant behavior related to positive reinforcement (accessing food); and (3) maternal defensive aggression that protects mothers' children from aggression from males and increase the inclusive fitness of the mother and is also based on operant conditioning related to negative reinforcement (removal of the aggressive male). In addition to different functional classes of aggression and violence, the literature often distinguishes normal aggression from abnormal, pathological violence with the possibility of different mechanisms for normal aggression and abnormal, pathological violence.

P. Sturmey, *Violence and Aggression*, Advances in Preventing and Treating Violence and Aggression, https://doi.org/10.1007/978-3-031-04386-4_5

5.1 Brain Structures

5.1.1 Types of Evidence

The literature attempts to identify brain structures that might underlie aggression, including abnormal violence, uses several approaches to infer causality. The first method studies abnormal brain structures that appear to be related to aggression. For example, tumors, medical conditions that imply structural problems, such as seizure disorders, and/or anatomical variations, such as unusually small or large brain regions, may correlate with differing levels of aggression and violence, differing levels of social behavior potentially, such as antisocial behavior, related to aggression, to violent criminal behavior, or to psychopathology related to aggression, such as intermittent explosive disorder. The second method involves tissue destruction. This can be done by psychosurgery, either directly for treatment of aggression and violence, such as lobotomies and removal of tumors putatively related to client aggression or for other conditions such as epilepsy which is otherwise uncontrollable, but which might also result in behavioral changes including reduction in aggression. In nonhuman animals, tissue destruction can also be done by stereotactic surgery for research, and other methods of tissue destruction, such as injection of alcohol, cauterization, or freezing tissue to destroy it. A third method is electrical stimulation of brain regions and observing behavior changes associated with brain stimulation of specific regions. This can be done in nonhuman animals in experimental studies, but can also be done in humans during psychosurgery when individuals undergoing brain surgery can report their experiences when they are awake during surgery. Deep brain stimulation has also recently emerged as a treatment for aggression and violence when some authors have used deep brain stimulation to treat aggression in humans (Franzini et al., 2013). (See Chap. 12, *Individual Biological Interventions for Violence and Aggression. II. Other Biological Treatments*). A fourth method is to study naturally occurring variations and naturally occurring abnormal physiological levels of neurotransmitters, hormones, and other substances. These can occur during medical conditions, such as hyperthyroidism-related aggression and irritability; and variations in aggression and related social behavior correlated with natural variations in hormones, for example, during different points of the menstrual cycle, changes in aggression and related hormones in pre- and post-menopausal women, and naturally occurring variations in testosterone in men and women. A fifth method that is also correlational, is to study correlations between groups of individuals with and without aggression-related psychopathology and then to study the potential biological correlates of aggression, such as testosterone and cortisol levels in psychopaths and nonpsychopaths. Sixth, psychopharmacological interventions to treat aggression and violence may also provide evidence of biological causes of aggression in that the changes in neurotransmission and other biological changes that occur when humans take psychotropic medications may be used as evidence of the causes in reductions of aggression and violence during treatment. The results of psychopharmacological treatments can also be used to impute the possible cause of aggression and violence prior to treatment. Finally, recent imaging techniques called optogenetics have been developed that permit studying the activities of living neurons in living and sometimes moving organisms. In optogenetics molecules are inserted into cells that convert light into electric impulses. This allows researchers to study specific neurons and pathways more locally and discretely than other methods. It also allows researchers to turn specific groups of neurons on and off and observe the effects on behavior. It should be noted that basic research on aggression is often done with an eye toward potential implications for prevention and treatment, especially, but not exclusively, in humans.

Some of these research methods are correlational, especially those with human subjects. Such correlational studies may sometimes have a lot of face validity, but it is difficult to infer causality due to the lack of experimental control. Other methods use experimental methods, but they are often with nonhumans or even in vitro living tissue samples rather than intact organisms, raising the issues of external validity. That is, it is unclear how the results might apply to understanding aggression in living humans. Research on the biology of aggression and violence involves many different species. These species may have significant anatomical and functional differences from one another, and, indeed, from humans as each species is adapted and evolved for particular ecological niches. Even among phylogenetically closely related species, such as other mammals and even primates, there may be important differences in the biological mechanisms that control aggression and violence. Finally, experimental studies can be done with humans, such as randomized controlled trials for psychopharmacology and psychosocial treatment of aggression. Whereas these studies enable conclusions about treatment efficacy to be made, inferring the mechanism(s) of action of effective treatment of aggression often requires other kinds of studies.

5.1.2 Brain Structures

In 1928, Hess reported that electrical stimulation of specific loci of the hypothalamus in the cat was associated with a wide range of behavioral phenomena, including aggressive behavior. Stimulation of other hypothalamic sites resulted in apparently fearful behavior in the cat. Around the same time, Bard (1928) reported that neurosurgical lesions in the cat that left the caudal hypothalamus intact, but removed the cerebral hemispheres, resulted in the so-called sham rage. That is, the cat appeared enraged, showed piloerection, had high blood pressure, and showed many of the behavioral and physiological characteristics of anger, even though there was no environmental trigger or victim for this behavior. These preliminary observations suggested that aggression and violence might be controlled by specific brain regions.

Since that time, numerous brain structures have been implicated in control of aggression violence including limbic and cortical structures. A very simplistic model suggests that the limbic system receives input from sensory organs and "attaches" emotional valence to those stimuli. This information is transmitted to cortical structures that appear to initiate or inhibit aggressive or violence behavior. For example, someone might see and hear a person with an angry face, waving their fist and threatening verbally, but a police officer might also be present nearby. The sense organs detect these stimuli and transmit impulses to the limbic system. The limbic attaches emotional significance, such as threat and fear, to the stimuli and electrical impulses are then transmitted to the cortical structures. These cortical structures may then inhibit aggressive behavior due to the person's prior learning, and history of punishment for aggressive behavior in the presence of police officers or similar stimuli. Such a model is a crude outline of some of the elements of brain structures involved in the control of aggression and violence. The next sections look at this question in more detail, specifically, the roles of the limbic and cortical structures.

5.1.3 The Limbic System

The limbic system sits above the spinal cord and brainstem structures and below the cortex. It is thought to be an evolutionary ancient structure sandwiched between the even more ancient brainstem structures responsible for movement and breathing, and the evolutionarily more modern structures in the cortex responsible for high functions such as integrating information from other regions of the brain and complex thinking in humans. The limbic system is implicated in many basic functions of the organism including olfaction; memory and learning; motivation, including hunger, arousal, sexual motivation; regulation of many homeostatic functions, including metabolism, and water and electrolyte balance; secretion and control of many hormones; and control of emotions including fear and aggression. The limbic system is a complex set of structures, nuclei, and pathways. Three main structures are most relevant to the regulation of aggression: The amygdala, hippocampus, and hypothalamus (Weiger & Bear, 1988).

The **amygdala** is a small almond-shaped structure with a complex structure with several nuclei each of which appear to perform different functions. Some parts receive input from visual, auditory, somatosensory, pain, olfactory, and taste sensory systems, suggesting it integrates multiple sources of sensory input. Outputs from the amygdala project to areas of the brain responsible for the expression of emotions, and instrumental behavior, such as avoidance and escape.

There are several sources of evidence illustrating the amygdala's role in aggression. Bilateral removal of the temporal lobes in dogs and monkeys results in a calming effect on behavior and reduces aggression. Stimulation of the amygdala evokes emotions of anger, violence, fear, and anxiety. For example, stimulation of the amygdala in typical humans may result in increased aggression and activation of the amygdala. Furthermore, people with various forms of psychopathology, such as veterans exposed to trauma, and adolescents with aggressive families, may also have structural changes in the amygdala and associated pathways. Psychosurgical studies of removing the amygdala that may reduce aggression (although this is controversial; See Chap. 12, *Individual Biological Interventions for Violence and Aggression. II. Other Biological Treatments*). These observations suggest that the amygdala plays a central role in control of aggression. For example, bilateral destruction of the amygdala results in the Kluver–Bucy syndrome which presents with a wide range of impairment including hyperorality (e.g. mouthing objects), hypersexuality, and *loss* of fear and anger. Human data on the amygdala are less clear than studies of nonhuman animals due to the difficulties in conducting experiments and the complexities of human behavior (Gouveia et al., 2019).

The **hypothalamus** is a small structure located under the thalamus which, like the amygdala, is structurally complex. The hypothalamus has many functions and regulates the autonomic nervous system by releasing hormones into the blood. It maintains homeostasis in the body by regulating the sympathetic and parasympathetic nervous systems. It regulates emotions such as "fight and flight" versus "rest and digest," and also regulates hunger, thirst, sleep, and digestion. It produces oxytocin and vasopressin, which regulate both physiological functions such as circadian rhythms and thermoregulation, but also broad behavioral effects including some social behavior which might be relevant to aggression (see below). The hypothalamus is connected to many of the structures mentioned here including the amygdala. Stimulation of some parts of the hypothalamus produces attack-like aggression in many species (Haller, 2020) and a wide range of other behavior, related to food, and social behavior. When engaged in aggressive behavior, various studies in nonhuman animals have shown that the hypothalamus is activated. Finally, other evidence for the importance of the hypothalamus in controlling aggression comes from psychosurgery studies in humans to reduce aggression in clinical populations (Gouveia et al., 2019). (See Chap. 12, *Individual Biological Interventions for Violence and Aggression. II. Other Biological Treatments*).

The **Periaqueductal Gray (PAG)** is a structure that surrounds the aqueduct between the third and fourth ventricles of the brain. Like the other structures described here, it has a complex structure controlling many different functions that interact one with another. The PAG regulates pain perception and analgesia, and female reproductive and maternal behavior among other functions. The PAG also influences glucose metabolism and emotional responses that might relate to physiological preparation for various behaviors including aggression. The PAG also influences production of facial expression that might be related to aggression. Electrical stimulation of the PAG can elicit affective and predatory aggression in cats (Haller, 2020).

The structures described above are presented for simplicity one by one, but the limbic control of aggression is a complex, dynamic system, rather than a series of isolated anatomical and / or structural units controlling functionally different forms of aggression. A study by Tulogdi et al. (2010) illustrates this point nicely. These authors investigated some of the brain mechanisms behind apparently emotionally cold, predatory aggression in rats that perhaps paralleled similar behavior in humans sometimes described as "callous-unemotional" violence. In the experimental rats they first limited glucocorticoid production by removing the adrenal gland and implanted glucocorticoid replacement to induce antisocial aggression. There was also a sham operation control group. They then subjected the rats to the resident/intruder test in which a smaller rat was placed in the cages with the individual rats from the experimental and control groups and observed behavioral and brain activation differences between the two groups. There were large behavioral differences between the groups. The experimental group showed a higher proportion of attacks to the intruder's vulnerable body parts with shorter latencies and less intentional signaling compared to the controls. In terms of brain mechanisms, the experimental group showed activation of the hypothalamus, PAG, and amygdala that resembled those controlling predatory aggression. The authors concluded that the similarities between this rodent model of predatory aggression and human emotionally cold, goal-oriented aggression were real both in terms of behavioral processes and brain mechanisms.

5.1.4 Cortical Structures

The second group of brain structures implicated in the control of aggression is the prefrontal cortex (PFC) (Weiger & Bear, 1988). The PFC is anatomically and functionally complex and has several different structures and functions associated with these structures. The PFC is involved in higher order executive functioning, planning and goal directed behavior, and behavioral inhibition, such as deferring accessing rewards, and inhibiting ongoing motor movements. Parts of the PFC are connected to the amygdala. Thus, the PFC is thought to be involved in the inhibition of aggressive impulses, for example, when there may be an ongoing cost to aggressing.

Damage to the PFC can result in increased aggression and emotional behavior thought to reflect the removal of inhibition of the activity of the limbic system. Further evidence for the involvement of the PFC in control of aggression comes from studies that have demonstrated that psychiatric patients with strong histories of aggression, such as those with intermittent explosive disorders, show evidence of increased risk taking, relative insensitivity to punishment (e.g. gambling losses) and other behavioral characteristics of PCF impairment, suggesting that impairment in the PFC predisposes toward aggression and violence (Best et al., 2002). This so-called inhibitory control or prefrontal deficit theory of control of aggression has recently undergone further refinement (Haller, 2020).

5.1.5 Summary

The studies reviewed here have identified specific brain structures that are part of the regulation and control of aggression and violence in humans and nonhumans and in humans with and without psychopathology. Thus, they form an important element in understanding aggression and violence, although limitations in the over simplistic idea that "the brain causes aggression" are discussed at the end of this chapter. Part of the biological mechanism of aggression and violence also related to neurotransmitters. The next section of this chapter goes on to consider this aspect of biological causes of aggression and violence.

5.2 Neurotransmitters

There appears to be some genetic basis for aggression. Since genes code for neuroanatomical structures and neurotransmitters, and given that specific neurotransmitters are associated with particular structures and pathways in the brain, some of which are associated with the control of aggression and violence, it seems unsurprising that there may be specific neurotransmitters that influence aggression and violence that would be an important part of a biological explanation of aggression and violence. Interest in neurotransmitters also comes from researchers and drug companies in the search for drugs that can alter neurotransmission and thereby perhaps treat aggression and violence in some contexts and populations.

Evidence for the involvement of neurotransmitters in aggression can come from several sources. Some of these sources include: (a) naturally occurring correlations between neurotransmitter levels and levels of aggression and violence; (b) naturally occurring differences in neurotransmitter levels between populations that are nonaggressive and aggressive, such as different groups of teenagers who are nonviolent or violent, or, groups of nonviolent and violent criminals of people with mental health issues, such as studies comparing people with intermittent explosive disorder and anxiety disorders; (c) experimental manipulations in nonhuman animals of neurotransmitter levels and their effects on aggression and violence in nonhuman animals; and (d) the effect of psychotropic medication which may change neurotransmitter levels and may thereby change aggression and violence. When considering the evidence that neurotransmitters affect aggression and violence all of these methods have limitations which should be considered. As noted above, some of these methods are correlational and are likely to involve confounds. Thus, it is difficult to be confident about any conclusions concerning causation. Experimental methods may demonstrate causality within the context of the experiment, but are often performed on nonhumans or in unusual circumstances and, hence, the external validity of the findings is often questionable. Finally, experimental drug studies in humans are also tricky to interpret in part because of measurement problems. For example, a common issue is that demonstration of a *selective* effect of a drug on aggression is challenging because it must be shown that there was not an overall suppression of behavior, but an effect specific only to aggression. Even when a clear demonstration of a drug effect can be shown, the mechanism of such an effect requires different kinds of studies. Hence, interpretation of the effects of neurotransmitters on aggression and violence, especially in humans, is challenging.

This section reviews the evidence for these causes of aggression. This section we will consider five main neurotransmitters related to aggression: Serotonin, dopamine, glutamate, norepinephrine, and gamma-Amino butyric acid (GABA.) When considering these individual neurotransmitters it is important to remember that simple, linear effects are unlikely. Neurotransmitters are likely to interact both with one another, and with other biologically active molecules including hormones, cytokines, enzymes, growth factors, and signaling molecules (Nelson & Chiavegatto, 2001). Finally, dietary intake of precursors of neurotransmitters such as serotonin may also affect neurotransmitter levels in intact living organisms (Gajos & Beaver, 2016; Hibbeln & Gow, 2014). (See Chap. 12, *Individual Biological Interventions for Violence and Aggression. II. Other Biological Treatments*, for a detailed discussion of serotonergic and other diets as a treatment for aggression).

5.2.1 Serotonin

Described by some as "the big dogma" (Olivier, 2004), it has widely been supposed that serotonin (5-hydroxytriptamine, [5-HT]) and aggression are reciprocally related, leading some such as Nelson and Chiavegatto (2001) to propose that serotonin is almost a common path for other biological mechanisms of aggression. Thus, they write " … the 5-HT data are most convincing. The 5-HT system generally dampens aggression in animals and violent behavior in humans" (p. 714). What is the evidence for such a conclusion?

Nelson and Chiavegatto (2001) cited the following evidence. First, aggression is correlated with the cerebrospinal fluid metabolite of serotonin 5-Hydroxy Indole Acetic Acid in humans and nonhuman animals. Second, procedures to increase serotonin, including loading with serotonin precursors and agents that inhibit reuptake of serotonin, results in decreased aggression. Third, when genes for the production of serotonin are missing, for example, in naturally occurring mutations in mice, then aggression is more likely. Finally, serotonin receptors are found in many regions of the brain associated with emotional control including aggression including hippocampus and PAG among other regions.

There are many complexities to this picture (Nelson & Chiavegatto, 2001; Nelson & Trainor, 2007). First, there are many different serotonin receptor types each with their own distribution and possible functions, thus it is unclear exactly how altering serotonin levels interacts with these different receptor types to modulate aggression. Second, serotonin levels are affected by numerous biological mechanisms including hormones, cytokines, enzymes, growth factors, and signaling molecules; hence, if serotonin levels are some common pathway that may results in aggression, these other variables may be very important in certain contexts (Nelson & Chiavegatto, 2001). For example, Chiavegatto and Nelson (2003) reviewed the evidence that serotonin effects on aggression depended upon the availability of nitrous oxide (NO). Third, the effects of serotonin may depend upon the different functional types of aggression. Thus, studies have examined the relationships of serotonin to reactive angry aggression compared to emotionally cold, operant aggression and to apparently normal versus pathological aggression (Nelson & Trainor, 2007). Finally, although modulation of serotonin may change aggressive behavior, it is unclear exactly how specific its effects are on behavior more generally. For example, some have concluded that serotonin affects aggression by modulating impulsiveness more broadly, rather than modulating aggression specifically (Nelson & Trainor, 2007).

5.2.2 Dopamine

Dopamine is a neurotransmitter that is spread widely throughout the body and has many functions. Dopamine is widely implicated in the control of reward in many contexts including food consumption, sexual and maternal behavior (Nelson & Trainor, 2007). Electrical stimulation of the brain resulting in the release of dopamine is subjectively experienced as pleasurable. Furthermore, nonhuman animals will emit high rates of operant behavior to obtain such stimulation; and, many drugs of addiction, such as cocaine and methamphetamine, result in the release of dopamine. Finally, dopamine is found in the mesolimbic areas of the brain which have been implicated in the control of aggression. These observations suggest a relationship between dopamine and operant behavior. Hence, it is unsurprising that it has been implicated in the control of operant aggression by affecting the reward value of aggressive behavior.

Yamaguchi and Lin (2018) reviewed the role of dopamine in aggression. They noted that there are two main lines of evidence implicating dopamine in aggression. First, anti-dopaminergic psychotropic medications, such as antipsychotic psychotropic medications, are often used to treat aggression in psychiatric patients and individuals with autism spectrum disorders. Furthermore, stimulants such as methamphetamine that result in release of dopamine can facilitate aggression in humans. These observations on the effects of psychotropic medication on aggression in humans require considerable qualification. For example, traditional antipsychotics such as haloperidol do not reduce aggression specifically, rather they suppress a wide range of behavior by sedating the person among other possible mechanisms. Similarly, the effects of risperidone on aggression in individuals with autism may not be specific to reducing aggression. Thus, it is unclear if the dopaminergic effects of antipsychotics actually do specifically mediate control of aggression. Similarly, dopamine agonists reduce aggression in nonhuman animals in some, but not all studies (Yamaguchi & Lin, 2018).

It is commonly believed that methamphetamine is associated with human aggression and violence, especially among chronic methamphetamine addicts; indeed, it is a contemporary established media trope. Thus, it is tempting to believe that this is due to methamphetamine's dopaminergic effects. In a critical review of human studies of methamphetamine and aggression, however, Tyner and Fremouw (2008) showed that this relationship is not so clear. Human methamphetamine-related aggression can occur in many circumstances. These may include: (a) during drug use; (b) as a result of drug withdrawal; (c) during acquisitive offenses to access methamphetamine directly or to obtain money or good in order to buy methamphetamine; and (d) during methamphetamine-induced psychosis (Hoaken & Stewart, 2003). Tyner and Fremouw pointed out that establishing a link between methamphetamine use and violence requires: (a) valid measurement of drug use; (b) valid measurement of aggression and violence; (c) demonstration of a parametric relationship between drug use and aggression and violence; (d) demonstration that drug use preceded aggression and violence; and (e) elimination of confounding factors that might also explain the aggression and violence. Thus, Tyner and Fremouw (2008) concluded that convincingly demonstrating that methamphetamine increases aggression a violence in humans through increasing dopamine is extremely difficult due primarily to the correlational and uncontrolled nature of human data.

Second, studies on nonhuman animals have shown that increasing dopamine can increase the reward value of access to and engaging in aggression. Paralleling the operant nature of some forms of aggression as shown in the classic studies of Azrin et al. (1965), studies of aggression in mice have shown that dopamine agonists increase the reinforcing value of access to the opportunity to engage in aggression, that is, mice will emit higher rates of operant behavior to access the resident-intruder test and aggress against a conspecific.

5.2.3 *Other Neurotransmitters*

Whereas serotonin and dopamine are the most frequently cited neurotransmitters involved in the control of aggression, there are many other neurotransmitters and biologically active substances that may have direct effects or influence aggression through influencing serotonin, dopamine, or other biological mechanisms influencing aggression. In this section we will consider some of the more commonly cited examples, namely, glutamate, norepinephrine, GABA, and nitrous oxide.

The first example is **Glutamate,** which is an amino acid and widespread throughout the body, including the nervous system and brain. It has an excitatory effect on neurotransmission. Agents that increase glutamate may also be associated with greater than usual reactivity to environmental stimulation, such as intruders or physical handling. Antiseizure medications, such as valproate, reduce glutamate and are also used for treatment of aggression (Miczek & Fish, 2005a). These observations suggest that glutamate increases the propensity toward aggressive behavior. Some studies have suggested that glutamate may increase aggression in part by stimulating release of dopamine (Miczek & Fish, 2005b).

A second example is **Norepinephrine** (Noradrenaline), which is both a neurotransmitter, a hormone (released by the adrenal glands into the blood), and as a medication injected into the body. Its numerous effects include readying the body for action. This is seen in its effects on increasing arousal; increasing attention and memory; increasing heart rate, and blood pressure; releasing glucose into the blood;

and diverting blood flow to the skeletal muscles. It is often described as being responsible for the "fight or flight" response. Thus, norepinephrine might be implicated in the control of aggression in a number of ways, some obvious, such as increasing arousal and readying the organism for agonistic social encounters, and others less obvious, such as interactions with other neurotransmitters and hormones.

Evidence for norepinephrine's role in aggression is slimmer than for other neurotransmitters. It includes observations that drugs such as propranolol, which reduce aggression in humans and nonhuman animals, also reduces synaptic norepinephrine levels. Other human and nonhuman animal studies have also found evidence that reductions in norepinephrine reduced aggression (Haller & Kruk, 2006; Marino et al., 2005). The relationship between noradrenaline and aggression is not consistent, however, perhaps because of differences in measurement procedures between studies (Miczek & Fish, 2005b).

Third, gamma-amino butyric acid (**GABA**) has also been implicated in control of aggression. GABA is the main inhibitory neurotransmitter, generally dampening down neural activity. GABA levels in rodents correlate positively with aggressive behavior, although some pharmacological agents may have an inverted-U shaped rather than linear relationship with aggression. The well-known effects of alcohol on increasing aggression may also be mediated through GABA in that alcohol may enhance the physiological effects of GABA on aggression. These effects may be tenuous due to individual differences in GABA receptors (Nelson & Trainor, 2007).

Finally, **Nitrous Oxide** is a free radical involved in blood vessel tone, immune response, and aggression. The effects of NO on blood vessel tone are well known through the use of sublingual nitroglycerin to relieve heat pain in angina pectoris by opening up the blood vessels supplying the heart muscles and through the recreational use of amyl nitrites ("poppers"). NO enzymes are located in brain structures associated with control of emotions (and also in other parts of the body associated with control of blood vessel tone and immune response). Studies have shown that NO may increase aggressive behavior in nonhuman animals through genetic and pharmacological manipulations. It appears that NO influences aggression mainly through influencing serotonin although the precise mechanism has not yet been elucidated (Nelson & Chiavegatto, 2001; Nelson & Trainor, 2007).

5.2.4 Summary

In this section on other neurotransmitters we have considered only a sample of biologically active agents that may be involved in neurotransmission. There are many other biologically active agents that interact with the apparently principal neurotransmitters involved in the control of aggression and this range of biologically active substances interact with one another in many ways, only some of which may be known today (Nelson & Chiavegatto, 2001; Nelson & Trainor, 2007; Yamaguchi & Lin, 2018).

The role of neurotransmitters in control of aggression and violence, especially in humans, is complex as is the relationship between psychopharmacology, presumably mediated by neurotransmitters, and human aggression. Thus, the relationships between neurotransmitters and aggression are sometimes more clearly demonstrated in nonhuman animals than in humans. These relationships are further complicated by generalizability of results from one species to another, but also between different types of aggression (predatory, territorial defensive male-on-male, maternal defending offspring, etc. [Haller, 2020]).

5.3 Hormones

This section will review some basic endocrinological concepts and their relationships to possible hormonal causes of aggression and violence and review evidence for commonly researched hormones that may be related to aggression, specifically, testosterone, thyroxine, steroids, and, oxytocin and vasopressin. Although some of the treatment implications of this work will be mentioned in passing, the reader is referred to chapter for a more detailed review of hormonal treatment of aggression and violence.

5.3.1 *Basic Endocrinological Concepts*

Hormones are naturally occurring substances secreted into the blood by ductless glands that regulate a wide range of bodily functions and that maintain homeostasis of many important biological functions. There are also synthetic versions of many of these naturally occurring substances. The system of organs that secrete these substances is referred to as the endocrine system and is distributed throughout the body. Whereas the nervous system acts through neurotransmitters and tends to act locally, quickly, and over short periods of time, the endocrine system transports hormones throughout the body by secreting them into the blood to transport them throughout the body. Hormones then bind at specific receptor sites. They act over relatively longer periods of time, for example, they are involved in regulation of diurnal patterns of behavior such as sleep, as well as temporal patterns of behavior over longer periods of time, such as puberty, the menstrual cycle, and menopause. Regulation of hormones often involves negative feedback loops to maintain homeostasis of the relevant bodily function.

Hormones can regulate a wide range of bodily functions including growth and development, metabolism, sexual function, reproduction, and mood. There are eight principal endocrine organs. Table 5.1 summarizes their hormones and common functions. Some hormones, such as testosterone, human growth hormone, thyroxine, and steroids, have been extensively studied with respect to aggression and violence both in terms of natural functioning, abuse of substances related to aggression,

Table 5.1 The principal endocrine organs, their hormones and functions. (Modified from Cancer Registration and Surveillance Modules, n.d.)

Organ(s)/ Location(s)	Hormones/functions
Pituitary gland	1. Human growth hormone (somatotropin)/controls growth, glucose metabolism, and protein synthesis 2. Thyroid stimulating hormone/release of thyroid hormone resulting in alteration in metabolic rate, body mass, and weight 3. Adrenocorticotropic hormone/stimulates secretion of cortical hormones 4. Gonadotropic hormones/regulates the development, growth, and function of ovaries and testes 5. Prolactin/development of glandular tissue in the female breast during pregnancy; stimulates postnatal lactation
Pineal gland	Melatonin/controls sleep, diurnal cycle, and reproductive behavior
Thyroid glands	1. Thyroxine and triiodothyronine/heart rate, blood pressure, body temperature, and weight 2. Calcitonin/regulation of blood calcium level
Parathyroid gland	Parathyroid hormone/regulates blood calcium
Adrenal gland	Secretes multiple hormones including: 1. Aldosterone/regulates water and sodium; 2. Cortisol/regulates blood sugar; 3. Male (androgens) and female (estrogens) sex hormones in minimal amounts/effects are usually masked by secretion of sex hormones by gonads, but effects may be apprent post menopause when androgens result in masculinization in the absence of estrogens in post menopausal women; 4. Epinephrine/in response to stress increases heart rate, cardiac activity, bronchodilation, and release of sugar to the blood; and, 5. Norepinephrine/released in response to stress and low blood pressure
Pancreas	Glucogens and insulin/regulation of blood sugar
Gonads: Testes	1. Testosterone / puberty in males; increased muscular-skeletal growth, secondary male sexual characteristics, such as body hair and increased sexual drive
Gonads: Ovaries	1. Estrogens/development of secondary sexual characteristics in females and development of uterus and vagina and regulation of menstrual cycle 2. Progesterone/thickening of uterine lining prior to pregnancy and regulation of menstrual cycle

and interventions based control of hormones, such as castration. (See Chap. 12, *Individual Biological Interventions for Violence and Aggression. II. Other Biological Treatments*). Other hormones, however, might also be relevant either on a case by case basis or in ways that have not yet been studied extensively.

Table 5.1 shows that hormones have a very wide range of functions, but it is important to note that none directly relate to aggression and violence: There is no hormone whose primary job it is to regulate aggression or violence. Rather, biological mechanisms that control aggression and violence result from the interaction of: (1) biological factors, such as genes, neurotransmitters, and hormones; (2) situational determinants such as the social context; and (3) the accumulated effects of the individual organism's learning (Brain, 1994). Thus, any simple model of a hormone causing aggression or violence is another foolish endeavor as it must necessarily be incomplete.

Brain (1977) and Brain (1994) noted that there were three main endocrinological systems that related to aggression and violence: First, is the hypothalamus—pituitary—gonadal (HGP) system. This involves luteinizing hormone and gonadotropins from the pituitary gland sex hormones from the gonads. Second, is the hypothalamus—pituitary—adrenocortical (HPA) system which involves ACTH and adrenal steroids. Third, are the thyroid and adrenal medulla.

It is important to note that the action of these hormone systems is dynamic. For example, some of these hormones are structurally very similar and may be readily transformed into one another. Second, their effects may be determined by the proportion of receptor sites that are available for binding. Third, the observed values of hormones observed may be difficult to interpret because peripheral levels may not tell us what the CNS levels are. Furthermore, it may be more important to consider if the hormone levels are increasing, decreasing or steady, rather than what their absolute values are. Finally, much of the research has been done on nonhuman animals, including both nonprimates and primates, and hence it is difficult to know with certainty whether these results can be generalized to understanding and treating human aggression and violence.

Hormones can influence aggression and violence in several ways. Hormones can influence aggression through the so-called early programming effects. This refers to the way in which exposure to hormones early in development influences the likelihood of adults responding aggressively. For example, exposure to testosterone in neonatal rodents may increase the likelihood of aggression in those animals as adults. Second, increased production of sex hormones during puberty is sometimes evoked as an explanation for the relatively low levels of aggression in prepubertal animals and the apparent increase in aggression in adolescents. Finally, studies have evaluated the effects of sex hormones on aggression in adults. (See the section below on testosterone on human aggression). Generally, these relationships have been demonstrated more clearly in nonhumans perhaps because more clear experimental manipulations can be made, for example, through neonatal castration and subsequent exposure to controlled amounts of testosterone. The evidence of these types of relationships in humans tends to be more ephemeral than in nonhumans (Brain, 1977, 1994).

5.3.2 *Testosterone*

If males are more aggressive than females and males have approximately five to ten times as much testosterone than females, it seems reasonable to ask, "Does testosterone cause aggression?" Reviews of this question have reached a more nuanced set of conclusions about the role of testosterone and aggression.

The existence of testosterone was implied in a study in 1849 by Berthold (cited in Haller, 2020). Berthold castrated male roosters resulting in morphological changes, such as elimination or reduction in secondary sexual characteristics, such as a smaller comb, and behavioral changes including elimination of crowing, copulation, and aggression. Moreover, reimplantation of the removed testes reinstated these morphological and behavioral characteristics. Since the reimplanted tests did not form neural connections, Berthold concluded that the tests secreted a then unknown substance that had caused these changes. In 1935, David et al. (cited in Haller, 2020) extracted and isolated testosterone from 200 pounds of bull testes. Testosterone was also synthesized around the same time allowing the development of experimental studies of the effects of testosterone. Testosterone has two main functions: Building muscle mass, and development of secondary male characteristics. In an experimentally more elegant demonstration, DeBold and Miczek (1981) demonstrated that during the intruder test, castration in rats reduced aggression toward intruders and replacement of natural testosterone with synthetic testosterone resulted in reinstatement of aggression toward intruders. These effects are dependent upon previous learning in some but not all species (Miczek & Fish, 2005b).

In a review by Carré et al. (2017) identified three lines of evidence that testosterone causes changes in aggressive and related behavior. First, males are generally more aggressive than females and they often have much higher levels of testosterone than females. Second, some studies have reported that violent offenders have higher testosterone levels than nonviolent offenders and that self-reports of aggression often correlate with testosterone levels in various human populations. Third, baseline testosterone levels are very moderately correlated with subsequent aggression in sports competition and contrived laboratory situations, such as various games apparently giving the opportunity to "aggress" against an unseen confederate by imposing fines or dispensing hot sauce to a putative victim. These effects are, however, of small to vanishing magnitude (Carré et al., 2017) and inconsistent (Albert et al., 1993). If this is so, what is the role of testosterone in aggression?

Carré et al. (2017) and Carré and Robinson (2020) offer a more nuanced account that may explain the lack of any simple relationship between testosterone and aggression. They noted: (a) the dynamic nature of baseline levels of testosterone; and (b) that there are bidirectional relationships between hormones and behavior. Thus, any experimental study of the effects of testosterone on aggression takes place against a background of potentially changing levels of testosterone both of which must be taken into account if the role of testosterone is to be understood. For example, in some animals naturally occurring levels of testosterone are low and stable during nonbreeding parts of the year, then gradually increase until they remain high during the breeding season and then wane back to levels similar to prebreeding levels. Social control of testosterone comes from studies of various challenges in nonhuman and human animals that have shown that: (a) testosterone levels increase after winning contrived and naturally occurring games such as competitive sports, including observing teams winning in sports fans; and (b) during subsequent challenges, this increase in testosterone is related to subsequent increases in aggression and winning. This latter observation has suggested to some that one of the functions of acute increases in testosterone is to maintain social dominance and other benefits of aggression.

Experimental studies of testosterone in humans have used several methods. Earlier studies evaluated the effects of chronic administration of testosterone on sexual functioning and aggression in men with low testosterone levels. Systematic reviews of these studies found that testosterone often resulted in improvements in sexual functioning, quality of life, and improved mood. Although some studies reported increased aggression and aggression-related constructs such as anger, these effects were inconsistent across reviewed studies (Kristensen et al., 2020; Zarrouf et al., 2009).

5.3.2.1 Anabolic Steroids and "Roid Rage"

Anabolic steroids are synthetic hormones which are the same as or similar to testosterone which are sometimes prescribed legally for delayed puberty in boys, the effects of low testosterone in adult men, and, building muscle mass in individuals who are HIV positive or who have other muscle-wasting conditions. Even their legitimate use has potential problems as they may cause a range of negative physical problems, including hypogonadism, heart disease (Payne et al., 2004), and psychiatric disorders, such as mood disorders.

Anabolic steroids result in increased aggression, hostility and unprovoked rage to apparently minimal provocation (Piacentino et al., 2015). Lurid case reports of uncontrollable rage and murder in publically known athletes and violent crime associated with steroid abuse have resulted in the cultural meme known as "roid rage" (Thiblin et al., 1997) suggesting an association between abuse of anabolic steroids and aggression and violence. As compelling as such vivid stories are, it is important to note that use and abuse of anabolic steroids by athletes and violent criminals involve confounds and selection biases. For example, athletes who abuse steroids also exercise enormous amounts, train hard, have unusual diets and lifestyles, and perhaps unusual personalities – and also abuse steroids. Hence, it is difficult to untangle the web of causation between aggression and steroids in this population.

5.3.3 Corticosteroids

One of the main corticosteroids is cortisol which is released in the HPA axis. It is sometimes called the "stress hormone" as it is often released from the adrenal glands in response to physical and psychological stress and results in increased blood sugar, use of sugar in the brain, and aids in tissue repair. Some studies have shown that high levels of cortisol are related to anxiety and social avoidance and low levels are related to reduced stress and social approach. Hence, cortisol appears to function to prepare the organism for threat, ready it for action, and to repair itself when damaged.

Given the defensive and protective functions of aggression, it is unsurprising that some studies have found correlations between cortisol levels and aggression. For example, as part of a longitudinal study of adolescent boys, Van Bokhoven et al. (2005) collected data on salivary levels of cortisol. They observed that: (a) boys with conduct disorder (CD) had higher cortisol levels than boys without CD; (b) boys with aggressive forms of CD had higher levels of cortisol than covert forms of CD; and (c) reactive aggression also predicted higher cortisol levels.

A systematic review of the literature on baseline cortisol levels and cortisol reactivity to stressors (Figueiredo et al., 2020) found mixed and contradictory results with some studies finding associations between aggression and externalizing behavior disorders related to aggression and others finding no relationship or even the opposite relationship. These contradictory findings might reflect studies using different populations and different measurement techniques, but perhaps, as with testosterone, there is no simple relationship between cortical and aggression.

5.3.4 The Dual Hormone Hypothesis

One response to this lack of a clear simple relationship between either testosterone alone or cortisol alone has been the so-called dual hormone hypothesis (Mehta & Josephs, 2010). Noting the lack of a robust simple relationship between testosterone and aggression, Mehta and Josephs postulated that this weak relationship masks interactions between different hormone systems that influence aggression. Specifically, they hypothesized that testosterone only increases social dominance when cortisol levels are low. By "social dominance," these authors mean a behavioral style that leads to high social status which is characterized by a variety of behavior, such as assertiveness, energetic social behavior, leader-like behavior, stereotypic masculinity, and an expansive posture. (Note that social dominance does not include aggression directly, but does include apparently similar behavior).

In their original paper, Mehta and Josephs (2010) reported two experiments. In the first, undergraduate volunteers were assigned to be either a leader or a follower in a block design task in which leaders instructed and evaluated followers in the task. After nine block designs participants switched roles. Consistent with other research, the authors found very small but statistically significant correlations between both testosterone and cortisol levels and social dominance. More importantly, when cortisol levels were high, testosterone had little impact on social dominance, and when cortisol levels were low, high levels of testosterone resulted in a large increase in social dominance, but when both cortisol and testosterone levels were low levels of social dominance were also low. The authors concluded that these data and those from their second experiment supported the notion that social dominance is the result of the interaction of the activity of two hormonal systems indexed by baseline levels of testosterone and cortisol.

Since the publication of Mehta and Josephs (2010), the dual hormone hypothesis has generated considerable amounts of research. In a systematic review of subsequent research and analysis of multiple data sets, Grebe et al. (2019) painted a less optimistic picture. These authors conducted a systematic search of the literature and found 55 effect sizes which were mixed: Some supported the dual hormone hypothesis, some found no effects, and some even found effects in the opposite direction! Grebe et al. suggested that this might be due to either lack of experimental power due in part to small sample sizes, and/or to a lack of precision in operationalizing predictions of the dual hormone hypothesis before experiments are conducted. In their own study these authors analyzed seven independent data sets (total $n = 718$) which also showed no robust evidence of testosterone x cortisol interactions on measures of status-striving personality traits.

Yet again, attempts to rescue simple models of hormones causing aggression using two-factor models disappoint. This makes the implications for treatment of human aggression especially challenging.

5.3.5 *Thyroxin*

Hyperthyroidism is the excessive secretion of thyroid hormones by the thyroid gland. This results in numerous physiological symptoms such as unexpected weight loss with apparently normal consumption of calories, heart racing and irregular heartbeat, sweating, sleep disturbance, and general irritability which sometimes includes aggression and violence. Hyperthyroidism treatment consists of treatments to reduce the secretion of thyroid hormones. This is most commonly performed by: (1) antithyroid drugs; (2) injection of radioactive iodine that is absorbed into the thyroid gland and destroys a sufficient number of thyroid gland cells to reduce the production of thyroxine; and (3) surgical removal of parts of the thyroid gland and supplementation with synthetic thyroid hormone to optimize the level of thyroxine. Hyperthyroidism has several causes but a common cause of Graves disease, an autoimmune disease which stimulates the production of thyroxine (Burch & Cooper, 2015).

That hyperthyroidism is associated with a broad range of psychiatric disorders, such as anxiety, depression, and irritability, has been known since the nineteenth century. Subjectively, individuals with hyperthyroidism report feeling chronically restless, emotionally out of control, impatient, angry, and unable to control emotional outbursts, while simultaneously feeling tired and exhausted from lack of sleep and poor quality sleep (Nexø et al., 2015). Their friends, families, and coworkers often report that the person appears on edge, driven, "hyper," overactive and over-reactive to minor social problems and daily frustrations. Thus, untreated or poorly regulated hyperthyroidism could easily predispose someone to aggressive and violent outbursts

There is also some evidence for an association between hyperthyroidism with violence and aggression and related traits in some psychiatric populations (Trifu et al., 2020), although the evidence is weak and suggestive, rather than conclusive.

For example, a couple of studies have found higher levels of thyroid hormones in violent criminals with antisocial personality disorders than in nonviolent patients and noncriminals (Evrensel et al., 2016; Stalenheim, 2004). There is also some evidence of an association between hyperthyroidism and aggression in nonhuman animals (Trepanier, 2006).

5.3.6 Oxytocin and Vasopressin

Oxytocin and vasopressin are two peptides which are synthesized in the hypothalamus and secreted by the pituitary gland. Oxytocin was originally implicated in the regulation of progression of labor and regulation of glucose and insulin. Hence it is used to treat diabetes. Vasopressin's main physiological function is to inhibit secretion of water by the kidneys and hence regulate the amount of water secretion by the kidney and ion concentration in the body. Hence, it is sometimes referred to as the "antidiuretic hormone."

Subsequent research has implied that oxytocin and vasopressin have much broader functions. These include roles in memory and broad regulation of social behavior including maternal behavior and scent marking in nonhumans including fish birds and rodents (de Jong & Neumann, 2017). For example, some have suggested that oxytocin not only promotes labor but also prepares the mother and facilitates prosocial behavior with infants (Young & Flanagan-Cato, 2012). These findings have potential implications for promoting human prosocial behavior and understanding and treating aggression and violence.

Evidence for the involvement of oxytocin in aggression includes the hypothesis that impairment in oxytocin systems may facilitate aggressive behavior and that administration of oxytocin often might both facilitate prosocial behavior and decrease aggression. De Jong and Neumann (2017) found somewhat stronger evidence for the anti-aggressive effects of oxytocin in nonhumans and equivocal and weak evidence for humans, but also noted that oxytocin might increase maternal aggression. These effects, however, may well depend upon numerous contextual factors. For example, oxytocin may only reduce aggression in individuals who are oxytocin deficient, perhaps only as measured directly by oxytocin levels in the CNS, rather than peripheral oxytocin levels. Furthermore, these effects in humans may only occur in individuals with weak prosocial behavior and/or in environments where prosocial behavior is not supported. Hence, detecting the effects of oxytocin on human aggression, if it is present, may be tricky.

Evidence for the involvement of vasopressin in aggression comes from both nonhuman animal studies and studies in psychiatric patients (Nelson & Trainor, 2007). For example, in rodents, stimulation of the hypothalamus results in release of vasopressin and increased aggression and blockade of vasopressin receptors results in decreased aggression (Ferris & Potegal, 1988). Some studies have found increased levels of vasopressin to be associated with impulsivity and aggressive traits in humans and that vasopressin levels may be related to social behavior more generally

(Albers, 2012). (See Chap. 12, *Individual Biological Interventions for Violence and Aggression. II. Other Biological Treatments*, for evidence of the effects of oxytocin and vasopressin in the treatment of human aggression).

5.3.7 Summary

The putative relationship between hormones is intriguing and plausible, it is a part of our cultural views of the causes of aggression and the basis for various individual biologically-based hormonal treatments such as castration. (See Chap. 12, *Individual Biological Interventions for Violence and Aggression. II. Other Biological Treatments*). Despite these factors, pinning down the evidence to support this view has proven difficult. Although sometimes dramatic effects can be seen in nonhuman animals demonstrating such relationships in humans have proven difficult. The possibilities of interactions between multiple factors—both endocrinological and environmental—has proven just as difficult to demonstrate robustly. Of course, testosterone, steroids, thyroxine, and vasopressin and oxycontin are not the only hormones that might be involved in the regulation of aggression and violence—they are just the most extensively studied at this time. Other hormones could easily be involved in control of aggression. Similarly, interaction models of hormone action, such as the dual hormone hypothesis, might simply be the first efforts in this field and other multifactor models might be more plausible.

5.4 Summary

This chapter and the preceeding chapter have reviewed a very wide range of evidence of biological causes of aggression and violence. Biological causes of aggression and violence refers to a wide range of potential causes. Some of this research does point to the interaction between environmental factors, primarily to the effects of experience interacting with genetics. To date however, there is little research on the interaction between genetic and cultural factors on aggression and violence.

References

Albers, H. E. (2012). The regulation of social recognition, social communication and aggression: Vasopressin in the social behavior neural network. *Hormones and Behavior, 61*(3), 283–292.

Albert, D. J., Walsh, M. L., & Jonik, R. H. (1993). Aggression in humans: What is its biological foundation? *Neuroscience & Biobehavioral Reviews, 17*, 405–425.

Azrin, N. H., Hutchinson, R. R., & McLaughlin, R. (1965). The opportunity for aggression in an operant reinforcer during aversive stimulation. *Journal of the Experimental Analysis of Behavior, 8*(3), 171–180.

Bard, P. (1928). A diencephalic mechanism for the expression of rage with special reference to the sympathetic nervous system. *American Journal of Physiology, 84*, 490–515.

Berthold, A. A. (1849/1944). Transplantation of testes. Translated by D. P. Quiring. *Bulletin of the History of Medicine, 16*, 399–401.

Best, M., Williams, J. M., & Coccaro, E. F. (2002). Evidence for a dysfunctional prefrontal circuit in patients with an impulsive aggressive disorder. *Proceedings of the National Academy of Sciences, 99*(12), 8448–8453.

Brain, P. F. (1977). Hormones and aggression. *Annual Research Reviews: Hormones & Aggression, 1*, 26.

Brain, P. F. (1994). Hormonal aspects of aggression and violence. In A. J. Reiss, K. A. Miczek, J. A. Roth, National Research Council, Division of Behavioral and Social Sciences and Education, Commission on Behavioral and Social Sciences and Education, & Panel on the Understanding and Control of Violent Behaviors (Eds.), *Biobehavioral influences* (Vol. 2, pp. 173–244). The National Academies Press. https://doi.org/10.17226/4420

Burch, H. B., & Cooper, D. S. (2015). Management of Graves disease: A review. *Journal off the American Medical Association, 314*(23), 2544–2554.

Cancer Registration & Surveillance Modules (n.d.). Retrieved March 01, 2021, from https://training.seer.cancer.gov/anatomy/endocrine/glands/thyroid.html

Carré, J. M., & Robinson, B. A. (2020). Testosterone administration in human social neuroendocrinology: Past, present, and future. *Hormones and Behavior, 122*, 104754.

Carré, J. M., Ruddick, E. L., Moreau, B. J., & Bird, B. M. (2017). Testosterone and human aggression. In P. Sturmey (Ed.), *The Wiley handbook of violence and aggression* (pp. 1–14). Wiley.

Chiavegatto, S., & Nelson, R. J. (2003). Interaction of nitric oxide and serotonin in aggressive behavior. *Hormones and Behavior, 44*(3), 233–241.

David, K., Dingemnanse, E., Freud, J., & Laquer, E. (1935). Uber Kristaliniches Mannliches Hormon aus Steerentestes. *Acta Breeve Neerland Physiol Pharmacol Microbio, 5*, 85–86.

de Jong, T. R., & Neumann, I. D. (2017). Oxytocin and aggression. In R. Hurleman & V. Grinevich (Eds.), *Behavioral pharmacology of neuropeptides: Oxytocin* (pp. 175–192). Springer.

Debold, J. F., & Miczek, K. A. (1981). Sexual dimorphism in the hormonal control of aggressive behavior of rats. *Pharmacology Biochemistry and Behavior, 14*, 89–93.

Evrensel, A., Ünsalver, B. Ö., & Özşahin, A. (2016). The relationship between aggression and serum thyroid hormone level in individuals diagnosed with antisocial personality disorder. *Nöro Psikiyatri Arşivi, 53*(2), 120.

Ferris, C. F., & Potegal, M. (1988). Vasopressin receptor blockade in the anterior hypothalamus suppresses aggression in hamsters. *Physiology & Behavior, 44*(2), 235–239.

Figueiredo, P., Ramião, E., Azeredo, A., Moreira, D., Barroso, R., & Barbosa, F. (2020). Relation between basal cortisol and reactivity cortisol with externalizing problems: A systematic review. *Physiology & Behavior, 225*, 113088.

Franzini, A., Broggi, G., Cordella, R., Dones, I., & Messina, G. (2013). Deep-brain stimulation for aggressive and disruptive behavior. *World Neurosurgery, 80*(3–4), S29–e11.

Gajos, J. M., & Beaver, K. M. (2016). The effect of omega-3 fatty acids on aggression: A meta-analysis. *Neuroscience & Biobehavioral Reviews, 69*, 147–158.

Gouveia, F. V., Hamani, C., Fonoff, E. T., Brentani, H., Alho, E. J. L., De Morais, R. M. C. B., … Martinez, R. C. (2019). Amygdala and hypothalamus: Historical overview with focus on aggression. *Neurosurgery, 85*(1), 11–30.

Grebe, N. M., Del Giudice, M., Thompson, M. E., Nickels, N., Ponzi, D., Zilioli, S., … Gangestad, S. W. (2019). Testosterone, cortisol, and status-striving personality features: A review and empirical evaluation of the dual hormone hypothesis. *Hormones and Behavior, 109*, 25–37.

Haller, J. (2020). *Neurobiopsychosocial perspectives on aggression and violence*. Springer.

Haller, J., & Kruk, M. R. (2006). Normal and abnormal aggression: Human disorders and novel laboratory models. *Neuroscience & Biobehavioral Reviews, 30*(3), 292–303.

Hibbeln, J. R., & Gow, R. V. (2014). The potential for military diets to reduce depression, suicide, and impulsive aggression: A review of current evidence for omega-3 and omega- 6 fatty acids. *Military Medicine, 179*(suppl_11), 117–128.

Hoaken, P. N., & Stewart, S. H. (2003). Drugs of abuse and the elicitation of human aggressive behavior. *Addictive Behaviors, 28*(9), 1533–1554.

Kristensen, T. T., Christensen, L. L., Frystyk, J., Glintborg, D., T'sjoen, G., Roessler, K. K., & Andersen, M. S. (2020). Effects of testosterone therapy on constructs related to aggression in transgender men: A systematic review. *Hormones and Behavior, 128*, 104912.

Marino, M. D., Bourdélat-Parks, B. N., Liles, L. C., & Weinshenker, D. (2005). Genetic reduction of noradrenergic function alters social memory and reduces aggression in mice. *Behavioural Brain Research, 161*(2), 197–203.

Mehta, P. H., & Josephs, R. A. (2010). Testosterone and cortisol jointly regulate dominance: Evidence for a dual-hormone hypothesis. *Hormones and Behavior, 58*(5), 898–906.

Miczek, K. A., & Fish, E. W. (2005a). Dopamine, glutamate, and aggression. In W. Schmidt, W. J. Schmidt, & M. E. Reith (Eds.), *Dopamine and glutamate in psychiatric disorders* (pp. 237–263). Humana Press.

Miczek, K. A., & Fish, E. W. (2005b). Monoamines, GABA, glutamate, and aggression. In R. J. Nelson (Ed.), *Biology of aggression* (pp. 114–149). Oxford University Press.

Nelson, R. J., & Chiavegatto, S. (2001). Molecular basis of aggression. *Trends in Neurosciences, 24*(12), 713–719.

Nelson, R. J., & Trainor, B. C. (2007). Neural mechanisms of aggression. *Nature Reviews Neuroscience, 8*(7), 536–546.

Nexø, M. A., Watt, T., Cleal, B., Hegedüs, L., Bonnema, S. J., Rasmussen, Å. K., … Bjorner, J. B. (2015). Exploring the experiences of people with hypo-and hyperthyroidism. *Qualitative Health Research, 25*(7), 945–953.

Olivier, B. (2004). Serotonin and aggression. *Annals of the New York Academy of Sciences, 1036*(1), 382–392.

Payne, J. R., Kotwinski, P. J., & Montgomery, H. E. (2004). Cardiac effects of anabolic steroids. *Heart, 90*, 473–475.

Piacentino, D. D., Kotzalidis, G., Del Casale, A., Rosaria Aromatario, M., Pomara, C., Girardi, P., & Sani, G. (2015). Anabolic-androgenic steroid use and psychopathology in athletes. A systematic review. *Current Neuropharmacology, 13*(1), 101–121.

Stalenheim, E. G. (2004). Long-term validity of biological markers of psychopathy and criminal recidivism: Follow-up 6–8 years after forensic psychiatric investigation. *Psychiatry Research, 121*, 281–291.

Thiblin, I., Kristiansson, M., & Rajs, J. (1997). Anabolic androgenic steroids and Behavioural patterns among violent offenders. *Journal of Forensic Psychiatry, 8*(2), 299–310.

Trepanier, L. (2006). Medical management of hyperthyroidism. *Clinical Techniques in Small Animal Practice, 21*, 22–28.

Trifu, S. C., Tudor, A., & Radulescu, I. (2020). Aggressive behavior in psychiatric patients in relation to hormonal imbalance. *Experimental and Therapeutic Medicine, 20*(4), 3483–3487.

Tulogdi, A., Toth, M., Halasz, J., Mikics, E., Fuzesi, T., & Haller, J. (2010). Brain mechanisms involved in predatory aggression are activated in a laboratory model of violent intra- specific aggression. *European Journal of Neuroscience, 32*(10), 1744–1753.

Tyner, E. A., & Fremouw, W. J. (2008). The relation of methamphetamine use and violence: A critical review. *Aggression and Violent Behavior, 13*(4), 285–297.

Van Bokhoven, I., Van Goozen, S. H. M., Van Engeland, H., Schaal, B., Arseneault, L., Séguin, J. R., … Tremblay, R. E. (2005). Salivary cortisol and aggression in a population- based longitudinal study of adolescent males. *Journal of Neural Transmission, 112*(8), 1083–1096.

Weiger, W. A., & Bear, D. M. (1988). An approach to the neurology of aggression. *Journal of Psychiatric Research, 22*(2), 85–98.

Yamaguchi, T., & Lin, D. (2018). Functions of medial hypothalamic and mesolimbic dopamine circuitries in aggression. *Current Opinion in Behavioral Sciences, 24*, 104–112.

Young, L. J., & Flanagan-Cato, L. M. (2012). Editorial comment: Oxytocin, vasopressin and social behavior. *Hormones and Behavior, 61*(3), 227.

Zarrouf, F. A., Artz, S., Griffith, J., Sirbu, C., & Kommor, M. (2009). Testosterone and depression: Systematic review and meta-analysis. *Journal of Psychiatric Practice, 15*(4), 289–305.

Chapter 6
Social Causes of Violence and Aggression: I. Cultural Evolution

This chapter considers the naturally occurring processes of cultural evolution of aggression and violence. All cultures, of course, do engage in some explicit interventions, such as education, laws requiring seat belts, speed limits, and taxation, in order to change the behavior of its members. Such attempts related to aggression and violence are usually unsystematic and intuitive, are often responsive to some immediate problem, such as an uptick in road rage, and, although they may have some specific immediate goal such as reduction in road rage, may have no overarching goal, such as increasing lifespan of the culture's members. An exception to this is intentional communities that are often religious and perhaps strive for some final utopian perfection here on Earth. These intentional communities, such as the Hutterites (see Chap. 14, *Cultural Interventions to Change Violence and Aggression. II. Broad-Based Intuitive Approaches*), may have explicit goals and explicit social mechanisms to achieve them. Often these goals do relate to social behavior broadly, including promoting prosocial behavior and control of aggression and violence.

This chapter will introduce Skinner's ideas on cultural evolution in more detail and consider one example of cultural evolution—cultural evolution in primates. The next chapter will extend consideration of these ideas to human cultures.

6.1 What Is Cultural Evolution?

6.1.1 Skinner and Cultural Evolution

As noted at the beginning of this book, Skinner (1953) proposed that behavior comes from three kinds of evolution: Biological evolution, cultural evolution, and evolution of operant behavior during the lifespan. Each of these processes is conceptually similar. There are variations in genetically controlled behavior, form, and physiology; variations in cultural practices; and variations in the behavior of

P. Sturmey, *Violence and Aggression*, Advances in Preventing and Treating Violence and Aggression, https://doi.org/10.1007/978-3-031-04386-4_6

individual organisms. The environment selects some forms of behavior but not others via biological evolution, cultural selection, and operant learning. Present complex behavior arises out of progressive and additive change which is often (but not always) slow. This tripartite history determines our current behavior, including aggression and violence. Thus, Skinner wrote extensively about cultures, cultural evolution, and the reasons for explicit cultural design (Skinner, 1953, 1961, 1965).

Skinner defined culture as follows: "the culture into which an individual is born is composed of all the variables affecting him which are arranged by other people" (p. 419) and Glenn (2004) defined a culture as "patterns of learned behavior transmitted socially, as well as the products of that behavior (objects, technologies, organizations, etc.)" (p. 139). So, how does a culture affect the behavior of its members and how can we design cultures to influence aggression and violence in ways that are considered desirable?

For Skinner one part of the answer to the question of how cultural practices evolve is quite simple. Some cultural practices come from the non-social environment. Designing a boat is in part a response to the local water conditions and agricultural practices are in part a response to the local weather and physical environment. The properties of the stream select some kinds of boats and not others and the properties of the soil select some kinds of plows and not others. Cultures that build bad boats drown and cannot grow enough food to support their population drown and starve and eventually these cultures extinguish. There are a myriad of examples of how the properties of the physical environment select certain properties of cultural artifacts, such as tools from the door handle to the computer keyboard, to enable the members of the culture to use such objects effectively to achieve personally or socially important outcomes that may contribute to the survival of the culture.

The most important source of cultural control of behavior, however, comes from differential social reinforcement and punishment from other members of the culture. Other people show us directly, or cultural artifacts such as books or other instructional media show us how to make the best boats and plows and reinforce following the rules related to that. Social practices like apprenticeship institutionalize and perhaps even ossify such practices with contingencies reinforcement, such as admiration of others for working with a master craftsman, and contingencies of punishment, such as punishment for making one's own ineffective or suboptimal products. Social control of cultural practices may include specific, narrows cultural practices. For example, when members of the culture reinforce compliance with dietary restrictions, eating the right food at the right time and place; and when people break the rules they may be castigated, shunned, expelled, or required to engage in effortful behavior, such as praying for forgiveness to rejoin a religious organization or undergoing retraining to rejoin a professional guild.

Skinner (1953) also suggests that conformity to social norms more generally is reinforced. He suggested that unusual behavior becomes aversive to a culture's members and they respond to remove that aversive stimulus with contingencies of punishment. Hence, wearing the wrong clothes to a party, talking too loudly in a meeting, and failure to observe the Sabbath may all elicit subtle or not so subtle social punishment. The important and interesting question is not that these social contingencies are the mechanism for transmitting and modifying cultural practices—they are obvious in everyday life, especially if a member of the culture persists and escalates in not following the rules. Rather, the interesting question is why these particular practices came to be the ones that the culture's environment selected and maintained.

It seems paradoxical that many cultural practices seem non-functional. For example, a food prohibition or admiration for having a large family might have been highly useful in previous, historical environments but the continued practice may appear useless and wasteful when refrigerators keep meat safe and when most children survive to adulthood and families no longer need large broods to look after its older members. Nevertheless, such apparently useless, inefficient, or even harmful practices continue due to the continued social contingencies and may have an interesting function in keeping members of a culture close to one another with other potential benefits.

Skinner (1953) noted that within a culture there are many variations in practices. Fashions and cultural memes come and go and immigrants arrive with different cultural practices. Within a culture are special groups, such as different social agencies with different practices, e.g., religious and secular organizations, and gangs, each with their own variations in cultural practices. Generational differences in cultural practices are also often apparent. Skinner noted that cultures can foster a wide range of behavior maintaining a certain level of work; including establishing or discouraging certain types of motivations, emotional dispositions; teaching certain repertoires, such as problem solving, social skills, self-control, and self-knowledge; and inducing or preventing neurotic behavior. Hence, an aggressive culture might foster anger and revenge as appropriate; fail to teach alternatives to aggression and teach aggressive behavior instead; and induce neurotic behavior that is avoided and escaped by aggression and violence. Alternatively, a pacific culture might foster a calm and patient disposition; teach good problem solving and social skills; explicitly teach children self-control from an early age; and prevent neurotic behavior and teach better ways than aggression and violence to deal with it.

Skinner (1953) noted that there are many similarities in behavior across cultures. Most people respond to contingencies mediated by the physical environment in a similar way: We all react to stubbing our toes and a cold day in broadly similar ways. Within a culture are large individual differences reflecting the different learning histories of each person and the somewhat different experiences of the same environment for each person. Hence, Skinner concluded that it is difficult to decide whether a particular cultural practice is directly related to the learning history of an individual member.

Skinner (1953) noted that cultural practices may arise adventitiously, for example, through the model of a highly regarded and admired Saint or world leader. Cultures experiment informally with cultural design constantly, but the outcomes may be pretty hit and miss. For example, congestion charging is sometimes quite effective in reducing congestion and pollution and in switching travelers to public transport, walking, and cleaner vehicles. In contrast, other attempts to change behavior can be quite ineffective or counter-productive. For example, laws passed to limit the use of restraint to manage dangerous behavior sometimes increase their use (Sturmey, 2015) or other forms of restraint are substituted. A TV chef may try through exhortation and education to induce school children to eat more healthily only to see the children's parents hand-feeding their children French fries through the school fence. Thus, since intuitive approaches to cultural design are often ineffective and sometimes harmful, Skinner went on to advocate that cultural design should be done explicitly rather than left to chance and should be informed by scientific methods rather than intuition and opinion as he could see no virtue in leaving such important matters to chance.

The person initiating a change in cultural practice does so because of their history of previous attempts to do so. In so doing, the cultural designer may generate variations in cultural practices that are similar to genetic mutations. The new practice will modify the behavior of members of the culture and the value of the change will be measured in terms of future survival of the culture.

Skinner (1953) only mentioned aggression and violence as cultural practices only once or twice in relation to aggression and violence. He wrote "Aggressive action has usually been most successful in promoting the survival of one group against another or of one individual against another" (p. 432). He mentioned this in the context of a culture having to decide to choose a practice that may be effective in promoting survival of the culture, but one which is against the espoused values of the culture. Such choices illustrate the need for cultures to find effective alternatives to aggression and violence that also promote their long-term survival.

Skinner's (1953) ideas on cultural evolution are readily applicable to prevention and treatment of aggression and violence through cultural design, not only through the general principles of treating cultural design as a form of experimentation to generate new possible ways of living, but also in terms of specific strategies. For example, teaching self-control in young children has been addressed in a recent line of research to teach children social and communication skills, requesting items and help, tolerating delays to reinforcement and prosocial behavior. This program was effective in teaching these skills and reducing a wide range of problem behavior in young children (Hanley et al., 2007).

There are also numerous explicit parallels between Skinner's writings in the 1950s and Jared Diamond's (2005) subsequent writings. Both take a functional approach to cultural evolution and extinction. Namely, both see cultural evolution as a product of adaptation to the physical and social environment, rather than some intrinsic or mystic property of people, such as intelligence or cultural values.

6.1.2 Cultural Evolution, Aggression, and Violence

Cultures pay a cost for aggression and violence, including loss of productive members, reduced fertility, and diversion of resources away from alternate perhaps better uses, such education, work, and looking after other members of the culture. There are many examples of cultures that have failed to defend themselves effectively from other cultures when these other cultures take or destroy their resources, including territory, food, land, workers, and fertile females. Many such cultures no longer exist, perhaps in part because of the lack of effective aggression and violence and other cultural practices to enable the survival of the culture.

In addressing the cultural evolution of human aggression and violence, Rumph and Ninness (2001) noted that modern humans are very recent species with a mere 90,000 years of recent probably relatively unchanging genetic endowment, about a 100,000 generations of hunting and gathering, 500 lifetimes of agriculture, a mere 10 generations of industrial society, and only three generations of computers and internet. They proposed that our genetic endowment being sensitive to the reinforcing properties of aggression and its consequences are highly functional and contributed to the survival of early human cultures. Early human societies may have benefited substantially from effective escape from and effective combatting against predacious and competing organisms from other species as well as groups of fellow humans. Note that intraspecific competition may be especially important since conspecifics are likely to compete for the same resources (Lorenz, 1966). The initial benefits of susceptibility to aggression's reinforcers may have served early human societies well in the short term, but today we may be paying for those initial benefits with long-term costs to human cultures.

Rumph and Ninness (2001) went on to speculate that in the past human cultures had limited exposure to other cultures. Hence, the opportunities for aggression between cultures were relatively limited by geographical proximity and lack of long distance transport and communication. Today, however, transport and communication allow groups to interact—sometimes aversively—with one another, thereby setting the occasion for aggression to be reinforced. Today, two groups of soccer players and their fans can meet up for a soccer match for the first time and fight one another in a novel location. Their fighting produced immediate stimulation, including observing other people suffering which may reinforce aggressive behavior (Skinner, 1953), but also result in immediate and long-term removal of rival team supporters. Their fans can watch the match around the world and fight each other in those locations too. Bringing disparate cultures together that have not interacted in the past may sometimes set the scene for these coercive and aggressive interactions to minimize aversive stimulation for both cultures. Thus, when considering cultures that already have a history of interacting aggressively, modifying the behaviors of members of each culture to have mutually beneficial outcomes using non-aggressive means is the challenge (Rumph & Ninness, 2001).

Jared Diamond's (1997) account of the European conquest of the Americas in *Guns, Germs, and Steel*, takes a similar functionalist approach to the cultural value of aggression, when he wondered why, when the conquistadors arrived in the Americas, that the Aztecs did not kill them, seize the Spaniards' ships, and go invade and conquer Spain? The bottom line here was that the Aztecs were insufficiently aggressive and lacked the technology to produce effective killer weapons to repel and kill the conquistadors. They paid the price of large-scale extinction of much of the indigenous cultures of the Americas mostly through disease, but also through direct conquest and fighting. The more effective aggression of the Spaniards and other Europeans (as well as their more effective technology and more lethal viruses) resulted in huge gains for the Spaniards in terms of access to food, goods, minerals, labor, etc. Additionally, they arrived in the Americas before many other Europeans arrived and so were also effective in excluding most other cultures from access to these resources.

The other side of the coin here is how did some cultures survive and expand for so long and what role did aggression and violence play in their survival? The persistence of aggression and violence within cultures is paradoxical since it is so costly. It seems that there is something to be explained here. One simplistic answer is that aggression and violence can make an important contribution to the survival of cultures through the maintenance or acquisition of resources such as land, mates, food, and water and can eliminate competition for those resources from other cultures. Rome and Carthage engaged in prolonged wars from 264 to 146 BC. When Rome finally had enough of fighting Carthage in 122 years of expensive and disruptive wars, they finally destroyed Punic culture by systematically burning the city to the ground, selling of the diminished surviving population into slavery and dispersing them, and sewing the fields of Carthage with salt to remove the possibility of further war by utterly destroying the city and its people. In so doing they eliminated a significant source of cultural competition. (The salting of the fields of Carthage might be a nineteenth century myth, but various conquering cultures have indeed salted the fields of the defeated cultures and several countries have used salting land as a punishment for certain crimes the outcome of which is destruction of future food sources for the local culture.)

Aggression and violence also are effective means of aversive control of members of society. Greater aggression and violence can be used to reduce lesser aggression and violence, but can also be useful in coercing members of society to pay taxes, refrain from a wide range of personally interesting behavior that might have costs for society generally, and even induce members of society to risk their own lives in war for the greater good of strangers. Schools can coerce children to come to school by threatening to imprison the children or their parents and can make them be still, quiet, and docile with the paddle or just plain physical violence from the teacher. Such an approach can create a quiet and orderly school, even if it is filled with children full of fear, but such approaches are unlikely to create a love of learning with children approaching schools and teachers at every opportunity.

Such coercion comes at a tremendous long-term cost. Skinner (1953) wrote that "punishment, unlike reinforcement, works to the disadvantage of both the punished organism and the punishing agency (p. 183)." In Sidman's (1989) classic, *Coercion And Its Fallout,* Sidman suggested that coercion might be justified in very rare circumstances, such as to control a dangerous behavior when no effective alternative strategies exist. Even its justified use, however, is problematic because its use is so reinforcing to the punishing agent by terminating other people's aversive behavior, that its use easily generalizes to many other unjustified applications. Punishment has several well documented negative side effects. These may include: (a) suppression of the problematic behavior may only be temporary; (b) the person punished may avoid the punishing agent by avoidance, lying, and duplicity; (c) punishment if painful might elicit aggression from the person punished; (d) the person punished may rapidly learn when punishment is not available and engage in the problem behavior at those times; (e) the person punished and others may learn through modeling to use punishment in other situations; and (f) punishment may have emotional negative emotional side effects on the person punished, such as fear which also inhibits adaptive behavior, and the person may learn to attenuate these negative emotional side effects through other problematic behavior, such as alcohol or drug abuse.

6.1.3 Summary

A functional view of cultural evolution sees cultural evolution as a process that involves selection of certain practices by the physical and social environments. These practices are primarily maintained through coercive social contingencies. If the practices are effective, they will contribute to the survival and expansion of the culture; if they are problematic, they may contribute to the eventual extinction of the entire culture.

6.2 Cultural Evolution in Non-Human Primates

6.2.1 Organized Violence in Primates

If we are interested in human culture as a cause of aggression and violence, there are at least two reasons why we should concern ourselves with other primate cultures. First, we tend to think of culture as something uniquely human, but if we look at Skinner's definition of culture above, we can see that many social animals have some form of culture that involves intergenerational transmission of behavior and customs such as tool use and communicative behavior. Although most non-human cultures may lack something like language as a means of cultural transmission, they may have other learning processes such as those reflecting their shared physical environment and social processes such as imitation, that may form the basis of transmission of culture. Second, some primatologists and others believe that there is some form of similarity, if not continuity, between non-human primate and some human cultures. Such similarity might be based on the effects of the same or similar variables controlling engaging in prosocial behavior, aggression, and violence in non-human and human cultures. Hence, studying cultural practices related to aggression and violence in non-human primate cultures might inform us about human cultures and at least some forms of human aggression and violence. For example, some forms of aggression in non-human primates closely related to humans appear similar to socially organized behavior which has similarities to organized human aggression and violence. In addition to being an interesting phenomenon, it appears that such organized social behavior waxes and wanes depending upon the environment and hence may be a model for changes in human aggression and violence.

The two closest human relatives are chimpanzees (*Pan troglodytes*) and bonobos (*P. paniscus*). All three species share a common ancestry some 5–8 million years ago (Gruber & Clay, 2016). Thus, these two species share much of their genome with one another and with humans and all three species must share much of their genome with the closest common ancestor. So, perhaps studying social behavior in general and aggression in particular in chimpanzees and bonobos might give us some insights into the evolution and nature of human aggression and violence.

6.2.1.1 Chimpanzees

Some of the earliest systematic observations of aggression in chimpanzees come from the seminal work of Jane Goodall who observed their behavior in the Gombe National Park, Tanzania, for over 50 years (Goodall, 2010). Chimpanzees are primarily herbivorous, eating large quantities of leaves daily, but also fruit, termites, etc. They live in socially organized troops. During the first 10 years of her work, Goodall observed that occasionally they also ate fresh meat after killing baboons, but appeared cautious about meat from dead animals. Other important observations included that chimpanzees would prepare and use. For example, they built nests by weaving fronds of vegetation together to make nests every evening high in the trees, and fashioned and used sticks to obtain termites from their nests and chewed leaves to make sponges to obtain water from tree hollows.

Early observations of aggression were unremarkable. They included fights over food both between chimpanzees and between chimpanzees and baboons. They also included male displays which consisted of noisy episodes, feet stomping, and throwing vegetation around, and in one case using empty oil cans from Goodall's camp to create noise. These displays appear to be costly in terms of energy and time. In some ways they seem to be behavioral analogs of the peacock's tail as they might indicate male genetic fitness to females. Aggression was also seen between mothers and their offspring as older offspring were weaned from breastfeeding when a new sibling was born. This nicely illustrates the competition between genes in the mother and her offspring and between the genes of her oldest and youngest offspring. (See Chap. 4, *Biological Evolution of Violence and Aggression. I. Evolution and Genetics*.) (This put the present author in mind of episodes of the TV series *SuperNanny* where parental efforts to extinguish children begging for food. This resulted in extinction busts in children and extinction bursts in frustrated parents whose pleas for patience in their children were also extinguished as the parents sometimes finally delivered snacks to their demanding brood.) Although novel and important, none of these observations of aggression in chimps in the Gombe National Park was very surprising.

But there were two big subsequent surprises. The first was that after the initial troop that Jane Goodall had been observing split into a large and a small troop, the large troop systematically raided the smaller troop. Groups of males from the first troop killed all the males and some of the females in the smaller troop. This was described as the "Gombe wars." Even more disturbing was observation and indirect evidence that one mother-offspring dyad killed at least 10 infants of other mothers in the troop. Goodall was perplexed and disturbed by such observations. At first she assumed that this murderous dyad must have been pathological. As we have seen elsewhere, however, killing conspecifics and even relatives to obtain territory, food, nests, and fertile females often makes evolutionary sense, as do infanticide and siblicide in certain circumstances. It seems that, like humans and other animals, chimpanzees kill their relatives and neighbors.

Other work on chimps comes from Wrangham's extensive research which describes how chimps are highly territorial and hierarchical (Wrangham et al., 1996). A few high ranking males have access to many females. Control of territory, including acquisition of territory from other troops, may result in acquisition of more fertile females either directly or indirectly through the acquisition of resources in the new territory resulting in more females later. Troops sometimes split and merge and sometimes members of one troop will kill most members of a neighboring troop resulting in acquisition of their territory and perhaps fertile females. In addition, they may eat and share the flesh of their victims.

Further evidence of cultural stability in non-human primates comes from Whiten et al. (1999) who reported stable differences in the behavior of seven different groups of chimps in long-term studies representing 151 years of observations. They coded the presence and intergenerational stability of 39 different behaviors, including tool use, grooming, and courtship. They found large, stable differences in the behavioral profiles of different groups of chimps which they described as "cultural variations." These observations seem similar to Skinner's (1953) and Glenn's (1988) definition of "culture" in that there were behavioral consistencies across members of the group and across generations and behavioral differences across groups of chimps suggesting that there were different social contingencies within each troop maintaining different profiles of behavior.

Male chimpanzees tend to stay in their natal areas and are often related to one another and cooperate in feeding and hunting for food. Male chimps appear to patrol territories in socially organized groups. They do so relatively silently, mostly avoiding members of other troops. They appear to listen for and to the calls of other troops.

Wilson et al. (2001) conducted "call back experiments" in which speakers placed in the wild played either nothing or calls from one or many other chimpanzees. Wilson et al. observed that when there were more than three patrolling males, all patrolling chimps vocalized loudly and approached the speaker irrespective of their social rank. When there were fewer than three patrolling males, they did not respond aggressively to calls and they approached the speaker more slowly than chimps in larger groups. If a patrol did encounter members of another troop, they vocalized together and may attack and kill males of other social groups. They did so in an apparently cooperative fashion in which some members hold down the victim while other members kill the victim (Goodall, 1979; Whiten et al., 1999).

What environmental circumstances make violence in chimps more likely? Pandit et al. (2016) noted a number of environmental influences that affect the likelihood of such chimpanzee aggression. These included: (a) abundant fruit; (b) the "imbalance of power," such that when the ratio of aggressors to intruders is greater than 3:1 that aggression is more likely; and (c) the safety of aggressors, namely aggressors rarely incurred serious injuries. Pandit et al. offered a functional explanation for increases in aggressive behavior when they wrote, "raiding by male chimpanzees can be understood as a consequence of their living in a male-bonded fission-fusion social system based on a mating system of joint resource defense polygyny. The fission-fusion element creates potentially large asymmetries in party sizes of neighboring communities. Raiding is perhaps facilitated by social hunting because it reduces the risk to the attackers" (Pandit et al., 2016).

Examples of cultural evolution and extinction, albeit on a small scale, from at least two examples of chimpanzee community extinctions (Goodall, 2010; Nishida et al., 1985) in which one troop took over the territory of a neighboring troop and killed many members of that troop. Some authors use the term "chimp wars" or "Gombe wars" to refer to such events. This work, however, is tainted by unbound anthropomorphism. The authors used terms such as "war," "genocide," "murder," "alliances," "patrols," "assault," "gang," and "The Gombe chimpanzee war." These authors seem to be able to divine the intent of chimpanzees when they describe the purpose of their behavior as "deliberately bit and attacks," "comfort"! and "defended him." These authors also have no problems reading the minds of these chimpanzees when they write that they are "interested .. Concerned .. [show] Empathy," "calculated," "ambitious", and even appear to be "cultivating awareness that the community had lost its moral compass."

6.2.1.2 Bonobos

The other close living relative of humanity is the bonobo monkey, which seems quite different behaviorally from Chimpanzees in many ways. Bonobos also live in troupes but although aggression does occur sometimes, disputes within the group seem to be settled through non-aggressive encounters. Male bonobos do not socialize with one another as much as male chimpanzees and do not support each other during aggressive episodes, but rather bond more with their mothers and other females. Bonobos do not maintain strict territories and hence do not engage in male patrols. Male to female aggression seems much less frequent than in chimpanzees and infanticide appears to be nearly absent.

Female bonobos form strong bonds with one another and engage in a lot of face to face social interaction including rubbing of vulvas. They cooperate in food defense and are tolerant of both stranger females and humans. Unlike chimpanzees, female bonobos eat the majority of meat. Females may also occupy high status positions in social hierarchies within a troup. Behaviorally, bonobos appear immature.

They have been described as "self-domesticating," that is they appear to maintain juvenile forms of behavior for much longer than chimpanzees. They also use tools less than chimpanzees (Gruber & Clay, 2016).

6.2.1.3 Comment

Chimpanzees and bonobos share much in common behaviorally. They live in large, mixed-sex groups as part of a fission-fusion society. Females move between groups. They travel together on a daily basis in subgroups with differing compositions. They have offspring who are dependent upon parental care for up to five years and have relatively long inter-birth intervals (Gruber & Clay, 2016). These differences between chimpanzees and bonobos are perhaps caricatures that hide large variation within each species and significant differences within species across geographical areas. Additionally there is much more research on chimpanzees than bonobos and so the picture remains incomplete.

What is notable is that the two living relatives of humans are quite different socially including in terms of aggression and gender differences in behavior. Additionally, there are large differences between subspecies and chimpanzees and populations in different geographical areas rendering firm conclusions about inter-species differences difficult. Attempts to extend knowledge of aggression from *Pan* species to humans are challenged by these facts.

6.2.1.4 Application to Humans

Aggression among our closest primate relatives provides different models of aggression and prosocial behavior. Can any of it be applied to humans? Pandit et al. (2016) noted some apparent similarities and possibly important differences between human and non-human primate aggression. A similarity Pandit et al. noted was that human hunter gatherers, like chimpanzees, also engage in "simple wars" consisting of small raiding parties on nearby neighbors. These authors argued that non-human primate and human hunter gatherer social organization and raids appear similar: Males hunt together in small groups and women forage and raise children together in a defended physical territory. Although human hunter gatherers patrol territories, a potentially important difference between human and non-human primates is that in human "simple wars" the men kill at a distance using slings, bows, and arrows with reduced risk of injury and death. The authors also argue that human "simple wars" are different from human "total wars" in three ways. Human "total wars": (a) involve the entire community rather than smaller groups of males; (b) are not opportunistic and do not wait for the best opportunity to kill; and (c) involve the killing of both dominant and non-dominant members of the other group, whereas "simple wars" tend to kill weaker member of the other group.

Manson et al. (1991) posed three questions related to the question of potential parallels between non-human primate and human socially organized aggression: (1) What conditions facilitate the evolution of male-on-male inter-group aggression? (2) Why is it that males are mostly involved? and (3) Do males compete over territory or females? In a comparison across various non-human primate species, they observed that females are most likely to participate in aggression when they raise their offspring within their birth groups. This is apparently unrelated to sexual dimorphism within the species, that is, it occurs in both sexually bimorphic and sexually non-bimorphic species. They suggested that a potentially important variable in determining aggression and violence is whether or not a resource is alienabile, i.e., cannot be moved or transferred from one group to another. Hence, fruit trees are alienable because they can be transferred from one territory to another. Likewise, females may be an alienable resource if they remain in the area or group in which they were raised. If females may transfer between groups, they are an inalienable resource, for example, if they sometimes move to other nesting sites by transferring to other troops. Mason and Wrangham found that in human hunter gatherer human societies, aggression was related to competition over females if resources were alienable, but aggression was more likely to be related to territory if resources were inalienable. Hence, Manson et al. concluded that: (a) attacks are lethal when the cost of aggression to aggressors is minimal; (b) males are more aggressive to other males when females remain in their birth groups and are unlikely to move; and (c) males are aggressive in order to compete over females rather than resources such as food.

6.2.2 *Limitations in Applications to Humans: Contemporary Human Wars*

These ideas may be applicable to aggression and violence in some human situations, such as early forms of human species and human hunter gatherer societies. Beginning some 12,000 years ago, the hunter gatherer organization of human society was progressively replaced by settled and progressively more organized and differentiated agricultural societies. Thus, today human hunter gatherer societies are almost completely extinct with a few surviving examples in Africa and the Amazonas. Thus, almost all humans are no longer foragers.

Today, human wars take place in highly organized, complex societies with settled urban communities, a great deal of labor specialization and hierarchical control of resources. Most contemporary human wars involve ideology and complex sets of beliefs shared by different groups within the society. So, perhaps such non-human primate models of cultural aggression largely no longer apply to human group aggression but only to early cultural evolution of human aggression rather than "complex" wars in hierarchical societies which are unknown in non-human societies.

6.3 Chapter Summary

Cultural evolution is another form of behavior change involving variations and selection. Skinner's other researchers' definitions of culture and description of how cultures may evolve can be applied to non-human primates including various aspects of social behavior including aggression and violence. Descriptions of chimpanzee and bonobo cultures illustrate the relationship between social behavior and the physical environment, but also demonstrate that species that are very closely related may be quite different from one another behaviorally. Attempts to extend understanding of primate aggression and violence to humans are limited and perhaps only applied very broadly to some aspects of human aggression and violence in non-urban societies. Human "total war" appears quite different because of the different nature of urban, human cultures.

References

Diamond, J. (1997). *Guns, germs and steel: The fates of human societies*. Vintage.

Diamond, J. (2005). *Collapse: How societies choose to succeed or fail*. Viking Penguin.

Glenn, S. S. (1988). Contingencies and metacontingencies: Toward a synthesis of behavior analysis and cultural materialism. *The Behavior Analyst, 11*(2), 161–179.

Glenn, S. S. (2004). Individual behavior, culture, and social change. *The Behavior Analyst, 27*(2), 133–151.

Goodall, J. (1979). Inter-community interactions in the chimpanzee population of the Gombe National Park. *The great apes*.

Goodall, J. (2010). *Through a window: My thirty years with the chimpanzees of Gombe*. HMH.

Gruber, T., & Clay, Z. (2016). A comparison between bonobos and chimpanzees: A review and update. *Evolutionary Anthropology: Issues, News, and Reviews, 25*(5), 239–252.

Hanley, G. P., Heal, N. A., Tiger, J. H., & Ingvarsson, E. T. (2007). Evaluation of a classwide teaching program for developing preschool life skills. *Journal of Applied Behavior Analysis, 40*(2), 277–300.

Lorenz, K. (1966). *On aggression*. Routledge.

Manson, J. H., Wrangham, R. W., Boone, J. L., Chapais, B., Dunbar, R. I. M., Ember, C. R., … Worthman, C. M. (1991). Intergroup aggression in chimpanzees and humans [comments and replies]. *Current Anthropology, 32*(4), 369–390.

Nishida, T., Hiraiwa-Hasegawa, M., Hasegawa, T., & Takahata, Y. (1985). Group extinction and female transfer in wild chimpanzees in the Mahale National Park, Tanzania. *Zeitschrift für Tierpsychologie, 67*(1–4), 284–301.

Pandit, S. A., Pradhan, G. R., Balashov, H., & Van Schaik, C. P. (2016). The conditions favoring between-community raiding in chimpanzees, bonobos, and human foragers. *Human Nature, 27*(2), 141–159.

Rumph, R., & Ninness, C. (2001). The promise of the design of culture. *Behavior and Social Issues, 11*(1), 25.

Sidman, M. (1989). *Coercion and its fallout*. Authors Cooperative.
Skinner, B. F. (1953). *Science and human behavior*. Norton.
Skinner, B. F. (1961). The design of cultures. *Daedalus, 90*(3), 534–546.
Skinner, B. F. (1965). *Science and human behavior*. Simon and Schuster.
Sturmey, P. (2015). *Reducing restraint and restrictive behavior management practices*. Springer.
Whiten, A., Goodall, J., McGrew, W. C., Nishida, T., Reynolds, V., Sugiyama, Y., … Boesch, C. (1999). Cultures in chimpanzees. *Nature, 399*(6737), 682–685.
Wilson, M. L., Hauser, M. D., & Wrangham, R. W. (2001). Does participation in intergroup conflict depend on numerical assessment, range location, or rank for wild chimpanzees? *Animal Behaviour, 61*(6), 1203–1216.
Wrangham, R. W., McGrew, W. C., & De Waal, F. (Eds.). (1996). *Chimpanzee cultures*. Harvard University Press.

Chapter 7
Social Causes of Violence and Aggression: II. Human Cultural Evolution

There is considerable evidence attesting to the evolution of aggression and violence in human cultures—mostly evidence that human cultures have become much less violent than in the past. This chapter reviews evidence that human cultures differ substantially in their rates of aggression and violence and over time. Some examples are given of naturally occurring examples of cultural evolution of aggression and violence such as historical reductions in homicide, aggression, and violence during evolution of the mafia, US southern culture of honor, the extinction of Greenland's Viking culture, and race riots and pogroms. The chapter concludes with consideration of what mechanisms might underlie these changes.

7.1 Differences in Aggression in Human Cultures

Evidence of cultural differences in aggression and violence comes from three sources. These include: (a) Data on large differences in homicide rates between cultures in both contemporary and historical data; (b) Changes over time within cultures, countries, cities, and neighborhoods, including both reductions and upward and downward spikes in homicide rates; and (c) Large reductions in homicide rates across the world, documented most extensively in European and North American cultures, over the last several hundred years (Eisner, 2003). Changes in the rates of homicide are fairly reliable but not completely so, and is often taken as an index of other changes in less lethal forms of aggression and violence which are even more difficult to measure. For every person who was killed there are probably some living survivors of similar other incidents of aggression and violence. These changes in homicide may have occurred in an unplanned manner. They may also have occurred in response to specific cultural actions, such as forms of cultural design including a change in laws or other planned social practices intended to reduce aggression and violence or promote other prosocial behavior, that may subsequently result in

P. Sturmey, *Violence and Aggression*, Advances in Preventing and Treating Violence and Aggression, https://doi.org/10.1007/978-3-031-04386-4_7

reductions in aggression and violence. Some of these kinds of evidence regarding reduction in homicide, such as differences between cultures and changes over historical periods of time, sometimes have also been analyzed to try and ascertain the mechanisms which might underlie these changes in rates of homicide.

7.1.1 Cultural Differences in Homicide

The WHO survey of global homicide in 2014 (Mikton et al., 2016) reported wide variation in homicide rates between different countries. In fact, homicide rates varied by a factor of around 200 from one country to another. Many countries having murder rates of about 1/100,000 population, yet Colombia and El Salvador had murder rates of 32 and 40/100,000, respectively. Homicide rates were somewhat lower in high than low income countries. Countries with very high murder rates tended to have problems with drug production, drug-related gangs, and weak central government (e.g. Colombia and El Salvador in 2014), alcohol abuse and poverty (e.g. Greenland), political instability and weak central governments (e.g. Nigeria, South Africa) or combinations of these factors (e.g. Venezuela).

Men and boys were victims of homicide in 82% of all homicides at a rate four times higher than girls and women, teenagers, and young adults (aged 15–44 years) were at especially high risk of being killed. For example, the rate of murder of boys and men aged 15–29 years was 18.2/10,000. Thus, men are much more likely to be both the perpetrators and victims of murder.

Mikton et al. (2016) also estimated that there had been a 16% *decline* in the murder rate between 2000 and 2012 and this trend was marked in high income countries. It is remarkable that such a large reduction took place in only 12 years. There were large differences between countries in terms of availability of national databases, policies, and enforcement of those policies. Thus, almost all countries had laws against statutory rape, but only just over half had effective enforcement of these laws whereas only approximately 40% of countries has laws against abuse of elders in institutions and only approximately 20% had effective enforcement of such laws. About one- to two-thirds of countries had various kinds of services in place for various groups of victims.

7.1.2 Historical Trends in Human Homicide

It is tempting to believe we live in violent times. The media serves up violent images, not merely every day, but continuously. The phrase “If it bleeds, it leads” continues to be true. The media often uses lurid violent images of apparently inexplicable individual violence, group violence, such as soccer violence and gang fights, bombs exploding and destroyed buildings and cities, including images of apparently innocent victims, and mass shootings. Media violence sells soap and bloody, violent images are effective clickbait.

7.1.2.1 Broad Historical Trends

Pinker (2011) provided data suggesting that, contrary to romanticized images of an idyllic, peaceful, pastoral past, that murder rates in hunter gatherer communities in humanity's distant past were much higher than in contemporary urban societies. For example, among the Jivaro, an indigenous people in Peru and Ecuador, Pinker estimated that male deaths by warfare were near 60% of all men and in other nonurban communities that murder rates were approximately 5–35% for males. In contrast, murder rates in contemporary Europe are very much less than 1%.

Pinker (2011) suggested that in the historical past that murder was legitimized and even encouraged. Evidence for this comes from many historical laws used the death penalty for minor crimes. Pinker noted that the biblical Old Testament required death as punishment for crimes such as picking up sticks on the Sabbath, and talking back to one's parents. The biblical Old testament described how God destroyed much of humanity in the flood and rained down fire and brimstone on Sodom and Gomorrah to utterly destroy these cities. The Bible's Old Testament is also full of exhortations for mass murder both for non-Jews and for sinners.

Many medieval European states also used a wide variety of methods of torture and other forms of violence to control and punish much behavior. This legitimized violence through institutions such as state-sponsored violence, religious persecutions, pogroms, and slavery. Such practices are now less common in many countries.

Eisner (2003) provided some very interesting data on the rate of homicide from historical records in England, other European countries, and colonial USA, indicating a massive 100- to 10-fold reduction in the homicide rate. For example, Gurr (1981) plotted an "S"-shaped curve to data on murder rates from parts of England from approximately 1200 to 2000 and found very large reductions in homicide. Subsequently, Eisner (2001, 2013, 2014) expanded this database considerably by adding data sets from more European cities and 18 countries to develop the *History of Homicide* database, leading to the conclusion that there has been a *99% drop* in male-on-male homicide during the period. Eisner estimated that around the year 1500 homicide rates in England were approximately 10–100/100,000 per annum population and fell continuously between 1500 and 2000 such that contemporary murder rates in England are generally now below 1/100,000. Eisner's analysis of murder rates in other European countries indicated similar reductions from approximately 30–40 to around 2 per 100,000 per annum during this time period.

This database also includes codes for perpetrator and victim characteristics. The database confirmed and refined Gurr's (1981) original observations. Very large reductions in homicide have occurred across all European countries during this timespan which are unlikely to be explained by artifacts or biases in the data. The findings in the database also refined Gurr's observations in two ways. First, there were large regional differences in when homicide rates fell. Murder rates first fell in England and The Netherlands well before 1800, with murder rates in England being below the rest of Europe even in medieval times. Murder rates in Scandinavia began to fall around the middle of the seventeenth century, but the decline in Italy did not begin until the eighteenth century but did not begin in Corsica and Sardinia until the mid-nineteenth century. Second, the characteristics of crimes changed considerably.

Early homicides were committed very often by males, many aged between 16 and 30 years, in public, with weapons such as sharp or blunt weapons, such as staffs. Today, although there are still many more homicides perpetrated by males than females, the ratio of males to females is much lower than in the past. Today, murders are more likely to be committed hands on rather than with weapons, and, more likely to be committed in nonpublic rather than public spaces. These observations suggest that murder had gone from a male-dominated, public act with blunt weapons to a more gender-equitable, private act at home between men and women. (These large reductions in the ratio of men to women do not obviate the fact that today the majority of intimate partner violence is perpetrated by men against women in domestic situations).

Pinker (2011) also presented data that today warfare is less common, shorter, and less lethal than in the past. Previous European wars have lasted decades and even centuries. In contrast, since World War II, for example, Europe has enjoyed an unprecedented 75 years of more or less peace. There have been relatively brief and localized wars in the Balkans, which were much less lethal than previous large scale, and local internal conflicts, such as The Troubles in Northern Ireland and Russian invasion of the Ukraine, and violent separatist or protest movements.

7.1.2.2 Recent Homicide Data

It is easy to dismiss broad historical trends as not reflecting what is happening today. When listening or looking at contemporary media, it is easy to see examples of recent events that might reflect upswings in homicide that seem to go against the conclusion that the world is a safer place. (Very recent increases in violence coinciding with COVID is one such example and upswings in violence during increases in temperature are also commonly observed. Another significant and worrying trend has been the upswing in mass shootings in the USA). Indeed, it would be unsurprising if there were not recent upswings in homicide somewhere in the world, but it is worth remembering that even the recent data from Mikton et al. (2016) reported recent reductions in homicide worldwide. Thus, such reductions in homicide also appear to be present in relatively recent historical data. For example, FBI crime statistic from 1992 through 2014 indicate reduction in murder rates from around 25 to just under 10/100,000 population per annum. Langton and Truman (2014) reported a similar large reduction in nonfatal violent crime in the USA between 1993 and 2014, in which the rates of nonfatal crime fell from approximately 80 to 20/1,000 persons. Although there are numerous examples in upswings in murder rates, for example, during the permissive age of the 1950s and 1960s in the USA, outbreaks of violence in specific American cities at certain times, and increases in intimate partner violence during the COVID pandemic, clearly, the recent reduction in homicide and violent crime is something to be explained.

7.1.2.3 How We Got Civilized

One explanation of these cultural trends comes from the work of Elias (1978) in his book, *The Civilizing Process*, a historical-sociological account of the development of civilization including the historical changes in behavior. Thus, rather than provide a history of events, *The Civilizing Process* provides a history of cultural evolution, behavior, manners and control of affect, and the processes that might account for these changes.

Elias argued that in medieval times people did not control emotions, including aggression, in the way people do today. It was not only that people were more aggressive, but that this was part of a general lack of inhibition of all emotions: In the past people were more joyous, more pious, and more murderous. Not only did people kill more, they also enjoyed killing and enjoyed public violence more: It was insufficient to kill your enemy—you had to put out their eyes and cut off breasts for the joy of it. Violence—hunting, injuring, torturing, and killing animals *and other people*—was something to be enjoyed and relished. Hangings, floggings, and burning of people and animals were spectacles to enjoy and entertain the public. People not only killed: First they mutilated. They cut off noses, hands, and breasts, not only to render their prisoners useless for war before being returned to the enemy, not only to avoid the expenses of feeding and maintaining prisoners, but also because people enjoyed mutilating other people as part of the joy of war.

Elias (1978) characterized early Europe as being a series of independent, small communities with less need for cooperation with other communities and each other. People engaged in disinhibited emotional behavior, which today we consider childish and uncivilized, but at that time people were even admired for doing so. It was unnecessary to cooperate with strangers very much and disputes with family and neighbors could legitimately be settled directly and if necessary with violence. Without powerful monarchs and other effective forms of national and supranational governments to control violence, chronic, local wars, and dynastic and family disputes could endure for decades and generations. Minor disputes and insults could be resolved by killing with little state control of murder. Think of *Romeo and Juliet* not as a beautiful, if tragic, love story, but as an intergenerational, murderous dispute in which the young members of the Montague and Capulet clans are sacrificed to appease the intergenerational hatred of their ancestors and senior family members: "*From ancient grudge break to new mutiny/Where civil blood makes civil hands unclean.*"

Today, state-sponsored violence is more impersonal. We do not know and have not met our enemies that we are about to kill. They are called "collateral" and "targets" rather than people. We kill only geometric images of people on a screen who are thousands of miles away with the press of a button, rather than face to face and with hand to hand combat with a known neighbor or family member. Today, the state must coerce unwilling warriors to risk their lives to kill strangers through money, services, propaganda, and high status and if that does not work, punishment if the warriors do not kill the states' enemies. In the past this was unnecessary. You knew and saw the enemy. You hated them. You killed your enemy up close and personally and you enjoyed doing so.

The lack of centralized power in early Europe permitted violence to take place, however, according to Elias (1978), as small city states progressively came under control of monarchs' power and control of violence became progressively centralized and individuals' behavior became more controlled. Elias argued that the royal court became the first place to control affect. Courtiers who showed their emotions and showed their hand lost the political game and lost favor with the Prince because they revealed their actions' true motivations. Controlling emotions—making them less intense and making emotional outbursts less frequent—meant you could win favor, power, and influence and achieve your own goals. Thus, violent knights and warriors became artfully inhibited and self-controlled politicians and bureaucrats; they learned to observe and read other people's behavior and emotions carefully; they learned not to rise to the bait; so, as effective members of the royal court, they became *courteous*—both polite but also devious.

Elias (1978) argued that over time the centralized state not only monopolized violence but also progressively controlled both violent behavior specifically but also expression of affect more generally. The joy of killing and torturing your enemy at close quarters was replaced by watching aggression and violence at a distance as entertainment and sports. The public could watch tournaments instead of battles, and sports such as boxing and animal baiting, which the state controlled in limited public arenas. Most people no longer engaged in dirty fist fighting with blows below the belt and gouging of eyes. Rather, they could watch a ritualized and partly sanitized version of a fight between people you did not hate as much as in the past, and from a distance separated from the combatants by ropes and a raised arena. Public affect was also controlled so that childish and impulsive spitting in disgust, coughing, and other noisome and public display of smelly bodily functions was discouraged. Touching, smelling, and directly staring at others were discouraged as uncivilized and crude, bad manners. This discourteous, uncivilized behavior was replaced by modest, genteel looking, glancing, and spectating—the "pleasures of the eye"—rather than sticking one's fingers and nose into things and people to touch and sniff them. In the past the behavior of the King and Queen of France who participated in the public burning of sacks of living cats and sometimes foxes for the general entertainment was considered a public joy and entertainment. Today, however, it is seen as a sign of psychopathy worthy of aversive social control, through incarceration in a mental health facility and required psychological treatment. An example of this trend has been the historically recent elimination of public state homicide: Public executions in England ceased in 1868 (McGowen, 1994), ceased in the US in the 1930s (Pitzulo, 2017), and the last use of public guillotine in France was 1939 (Taïeb, 2020).

A final driver to reduce violence, according to Elias (1978), was progressive specialization of labor in the nation state. Centralized control of violence not only allowed safer commerce between strangers in public spaces, but societies that could control public affect benefited economically from the subsequent growth of trade.

These civilizing societies also allowed labor to become more and more specialized creating even greater interdependence between strangers including very distant strangers. Progressively longer chains of interdependent people developed who were united by economic tasks and their final products. Disruption of these interdependent chains of people might mean harm to many people in the chain and the economies of many places. Cultures that could solve these problems most effectively and quickly would prosper more and eventually displace other cultures that did not do so or did so less effectively.

Eisner's (2014) essay, *From Swords to Words*, influenced by Elias' (1978) *The Civilizing Process,* offered the following similar explanatory account. To obtain the long-term benefits of trade and exchange of knowledge, societies must enforce self-control of impulsive violence with its immediate, short-term benefits, such as stealing other people's goods and money, and replace impulsive violence with behavior with long-term benefits, such as international trade and increased wealth. Societies that endure can do so by disciplining, rationalizing, and organizing participants' behavior. They promote self-discipline through child-rearing practices that teach children to inhibit their affect, including aggressive and violence behavior, through religious instruction, and education emphasizing virtues and patience for long-term beneficial outcomes, or through other mechanisms that are socially mediated, such as judicial and institutional discipline. Dutch sociologists have hypothesized that, at various times, that societies have initiated control of civilizing offenses through mechanisms such as religious reforms to control expression of emotions, libidinous behavior, dancing, fairs, and alcohol consumption. This was done in part to reduce violence, but also to increase civilized behavior gradually for the broader, delayed benefits of society and its members.

Eisner (2014) also emphasizes the importance of literacy and education in reducing violence. Historically, both book production and increased literacy are correlated with reductions in violence and murder. Eisner hypothesized that reading promoted self-control both directly and indirectly. It involves prolonged periods of learning, relative immobility, speaking quietly or not speaking at all, a controlled posture, and control of hands, fingers, head, and eyes. The content of many books too promoted personal virtues. Biographies and the lives of the saints provided models for long-sighted behavior, such as patient studying, acquiring skills, planning one's life, forbearance, and suffering, etc. for a better future, perhaps not even on this Earth. Reading and learning to read themselves often take place in the most common vehicle of self-control: Schools and families. Schools and families promote these practices and, by doing so, promote self-control and its long-term benefits. If the entire family or school work together effectively and efficiently, everyone will have a quiet, productive, and orderly, shared life with long-term benefits for all present and future members of the group. Families and schools that do not do this benefit less and their cultures are more maybe more likely to disappear over time.

7.1.2.4 Conclusions

What data we have on broad and historic trends on homicide converge on a clear conclusion: Most of us live in times that are much less violent than early humanity, medieval times, and much less violent than even a few decades ago. Of course, this will not be true if you live in an area with a recent uptick in drug-related violence, if you live in The Philippines where the state has publically embarked on a campaign of extra-judicial murdering of drug-related crime since 2016 or if you live in Venezuela where in 2018 money is nearly worthless and failure to fight for your food might result in you and your family dying. Yet, the overall historical trend is clear: Despite local upsurges in violence and dramatic bloody incidents, we are living in times that are much more peaceful than in the historical past.

Reviewing Eisner's and related work is interesting because of the parallels with other functional explanations of behavior, especially behavior analysis's account of cultural evolution (Skinner, 1965). Eisner's work emphasized social control of behavior and the gradual elimination of the most heinous forms of punishment—arbitrary violent and disgusting punishment from other people in public places, state use of torture and humiliation, etc.—with progressively less heinous and offensive forms of social control of behavior. Furthermore, these historians make great play of the acquisition of self-control to prevent violence, learning to control expression of emotions, and engage in alternate peaceful ways of behavior, such as politeness, respectful conflict resolution, and avoidance and control of alcohol and weapons, etc. These authors mention societal control which encourages members to engage in self-control, but do not mention the specific learning strategies that result in acquisition of self-control repertoires in individuals. Behavior analysis provides an account of how this works at the level of the individual and culture (Skinner, 1965). (See Chap. 6, *Social Causes of Violence and Aggression: I. Cultural Evolution,* for details).

7.1.3 *US Southern Culture*

Other evidence for cultural evolution of aggression and violence—this time cultural evolution of higher rates of aggression and violence than elsewhere—comes from work on the so-called US Southern culture of honor and violence. Rates of aggression and violence, including murder are higher in the southern than northern parts of the USA. For example, the CDC (2020) reported that in 2018 the following states had the highest rates of murder: Mississippi, 13.4/100,000), Lousiana (13.3/100,00), Alabama (12.2/100,000), Missouri, 11.4/100,000), and New Mexico (10.8/100,000). In contract the following states had the lowest rates of murder in 2018: Main (0), Rhose Island (0) Vermont (0), New Hampshire (1.8/100,000), and Nebraska (1.9/100,00). As well as higher rates of murder than the US South, some have observed higher use of violent TV, more hunting licenses per person, and more executions and lynchings in the South than in the North. In addition, there are more legal defenses to injure or kill another person in terms of justifications to do so because one's honor was offended (Copes & Vieraitis, 2009). There is also some evidence from experimental social psychology to confirm these impressions. In an experiment an experimental stooge bumped into the participant and called them "an asshole." Individual men from the US South were more likely to react to the insult, believing that their masculinity was impugned, were more upset and react with higher cortisol levels, show a larger rise in testosterone, and show more aggressive cognitions and behavior than US Northerners (Cohen et al., 1996; Cohen & Nisbett, 1994).

What explanations are there for these cultural differences in aggression and violence? Some have attempted to explain these regional differences in murder rate in terms of higher temperature in the South than the North of the US (Anderson 1989, 2001), whereas others have suggested that cultural differences are also at play, leading to evoke the term the *Southern Culture of Violence*. Nisbett and Cohen (1999) suggested that when the US South was settled, it was settled by British herdsmen from the fringes of Britain where state control of behavior was weak and who were used to defending their cattle from marauders who might steal them. Failure to protect their cattle and to train one's children and peers to do so was a cultural practice likely to lead to cultural extinction. Upon arrival to the US south, population density was low until the mid-nineteenth century and rule of law was hard to find. Again, failure to defend one's cattle and land effectively and aggressively would result in loss of cattle and cultural extinction. Being able to project clearly to others than one would indeed aggressively defend one's property, family, and honor would pay off in many ways. Responding vigorously and aggressively if one's honor was questioned was important and southerners were more likely to be empathic toward the killer than the victim because the killer was justified and admired for defending his honor.

Nisbett and Cohen (1999) suggested that there were several cultural mechanisms for maintaining high levels of aggression and violence. These included: (a) parents who trained their children to fight early on and that violent entertainment, such as no holes barred fights; (b) the adoption of more punitive forms of Protestant Christianity with emphasis on sin, punishment, and retribution in this life not the next; (c) slavery, an institution which justified violence toward other people, again with strong instrumental overtones, such as protecting what was then one's property; and (d) a legal system with justifications to kill when standing one's ground, defending one's property, and defending one's honor, if a man found his wife with another man in bed. These cultural practices were important vehicles for transmitting and maintaining a culture of honor across generations. This type of aggression was a type of instrumental aggression in keeping one's possessions and maintaining one's social status.

More recently, several authors have been critical of these explanations of cultural differences in aggression and violence. Anderson (1989, 2001), for example, pointed out this hypothesis is at least incomplete as: (a) it does not explain changes in aggression and violence related to time, temperature, and other factors; (b) some data have shown that, compared to temperature and poverty, southerness has negligible, independent effect on violent crime; and (c) "the simple heat hypothesis however, accounts for all of these effects" (p. 70). Other more recent researchers have also failed to provide empirical support for the culture of southern violence hypothesis and defensive use of guns in particular (Copes & Vieraitis, 2009), perhaps indicating changes in cultural values or changes in the population of the south in recent years, although the data cited at the beginning of this section suggests this is actually not true.

7.1.4 The Cosa Nostra

When thinking about the mafia, most people think about media representations such as Al Capone, *The Godfather* movies, and *The Sopranos* on television. As Dickie (2005) and other scholars have shown, however, there is much more to mafias than public massacres and the olive oil business. The mafias have a two hundred year history and have evolved and adapted over time, sometimes being very violent and more recently much less so, choosing largely to disappear below the surface of society. The history of the mafia is quite well documented in both the academic and yellow press and hence is an instructive case study as to how aggression and violence are selected and maintained and how they wax and wane depending upon the current social context.

There are four large mafias in Italy (Dickie, 2005). The Cosa Nostra ("our thing") is Sicilian which linked to but independent from the American mafia; the Camorra is the mafia based in Naples (Saviano, 2008); the Ndrangheta is a family-based mafia centered in Calabria (Sergi & Lavorgna, 2016; Serenata, 2014); and the more recently founded Sacra Corona Unita in Puglia (Serenata, 2014; Sergi & Lavorgna, 2016). This section will only deal with the evolution of the Cosa Nostra.

Dickie (2005) charts the beginning of the Cosa Nostra back to around 1860–1876. At that time Sicily was part of the Kingdom of the Two Sicilies with its capital across the sea in distant Naples, until it was incorporated into the modern Italian state in 1861 when the capital moved even further away to Rome. At that time, rural parts of Sicily was still quasi-Feudal with many impoverished, agricultural workers effectively indentured to local land owners on latifundia (very large landed estates run with effective slave labor). Latifundia were not completely abolished until the 1950s. Justice and policing were weak or absent for many. You may also recall that Sicily was one of the last parts of Europe to experience reductions in homicide, reflecting, perhaps, weaker government, and more ineffective formal systems of justice than other parts of Europe.

In the early nineteenth century, Sicily produced and exported large quantities of citrus, including lemons to the British navy and the USA, oranges, and bergamots for Earl Grey tea; nineteenth century Palermo high society had a significant British presence. Production and banking were mainly based around Palermo's Conca d'Oro. Citrus production, however, is risky as it involves several years' investment of money and labor without return during which the crop remains vulnerable to lack of water supply and vandalism to the lemon trees. Sicily was also a major producer of sulfur exported to the new industries in the United Kingdom and France. In the nineteenth century, the demand for these products increased greatly, making their production and associated transportation something worth defending against bands of brigands who operated in rugged territory where there was no effective state protection for landowners.

In the 1860s early individual local Cosa Nostra "cosce" (cells or clans) formed. The local cosce used threats, theft, violence, murder, and corruption of local police forces, professionals, and the Catholic churches to provide protection to local citrus and sulfur industries, obtain justice and power, get their friends hired, take bribes, and effectively tax local citrus production for their own benefit. Such organizations evolved as financially profitable rackets corrupting citrus production that got their members, friends, and family members jobs when the early Italian state was untrustworthy and ineffective in Sicily.

Buonanno et al. (2012) have provided an interesting empirical study of the growth of the mafia in Sicily. Using historical demographic, economic and geographical data, they reported that the mafia was concentrated initially in Western Sicily around Palermo. Using multiple regression to predict the degree of mafia activity, they found that geographical variables related to citrus production, such as land fertility and the presence of river water, the presence of sulfur mines, and territorial ruggedness predicted the degree of mafia activity. In western Sicily the presence of sulfur mining and population density correlated negatively with mafia activity, suggesting the importance of protecting transportation to Sicily's main port. In eastern Sicily, the presence of river water and farming predicted the extent of mafia activity, suggesting the importance of protecting agriculture as a driver of mafia activity. This is a specific example of the more general view that the geographical environment can promote certain kinds of human behavior, in this case by weakening governance and facilitating illegal behavior (Cf. Diamond, 1997).

Buonanno et al. (2012) framed these findings in a functionalist view of the cultural evolution of violence. That is, the protection of property and goods through violence is a valuable commodity or service that citizens purchase—a kind of insurance policy. It is a commodity that is most needed when there is something valuable to protect and physical geographical factors, such as distance to ports and dangerous terrain, and social factors, such as the presence of gangs of brigands stealing the goods, make it difficult to protect one's valuable property and goods by other means. Absent legitimate, effective state violence funded through taxes, citizens who cannot defend their own property purchase this commodity of violence through other social agencies such as the mafia. This perspective sees the nineteenth century mafia primarily as an organization providing protection of valued goods, even collaborating with brigands to stimulate demand for this product.

7.1.4.1 Selecting and Training Murderers

The early Cosa Nostra used several mechanisms to ensure loyalty from members and that their members were violent as needed. First, they selected potential members carefully who were both self-controlled, discrete, and had weak family ties. They then required loyalty to the Cosa Nostra through initiation rituals similar to Masonic rituals. New members took loyalty oaths and *omerta* (oaths of silence) and gained substantial social, financial, family, and other personal benefits of membership. Loyalty and murder were sometimes enforced by requiring members to kill first before joining the organization on some occasions and murdering disloyal members and their family members; merely observing these practices probably encouraged loyalty and murdering others when required to.

The Cosa Nostra of the 1860s was not a corporation or state, but rather a group of many independent coscas, each headed by their own capo ("boss"). The way in which individual cosca became corporatized into one organization is unclear, but during the late nineteenth century the Cosa Nostra in Palermo, the most prosperous city in Sicily, may have begun to incorporate individual coscas through violent conflict and became a more centralized organization. Subsequently, the Cosa Nostra expanded into other Sicilian cities and diversified into other industries. The cosa nostra then expanded its political corruption and influence nationally into Rome in part by delivering votes to politicians at a time when only 2% of the male population of Italy could vote and hence could easily be controlled. By 1900 the Cosa Nostra, riding the massive wave of Sicilian immigration to the US, expanded across the Atlantic to the port cities of New York and New Orleans and other new large American cities, until an American mafia developed which had close ties to the Sicilian Cosa Nostra, but was distinct from it organizationally. A twentieth century gift to the Cosa Nostra was American prohibition, which provided massive untaxed income for a highly popular product. After effective but temporary suppression of the Cosa Nostra by Mussolini in the 1930s, the allies' conquest of Sicily at the end of World War II resulted in naive American political and economic cooperation with the Cosa Nostra. Finally, the Cosa Nostra expanded into illegal drugs in the 1950s and later into the European Community, former USSR republics, and the Balkans.

During each of these expansions, violence was useful in eliminating economic rivals and intimidating producers of goods to cooperate. The functional significance of such expansion is pedestrian: It resulted in enormous economic and political benefits to the cosa nostra and its economic associates and political friends. Failure to cooperate might lose your business, ensure you are not re-elected, or get you, your friends and family killed.

The benefits of violence changed during the two mafia wars of 1962–1969 and 1983–1992. The first mafia war began as a series of retaliatory killings for a drug deal gone wrong, which included murders and gun fights in the streets of Milan but ended up with car bombings and the subsequent arrest of 2,000 mafiosi. The effects were to bring the Cosa Nostra to unwanted national and international attention and subsequently resulted in a decade or so of relative peace. In April 1981 violence increased dramatically as a series of revenge killings and fights between different mafia factions in Sicily resulted in over 200 murders; some days 12 people were murdered in different incidents. This period is sometimes referred to as the *mattanza* ("slaughter"), drawing the analogy with the rounding up and killing of entire schools of tuna fish in Sicily in an ocean of blood. There were mafia bombings of public figures in Rome, Milan, and Florence. At the same time the left wing radical Red Brigades were also bombing, kidnapping, and murdering across Italy; it must have appeared that Italian society had fallen apart.

After hundreds of mafiosi were arrested, 474 were placed on trial—the so-called maxi-trial—and, apparently illustrating the availability of due legal process, 144 were acquitted. The change in the effectiveness of the Italian judiciary reflected the retirement of old school judges and appointment of new judges from an improving Italian education system, and tremendous individual bravery and hard work on the part of many individuals. Subsequently, the Cosa Nostra murdered the trial judge, Giovanni Falcone, his wife, and three members of his police escort by bombing their car. There was evidence that the elected Italian Prime Minister Guilio Andreaotti probably had ties to the mafia throughout his career. He went to court, dissimulated ignorance of the mafia, and was found not guilty. The Falcone murder and trial of the prime minister Andreaotti produced a large backlash in public opinion, including antimafia protests by citizens. There were no winners; the Cosa Nostra probably murdered Andreotti's best friend, perhaps when Andreotti did not deliver on certain promises.

The publicity and public protest was not good for the Cosa Nostra business and absorbed enormous resources. Subsequently, there was an about face. The Cosa Nostra embraced a kind of peace in which the worst public violence was no longer practiced so frequently, but the Cosa Nostra instead decided to "sink down," mostly out of public view, involving itself in corruption in construction, garbage disposal, and recycling. Some sources estimate that the combined mafias account for approximately nearly a tenth of Italian gross domestic product in 2007 (Barone & Narciso, 2015) and suppressed productivity by a fifth in Southern Italy (Pinotti, 2011). Additionally, there is evidence that as the mafia has expanded to other regions of Italy, murder increases that its presence suppresses economic productivity in part by protecting favored entities from competition, discouraging private investment, and encouraging public investment which can be skimmed and corrupted more readily (Pinotti, 2013).

The Cosa Nostra is less visible than in the past but it has not vanished; numerous relatively minor crimes—such as threats, fires, and vandalism reminiscent of the property destruction of nineteenth century lemon groves—still indicate that violence may be available as needed. Dickey's book only provides a history of the cosa nostra up to 2006 and the practice of mafia violence continues to evolve. In 2017, 25 years after the Falcone bombing, Giuseppe Dainottithe, the current leader of the mafia, was gunned down in the streets of Palermo, which suggests that mafia violence has not disappeared.

This brief history of violence shows how external factors can select either violence or relative peace, even in the cosa nostra. The most obvious function of violence in this case is an acquisitive one—organized crime acquires resources and eliminates legal and illegal rivals for resources. The short-term costs of organized violence are clear, but the long-term benefits are large and enduring.

Dickie's account of the mafia echoes the work of Eisner (2003) and Pinker (2011): The initial nineteenth century evolution of the mafia occurred when there were still large differences in murder rates between north and south Italy. The growth of the mafia reflected the differences that Pinker identified, namely, the lack of state appropriation of violence from other actors and the lack of education, reading, and schools as society's vehicles to induce control and take the long-term view of one's own behavior. The reduction in mafia violence and its apparent submersion from public view is seemingly paradoxical. During the first and second Mafia wars the Italian state finally was able to assert itself, develop a more effective judiciary, arrest and convict some murderers, and do real damage to the mafia by eliminating its workforce and confiscating significant parts of its property and money. In addition, public opposition to the mafia, in terms of protests and boycotting businesses that pay taxes to the cosa nostra, may all have added to the financial and other costs of continuing violence relative to nonviolence. The so-called *addiopizzo* movement is a grassroots movement which refuses to pay mafia extortion money and advertise themselves as such to encourage patronage of their businesses.

7.1.5 Race Riots and Pogroms

Just as group violence from the mafia waxes and wanes over time, many societies experience sudden flare ups of mass violence. These flare ups are often intense but may be brief, take place over a discrete period of time, or may be recurrent over more extended periods of time. Thus, one interesting behavioral phenomenon related to violence and aggression is situations where one group engages in violence against a second group often in explosive and intense outbursts of violent behavior accompanied by theft, looting, property destruction, violence, and murder based on the group identify of the victims. Examples include race riots, such as those in the USA against African Americans and, in fewer cases, Latinos and Irish; race riots by American troops stationed abroad in World War II (Krugler, 2014); and antisemitic race riots and pogroms in the United Kingdom, Australia, Imperial Russia and the USSR, and the middle east.

7.1.5.1 Tulsa Race Riot 1921

Following the Civil War and emancipation, many African Americans were suppressed economically and socially by Jim Crow laws limiting and preventing voting and economic prosperity, although some African American communities did indeed begin to prosper. Later, when African American troops returned from World War I, many African American veterans were dismayed and frustrated that, after laying their lives on the line for the United States, they continued to be subjected to discrimination after serving their country. Indeed some African American soldiers were subject to violence by members of the American public while serving. For example, there were incidents in which some African American men were beaten up, stripped of their uniforms, and run out of town in the USA. Lynchings of African Americans by Whites increased post World War I with at least 61 lynchings of African Americans in 1919. In addition, the Klu Klux Klan and many local newspapers stoked racial hatred across the USA. The year 1919 is called by some as the "red summer," to refer to more than 25 race riots across the USA, not only in the South. For example, in the 1919 Chicago race riot, 38 African American and Whites were murdered and 537 people injured requiring military intervention to quell the violence (Krugler, 2014.)

One well documented example is the race riots in Tulsa, Oklahoma in 1921 (Johnson, 1998, 2014). Tulsa had an economically and culturally thriving African American neighborhood, Greenwood, which was geographically, economically, and culturally segregated from the majority White part of Tulsa. Two factors boosted the economic status of African Americans in Tulsa. First, some African Americans were members of American-Indian tribes who received land following treaties between the American government and Muskogee (Creek) tribes. Second, Tulsa had an oil boom in the 1900s, boosting the economy for everyone. Despite these positive features, Tulsa's population at this time was booming which resulted in social instability. The Klu Klux Klan was active and expanding in Oklahoma, there were lynchings of African Americans, the local yellow press stoked racial hatred, and the police and justice system was weak resulting in widespread lawlessness. These factors stoked White resentments against African Americans and permitted lawlessness and mob justice for years before the riot, including even public lynchings of a White man for alleged murders.

On Monday, May 30 1921, Dick Rowland, a 19-year-old African American, was riding in an elevator. The elevator car lurched and he fell against the elevator operator, a White woman, Sarah Page. She screamed. A clerk ran to her aide and Dick Roland ran away. She accused him of assault and he was arrested the next day. When interviewed by the police, Sarah Page recanted and corroborated Dick Roland's story of innocence. Rumors and the local yellow press fueled the story that there had been an attempted rape. That night a White mob assembled outside the jail, probably planning to lynch Dick Rowland, and a group of armed African American men from Greenwood appeared to protect him. When a White policeman attempted to disarm an African American veteran, shots were fired and chaotic gunfire broke out. The police and mobilized armed guard deputized White members of the mob who looted stores, stealing guns and bullets.

By 6 am a White mob of 15,000 people, including armed boys as young as 10 years old invaded the Greenwood neighborhood. White men and women ransacked businesses and houses stealing valuables, then murdered the occupants and set houses on fire. Witnesses reported that planes from a local airport were used to bomb Greenwood using nitroglycerine and kerosene to start fires and to spray bullets into the streets. The mob burned buildings and murdered and burned people at will without intervention from law enforcement. By noon, state troops arrived, martial law was declared, and the violence largely ceased. Subsequently, hundreds of African American citizens were rounded up, marched through the streets of Tulsa at gunpoint and interred for weeks in jails and the city convention center; more African men were shot, even when unarmed and with their hands raised. Some African Americans were allowed out of being interned, if their White employers vouched for them. The Red Cross was active in providing relief to many African Americans and some White employers protected their African American employees.

By the end of the riot around 300 people were murdered, and 35 city blocks of Greenwood were destroyed. The local press blamed African Americans for the riot. The City of Tulsa then passed laws limiting economic regeneration of Greenwood and the mass grave of many African Americans remained unmarked for decades.

7.1.5.2 Pogroms

Pogroms are defined by persecution, violence, and murder of members of an ethnic and/or religious group by another group. The prototypical and most commonly documented example is antisemitic pogroms, although some have also described other outbreaks of violence against specific and identifiable groups of people as pogroms. For example, the so-called Belfast pogroms in 1920–1922 by Protestants against Catholics took place following the expulsion of Catholics from working in the shipyards in July 1922. Around 450 hundreds of mostly Catholics civilians were murdered, around 1100 injured and around 650 houses and businesses were destroyed (Lynch, 2008.)

Antisemitic pogroms date back to early Roman times. Hundreds have been documented. Many are associated with traumatic, economic, and political events. Examples of these events include economic downturns; the Black Death; the unification of Spain and the subsequent expulsion of Jews from Spain and Portugal in 1492; the incorporation of parts of current Poland and Ukraine into the Russian Empire in the late nineteenth century; the end of World War I in parts of Europe and Argentina; Nazi Germany (infamously, Kristallnacht); and, throughout Europe including the United Kingdom and the Mediterranean basement, during and after World War II. Pogroms involve both violence and murder against a defined and visible group and their property, but also looting, theft and destruction of businesses and homes by other groups of people.

The Farhud is one example of an antisemitic pogrom which took place in Bagdad, Iraq between June 1 and 2, 1941. After World War I and the collapse and breakup of the Ottoman empire, the British government controlled Iraq as a Mandate which became independent in 1932 but with considerable British influence.

In early 1941, there was a pro-Nazi coup which attempted to further minimize British influence in Iraq. Thus, in addition to historical antisemitic violence, there was an increase in antisemitic propaganda and the government ordered Jews to stay at home apparently for their protection prior to the pogrom. A trigger event occurred on June 1 when a Jewish delegate to the local government was attacked by an Arab mob. Later that day, antisemitic violence spread across several neighborhoods of the city with police joining in the violence; some Arabs protected Jewish neighbors. After 2 days of riots, British troops imposed a curfew and shot violators on sight, indiscriminately shooting into crowds of looters. Around 200 Jewish Iraqis were killed, although more people may have been buried in a mass grave. Hundreds more were injured and hundreds of homes and businesses were looted or destroyed. Between 1949 and 1952 there was an Exodus of Iraqi Jews to Israel (Black, 2010.)

7.1.5.3 Race Riots and Pogroms: Similarities and Differences

Although we refer to race riots and pogroms with different words there are some probably important similarities. First, there are groups of people who are readily identifiable through physical appearance; culture-specific behavior, such as language or religious practices; and cultural symbols, such as clothes, names, and buildings. They are often identifiable because of geographical or social separation, such as living in certain neighborhoods, working in certain occupations, or conducting certain activities at specific times and places. Second, there is often a long history of discrimination, resentment, and mistreatment that is a part of the dominant culture prior to the race riot or pogrom. Third, following a recent heating up discrimination, perhaps reflecting cultural changes, such as soldiers coming back from war or economic problems, a trigger event or series of events occur and seems to precipitate the immediate violence. During the heated phase of the violence there is often social contagion: Following the instigation of the violence by a smaller number of people, many others who are not involved directly in the initial violence, join in the violence, and engage in looting, murder, and destruction. This social contagion suggests that modeling from other similar people, obtaining goods and money, and destroying dissimilar people's property and harming members of the other group, are important consequences at the time for many rioters. In addition, often at the time of the outbreak of violence, the social forces that normally suppress such violent behavior such as the police or justice system—which may already be relatively weak—are now ineffective, absent, or indeed join in the rioting and violence. Racism that was previously kept in check boils over when factors keeping it under control are removed. During the heated violence, there are many consequences such as encouragement from others to participate, but an important part of those consequences are also economic: The rioters gain goods, property, and money, destroy economic rivals, often resulting in the removal of the minority group through murder or because the minority group flee danger and leave the area.

Why do people not riot against identifiable groups and kill them at other times? The changes in circumstances that precede or accompany race riots and pogroms may be those that involve the removal of the usual breaks on violent behavior. Weakening of social control of violence, such as transition from one government to another, and such as the end of a war when temporarily there is little no effective social control of violence, may predispose to an outbreak of violent behavior. A second factor might be that sometimes the changes that precede the outbreak of violence make violence a more attractive choice, as when there is an economic downturn and people cannot support themselves and their families through legitimate means.

Why do people not victimize other groups? The victimized group is not chosen arbitrarily. Rather there is often a long history of discrimination, mistreatment, and perhaps historical violence that includes historical outbreaks of largescale violence or individual violent acts. These factors provide models and legitimization for the violence. Economic resentment and jealousy from a deprived, poor majority group against a minority prosperous group also is sometimes a factor. Finally, the minority victim group may be relatively unable to defend themselves against the majority group either because the majority group is simply too numerous, better organized, and better armed.

7.2 Cultural Extinction and Homicide

Violence and aggression can contribute to both cultural survival and extinction in a variety of ways. Aggression and violence can protect current resources and result in acquisition of resources and elimination of rival consumers of those resources. Jared Diamond's (1997) functional approach to cultural evolution and cultural extinction is an interesting example of this approach as shown in his two volumes *Guns, Germs, and Steel* (1997) and *Collapse* (2005). Diamond's account of the extinction of cultures is a general one and does not focus on the role violence and aggression but mentions it frequently.

Violence and aggression may have short- and long-term benefits which may contribute to cultural evolution. Diamond (2005) described that first encounters between different cultures can be scary, frightening, dangerous, and often involve threats and demonstrations of violence and aggression which seem effective in protecting each party's resources. Indeed, Cabeza de Vaca's's (1542/1961) *Adventures in the unknown interior of America* is an account of the first conquistadores' journey across the US is full of mutually terrifying accounts of first encounters of the conquistadores with the indigenous peoples of the Americas. Other examples of functional use of violence comes from some cultures which practice abortion, infanticide, and suicide as measures explicitly recognized by the culture to control population growth so that the population does not exceed what the environment can maintain without cultural extinction (Glenn, 1988). Diamond argued that such practices are effective in controlling population growth in marginal environments and hence are useful in maintaining the culture when other cultural practices might readily lead to cultural extinction. Diamond also listed five features of cultural extinction some of which are related to aggression and violence in several ways. They are: (a) negative human impacts on the environment; (b) adverse climate change; (c) loss of friendly neighbors; (f) arrival of hostile neighbors; and (e) dysfunctional cultural attitudes and practices.

7.2.1 *Viking Extinction in Greenland*

Diamond's (2005) prototypical example of cultural extinction is the extinction of Viking culture in Greenland. Following an increase in population density in the Viking homelands, there was a general eastward expansion of the Viking population to The Shetland, Orkney and Faroe islands, Iceland, Greenland, and a few parts of North America, such as Baffin Island, Newfoundland, and perhaps elsewhere. The Vikings arrived in Greenland around 980 CE but by approximately 1400 or so, all Vikings in Greenland had probably died in spite of 450 years of expansion and relative success in a challenging environment for hundreds of years.

Diamond (2005) considered several factors that contributed to this cultural extinction. He noted that the Vikings unintentionally caused soil erosion and deforestation in a marginal ecosystem by reckless and unsustainable farming and logging practices which got them food and fuel in the short term, but deprived them of both food and charcoal in the long term. In contrast, Vikings in Iceland also faced a marginal environment and problems with soil erosion, but changed their farming practices early on to preserve topsoil. In addition, Greenland was a cold and hostile environment with relatively few patches of useful land and limited transport options, as overland travel between the habitable areas is difficult or impossible, making access to the sea via fiords essential for trade and cultural activities with other Viking communities within and beyond Greenland. The Vikings also had the misfortune to arrive when the climate was a little better than average, and so in the late 1300s when the climate became colder there was loss of food sources and increased blocking of fiords with ice flows, making access to seals as food difficult. An additional geographical factor was the rise in sea level, which may have inundated a significant portion of the already limited area available for agriculture (Mikkelsen et al., 2008).

During the first part of Viking settlements, the Vikings living in Greenland were dependent on their friendly, but distant fellow Scandinavians. They were a regular source of supplies of resources and cultural artifacts unavailable in Greenland, such as books, expensive cloth, iron, wood, and brass church bells. These were exchanged for local valuables from Greenland, such as Walrus ivory and furs. Over time, however, ships from Norway became fewer, more intermittent, and eventually stopped coming altogether. The result was that the Vikings of Greenland had limited access to these valuable supplies of iron. So, they were forced to sail to Baffin Island and Newfoundland to search for timber and iron. Evidence recently emerged confirming that Vikings were in Anse Les Meadows, Newfoundland in 1021 (Kuitems et al., 2021) where they established a camp with buildings, perhaps overwintering there, and maybe used it as a base for further exploration to the south. Due to the failing supplies of these basic materials, eventually the Vikings of Greenland had to recycle existing iron, wood, and other limited resources, eventually using wooden rather than metal nails in buildings and recycled existing wooden panels in new construction.

The Vikings of Greenland seemed to regard their local Inuit neighbors with contempt and referred to them as *skraelings* ("wretches") to be killed. Unfortunately, for the Vikings these wretches were in fact highly skilled and well adapted to the hostile environment of Greenland. The Inuit had many effective cultural practices, which still endure today. They include efficient and technically demanding hunting of seals, birds, and whales in kayaks that Diamond described as highly designed technical marvels skillfully designed to protect the person using the kayak from extreme cold and providing them with an armamentarium to safely kill whales. In addition, through Inuit cultural contingencies and individual learning histories, they learned to use a kayak when young which enabled Inuits to gain hyperextension of their arms, such that they could throw lances and kill whales even more efficiently.

Finally, the Greenland Vikings persisted in dysfunctional cultural practices which were destructive in the long term. These included destruction of the limited supply of trees, turf, and topsoil by deforestation and grazing with sheep which eventually also resulted in loss of the associated food production. Building practices, such as using turf for walls and roofs as insulation, required further destruction of vegetation holding in the topsoil. Cultural inflexibility, such as grazing sheep, raising cattle, and building cathedrals with associated cultural paraphernalia that could not be produced locally, also limited their adaptation to a challenging environment. It seems perverse that the Vikings did this in plain sight of highly effective Inuit cultural practices which they failed to adopt. Thus, they did not learn to use kayaks or Inuit specialized tools for hunting, even in the face of increasingly colder weather, loss of fertile land, and other dwindling resources. This clear example of cultural inflexibility, failure to experiment and ossification of cultural practices that were no longer adaptive resulted in extinction of the Viking culture.

Some three hundred years after the Vikings in Greenland had disappeared, other Europeans arrived in Greenland to find the Inuit doing just fine. In the end, it seems that the Viking community became progressively smaller, due to dwindling food and other supplies, loss of land fertility, ecological destruction, and failure to adapt to a new environment. Diamond concluded that the Inuit probably killed the Vikings, first by blocking access to fjords thereby limiting access to food sources such as seals, and probably eventually killed the remaining Vikings as they starved surrounded by food sources.

7.2.1.1 The Role of Aggression and Violence in Cultural Extinction

Diamond's (2005) descriptions of the cultural extinction of the Vikings in Greenland and elsewhere make frequent references to aggression and violence. Diamond concluded that from the get go, the Vikings were hostile to the Inuit and probably killed most of those they first encountered: Regarding them as "wretches" did not bode well for a future, mutually beneficial productive and peaceful relationship with the new neighbors.

The Greenland Vikings were unable to defend themselves in the long run from neighbors who probably became hostile to them. Diamond speculated that in the end, the Viking culture in Greenland may have looked like a sinking liferaft. He suggested that the remaining Vikings lost all respect for former civil and religious authorities. These now useless authorities could no longer promise them the blessings of an afterlife, since there were no Bishops coming from Norway, and hence no baptisms. Furthermore, these authorities could no longer feed them or keep them safe in the here and now. So, perhaps the end was some combination of in fighting and murder among the Vikings themselves often related to acquisition of remaining and dwindling valuable resources and the Inuit turning up later to finish off the last survivors and grab some useful artefacts.

Diamond's (2005) *Collapse* has many other examples of the prominence of violence during cultural extinction, mostly in island cultures. These examples are often related to violence between newly arrived communities with indigenous communities such as land, etc., which set the arriving community for long-term failure. Aggression and violence also commonly occur near the end of cultural extinction within a community over access to dwindling resources needed for survival and sometimes are related to violence with members of other communities, again often over precious and ever-dwindling resources.

7.3 Functional Explanations of Cultural Differences and Evolution

While it is easy to provide tantalizing illustrations of cultural differences and changes in time in aggression and violence, it is more difficult to explain them. Is it simply that members of different cultures are biologically different kinds of people or, is there something else going on?

One approach to understanding culture behavior analysis which uses the notion of meta-contingencies to explain cultural practices. This notion is compatible with other approaches such as cultural materialism (Glenn, 1988). As Skinner (1965) noted, identifying the contingencies that account for cultural differences is relatively obvious and trivial. People in one culture often reinforce aggression in members of the police against some members of the public and reinforce sport players for aggression against each other and often punish aggression by teachers and parents toward children, etc. The important question is how such contingencies are maintained within a culture. Glenn (1988) evoked the notion of a meta contingency which she defined as "the unit of analysis encompassing a cultural practice and all of its variations, and the aggregate outcome of all the current variations." These outcomes of cultural practices must be specified but can include anything from milk production and number of children. The behavior of the individual members of the culture is controlled by the individual interlocking contingencies, but the selection of the cultural practice is maintained by its effect on the survival or nonsurvival of the entire culture.

Glenn (1988) makes only passing references to violence, interestingly related to violent infanticide in humans. She noted that control of reproduction is essential to the survival and success of a culture. Failure to produce enough children will ensure the extinction of a culture. Similarly, overproduction of children may also result in cultural extinction, if the number of children is so great or the number of children are so tightly spaced that caring for them interferes with other activities necessary for the survival of the culture, such as caring for current living children.

Glenn (1988) noted that during human paleolithic history, birth control was possible through continued lactation, herbal remedies, and separation of the sexes. When these methods failed, then traumatic abortion, such as pounding the pregnant woman's stomach while pregnant, was practiced. Some data from near contemporary paleolithic-like cultures indicate that approximately 30% of infant deaths come from infanticide, especially of girls (Hill et al., 2007). Thus, Glenn implied that cultures that use violence or other methods to adjust the number of children to below the maximum that their environment can support will be more likely to survive than those that do no. Obviously control of the number of children, in part through violence, is just one part of a web of cultural practices that contribute to the survival or nonsurvival of the culture. This observation suggests a functional explanation at the level of cultures for one form of violence.

7.3.1 Nonviolent Protest

There have been many attempts to design peaceful yet effective cultures and cultural practices. One such practice is nonviolent protest. (See Chap. 14. *Cultural Interventions to Change Violence and Aggression. II. Broad-Based Intuitive Approaches*, for a discussion of several examples of Utopian communities that sometimes share some features with nonviolent protest.)

It is tempting to assume that nonviolent protest must fail to restore social injustices and prevent violence and the failure of nonviolent protest is one of the factors that necessarily underlies subsequent violent and effective protest, but, perhaps this is not so. Some might point to the examples of Mahatma Gandhi, in liberating India from British colonial rules through nonviolent practices, such as making one's own salt, boycotting British products, such as cloth from Manchester, and spinning one's own cloth. Other might also give the example of Martin Luther King, whose organized boycotts of buses, other economic boycotts, peaceful marches, and reaching out to diverse, disadvantaged communities as a call for unity and common purpose, etc. which were highly effective in achieving civil rights, such as the right to vote, for African Americans, and economic justice for some low paid workers in the USA. Indeed, until recently, scholarship has largely focused on violence rather than its alternatives perhaps because violence is seen as a more immediate and pressing problem, because it is difficult to quantify nonviolent protests and there is confusion over the term nonviolent protest, as it can be confused with weakness, pacifism, or activism (Chenoweth & Cunningham, 2013.)

This assumption and approach were challenged by the work of Erica Chenoweth and colleagues. Stephan and Chenoweth (2008) defined nonviolent civil resistance methods to include "boycotts, strikes, protests and organized noncooperation to challenge entrenched power and exact political concessions" (p. 8), symbolic protests, and social and political noncooperation; elsewhere they have noted there are several hundred such practices. These nonviolent approaches to change usually take place outside the channels of conventional politics, such as elections, courts, and legislative processes. Nonviolent civil resistance can be contrasted with other nonviolent events, such as once off strikes or protests and refusals to pay unjust taxes, because it is relatively organized, purposeful, coordinated, takes place over sometimes extended time periods, has grievances that are shared by a large group of people, and shared goals. Finally, nonviolence can be principled or instrumental. Principled nonviolence can be associated with ethical and religious commitment to nonviolence in all circumstances, as illustrated by the beliefs and practices of Hutterites, Quakers, and other such groups. Nonviolence can also be an instrumental strategy, one which is chosen because it is likely to be effective in obtaining one's political aims, as when a violence campaign adopts nonviolent methods because they appear more likely than violent methods to succeed (Chenoweth & Cunningham, 2013.)

One of the main contributions to studying effective, nonviolent cultural practices comes from a study by Stephan and Chenoweth (2008) who constructed a database of 323 violent and nonviolent conflicts from 1900 to 2006. They coded each conflict as: (a) successful (largely or completely achieving their aims); (b) partially successful (achieving some concessions); and (c) nonsuccessful in achieving their aims. They also coded a number of variables for each conflict and used these variables to predict the outcome of nonviolent and violent campaigns.

Surprisingly, and in contrast to the conventional wisdom, Stephan and Chenoweth (2008) found that 53% of nonviolent campaigns but only 26% of violent campaigns were successful. The main predictors of success in nonviolent campaigns were: (a) nonviolent civil resistance that was violently suppressed was six times more likely to succeed completely; (b) repressive regimes were 12 times more likely to make partial concessions to nonviolent groups; (3) defections from the violent group increased the chances of success of the nonviolent campaign four-fold; (d) nonviolent campaigns that received external support were three times more likely to succeed, but international sanctions were not associated with success; (e) longer nonviolent campaigns were more likely to achieve partial success; (f) repression of nonviolent campaigns resulted in a greater chance of their success; and (g) defections from security forces were associated with a 46-fold increase in the success of nonviolent civil resistance. Stephan and Chenoweth attributed the success of nonviolent campaigns to two groups of factors. First, nonviolent protest harnesses broad legitimization both nationally and internationally which increases population-wide participation, alienation of the target group, and undermining of the target group's legitimacy and economic, political, and military power. Second, nonviolent protest makes it difficult for the dominant group to justify their own violence. Violent suppression of nonviolent protests can readily backfire against the dominant group as when it increases solidarity within the protest movement, and increases broad support and sympathy for the nonviolent group. It may also make it more likely that members of the target group defect and lose domestic and international support. If the nonviolent group remains nonviolent, accrues martyrs, for example, during violent and/or illegal suppression of peaceful, unarmed protesters or rounding up and imprisonment of civil leaders, and gains international support for its legitimacy and goals through expressions of support and material and financial aid, then it can make it more likely that the violent group will have to make concessions to the nonviolent group.

Chenoweth's work is interesting and makes a unique contribution to understanding the cultural evolution of violent and nonviolent protest. Its main contribution is the focus on the characteristics of effective nonviolent civil resistance as an alternative to violent campaigns. These ideas are compatible with functional views of cultural evolution of effective nonviolence and violence. Societies may select effective, nonviolent practices over time when nonviolent cultural practices address a goal or set of goals shared by a large group of the population, when that group has no effective and legitimate alternatives in achieving those goals, and when the suppressing violent group loses internal and external support and resources. In contrast, cultures may select and continue to engage in violent practices when those practices are effective in achieving the political, material, and cultural goals of the violent group and no effective nonviolent or violent opposition to violent behavior control exists.

7.4 Summary

This chapter and the preceding chapter have documented the importance of cultural factors in the evolution and maintenance of aggressive behavior. The evidence is that cultural differences in violence and aggression are large, that violence has reduced considerably and the world is probably less violent than it used to be. We can use notions from diverse intellectual sources, including metacontingencies to help explain these differences and perhaps suggest interventions based on cultural metacontingencies supporting violence.

References

Anderson, C. A. (1989). Temperature and aggression: Ubiquitous effects of heat on occurrence of human violence. *Psychological Bulletin, 106*(1), 74–96. https://doi.org/10.1037/0033-2909.106.1.74

Anderson, C. A. (2001). Heat and violence. *Current Directions in Psychological Science, 10*(1), 33–38.

Barone, G., & Narciso, G. (2015). Organized crime and business subsidies: Where does the money go? *Journal of Urban Economics, 86*, 98–110.

Black, E. (2010). *The Farhud: Roots of the Arab-Nazi alliance in the Holocaust*. Dialog Press.

Buonanno, P., Durante, R., Prarolo, G., & Vanin, P. (2012). *Poor institutions, rich mines: Resource curse and the origins of the sicilian mafia*. University of Bologna: Quaderni Working Papers

DSE No. 84. Retrieved October 8, 2021, from https://www.siecon.org/sites/siecon.org/files/oldfiles/uploads/2012/08/Buonanno-Durante-Prarolo-Vanin.pdf

CDC. (2020). *Homicide mortality by state*. Retrieved October 8, 2021, from https://www.cdc.gov/nchs/pressroom/sosmap/homicide_mortality/homicide.htm

Chenoweth, E., & Cunningham, K. G. (2013). Understanding nonviolent resistance: An introduction. *Journal of Peace Research, 50*(3), 271–276.

Cohen, D., & Nisbett, R. E. (1994). Self-protection and the culture of honor: Explaining southern violence. *Personality and Social Psychology Bulletin, 20*(5), 551–567.

Cohen, D., Nisbett, R. E., Bowdle, B. F., & Schwarz, N. (1996). Insult, aggression, and the southern culture of honor: An experimental ethnography. *Journal of Personality and Social Psychology, 70*(5), 945–960.

Copes, H., & Vieraitis, L. M. (2009). Bounded rationality of identity thieves: Using offender-based research to inform policy. *Criminology & Public Policy, 8*(2), 237–262.

de Vaca's, C. (1542/1961). *Adventures in the unknown interior of America*. University of New Mexico Press.

Diamond, J. (1997). *Guns, germs and steel: The fates of human societies*. Vintage.

Diamond, J. (2005). *Collapse: How societies choose to succeed or fail*. Viking Penguin.

Dickie, J. (2005). *Cosa Nostra: A history of the Sicilian mafia*. Macmillan.

Eisner, M. (2001). Modernization, self-control and lethal violence. The long-term dynamics of European homicide rates in theoretical perspective. *British Journal of Criminology, 41*(4), 618–638.

Eisner, M. (2003). Long-term historical trends in violent crime. *Crime and Justice, 30*, 83–142.

Eisner, M. (2013). What causes large-scale variation in homicide rates? In *Aggression in humans and other primates biology, psychology, sociology* (pp. 137–163). de Gruyter.

Eisner, M. (2014). From swords to words: Does macro-level change in self-control predict long-term variation in levels of homicide? *Crime and Justice, 43*(1), 65–134.

Elias, N. (1978). *The civilizing process*. Blackwell.

Glenn, S. S. (1988). Contingencies and metacontingencies: Toward a synthesis of behavior analysis and cultural materialism. *The Behavior Analyst, 11*(2), 161–179.

Gurr, T. R. (1981). Historical trends in violent crime: A critical review of the evidence. *Crime and Justice, 3*, 295–353.

Hill, K., Hurtado, A. M., & Walker, R. S. (2007). High adult mortality among Hiwi hunter-gatherers: Implications for human evolution. *Journal of Human Evolution, 52*(4), 443–454.

Johnson, H. B. (1998). *Black Wall Street. From riot to renaissance in Tulsa's historic Greenwood district*. Easking Press.

Johnson, H. B. (2014). *Tulsa's historic Greenwood district*. Arcadia Publishing.

Krugler, D. F. (2014). *1919, the year of racial violence*. Cambridge University Press.

Kuitems, M., Wallace, B. L., Lindsay, C., et al. (2021). Evidence for European presence in the Americas in AD 1021. *Nature*. https://doi.org/10.1038/s41586-021-03972-8

Langton, L., & Truman, J. L. (2014). *Socio-emotional impact of violent crime*. US Department of Justice, Office of Justice Programs, Bureau of Justice Statistics.

Lynch, R. (2008). The people's protectors? The Irish Republican Army and the "Belfast pogrom," 1920–1922. *Journal of British Studies, 47*(2), 375–391. https://doi.org/10.1086/526757

McGowen, R. (1994). Civilizing punishment: The end of the public execution in England. *Journal of British Studies, 33*(3), 257–282. https://doi.org/10.1086/386055

Mikkelsen, N., Kuijpers, A., & Arnemorg, J. (2008). The Norse in Greenland and late Holocene sae-change level. *Polar Record, 44*(1), 45–50.

Mikton, C. R., Butchart, A., Dahlberg, L. L., & Krug, E. G. (2016). Global status report on violence prevention 2014. *American Journal of Preventive Medicine, 50*(5), 652–659.

Nisbett, R. E., & Cohen, D. (1999). Violence and honor in the southern United States. In J. E. Dizard, R. M. Muth, & S. P. Andrews (Eds.), *Guns in America: A reader* (pp. 264–274). NYU Press.

Pinker, S. (2011). *The better angels of our nature: The decline of violence in history and its causes*. Penguin.

Pinotti, P. (2011). *The economic consequences of organized crime: Evidence from Southern Italy*. Working Paper 868. Bank of Italy.

Pinotti, P. (2013). Organized crime, violence, and the quality of politicians: Evidence from southern Italy. In P. J. Cook, S. Machin, O. Mariue, & G. Mastrobuoni (Eds.), *Lessons from the economics of crime: What reduces offending* (pp. 175–188). MIT Press.

Pitzulo, C. (2017). The skirted sheriff: Florence Thompson and the nation's last public execution. *The Register of the Kentucky Historical Society, 115*(3), 377–410.

Saviano, R. (2008). *Italy's other mafia*. Macmillan.

Serenata, N. (2014). *The 'Ndrangheta and sacra Corona Unita. The history, organization and operations of two unknown mafia groups. Studies of organized crime, 12*. Springer.

Sergi, A., & Lavorgna, A. (2016). *'Ndrangheta: The glocal dimensions of the most powerful Italian mafia*. Springer.

Skinner, B. F. (1965). *Science and human behavior*. McMillan.

Stephan, M. J., & Chenoweth, E. (2008). Why civil resistance works: The strategic logic of nonviolent conflict. *International Security, 33*(1), 7–44.

Taïeb, E. (2020). *Hiding the guillotine: Public executions in France, 1870–1939*. Cornell University Press.

Chapter 8
Social Causes of Violence and Aggression: III. Evolution of Aggression over the Lifespan

Whatever biological and cultural factors may be at play in causing aggression and violence, the role of the individual organism's learning during their lifetime is clear and robust. This gives one cause for optimism. Learning can be the basis for prevention and what is learned may be unlearned or, if not, maybe even a persistently violence person can learn to control their behavior by learning to do something else or at least learning to avoid the triggers and situations for violent behavior. Even in studies reporting high heritability of aggression and violence, not all identical twins are concordant for aggression and violence; even identical twins where one is psychopathic may be discordant for psychopathy and often so-called concordant twins are quite different too suggesting that something other than genetics is at play. Not all individuals with XYY syndrome are violent; some individuals with Down syndrome, despite the apparent "Down syndrome advantage" of prosocial behavior, are aggressive; and some individuals with Autism Spectrum Disorders are not aggressive at all, despite autism being a risk factor for aggression. In a similar vein, some authors have used forensic case formulation to explain the acquisition of extreme violence including extreme cases, such as multiple murder, in a way that is appealingly rational and uses learning during the lifespan as a key component of understanding the evolution of extreme violence (Gresswell & Hollin, 1992; Hart et al., 2011; Sturmey & McMurran, 2011). All this suggests that the environment plays an important role in selecting, elaborating, and maintaining nonaggressive or aggressive behavior within each individual's lifespan.

This chapter reviews the evolution of aggression and violence over the lifespan of an individual organism. It focuses on behavior analytic accounts of the mechanisms behind such changes based on learning, principally operant learning. It begins by considering the role of classical conditioning in learning aggression and then considers operant learning processes including modeling. The next section considers behavioral accounts of emotion and how it applies to the emotion of anger. A key concept in behavior analysis is self-control. This section relates acquisition of self-control to the initial acquisition and interventions based on self-control. Finally, the

P. Sturmey, *Violence and Aggression*, Advances in Preventing and Treating Violence and Aggression, https://doi.org/10.1007/978-3-031-04386-4_8

next section considers the role of verbal behavior and culture in the evolution of aggression during the lifespan.

8.1 Classical Conditioning

Skinner (1953) noted that biological evolution prepares future organisms for aspects of the environment that are relatively permanent, such as gravity; common threats to the integrity of the organism, such as choking and some kinds of injury and other aspects of behavior which directly relate to survival, such as infants suckling, digestion, and sexual behavior. Other aspects of the future environment are difficult to predict. Although gravity is permanent, the exact stimulus properties of walking hazards—snow and ice, and slippery, waxy floors made of the latest material, etc.—are unpredictable from an evolutionary timescale. We may never know where the patch of ice or other hazard might be and exactly how it might look. Foods too are unpredictably dangerous, and omnivores like humans should be very sensitive to acquiring new food reinforcers quickly when it is advantageous to do so and avoid poisonous and potentially lethal foods quickly. Thus, reflexive aggression in response to pain which signals a threat to the organism's integrity from a predator, another organism, or an environmental threat is likely to be highly advantageous to the survival of the organism's genes.

Important learning may take place through stimulus pairings. Initially, an unconditioned stimulus (US, e.g., pain) elicits an unconditioned response (UR, e.g., aggression). Sometimes a neutral stimulus (NS, e.g., predator squawking and approaching with threatening gestures) is paired with an unconditioned stimulus (e.g., pain from bites). After a number of pairings, the NS now comes to elicit a conditioned response (CR, e.g., aggression) that is similar to the UR. Stimulus generalization may occur without direct pairing such that novel stimuli that have physically similar properties to the CS may also come to elicit the CR. For example, a physically or behaviorally similar novel predator's approach might elicit the CR, even though this particular stimulus has never been paired previously with the US. Classical conditioning has been demonstrated for many behaviors related to aggression in many species including classical conditioning of a wide range of emotional responses, such as anger, fear, and anxiety.

The observation that human and nonhuman animals lash out when injured or when danger of being injured is impending is not new: It has been studied systematically since the 1930s. It is a well-known everyday phenomenon, as anyone who has had the bad luck to stand next to someone who gets unexpectedly hurt unfortunately sometimes knows. If the person or pet next to you gets hurt, if you are lucky you only get yelled at or even pushed, even though you are not responsible for the harm: You are merely the unfortunate Pigeon B (see below.)

Basic research with nonhumans in the 1960s demonstrated classical conditioning of pain-elicited aggression. For example, Vernon and Ulrich (1966) paired a neutral tone with pain and, subsequently, the neutral tone alone elicited aggression between pairs of rats. Farris et al. (1970) showed that this could occur early in the development of rats. Further, studies have shown arbitrary stimuli can elicit conditioned aggression in Siamese fighting fish when the image of an intruder male is used as the unconditioned stimulus (Thompson & Sturm, 1965.) Lyon and Ozolins (1970) also demonstrated this in rats. Presentation of the CS prior to an encounter with an actual intruder male increases the intensity of aggression when the actual intruder is presented (Hollis, 1984, 1990) (see Ulrich, 1966 for an early review of these studies).

This phenomenon was studied extensively in the 1960s in a series of classic studies by Nathaniel Azrin and colleagues and numerous other independent groups of researchers since in many human and nonhuman species. The typical paradigm involved delivering painful, behavior-independent stimuli, such as electric shocks or tail pinches to one organism (Pigeon A) and having a conspecific (poor pigeon B!) who will get hurt or an object such as a suspended ball near the animal that the animal can aggress against. Following the painful stimulus, the hurt animal will typically aggress to another nearby organism or object.

In an early study, Azrin et al. (1965) observed the rates of aggression in six squirrel monkeys. They investigated the effects of the intensity of tail pinch on aggression measured by observing the probability of attacking a cloth covered ball in a cage. The monkeys were in Skinner boxes. The authors found that for all participants aggression was more likely with higher intensities of tail pinches and that there were also large individual differences between participants which were of unknown origin. Similar data have been reported in other species including primates. It is important to note that in this situation, access to the opportunity to aggress is reinforcing, i.e., accessing the opportunity to aggress is an operant behavior. For example, Azrin et al. (1965) conducted an experiment, again with squirrel monkeys, which showed that squirrel monkeys emitted operant behavior by pulling a chain to obtain access to pull chain to obtain an object to aggress against. This chain pulling response was reinforced and extinguished, demonstrating that this behavior was an example of operant behavior. (One may think of this as an analog of chasing after someone when they have hurt you to aggress against them.)

Dos Reis Soares and Goulart (2016) conducted a systematic review of the *Journal of the Experimental Analysis of Behavior* to identify publication on pain-induced aggression. Most studies were published between 1962 and 1981 with little activity thereafter and usually involved pigeons, rats, and monkeys, but not humans. These early studies spawned a considerable series of experiments on variables that influenced shock-induced aggression. In these studies, aggression was a function of shock intensity (stronger shocks were more likely to induce aggression than weaker shocks) and shock density (more shocks per unit time were associated with more aggressive behavior). Further, punishment of pain-induced aggression with electric shock reduced aggression, suggesting that although pain-elicited aggression was acquired through respondent mechanisms, it was also sensitive to operant processes. Generally, these results were not a function of species and were replicated across rats, pigeons, various monkey species, and humans. It is notable that this particular line of research with nonhuman animals largely ceased after 1982, presumably because of the ethical objections to deliberately causing pain to animals and the diminishing returns of such studies to our knowledge of aggression.

There have been a number of interesting reactions to this line of research, especially in the area of its implications for understanding human aggression. Archer (1995) linked pain-induced aggression to ethology by questioning whether pain-induced aggression was a general model of aggression and how it related to aggression that occurred in the wild. He noted that ethological studies of aggression in rats differed in topography from that studied in typical studies. Namely, when rats aggress in the wild, they attack intruder rats by biting specific body sites, a detail absent from these behavior analytic studies. Archer understands pain-induced aggression to be analogous to fear-related aggression similar to that induced when a rat is cornered by an aggressor but not of other types of aggression.

A different reaction to pain-induced aggression came from Berkowitz (1993) who put forward the thesis that there are both important similarities and differences between pain-elicited aggression in humans and nonhuman animals. Berkowitz emphasized that, although there may be substantial similarities in pain-induced aggression between humans and nonhumans in some circumstances, key differences might arise due to cognition. Berkowitz extended the notion of "pain" in human organisms to include milder forms of physical stimulation, such as uncomfortable temperatures, and even "unpleasant events of any kind" (p. 1135). He noted that pain in animals may induce both attack but also fleeing and that perhaps some aspects of pain-induced experiments might be odd in that they do not permit the option to flee, although some experiments on pain-induced aggression in nonhumans might include an operant procedure of terminating the pain or related CSs that might be analogous to fleeing (see below.)

Studies of pain-induced aggression also have interesting implications for the treatment of aggression. In attempts to reduce some human aggression we can consider either reducing or eliminating aversive stimulation, for example, by using air conditioning to reduce discomfort or teaching alternate behavior such as social skills and parent training to remove aversive social stimuli. Based on this literature, we might provide other objects of aggression, teach alternate behavior, or punish pain-induced aggression (although the ethics of this latter approach would usually be questionable in many circumstances). If pain-induced aggression is sometimes an example respondent behavior, then we could also consider ways to modify the CS, such as by pairing it with positive stimuli, for example, increasing the number of pleasant activities between couples who engage in interpersonal violence, or to use respondent extinction by repeatedly presenting the CS, for example, repeatedly exposing the angry person to triggers such as insults until they no longer respond in an angry fashion. Some of these procedures appear analogous to certain components of anger management.

This phenomenon of pain-induced aggression might be a model for human angry, "hot" aggression in that arbitrary stimuli (mean and dirty looks and minor insults) may come to elicit both angry emotions and aggressive behavior in some people. Evidence for such a view comes from studies that have shown that anger can also be classically conditioned and extinguished in humans (Brondolo et al., 1997; DiGiuseppe et al., 2007).

8.2 The Selecting Environment

Although unconditioned and conditioned reflexes that allow organisms to respond quickly to threats such as injury, pain, and loud sounds by withdrawing quickly, these reflexes are limited when it comes to responding to the highly variable stimuli which might predict threat or be dangerous, such as someone walking toward you with a knife or a drunken person threatening you. Evolution cannot prepare flexible organisms for most eventualities other than granting the organism the susceptibility to rapid and efficient operant learning. How does this occur?

Behavioral theorists evoke the notions of variation, selection, and extinction, borrowed from biological evolution to explain learning simple and complex behavior (Glenn, 1988; Skinner, 1953). From birth, if not earlier (March et al., 2009), behavior is modified by its consequences. Newborn infants turn and cry; move their arms, legs, and head; and grab, hold, and mouth objects. The elementary behaviors may be followed by consequences which strengthen or weaken the future likelihood of these behaviors. Both the physical and social environment may select and elaborate such behavior. Grasping and manipulating an object with one kind of grip will produce stimulation that strengthens the behavior in the future and grasping the object another way will cause it to fall and weaken that form of grasping in the future. Over time attempts to scoop up a pea on a table surface with an open palm are replaced with a delicate pincer movement of a finger and thumb. When an infant is uncomfortable, turning one way will relieve the discomfort and turning another will increase discomfort. The social environment of infants also selects certain behavioral variants including both relatively modest behavioral change, such as acquisition of phonemes, and broad changes, such as the broad rate of language acquisition (Hart & Risley, 1995).

Differential reinforcement and punishment can account for changes in the rate, intensity, latency, etc. of specific existing responses, but where does new and complex behavior come from? There are several processes that can account for the acquisition of complex behavior, including complex behavior relating to aggression. Shaping of progressively more complex forms of behavior and extinction of earlier, more rudimentary forms of behavior can give rise to complex motor, verbal, and social behavior. Behavior can appear more complex when it is brought under more varied stimulus control, as when an infant learns to orient to the nearby location of a sound and later learns to orient to near and far sounds. Extinction of operant behavior is often accompanied by response variation including novel responses, and sometimes the environment might strengthen such new behavior through reinforcement. Complex sequences of behavior can arise through chaining as when a child learns to reach for a container, open it and obtain a preferred toy or snack. Imitation and following instructions, a more complex form of acquisition of stimulus control, can both lead to much novel behavior. Finally, behaviors acquired under one set of schedules of reinforcement can be combined under new schedules of reinforcement (so-called contingency adduction; Andronis et al., 1997). These learning processes and combinations of these learning processes might account for acquisition of complex behavior across the lifespan (Epstein et al., 1980).

8.2.1 Reinforcement and Shaping of Aggression

There is considerable evidence that aggression can be shaped and maintained by differential reinforcement in domestic and wild nonhuman animals and in humans. Shaping involves two processes: (a) extinction of previous behavior, which often induces response variability and side effects, which include emotional behavior including aggression; and (b) differential reinforcement of other behavior. During the course of extinction and differential reinforcement, complex behavior can be learned through shaping even including pathological behavior such as self-injury in both humans and nonhumans (Layng et al., 1999; Schaeffer, 1970).

The phenomenon of differential reinforcement of aggression is nicely illustrated by an anecdotal letter by John Bailey (1987) concerning the *Terrorist Geese of Tallahassee*, responding to another letter from a member of the public who described how "These creatures were totally crazy and so insolent in begging for food that we couldn't do anything but leave a loaf of bread for them and escape to our car as quickly as possible" (p. 203). The author advocated for removal of the geese to a desert lake and replacement with swans that were more beautiful. In reply, Bailey, noted that, as beautiful as the swans would probably be, they too would become aggressive if members of the public shaped aggressive behavior. Rather than relocate the geese, Bailey suggested that members of the public throw their bread over the backs of the geese, stand their ground, and do not drop any bread. Bailey reported that he tried this the morning of writing the letter and it worked just fine. Differential reinforcement of aggressive behavior has been implicated in aggression in many species including sheep, primates, dogs, and of course humans and appears to be a robust phenomenon (Fernandez, 2020).

Although differential reinforcement of aggression increasing aggressive behavior is well documented, naturally occurring shaping of aggressive behavior over the lifespan is difficult to demonstrate. There are several good reasons for this. One is that it might take a long time to do such studies. Second, it might be difficult to predict what current behavior and which current consequences might be involved in shaping aggressive behavior which has not yet occurred; hence, it might be difficult to detect the exact developmental precursors to aggression before aggressive behavior itself is emitted. Third, there would be ethical problems in knowingly failing to intervene before aggression was acquired.

8.2.2 Chaining

A stimulus–response chain is a sequence of stimuli and responses in which each member of the chain acts as the discriminative stimulus for the next response and the conditioned reinforcer for the preceding member of the chain. The chain ends with some terminal reinforcer. For example, one might search for one's phone, turn it on, open the text message function, and see a text or picture. Chaining of behavior can occur both naturally or automatically and is called "autochaining," and chaining can also occur programmatically. For example, autochaining might occur if there is an aversive stimulus and the behavior chain of getting up, looking for the source of the noise, and closing the window terminates the aversive noise. Programmatic chaining occurs when the environment is designed to teach a stimulus–response chain, as when someone teaches another person to close the window when it gets noisy before they become angry.

Chaining has been implicated in aggression in a number of ways. Everyday examples of escalation of problem behavior may well be examples of stimulus–response chains. For example, a typical sequence might involve: (a) weak and/or ineffective adaptive behavior that is extinguished, such as a quiet request that is ignored; (b) precursor behavior, such as increasingly loud speech and threats which may be reinforced intermittently; and (c) aggressive behavior with final access to the reinforcer maintaining aggression such as being removed from an unpleasant environment. It is possible that caretakers inadvertently chain behaviors including aggression in various ways. Such stimulus–response chains can be seen in many populations and contexts, but is documented most rigorously in individuals with developmental disabilities. It is also an important component of positive interventions that teach progressively more complex adaptive behavior as alternatives to aggression (e.g., Ghaemmaghami et al., 2018.)

An early study which demonstrated stimulus–response chains and aggression is by Murphy et al. (2000). The participant was Ms. G., a 38-year-old woman with severe intellectual disabilities and autistic features who had lived in institution settings between the ages of 9 and 33 years and currently lived in a fully staffed community home. Her group home staff referred her because of problems with managing chronic stereotypies and aggression toward staff. Her aggression had failed to respond to haloperidol. Initial staff interviews revealed that she spent much time acquiring paper items, tearing them up into tiny pieces, and flushing them down the toilet. (She also showed other stereotypies.) Staff reported that: (a) a common antecedent was a piece of paper on the floor; (b) aggression was most likely to occur when the opportunity to flush paper was blocked; and (c) she became agitated when she could not flush the items away. The authors conducted a functional analysis in which they compared four conditions: access to flushable items that was either unblocked or blocked by the experimenter and access to unflushable polystyrene that was either blocked or unblocked. Hitting the experimenter occurred only in the condition in which the experimenter blocked access to the flushable items. The authors concluded that "the behavior of running to a toilet, ripping and flushing formed a chain, which when the putative reinforcer (disappearance of the paper) was no longer presented, resulted in an extinction burst (of the behaviors which formed the chain) and that the aggressive behavior was occasioned by the prevention from flushing or removal of the putative reinforcer (i.e., 'prevention' was an establishing operation for aggression)" (p. 56). Subsequent studies by Hausman et al. (2009) and White et al. (2011) have also found similar evidence of response chains and access to ritualistic behavior being implicated in environmental control of aggression.

The notion of stimulus–response chains might be useful in understanding aggression and violence in many contexts and populations. Examples of violence involving stalking and seeking out victims, acquisitive violent crime which often involves a series of responses that result in acquisition of goods or money, and escalating anger which results in another person suffering or removal of an aversive stimulus all appear similar to stimulus–response chains. This might be useful as it directs attention to intervening early in the response chain and identifying early members of the chain that might include adaptive behavior that could be strengthened during intervention.

8.2.3 Generalized Imitation

Many have recognized that generalized imitation is both a naturally occurring phenomenon and an important skill that can be learned. When organisms cannot imitate, their learning is greatly impaired. Further, teaching becomes difficult as the available options are limited to procedures such as differential reinforcement, shaping, and prompting and fading which are often much less efficient than showing what to do. Many researchers have also recognized imitation as a key mechanism in acquisition of a wide range of socially important behavior, including language and prosocial behavior (Schlinger, 2002)

People have also long expressed concerns over the transmission of aggression from models, such as parents who engage in aggressive and coercive child management practices, siblings, peers, and media, such as the news, movies, cartoons, and violent pornography. Further, these sources often show aggression from high status characters with reinforcement for aggressive behavior and minimization of its negative consequences. Environments where aggression and violence are common, such as prisons, youth facilities, and some residential services, are often characterized by authoritative figures who themselves engage in violence, thereby apparently legitimizing its use for everyone. Bandura's work amplified these concerns when he noted that in some circumstances caregivers explicitly use imitation and exposure to aggressive models to teach their children, especially their boys, to "stand up for themselves," "man up," and "not be a sissy" (Bandura, 1973).

To understand these processes, we should first carefully consider what we mean by imitation. People say, "monkey see, monkey do," but maybe there is more to imitation than mere exposure to models. It is commonplace to see people apparently copy each other's behavior in many circumstances. Caretakers, siblings, and peers may instruct children directly and indirectly to imitate certain models. They also formally and informally expose them to models of appropriate behavior: "If you lie down with dogs, you get up with fleas" says the proverb published by Benjamin Franklin. In addition, caretakers praise correct imitation, for example, when they say, "You are a big boy now, just like your brother," and punish imitation of inappropriate models, for example, a caretaker might say, "why are you behaving like *her* now?" Hence, some of the antecedents and consequences for imitation seem readily apparent to the casual observer.

Individuals, however, often imitate many different models, including models for which they have never been reinforced. Thus, an important concept is generalized imitation. Young et al. (1994) defined generalized imitation as "a class of responses that temporally follow and are topographically similar to behavior modeled by another person" (p. 685). Generalized imitation is a functional response class because reinforcement of some imitative responses may increase other imitative responses that are not directly reinforced and extinction of some imitative responses may decrease the probability of other members of the response class (Baer & Deguchi, 1985).

Research has identified several variables that influence the occurrence of generalized imitation. These include reinforcement for imitation (Young et al., 1994), similarity between the model and the person learning (Marcus & Wilder, 2009), and reinforcement or punishment of the model for engaging in the modeled behavior (Wheeler & Smith, 1967). Further, individuals do not imitate all models all the time. We look at others to determine which fork to use during each course of a meal and avoid the scowl of social disapproval by not breaking the rules of social niceties by behaving like everyone else around us (Brown et al., 2008). Some forms of generalized imitation are called cheating and other forms of generalized imitation are called teamwork! Thus, imitation is under the stimulus control (DeQuinzio et al., 2007; Furnell & Thomas, 1976). Imitation by others may also be a positive reinforcer for the behavior of the individual imitated (Pelaez et al., 2011). Thus, we have a good behavior analytic conceptual framework of what generalized imitation is and the variables that influence imitation.

8.2.3.1 Bandura and Bobo Dolls

The classic studies on modeling and aggression come from Albert Bandura who made an enduring contribution to the understanding of aggression, both from his classic "bobo" dolls experiments and from several books addressing this topic more broadly (Bandura, 1973; Bandura & Walters, 1963). In the 1950s and 1960s, influenced by psychoanalytic models of aggression, it was generally believed that observing aggression and violence was cathartic, that is, observing aggression resulted in *reduced* aggression in the observer. Thus, many researchers at that time believed that the idea that observing aggression would decrease aggression.

There were three original bobo doll studies. In Bandura and Huston (1961), the authors assigned 24 boys and 24 girls aged 45–61 months from a university nursery to three conditions. Prior to imitation learning, there were two, 15-minute sessions scheduled over 5 days. In the nonnurturant condition, the experimenter placed toys out, but sat at a desk busying herself with paperwork and did not interact with the children, but in the nurturant condition, the experimenter sat on the floor near the children and interacted warmly with them. In a subsequent session, the children were then given a distracting task, and the authors manipulated the behavior of the model including modeling aggression. A control group of children did not participate in any of these sessions prior to imitation learning. Children in the experimental groups were subsequently more aggressive than the control group; this was unaffected by whether the children had observed a nurturant or nonnurturant model. This was a surprising result at the time suggesting that imitation and stimulus properties of the model were important determinants of child aggressive behavior.

Bandura et al. (1961) extended these results in several ways. In this study, children were exposed to adult models who were either aggressive or nonaggressive to bobo dolls. Second, in order to evaluate generalization of modeling, observation of the children's behavior took place in a different room. Following the model, the children were given a frustrating task (loss of preferred toys) to increase the likelihood of subsequent aggression. The children exposed to the aggressive model were much more likely than children in the other groups to show aggression. In contrast, children exposed to the quiet model showed little aggression, but were generally much quieter and inhibited. Subsequently, Bandura et al. (1963) showed similar effects using movies with aggressive behavior or no aggressive behavior, replicating an earlier similar study by Lovaas (1961). These latter two studies also reported some data that children were more likely to imitate male models and models of the same gender as the child.

These studies were highly important because they demonstrated that typical children showed *more* rather than less aggressive behavior when exposed to adult models, thus undermining the psychoanalytic, catharsis theory of modeling and aggression. Further, these studies demonstrated that aggressive behavior in the media, including children's cartoons, had the capacity to increase children's aggression, at least in the short term. These studies did not demonstrate the potential link between early exposure to aggressive models during early childhood, for example, from caregivers and peers, and subsequent development of more severe and persistent forms of aggression, delinquency, and criminal aggression and violence.

Bandura (1973) saw the role of modeling in the acquisition and maintenance of aggression very broadly; for him it was not limited to only immediate imitation of aggressive models, but was rather sometimes part of a much broader process of increases in aggression and violence. For example, he discussed the role of modeling in group violence, such as was seen in the student protest movements of the 1960s and American anti-Vietnam war violent protest movements like the Weathermen Underground. He speculated that exposure to progressively more assertive behavior, passive protests, vandalism, and property destruction eventually culminated in modeling of violence, such as bombings and fire settings.

Bandura also saw modeling as a key mechanism in the transmission of aggression and violence in highly aggressive cultures and in the transmission of pacific behavior in peaceful cultures. For example, he gave the example of how the Dani culture, which engaged in highly ritualized violence between neighboring groups, progressively trained its children using modeling to engage in culture-specific forms of aggression through escalating games, symbolic aggression, training in specific fighting skills, and observation of actual war. A final example of the broad impact of modeling on aggression and violence for Bandura relates to how the military, police, and other violent organizations train their members in part through modeling of progressively more violent behavior; for example, the trainee soldier is exposed to progressively more aggressive models and practices progressively more aggressive acts many times until the soldiers are fluent in aggression and inured to the impact of their own aggression and violence.

Bandura's ideas are still very much alive today. They are influential in research and public policy concerning the role of media in promoting prosocial or aggressive behavior and the role of violent pornography in promoting sexual violence toward women and children. Modeling is also a component of several theories of aggression such as social learning theory and treatments for aggression.

8.2.4 Schedule-Induced Aggression

Schedules of reinforcement induce a very wide range of behavior changes, not only changes in the target behavior (Andronis et al., 1997; Frederiksen & Peterson, 1977). For example, increases in the target behavior during a schedule of reinforcement is often accompanied by the reduction of other behavior and a general orderliness of behavior as the organism emits operant behavior, meets the schedule requirement, goes to obtain its reinforcer, interacts with or consumes it, and goes back to the operant manipulandum resulting in less behavioral variability than in baseline. Changes in behavior that are not directly reinforced are common. Such adjunct behavior is neither directly reinforced by the schedule of reinforcement for the other behavior nor elicited by stimuli associated with that schedule, rather they are changes in other existing responses within the organism's existing repertoire. There are many examples, but a commonly studied change in adjunctive behavior is drinking and/or polydipsia (Falk, 1971), but other adjunctive behavior in nonhuman animals can include eating, wheel running, airlicking, escape, and, indeed, aggression (Foster, 1978). Although there is an extensive literature studying adjunctive behavior in nonhuman organisms, Foster argued that the phenomenon might be greatly understudied and underreported in the applied literature because it may be difficult to detect and predict its occurrence.

In a comprehensive review of this phenomenon, Frederiksen and Peterson (1977) defined schedule-induced aggression as "aggress[ing] against other organisms or inanimate objects during a schedule of reinforcement in which reinforcement is delivered for emitting another unrelated response" (p. 57). The prototypical example of schedule-induced aggression is that which sometimes occurs during extinction of operant behavior. That is, during extinction not only does the target behavior decrease, but behavior becomes more variable and the organism may emit "emotional" behavior which sometimes includes aggression (see Lerman et al., 1999 reviewed below).

Frederiksen and Peterson's (1977) review looked at both the nonhuman and human literatures and broadly found similar findings. Operant extinction refers to the effect of removing the contingency between a reinforcer maintaining an operant behavior and the behavior itself. This is usually done by identifying the reinforcer maintaining the behavior and preventing access to the reinforcer, but can also be done by presenting that reinforcer independent of the behavior, for example, on a fixed-time schedule. Operant extinction has four main effects on behavior: reduction in the frequency of the operant behavior, temporary increase in the operant behavior, increased response variability, and "emotional side effects" which can include a variety of emotional behavior, including aggression.

An interesting demonstration of the effects of extinction on aggression comes from Knutson (1970), who ran two rats concurrently. Aggression to a conspecific occurred when either rat's behavior was extinguished, but was most likely when both rats' behavior was simultaneously extinguished. Thus, this demonstrated the effects of interactions of schedules of reinforcement for the behavior of more than one organisms on inducing aggression.

An example of schedule-induced aggression with typically developing humans comes from Frederikson and Peterson (1974). They placed 16 typically developing nursery children in an experimental room with an operant device with which they earned pennies and access to a bobo doll. The authors compared the effects of three conditions—baseline, continuous reinforcement of bar pressing, and extinction—on the children's aggression and rate of bar pressing behavior. Aggression to the bobo doll *never* occurred during the continuous reinforcement schedule for *any* child. Aggression occurred only and reliably during the extinction schedule and baseline. This was equally true for both boys and girls. The authors concluded that this was an example of schedule-induced aggression.

A second applied demonstration of extinction-induced aggression comes from Lerman et al. (1999) who examined the frequency of extinction bursts and aggression during extinction of self-injurious behavior with and without other behavioral procedures. The authors analyzed 41 data sets collected between 1989 and 1997 at a regional center involved in the treatment of self-injurious behavior (SIB). All data sets involved extinction of SIB in 30 individuals with moderate to profound Intellectual Disabilities, aged 5–54 years. All but one were aged over 18 years. In some cases the authors used extinction alone, and in others they used extinction in combination with other treatments such as NCR or differential reinforcement of other, incompatible, or alternate behavior. The authors reported prevalence of extinction bursts and extinction-induced aggression. Extinction bursts occurred in 39% of all cases, of which 22% of extinction bursts involved aggression. There were large differences in the likelihood of aggression when extinction was used alone or in combination with other procedures. When the authors used extinction alone, extinction bursts occurred on 625 of occasions. Of these occasions 29% of extinction bursts involved aggression as a side effect of extinction. When the authors combined extinction with other treatments, extinction bursts occurred during about 15% of applications. Of these applications, only 15% of extinction bursts involved aggression as a side effect of extinction.

Strangely, schedules of reinforcement may have aversive properties, especially related to periods of reduced reinforcement, such as immediately after the delivery of a reinforcer on a variable ratio or interval schedule and during periods of long runs prior to reinforcement on thin schedules of reinforcement. Thus, schedule-induced aggression also occurs during schedules of positive reinforcement under certain circumstances (Looney & Cohen, 1982). This occurs in a predictable manner. For example, increases in aggression during extinction occur shortly, but not immediately after the onset of extinction. In nonhumans, access to the opportunity to aggress may also be a positive reinforcer in some circumstances. These effects are accompanied by large individual differences in the rate of aggression (e.g., Azrin et al., 1966), but the source of these large individual differences are unknown (Frederiksen & Peterson, 1977).

The phenomenon of aggression as a side effect of extinction has several treatment implications. Some of these include: (a) avoiding using extinction alone by combining it with other treatments; (b) using punishment of aggression during extinction (although the ethics of such an approach are dubious, given the availability of alternative nonpunitive, effective treatments); (c) teaching self-control during extinction (Skinner, 1953); and (d) changing the environmental events, including changes in timing of activities such as working and leisure activities, that are responsible for schedule-induced aggression, rather than direct contingencies for the problematic schedule-induced behavior (Foster, 1978).

It is interesting to note that operant extinction-induced aggression is very similar to Dollard, et al.'s (1939) frustration-aggression hypothesis. They suggested that aggression occurs when goal-directed behavior (i.e., previously reinforced behavior) is blocked (i.e., no longer reinforced) such as not winning money or failing to obtain a grade, then aggression may occur. Unfortunately, "frustration" is not readily operationalized. Thus, an alternate approach is to conceptualize such phenomena in terms of schedules of reinforcement.

8.2.5 Summary

There are several relatively simple operant processes that could account for the acquisition and maintenance of aggression and violence. Their cumulative and combined effects, in part or as a whole, could account for the life-long ebb and flow of aggressive and violent behavior. Let us now move on to another, less frequently acknowledged aspect of a behavioral account of aggression and violence, emotionally based angry aggression.

8.3 Anger and Aggression

Behavior analysis is often critiqued for denying people's private lives of thoughts and feelings. This comes in part from a misunderstanding of behavior analysis' account of private behavior including emotional behavior. Behavior analysis addresses emotions differently from other approaches because it rejects emotions as the *cause* of behavior. Rather, private behavior such as emotion and other private behavior related to emotions are behavior to be explained. Explaining behavior through emotions ("She hit him because she was angry") is a distraction from the environmental control of emotions. Thus, it is better to say that "When he insulted her (environmental event), her face became red, her blood pressure rose, and her stomach turned over (physiological response), she said to herself that she felt mad as hell, and thought 'I'll teach this jerk a lesson' (covert angry behavior), and she balled up her fist and threatened him verbally (overt angry behavior). When he cowered, apologized and backed away (the consequences of her anger included seeing him suffering, removing his insult), she started to calm down (the man's behavior is negatively reinforced by removal of her angry behavior." The emphasis on emotions as a cause of behavior also makes it less likely that we will attend to the practical use of induction of emotions through environmental design and behavioral self-management (see below). Behaviorism also values precision in language, and many words used to refer to emotions are often undefined and imprecise and defined in a circular fashion (Friman et al., 1998.)

8.3.1 Skinner, Emotions, and Aggression

In a behavioral approach to emotions, including anger, Skinner (1953) argued that although emotions have their roots in part in biology that in part characterize angry behavior, learning and culture are also essential in modifying any biological predispositions and reflexes. For example, snarling, growling, and clawing in response to threatening stimuli may well be behavior that is initially unlearned, and organisms may be predisposed to have such behavior reinforced by tissue damage in other organisms. But, people learn to show anger in ways that are specific to one's own culture through modeling and the consequences of showing culturally specific appropriate forms of anger. Each of us learns to threaten, show contempt, and look down our noses, for our fellow humans in ways judged appropriate to our own culture. Each of us learn to engage in angry behavior and to the triggers and insults appropriate and sometimes specific to our own culture. Children early on learn to tease, take toys, destroy others' work, and call names in ways specific to their culture that cannot come directly from their biology. Actors also learn to behave in emotional ways as when they portray the angry person while acting.

Skinner (1953) argued that the everyday words describing emotions describe a predisposition to act in certain ways under certain circumstances. The angry person "shows an increased probability of striking, insulting, or otherwise inflicting injury and a lowered probability of aiding, favoring, comforting, or making love" (p. 162). Emotions vary by their breadth of impact on the behavioral repertoire of the organism. A mild emotion such as nostalgia describes a narrow effect, whereas anger and anxiety describe emotions with a broad impact. The responses that characterize an emotion are held together by common reinforcers. Angry behavior is reinforced by observing the suffering of others. The angry person not only rejoices when the object of their anger "gets what is coming to them" but also enjoys talking about and observing others' suffering as when they watch a good fight in a violent movie or hear about people getting their just deserts.

According to Skinner (1953), the main error in talking about emotions is to elevate them to the cause of behavior. Many everyday examples suggest environmental control of emotional behavior as when the lights go out and we become scared and when we receive an insulting email that makes us engage in angry behavior. Skinner noted that emotional behavior may be discrete and short-lived, as when even the kindest person eventually becomes angry for a short period of time, or broad and chronic, as when the perpetually angry person shows medical problems relating to a lifetime of angry behavior, including physiological responses, resulting in ulcers. Finally, Skinner noted that emotions are predispositions that may only result in observable behavior under the appropriate context. Thus, the angry person may behave in a mildly angry fashion in circumstances similar but not the same as when he will become violent. When we say, "Watch out, the boss having a fit again," we are saying in part that if you annoy the boss today, you will provoke an angry response.

In the last section of Skinner's (1953) chapter on emotions, he discussed the practical use of emotions including anger. We cannot evoke respondent behavior, but we can arrange the environment to present the stimuli that evoke such conditioned behavior in others and ourselves. Artists and media of all kinds design stimuli to change emotional behavior in members of their audience. Fake wrestling exaggerates participants' suffering to rile up the audience to keep them watching and paying. A good novel is designed to keep us turning the pages because it delivers the looked for response in their reader when the first chapters of the book induced us to hate the bad guy in the plot, and we are happy and relieved when the bad guy receives their deserved and righteous retribution in the last chapter. To eliminate inappropriate emotions in others and ourselves, we learn to eliminate the antecedent stimuli for anger as when we walk away from an annoying person or present antecedent stimuli for incompatible behavior as when we take a deep breath. This is the basis of many anger management programs.

We teach the angry person to walk away from the annoying other person or to ask them to leave or teach them to use relaxation training or pinch themselves when they begin to emit angry behavior. At other times, it is advantageous for us to become angry or to induce anger in others. We can all benefit from becoming a little angry if we want to induce behavior in others if alternatives are ineffective or unavailable. Screwing up our courage by rehearsing our grievances might make it more likely that we can engage in the right behavior to induce our supervisor to give us a raise. Actors also learn to induce their own emotions such as anger using the Stanislavsky method in which they rehearse an emotionally laden memory to induce their own emotional behavior on stage to become a more skilled and convincing actor to their public. The coach may want his team to fight better by giving a pep talk and rehearsing aggressive responses to induce an aggressive emotion. The prosecutor may induce the witness to say what the prosecutor wants them to say or do. The prosecutor does so by repeatedly presenting anger-inducing stimuli to provoke their witness's anger so the jury can see that the accused is really an angry violent person. Society induces violent behavior and killing in the members of their armed forces with stories of the enemy's atrocities and later attempts to induce nonviolent behavior by presenting stimuli to induce nonviolent behavior as is Truth and Reconciliation procedures.

8.3.2 Novel Aggressive and Angry Responding

Aggression may occur repeatedly in the same context: The bully approaches the victim, threatens, and takes their money every morning at school, and every time the angry person's significant other makes a negative remark about their work, they boil over and strike the other person. The former appears to be operant behavior, and the latter appears to be respondent behavior. In both the cases, aggressive behavior reoccurs because the same relevant stimuli are represented. In the first example, the victim is the discriminative stimulus for an operant response and each time the money reinforces the bully's aggression. In the second example, the angry person's partner is the conditioned stimulus that elicits the conditioned response of angry aggression, and, as in two factor theory, there is also negative reinforcement of angry aggression when the partner no longer presents criticism and positive reinforcement for the aggressor's behavior when the partner is seen to suffer.

An important question is where novel aggressive behavior comes from. In the case of operant aggression, novel topographies of aggression and novel stimulus control of aggressive behavior could be learned in various ways: shaping could produce novel aggressive topographies, physical stimulus similarity could produce novel stimulus control, and modeling could produce both. In the case of respondent, angry aggression, novel angry-aggressive responding could be produced by at least two processes. The first includes physical stimulus similarity between the conditioned stimulus and other stimuli. The second could be higher order conditioning. Both of these phenomena are difficult to detect in individual cases since it may be difficult or impossible to divine an individual's learning history due to the incompleteness and inaccuracy of self and other reports. Nevertheless something else that requires explanation is the phenomenon of when a stimulus elicits an emotional response without direct conditioning. How can behaviorism account for aggression in completely novel situations when these other mechanisms do not explain this phenomenon?

A more recent iteration of behavior analytic accounts of emotional behavior comes from work on stimulus equivalence and transfer of emotional function. Friman et al. (1998), in a discussion of anxiety that has more general application to behavior analytic accounts of emotional behavior, pointed out that since Skinner's (1953) earlier discussions of emotion, a new and very large amount of data has accumulated relevant to emotions that uses the ideas of stimulus equivalence and transfer of stimulus function among members of a stimulus class. Equivalence is shown when two or more stimuli elicit the same response. Thus, an equivalence class is a group of stimuli that show reflexivity, symmetry, and transitivity. In the context of matching tasks, consider three stimuli, A, B, and C, such as the written word "cat" (A), a picture of a cat (B), and the sound "meeow" (C). Reflexivity is shown when the person matches each stimulus with itself (A-A, B-B, etc.). Symmetry is shown when after choosing B, given A (A-B), the person also choses A, given B (B-A). Transitivity is shown when having learned A-B and B-C, the person choses C given A (A-C). Finally, equivalence is shown when given C, the person choses A (C-A). Members of a stimulus equivalence class may be physically different as in the example of a printed word, line drawing, and sound. Classes may have more than three members and can be expanded. For example, if a further stimulus such as the written word "gatto" (D) becomes a member of the class, for example, by teaching D-A, not only do the symmetrical relationships A-D emerge but also all other factorial combinations of relationships between D and other members of the class (i.e., D-B, B-D, D-C, C-D). Existing classes can also be combined with one another. For example, classes of "cats," "dogs," "sheep," etc. could be combined into a superordinate class of "domestic animals" (Fields & Verhave, 1987).

Members of a stimulus equivalence class may also elicit other responses such as conditioned responses, such as conditioned fear responses. Importantly for this discussion, Dougher and Markham (1994) demonstrated that when a fear response was conditioned to one member of a stimulus class, other members of the class also came to elicit fear without direct training, a phenomenon called transfer of function. Researchers have considered the possibility that transfer of emotional functions among members of a stimulus equivalence class might be another important explanation for the acquisition of psychopathology when direct conditioning, stimulus generalization, higher order conditioning, etc. are not likely explanations for generalized emotional responding (Friman et al., 1998). Transfer of function has been related to psychopathology to account for generalization of learning events, for example, perhaps a painful wasp sting paired with wasps, the sound of buzzing, and a physical location will produce generalization through transfer of stimulus function to other members of various stimulus classes such as "insects" and "annoying noises," so that a picture of a wasp and the sound of a fly both come to elicit anxiety.

Friman et al.'s (1998) review made few references to anger directly, but did suggest the possibility that this model may be applicable to any emotional response, not only anxiety (which was the focus of their review). Thus, if in some cases angry responses are learned in one context might generalize through transfer of stimulus function within stimulus classes. For example, being humiliated and physically abused when alone with a parent might establish the parent as a conditioned stimulus to elicit conditioned anger. Generalization might occur by the other methods discussed above, such as stimulus generalization to being alone with adults and people who look like the abusive parent. Generalization could also occur through higher order conditioning, for example, the abusive parent might pair themselves with an older sibling and with words like "Mind what your brother says, you better respect him!" But generalization could also occur through transfer of stimulus function within a stimulus class. If the child acquires a class that we might loosely call "Authoritative people," the angry/aggressive response to one member of a class might result in transfer of function to other members of the class such as the coercive math teacher. Over the course of a lifetime, these and other learning mechanisms might for some individuals establish angry behavior as a part of an individual's behavioral repertoire in many contexts far removed from the original acquisition with the abusive parent many years before.

An interesting twist in the analyses of emotional behavior and aggression and violence is the interplay between different emotions. At one point, Friman et al. (1998) mentioned anger in relation to obsessive compulsive disorder and fear of harming someone. People with obsessive compulsive disorder typically spend much time engaged in private verbal behavior, such as worrying about contamination, such as having clean hands whether or not they have performed something important such as unplugging the iron and socially unacceptable behavior, for example, profanity or harming others sometimes through aggression. These thoughts as well as exposure to relevant physical stimuli often elicit anxiety. Thus, avoidance of these thoughts and stimuli including those related to anger is often negatively reinforced by reduction in anxiety. (See McAndrew, 1989 for an example of a functional analysis of such a case.)

8.4 Behavioral Self-Control

In everyday language, we often describe individuals who are aggressive or violent as impulsive; instead of walking away or being assertive, the aggressive individual cannot wait but takes what they want or immediately stops the unpleasantness that they cannot tolerate by being aggressive. This seems analogous to the infamous, so-called marshmallow test in which children can choose to eat the marshmallow when the experimenter leaves the room or wait until the experimenter returns and obtain two marshmallows. Even at the age of 4 years, some children are skilled at waiting in the face of temptation and have skills such as distraction, singing, counting aloud, etc. to help themselves do so. There are others who do not know how to wait and give into temptation quickly. In the original research, Mischel and Ebbesen (1970) showed strong environmental influences on choice of immediate low-quality reinforcer versus delayed better-quality reinforcer. That is, children waited longest when the reinforcers were not immediately visible, but if reinforcers were in sight, they were more likely to behave more impulsively. Since this was an experiment with randomization of participants to conditions the differences could not have been due to participant characteristics, but rather due to the environmental manipulation.

Subsequently, Mischel et al. (1972) reported three experiments that both replicated these effects and also showed that the strategies that the children used also affected impulsive behavior. For example, instructions to think about fun things rather than sad things decreased impulsivity, and instructions to think about the future reinforcers increased impulsive behavior. Mischel and colleagues emphasized the importance of covert behavior, such as shifting attention away from the reinforcers and thinking about other things; however, the importance of cognition in self-control is challenged by subsequent research that showed that nonverbal animals such as rats (Chelonis et al., 1994; Logue et al., 1992), parrots (Vick et al., 2010), and other nonhumans (Beran, 2002) can learn to wait for bigger better reinforcement, if the environment is designed to facilitate that learning.

The marshmallow test studies have been highly cited as a part of the popular accounts of psychology in the media. This is in part because of the ability of the marshmallow test to predict that a broad swath of behavior later in life is as important as it is impressive because even this small sliver of early behavior predicts a wide range of adolescent and adult functioning, adaptive behavior, and mental health outcomes (e.g., Falk et al., 2019; Mischel et al., 1988). The interesting question is how organisms learn to wait either for larger delayed positive reinforcers and to tolerate aversive stimulation and how this can be applied to prevent and treat aggression and violence.

8.4.1 Skinner and Self-Control

Skinner considered self-control as a repertoire consisting of a controlling behavior and a controlled behavior. The controlling behavior (e.g., taking a deep breath) changes the probability of the controlled behavior (making a nasty comment to someone who is annoying.) Self-control was a special form of social behavior, but one in which only one person is involved. Thus, in socially mediated behavior, a friend might tell someone, "Take a deep breath and chill" which reduces the probability of the conversation partner making a nasty comment. In self-control, the process is the same, but only one person emits the controlling and controlled behavior. Self-control can be seen as a choice between an immediate, smaller, low-quality, easily available reinforcer—striking the annoying other person to terminate the aversive stimulus immediately—and a larger, better-quality, more delayed reinforcer requiring an effortful response—making an excuse in a calm and assertive manner to walk away in order to have a better relationship with the person later on. Impulsivity is the choice of the immediate, smaller low-quality reinforcer over a delayed, larger better-quality reinforcer. It is possible that in many situations, one can think of aggression as a problem of self-control, that is, a choice between an immediate reinforcer obtained through aggressive behavior and a delayed and sometimes larger better-quality reinforcer obtained through prosocial behavior.

Table 8.1 Skinner's seven types of self-control strategies and examples applied to self-control of aggression

Technique of control	Description and example
1. Physical restraint and physical aid	Present self to a jail or institution to prevent future aggression; place an ankle bracelet on to stop oneself leaving the house and looking for a fight; put hands in pocket to prevent pointing a finger toward someone's face; remove weapons and alcohol and lock them up to make violence less likely; place snacks and photographs our prior to a visit in order to make the visit with a potentially annoying relative involve more prosocial behavior
2. Changing the stimulus	Remove stimuli that precipitate aggression, e.g., look away from an annoying person to see something else or place your hands over your ears and say "la la!" so you do not hear what the other person says. Present a stimulus to do something else, such as reading a book or take out a written reminder describing how to handle an annoyance without aggression
3. Depriving and satiating	If seeing others suffer is a reinforcer for aggression, beating up a pillow or mattress until exhausted might be a form of satiation to prevent aggression
4. Manipulating emotional conditions	Induce a mood that is incompatible with anger, for example, by listening to relaxing music or rehearse previous appearances in court to "scare oneself to death" by thinking of the damage or problems to do with reoffending if one might commit another act of aggression
5. Using aversive stimulation	Make a public resolution which programs aversive consequences for failure to keep a promise, for example, publicly commit to going to therapy once a week for 3 months; make a cash deposit or deposit of favorite possessions which can only be returned when the therapy is complete; share a goal to engage in prosocial behavior with another person and agree to review progress with them on a fixed date
6. Drugs	Take a prescribed anxiolytic to induce relaxation and thereby reduce the likelihood of anger aggression; comply with taking medication to reduce sexual drive in order to reduce the future probability of violent sexual aggression
7. Operant conditioning	Associate with a friend who prompts and reinforces prosocial behavior; schedule your own reinforcers for prosocial and/or preventative behavior; schedule self-punishment for not reaching a goal, for example, making a cash deposit that will be lost if a goal is not met
8. Punishment	Associate with a friend who punishes precursors to aggression; confess your sins and engage in penance after being aggressive
9. Doing something else	Change the topic of conversation away from something annoying and anger-inducing to something positive in order to reduce arousal and anger; walk away when someone starts to make you angry

Thus, learning to wait is a key self-management skill both in passing the marshmallow test and in reducing aggressive behavior.

Skinner (1953) described nine common self-control strategies which are listed with examples related to aggression in Table 8.1. Many of these are very familiar to all of us. They also are analogous to many mental health interventions in which a therapist teaches a client to behave in a different way to prevent or ameliorate an actual or potentially aggressive behavior.

Skinner (1953) hypothesized that we learn these self-control skills naturally and primarily through caregiver contingencies of punishment. For example, a child who wriggles too much in public and embarrassed their caregivers may be scolded and told to sit up straight with their arms folded. Subsequently, their caregivers no longer receive scowls from other people, and the caregivers are no longer embarrassed. With a sufficient number of trials, the child may learn to avoid the scolding by folding their arms or sitting on their hands in situations in which their behavior has previously been punished and their caretakers will no longer be embarrassed. These approaches to managing one's own behavior are often quite acceptable and give the appearance that the person is autonomously choosing to engage in nonviolent behavior. Skinner, however, is clear that the ultimate source of control is the social and physical environment that teaches these self-management skills.

8.4.2 Interventions to Reduce Impulsivity

One approach to thinking about self-control is to consider it as a choice between two or more available consequences. An impulsive choice is the selection of an immediately available, low-quality reinforcer through a low effort response, and a self-controlled choice is the selection of a delayed, better-quality consequence through a more effortful response. The converse is true for punishers as when one trades off the painful visit to the dentist for a delayed much more painful future experience. In nonhuman animals and humans, these effects can be investigated directly by giving the organism an actual choice between two or more consequences, and in humans, these are sometimes investigated by eliciting verbal responses to hypothetical choices, for example, choosing between \$1 now and \$10 at some point in the future. Studies of different groups of individuals with psychopathology or social problems, such as drug addiction, attention deficit hyperactivity disorder, and suicidality, are often characterized by impulsivity (Critchfield & Kollins, 2001).

It is easy to consider aggression as a form of impulsive behavior in some individuals and situations. Aggression may result in immediate access to reinforcers, whereas alternate prosocial responding may result in delayed access to reinforcers. Some data support such as view. Bridge et al. (2015), in the point subtraction analog experimental paradigm, found that individuals with histories of suicidality and who were unmedicated were more likely to subtract points from a hypothetical confederate than individuals with histories of suicidality and who were medication and than match controls without histories of suicidality. Propensities toward aggression was correlated with family histories of suicidality, suggesting that aggression is associated with delay discounting at least in this population.

Some studies have found that ratings of impulsivity in preschoolers predict a wide range of social and academic outcomes in adolescence and ability to deal with frustration (Mischel et al., 1988), and adolescent Scholastic Aptitude Test scores and ratings of self-control (Shoda et al., 1990) suggest the importance of acquisition of self-control skills generally in adaptive behavior, and at least indirectly related to some aggression-related outcomes such as frustration tolerance. If this is true, it suggests the importance of early intervention to prevent the development of impulsivity and interventions for existing aggression and violence based on models of self-control.

A number of approaches have been used to promote nonimpulsive choices, including some research relating to aggression. The most common approach which can be construed in terms of delay discounting is functional communication training (FCT). FCT can be considered a choice between a problem behavior, such as aggression, and a communication response, such as asking for a person to go away. In the natural environment, aggression is often highly effective and a fluent, practiced response, whereas the communication response is sometimes absent, effortful, not responded to, or responded to infrequently and results in smaller delayed consequences.

There are several approaches to reducing impulsivity. In a study of typical children, Schweitzer and Sulzer-Azaroff (1988) demonstrated that gradual fading (increases) of delays resulted in waiting for larger reinforcers over immediately available smaller reinforcers. FCT interventions often include a variety of interventions that include procedures such as: (a) immediate reinforcement for the communication response; (b) extinction of the problem behavior; (c) manipulations of effort to make communication low effort and problem behavior effortful; (d) teaching delay tolerance by increasing the time between the communication response and its reinforcer, teaching waiting skills sometimes as a form of DRA; and (e) bringing the communication response under appropriate stimulus control that indicates availability of reinforcement for the communication response at only certain times, implying waiting when the reinforcer is unavailable (Tiger et al., 2008). Others have shown that signaling the delay for larger, delayed reinforcers (Vessells et al., 2018), for example, through a hand gesture or a timer during waiting, can promote mands (the nonimpulsive response) and low levels of aggression (the impulsive response) compared to unsignaled waiting (Vollmer et al., 1999). In a study of three typically developing children aged 3–5 years, Newquist et al. (2012) compared three strategies to reduce impulsive choices. They found that briefly stating rules or a countdown timer were ineffective, but providing highly preferred toys during the wait interval was an effective self-control strategy to promote choice of larger delayed reinforcers. These procedures parallel descriptions of self-control in Skinner's (1948) *Walden II*. Finally, Ninness et al. (1995) reported an experiment involving adolescents with serious emotional disturbances and high levels of aggressive behavior in community settings. The authors demonstrated that a treatment package of instructions, modeling, and role play for peer-to-peer aggression resulted in large reductions in peer-to-peer aggression. The authors then used a package of self-management strategies including self-assessment, self-recording, and self-reinforcement to maintain change in the absence of direct supervision which proved effective.

One more comprehensive program to teach self-management skills to typical young children comes from Hanley et al. (2007), who developed and evaluated a classroom-wide preschool life skills program. Many children are now raised less often by their parents than in the past; hence, preschool classroom programs to teach self-management and emotional-behavioral skills may now be more important. In the first study, Hanley et al. worked with 16 preschool children, one of whom had a developmental delay, in a university preschool. The curriculum consisted of four units which included instruction following (responding to name, single/multi-step instruction); FCT (requesting assistance, attention); delay tolerance (with the mediating response "when I wait quietly, I get what I want"); and friendship skills (such as saying thank you and making compliments). They taught these skills directly and explicitly by the classroom teacher who used instruction and modeled for one specific skill at the beginning of class every day. Thereafter, the teacher created situations in the classroom to practice the skill throughout the day and taught each skill using behavioral skills teaching. The program took 6, 8, 4, 8 school days to teach for each of the four units, respectively. The authors reported that children learned many of the skills rapidly and reduced problem behavior including aggression without direct intervention. Subsequently, Hanley et al. (2014) reported a systematic replication study this time with children in a Head Start program.

These studies showed that children can learn self-control/social-emotional skills in typical preschool settings when typical classroom staff deliver the intervention with consultation support. The program results in children acquiring skills acquisition and reducing challenging behavior including aggression. The authors reported highly socially valid ratings. One limitation was that the authors have not yet reported data on generalization and maintenance and long-term effects (Fahmie & Luczynski, 2018; Luczynski & Fahmie, 2017).

8.4.3 Summary

Self-control is a useful framework for thinking about and intervening with aggression. First, it suggests interventions for primary prevention of aggression and violence in typically developing, nonaggressive young children, namely teaching a broad and flexible repertoire of appropriate communication and social behavior and tolerance of delays to reinforcement. These strategies can be implemented in typical service settings. Second, self-control suggests practical solutions for existing problems of aggression including strategies for learning to wait and increasing tolerance for delayed reinforcers. For some people, learning and using self-control repertoires can become a way of life that is instrumental in becoming productive in the things that are important to each person and leading a happy life. This approach helps both avoiding everyday frustration and dealing with them more effectively (Epstein, 1997).

8.5 Verbal Behavior

Much human behavior derives from our repertoire of verbal behavior that allows us to participate in our current and past cultures and benefit from the experiences of other people through rule-governed behavior. Rule-governed behavior is a statement which explicitly or implicitly describes a contingency that results in behavior that is relatively insensitive to its contingencies (Skinner, 1953; Zettle & Hayes, 1982). For example, a parent might teach their child "Don't be a patsy. Hit the other kid first before they hit you. Real men stand up for themselves." If the parent or others have already taught their child to follow rules, the child might well follow this rule too. They may never experience the consequences of not following this rule, and the parents might maintain rule governance generally "You are a smart kid, you know how to do the right thing!" and specifically "I told you so, you hit him before he could hit you!" Others have added that rules may also change the function of stimuli as when someone says, "This is the winning lottery ticket," thereby increasing the reinforcing value of grabbing the winning ticket.

Zettle and Hayes (1982) distinguished three classes of rules. A ply (or "pliance") is "rule-governed behavior under the control of apparent socially mediated consequences for a correspondence between the rule and relevant behavior" (Hayes et al., 1989, p. 203). For example, if a child is told "good children wait for cookies, bad children hit their friends," the child patiently waits, and the caregiver tells the child "You are such a nice boy!" then this would be a ply. Tracking is "rule-governed behavior under the control of the apparent correspondence between the rule and the way the world is arranged" (Hayes et al., 1989, p. 206). An example of a track would be "Go tell your grandma she is sweet and she will give you a cookie," the child then does this, and gets the cookie. Augmenting is "rule-governed behavior under the control of apparent changes in the capacity of events to function as reinforcers or punishers" (Hayes et al., 1989, p. 206). An example of augmenting might be if an offender is told "When you go from 'green' status to 'yellow' status on your warning system, you will probably end up back in court and will have to pay more fines' and subsequently the offender avoids relapse related behavior that would place them on 'yellow' status."

Despite the notion of rule-governed behavior being discussed for over 60 years, there is relatively little empirical research linking it directly to aggression and violence. Rule-governed behavior related to prosocial behavior and compliance with rules to reduce aggression might, however, be two examples. Many practitioners and parents attempt to use rules, as when a teacher instructs a child "If you have safe hands, you get to play on the computer" and when a therapist tells an emotionally cold psychopath, "Maybe people are stupid, but if you start treating them better you might get out of the locked ward faster." Perhaps if attempts to use rule-governed behavior were better informed by how rule-governed behavior can be taught, these attempts might be more successful.

8.6 Summary

The material reviewed in this chapter shows that learning during the lifespan plays a key role in the development of aggression and violence, irrespective of whether the organism has minimal or strong biological and cultural influences toward aggression and violence. ABA has focused on operant aggression. Although it provides an account of emotions such as anger and rule-governed behavior, there is less empirical research in this area.

Behavior analytic studies of aggression are often conducted over a relatively short period of time, and they often study naturally high rate of aggression or aggression which can readily be evoked for the purpose of functional assessment and analysis. This might lead one to ask, can extreme, low-frequency violence be learned? It is tempting to think that it cannot because it seems irrational, uncontrollable, and inexplicable. But, perhaps if we cannot explain something, it might be worthwhile considering that we might be able to do so. Sometimes when we see a single extreme incident of violence, it is tempting to pathologize the behavior for at least two reasons: first, it is so intense; and, second, because it is the first instance of violence, it could not have been reinforced previously. However, there is a literature on forensic case formulation which addresses extreme violence and attempts to explain this through lifelong learning (Gresswell & Hollin, 1992; Sturmey & McMurran, 2011). Further, even though the first occurrence of extreme violence logically could not have been previously reinforced, it may be useful to consider it as just another example of the problem of the first incident of any behavior—the child's first step or first word—for example. Later chapters will address the extension of behavioral explanations of the development of extreme violence.

References

Andronis, P. T., Layng, T. J., & Goldiamond, I. (1997). Contingency adduction of "symbolic aggression" by pigeons. *The Analysis of Verbal Behavior, 14*(1), 5–17.

Archer, J. (1995). What can ethology offer the psychological study of human aggression? *Aggressive Behavior, 21*(4), 243–255.

Azrin, N. H., Hake, D. F., & Hutchinson, R. R. (1965). Elicitation of aggression by a physical blow. *Journal of the Experimental Analysis of Behavior, 8*(1), 55–58.

Azrin, N. H., Hutchinson, R. R., & Hake, D. F. (1966). Extinction-induced aggression. *Journal of the Experimental Analysis of Behavior, 9*(3), 191–204.

Baer, D. M., & Deguchi, H. (1985). Generalized imitation from a radical-behavioral viewpoint. In S. Reiss & R. R. Bootzin (Eds.), *Theoretical issues in behavior therapy* (pp. 179–217). Academic Press.

Bailey, J. S. (1987). Comment from readers. *Journal of Applied Behavior Analysis, 2*, 203–204.

Bandura, A. (1973). *Aggression: A social learning analysis*. Prentice-Hall.

Bandura, A., & Huston, A. C. (1961). Identification as a process of incidental learning. *The Journal of Abnormal and Social Psychology, 63*(2), 311–318.

Bandura, A., Ross, D., & Ross, S. A. (1961). Transmission of aggression through imitation of aggressive models. *The Journal of Abnormal and Social Psychology, 63*(3), 575.

Bandura, A., Ross, D., & Ross, S. A. (1963). Imitation of film-mediated aggressive models. *The Journal of Abnormal and Social Psychology, 66*(1), 3–11.

Bandura, A., & Walters, R. H. (1963). *Social learning and personality development*. Holt, Reinhart, & Winston.

Beran, M. J. (2002). Maintenance of self-imposed delay of gratification by four chimpanzees (Pan troglodytes) and an orangutan (Pongo pygmaeus). *The Journal of General Psychology, 129*(1), 49–66.

Berkowitz, L. (1993). Pain and aggression: Some findings and implications. *Motivation and Emotion, 17*(3), 277–293.

Bridge, J. A., Reynolds, B., McBee-Strayer, S. M., Sheftall, A. H., Ackerman, J., Stevens, J., … Brent, D. A. (2015). Impulsive aggression, delay discounting, and adolescent suicide attempts: Effects of current psychotropic medication use and family history of suicidal behavior. *Journal of Child and Adolescent Psychopharmacology, 25*(2), 114–123.

Brondolo, E., DiGiuseppe, R., & Tafrate, R. C. (1997). Exposure-based treatment for anger problems: Focus on the feeling. *Cognitive and Behavioral Practice, 4*(1), 75–98.

Brown, A. K., Brown, J. L., & Poulson, C. L. (2008). Discriminating which fork to use: Teaching selective imitation to people with autism. *Research in Autism Spectrum Disorders, 2*(2), 199–208.

Chelonis, J. J., King, G., Logue, A. W., & Tobin, H. (1994). The effect of variable delays on self-control. *Journal of the Experimental Analysis of Behavior, 62*(1), 33–43.

Critchfield, T. S., & Kollins, S. H. (2001). Temporal discounting: Basic research and the analysis of socially important behavior. *Journal of Applied Behavior Analysis, 34*, 101–122.

DeQuinzio, J. A., Townsend, D. B., Sturmey, P., & Poulson, C. L. (2007). Generalized imitation of facial models by children with autism. *Journal of Applied Behavior Analysis, 40*(4), 755–759.

DiGiuseppe, R., Cannella, C., & Kelter, J. (2007). Effective anger treatments require a functional analysis of the anger response. In A. Cavell & K. T. Malcolm (Eds.), *Anger, aggression and interventions for interpersonal violence* (pp. 55–76). Lawrence Erlbaum Associates Publishers.

Dollard, J., Miller, N. E., Doob, L. W., Mowrer, O. H., & Sears, R. R. (1939). *Frustration and aggression*. Yale University Press.

dos Reis Soares, P. F., & Goulart, P. R. K. (2016). Induced aggression in JEAB: A survey of publications. *Revista Brasileira de Análise do Comportamento, 11*(1), 11–15.

Dougher, M. J., & Markham, M. R. (1994). Stimulus equivalence, functional equivalence and the transfer of function. In S. C. Hayes, L. J. Hayes, M. Sato, & K. Ono (Eds.), *Behavior analysis of language and cognition* (pp. 71–90). Context Press.

Epstein, R. (1997). Skinner as self-manager. *Journal of Applied Behavior Analysis, 30*(3), 545–568.

Epstein, R., Lanza, R. P., & Skinner, B. F. (1980). Symbolic communication between two pigeons, (*Columba livia domestica*). *Science, 207*(4430), 543–545.

Fahmie, T. A., & Luczynski, K. C. (2018). Preschool life skills. Recent advancements and future directions. *Journal of Applied Behavior Analysis, 51*(1), 183–188.

Falk, A., Kosse, F., & Pinger, P. (2019). Re-revisiting the marshmallow test: A direct comparison of studies by Shoda, Mischel, and Peake (1990) and Watts, Duncan, and Quan (2018). *Psychological Science, 31*(1), 100–104. https://doi.org/10.1177/0956797619861720

Falk, J. L. (1971). The nature and determinants of adjunctive behavior. *Physiology & Behavior, 6*(5), 577–588.

Farris, H. E., Gideon, B. E., & Ulrich, R. E. (1970). Classical conditioning of aggression a developmental study. *The Psychological Record, 20*(1), 63–67.

Fernandez, E. J. (2020). Training petting zoo sheep to act like petting zoo sheep: An empirical evaluation of response-independent schedules and shaping with negative reinforcement. *Animals, 10*(7), 1122. https://doi.org/10.3390/ani10071122

Fields, L., & Verhave, T. (1987). The structure of equivalence classes. *Journal of the Experimental Analysis of Behavior, 48*(2), 317–332.

Foster, W. S. (1978). Adjunctive behavior: An under-reported phenomenon in applied behavior analysis? *Journal of Applied Behavior Analysis, 11*(4), 545–546.

Frederiksen, L. W., & Peterson, G. L. (1974). Schedule-induced aggression in nursery school children. *The Psychological Record, 24*(3), 343–351.

Frederiksen, L. W., & Peterson, G. L. (1977). Schedule-induced aggression in humans and animals: A comparative parametric review. *Aggressive Behavior, 3*(1), 57–75.

Friman, P. C., Hayes, S. C., & Wilson, K. G. (1998). Why behavior analysts should study emotion: The example of anxiety. *Journal of Applied Behavior Analysis, 31*(1), 137–156.

Furnell, J. R. G., & Thomas, G. V. (1976). Stimulus control of generalized imitation in subnormal children. *Journal of Experimental Child Psychology, 22*(2), 282–291.

Ghaemmaghami, M., Hanley, G. P., Jessel, J., & Landa, R. (2018). Shaping complex functional communication responses. *Journal of Applied Behavior Analysis, 51*(3), 502–520.

Glenn, S. S. (1988). Contingencies and metacontingencies: Toward a synthesis of behavior analysis and cultural materialism. *The Behavior Analyst, 11*(2), 161–179.

Gresswell, D. M., & Hollin, C. R. (1992). Towards a new methodology for making sense of case material: An illustrative case involving attempted multiple murder. *Criminal Behaviour and Mental Health, 2*(4), 329–341.

Hanley, G. P., Fahmie, T. A., & Heal, N. A. (2014). Evaluation of the preschool life skills program in head start classrooms: A systematic replication. *Journal of Applied Behavior Analysis, 47*(2), 443–448.

Hanley, G. P., Heal, N. A., Tiger, J. H., & Ingvarsson, E. T. (2007). Evaluation of a classwide teaching program for developing preschool life skills. *Journal of Applied Behavior Analysis, 40*(2), 277–300.

Hart, B., & Risley, T. R. (1995). *Meaningful differences in the everyday experience of young American children*. Paul H. Brookes Publishing.

Hart, S., Sturmey, P., Logan, C., & McMurran, M. (2011). Forensic case formulation. *International Journal of Forensic Mental Health, 10*(2), 118–126.

Hausman, N., Kahng, S., Farrell, E., & Mongeon, C. (2009). Idiosyncratic functions: Severe problem behavior maintained by access to ritualistic behaviors. *Education and Treatment of Children, 32*(1), 77–87.

Hayes, S. C., Zettle, R. D., & Rosenfarb, I. (1989). Rule-following. In S. C. Hayes (Ed.), *Rule-governed behavior* (pp. 191–220). Springer.

Hollis, K. L. (1984). The biological function of Pavlovian conditioning: The best defense is a good offense. *Journal of Experimental Psychology: Animal Behavior Processes, 10*(4), 413.

Hollis, K. L. (1990). The role of Pavlovian conditioning in territorial aggression and reproduction. In D. A. Dewsbury (Ed.), *Contemporary issues in comparative psychology* (pp. 197–219). Sinauer Associates. https://doi.org/10.1037/11525-009

Knutson, J. F. (1970). Aggression during fixed-ration and extinction components of a multiple schedule of reinforcement. *Journal of the Experimental Analysis of Behavior, 13*(2), 221–231.

Layng, T. J., Andronis, P. T., & Goldiamond, I. (1999). Animal models of psychopathology: The establishment, maintenance, attenuation, and persistence of head-banging by pigeons. *Journal of Behavior Therapy and Experimental Psychiatry, 30*(1), 45–61.

Lerman, D. C., Iwata, B. A., & Wallace, M. D. (1999). Side effects of extinction: Prevalence of bursting and aggression during the treatment of self-injurious behavior. *Journal of Applied Behavior Analysis, 32*(1), 1–8.

Logue, A. W., Tobin, H., Chelonis, J. J., Wang, R. Y., Geary, N., & Schachter, S. (1992). Cocaine decreases self-control in rats: A preliminary report. *Psychopharmacology, 109*(1), 245–247.

Looney, T. A., & Cohen, P. S. (1982). Aggression induced by intermittent positive reinforcement. *Neuroscience & Biobehavioral Reviews, 6*(1), 15–37.

Lovaas, O. I. (1961). Effect of exposure to symbolic aggression on aggressive behavior. *Child Development*, 37–44.

Luczynski, K. C., & Fahmie, T. A. (2017). Preschool life skills: Toward teaching prosocial skills and preventing aggression in young children. In P. Sturmey (Ed.), *The Wiley handbook of violence and aggression* (pp. 1–12). Wiley.

Lyon, D. O., & Ozolins, D. (1970). Pavlovian conditioning of shock-elicited aggression: A discrimination procedure. *Journal of the Experimental Analysis of Behavior, 13*(3), 325–331.

March, S. M., Abate, P., Spear, N. E., & Molina, J. C. (2009). Fetal exposure to moderate ethanol doses: Heightened operant responsiveness elicited by ethanol-related reinforcers. *Alcoholism: Clinical and Experimental Research, 33*(11), 1981–1993.

Marcus, A., & Wilder, D. A. (2009). A comparison of peer video modeling and self video modeling to teach textual responses in children with autism. *Journal of Applied Behavior Analysis, 42*(2), 335–341.

McAndrew, J. F. (1989). Obsessive-compulsive disorder: A behavioral case formulation. *Journal of Behavior Therapy and Experimental Psychiatry, 20*(4), 311–318.

Mischel, W., & Ebbesen, E. B. (1970). Attention in delay of gratification. *Journal of Personality and Social Psychology, 16*(2), 329–337.

Mischel, W., Ebbesen, E. B., & Raskoff Zeiss, A. (1972). Cognitive and attentional mechanisms in delay of gratification. *Journal of Personality and Social Psychology, 21*(2), 204.

Mischel, W., Shoda, Y., & Peake, P. K. (1988). The nature of adolescent competencies predicted by preschool delay of gratification. *Journal of Personality and Social Psychology, 54*(4), 687–696. https://doi.org/10.1037/0022-3514.54.4.687

Murphy, G., MadDOnald, S., Hall, S., & Oliver, C. (2000). Aggression and termination of "rituals": A new variant of the escape function for challenging behavior? *Research in Developmental Disabilities, 21*(1), 43–59.

Newquist, M. H., Dozier, C. L., & Neidert, P. L. (2012). A comparison of the effects of brief rules, a timer, and preferred toys on self-control. *Journal of Applied Behavior Analysis, 45*(3), 497–509.

Ninness, H. C., Ellis, J., Miller, W. B., Baker, D., & Rutherford, R. (1995). The effect of a self-management training package on the transfer of aggression control procedures in the absence of supervision. *Behavior Modification, 19*(4), 464–490.

Pelaez, M., Virues-Ortega, J., & Gewirtz, J. L. (2011). Reinforcement of vocalizations through contingent vocal imitation. *Journal of Applied Behavior Analysis, 44*(1), 33–40.

Schaeffer, H. H. (1970). Self-injurious behavior: Shaping "head banging" in monkeys. *Journal of Applied Behavior Analysis, 3*, 111–116.

Schlinger, H. D. (2002). Concepts in behavioral development. *Behavioral Development Bulletin, 11*(1), 1.

Schweitzer, J. B., & Sulzer-Azaroff, B. (1988). Self-control: Teaching tolerance for delay in impulsive children. *Journal of the Experimental Analysis of Behavior, 50*(2), 173–186.

Shoda, Y., Mischel, W., & Peake, P. K. (1990). Predicting adolescent cognitive and self-regulatory competencies from preschool delay of gratification: Identifying diagnostic conditions. *Developmental Psychology, 26*(6), 978.

Skinner, B. F. (1948). *Walden Two*. McMillan.

Skinner, B. F. (1953). *Science and human behavior*. Free Press.

Sturmey, P., & McMurran, M. (Eds.). (2011). *Forensic case formulation* (Vol. 49). John Wiley & Sons.

Thompson, T., & Sturm, T. (1965). Classical conditioning of aggressive display in Siamese fighting fish. *Journal of the Experimental Analysis of Behavior, 8*(6), 397–403.

Tiger, J. H., Hanley, G. P., & Bruzek, J. (2008). Functional communication training: A review and practical guide. *Behavior Analysis in Practice, 1*(1), 16–23.

Ulrich, R. (1966). Pain as a cause of aggression. *American Zoologist, 6*(4), 643–662.

Vernon, W., & Ulrich, R. (1966). Classical conditioning of pain-elicited aggression. *Science, 152*(3722), 668–669.

Vessells, J., Sy, J. R., Wilson, A., & Green, L. (2018). Effects of delay fading and signals on self-control choices by children. *Journal of Applied Behavior Analysis, 51*(2), 374–381.

Vick, S. J., Bovet, D., & Anderson, J. R. (2010). How do African grey parrots (Psittacus erithacus) perform on a delay of gratification task? *Animal Cognition, 13*(2), 351–358.

Vollmer, T. R., Borrero, J. C., Lalli, J. S., & Daniel, D. (1999). Evaluating self-control and impulsivity in children with severe behavior disorders. *Journal of Applied Behavior Analysis, 32*(4), 451–466.

Wheeler, L., & Smith, S. (1967). Censure of the model in the contagion of aggression. *Journal of Personality and Social Psychology, 6*(1), 93.

White, P., O'Reilly, M., Fragale, C., Kang, S., Muhich, K., Falcomata, T., … Lancioni, G. (2011). An extended functional analysis protocol assesses the role of stereotypy in aggression in two young children with autism spectrum disorder. *Research in Autism Spectrum Disorders, 5*(2), 784–789.

Young, J. M., Krantz, P. J., McClannahan, L. E., & Poulson, C. L. (1994). Generalized imitation and response-class formation in children with autism. *Journal of Applied Behavior Analysis, 27*(4), 685–697.

Zettle, R. D., & Hayes, S. C. (1982). Rule-governed behavior: A potential theoretical framework for cognitive–behavioral therapy. In P. C. Kendall (Ed.), *Advances in cognitive–behavioral research and therapy* (pp. 73–118). Academic Press.

Chapter 9
Psychological and Sociological Theories of Violence and Aggression

This chapter reviews psychological and sociological theories of aggression and violence, of which there are many. Bjørkly (2006) identified at least 104 theories of aggression some 20 years ago; surely there are many more today! Theories are attempts to go beyond prediction of a phenomenon but rather attempt to explain the observations in terms of some other type of variables. Thus, behavior may be explained in terms of various physiological variables, such as neurotransmission; mental events, such as expectations; and, cognitive theories, where hypothetical structures and cognitive processes explain behavior (Skinner, 1950).

Theories of aggression and violence vary tremendously in their content and scope. They can be focused on specific contexts, issues, and populations such as theories of workplace violence, intimate partner violence or school violence, or may be broad and encompassing such as the General Aggression Model (Allen et al., 2018) and Social Role Theory (SRT) (Eagly & Wood, 2011). These broad theories attempt to integrate all information we have about aggression from psychology, sociology, criminology, and evolutionary biology into one comprehensive, all embracing model. Theories of aggression also vary tremendously in terms of the explanatory variables they consider to be relevant and pertinent. Thus, sociobiology attempts to explain aggression in terms of inclusive fitness, the survival benefit to all copies of one's genes (McKenna, 1983), and feminist theories of aggression attempt to explain violence against women in terms of concepts such as patriarchy and imbalance of social power (McPhail, 2016).

This chapter will selectively review several representative types of psychological and sociological theories of aggression and violence. The reader should note that there is no attempt to review these theories of aggression comprehensively; such a task would be almost impossible within the scope of this chapter. The reader is referred to Chap. 4, *Biological Evolution of Violence and Aggression. I. Evolution and Genetics,* and Chap. 5, *Biological Evolution of Violence and Aggression. II: Brains, Neurotransmitters, and Hormones* for a more detailed review of various biological theories of aggression.

P. Sturmey, *Violence and Aggression*, Advances in Preventing and Treating Violence and Aggression, https://doi.org/10.1007/978-3-031-04386-4_9

9.1 Psychoanalytic Theory

Psychoanalysis is not a single unified field, and it is one that has been around for more than a hundred years. Hence, there is really no single psychoanalytic account of aggression. In this section, we will consider three examples of such theories: the work of Sigmund Freud, Anna Freud, and a more contemporary contributor Peter Fonagy.

9.1.1 Sigmund Freud's Theories

Sigmund Freud's early theory of aggression proposed that aggression was caused by two innate, instinctual, and evolutionarily derived drives: *Thanatos*, the destructive energy or drive for aggression, and *Eros*, the drive for pleasure. Both are a part of the hidden *Id*. Thanatos is expressed through aggression to others and to oneself. The Id negotiates with reality through the rational Ego and the idealistic Superego. The hidden conflicts between Thanatos and Eros cause inner tension and discomfort. The rest of the individual's personality structure either uses defense mechanisms to reduce tension or blocks conscious awareness of the discomfort. Thanatos is also the source of vitality, energy, autonomy, exploration that contribute to the survival of the organism. According to psychoanalysis, the aggressive drive cannot be really treated, but the psychic energy can be channeled or controlled in various ways which may be adaptive or problematic. Thus, aggression and violence are seen as potentially useful in that they regulate the individual's personal life and interpersonal relationships.

According to Sigmund Freud, a person's current aggression and violence reflect three factors: the expression of an unconscious conflict between parts of the psyche, impaired personality functioning, and dysfunctional relationships. Early experiences get copied through the process of identification (behaving like significant others, such as a parent), recapitulation (behaving as being in control of the other), and introjection (treating the self in the same way as others treated them in the past). Aggression and violence are developmental problems that at first are instinctual and unconscious, but, by the time the child is 4 years old the child can have a mental picture of others and their mental states (i.e., can "mentalize"). That is the child can imagine the other person's mind and mental state and can plan and act out a plan to make the other person suffer in the future. The ability to mentalize can also inhibit one's aggression and violence when the picture of the other person's suffering stops one from acting on the impulse (Taubner et al., 2017). (See below for further discussion of mentalization and aggression.)

9.1.2 Anna Freud's Contribution

Sigmund Freud's daughter, Anna Freud, also made important contributions to psychoanalytic models of aggression by extending and elaborating on her father's work. Early in her career, she was trained as a teacher of young children and later was personally involved in the care of a friend's children. These experiences gave her direct experience of working with many different children. After training as a psychoanalyst, she established the first clinics for child psychoanalysis in Vienna. Then, after the Anschluss, Sigmund Freud paid a large ransom to the Nazis, allowing some but not all members of his immediate family safe passage to London. Sigmund Freud lost four sisters murdered in the Holocaust. In London, Anna Freud went on to set up her own institute and clinic and worked with children, including refugee children from the Holocaust, many of whom had lost both of their parents as well as other friends and family members. This institute provided some of the earliest child psychology services in the United Kingdom and was a center for professional training and development for psychoanalysis in the United Kingdom and elsewhere (Young-Bruehl, 2008).

Following her work with these children, Anna Freud (1947) wrote about the development of normal and abnormal childhood aggression and related aspects of child development and personality. Anna Freud saw childhood not as a peaceful idealized phase of life but as one characterized by conflicts between innate instincts which eventually resulted in the child's ability to negotiate the conflicts between instincts requiring immediate gratification and the limitations of the real world. The ego can resolve these conflicts adaptively through denial, repressing the instincts, submitting to the environment (becoming the "good child"), or submitting to their instincts (becoming the bad, rebellious child).

According to Anna Freud (1947), there are two main instincts. The sexual instinct derives pleasure through touching and stimulation of the skin, mouth, and anus. Kissing, touching, and looking are apparently normal forms of expression of the sexual instinct and abnormal expressions include what Anna Freud regarded as sexual perversions or immature behavior, such as cunnulingus, etc. The aggressive instinct includes aggression turned toward oneself and destructiveness. Aggression turned inward included head banging and hair pulling in typically developing children, and more severe forms of self-injury as sometimes seen in children with autism and intellectual disabilities. The aggressive instinct is expressed through destructive love such as a child destroying one's teddy bear. In normal development, the sexual and aggressive instincts fuse and result in a healthy manner of dealing with the world and a healthy personality.

Pathological aggression, according to Anna Freud (1947), is seen in children who have been orphaned, institutionalized, or otherwise traumatized. It is expressed in apparently senseless aggression toward physical objects and people, which is cruel and hurtful, and the children are difficult to handle. Anna Freud concluded that direct treatment of the aggressive behavior would be useless and possibly harmful, and, rather, aggression should be treated by attempts to fuse the sexual and aggressive drives. She put forward the idea that in response to aggression from others, including parents and other caretakers, children identify with the aggressor. Recognizing aggression as a problem in some of these children, Anna Freud recommended that they should form a close relationship with a particular staff as a kind of substitute parent. Anecdotally, she reported that this helped with aggression. Note that many of these ideas were widely current in the 1940s onward and are reflected in the work of people such as Bettelheim in the field of autism.

9.1.3 *Fonagy, Mentalization, and Aggression*

Psychoanalysis has continued to be active in the field of explaining and treating aggression. For example, Fonagy (2003) posited that it is the development of mentalization skills during early child development that accounts for the well-documented decrease in aggression in early childhood. Further, he hypothesized that aggression and violence interfere with human's ability to imagine other people's minds and thereby promotes aggression. Fonagy suggested that failure to acquire mentalization, skills through inability to read facial expression and vocal tones, results in individuals who are aggressive and whom we call psychopaths.

Fonagy (2003) also hypothesized that some people fail to acquire mentalization because an attachment figure makes thinking about others' minds anxiety-provoking and so, rather than using prosocial behavior, the person uses aggression to control other people's behavior. Recovery from the path toward adult aggression and violence in adolescence is mediated by the development of attachment to others and the development of healthy social relationships, and in this way, the adolescent may develop mentalization that diverts them from the path to adult violence.

Mentalization also could have implications for early intervention, prevention, and treatment of aggression. For example, in the dyadic situation between caretaker and infant or young child, the reading or misreading of each other's mental states and/or intentions might prevent or predispose a child toward aggression. Interventions such as training to increase the sensitivity and accuracy of caretakers' mind reading and interaction skill with the child might be effective in preventing subsequent aggression. Unlike Sigmund and Anna Freud and many other psychoanalysts, Peter Fonagy has conducted conventional group research to evaluation psychotherapy.

9.1.4 *Comment*

Psychoanalysis is a very broad field with a great variety of theoreticians and writers working for over a hundred years. Hence, making generalizations is difficult. But, two kinds of limitations should be considered. The first is that psychoanalysis uses many nonobservable hypothetical variables to explain aggression and other social behavior and hence is difficult to test. The second is that the empirical support for psychoanalysis, including some predictions from psychoanalytic theory and the outcomes of some psychoanalytic treatments for aggression, is weak or has been directly disconfirmed.

One first challenging issue of psychoanalysis is use of many observables to explain aggression. For example, concepts such as drives, conflicts between and fusion of drives, and identification with the aggressor as explanations are not directly observable. The use of such explanations of aggression requires, at a minimum, reliable and valid measurement of such concepts. Do two independent psychoanalysts agree if the sex and aggression drives are fused or not? Further, given that observation of the alleged outcome of the changes in the constructs used to explain the change in aggression is necessarily confounded with the change in observed aggression, it is difficult to imagine how these two sets of variables could be measured independent of one another. For example, if, as a child grows older and becomes less aggressive, this knowledge may well influence any judgment as to whether or not the hypothetical sexual and aggressive drives have fused. That is, can two psychoanalysts independently agree both that the drives did not fuse, although aggression did decrease?

Second, some testable empirical predictions from psychoanalysis have been disconfirmed. Sigmund Freud and other psychoanalysts predicted that engagement in cathartic activities should reduce future aggression. For example, Freud wrote to Einstein "forces are turned to destruction in the external world, the living creature will be relieved and the effect must be beneficial" (Freud, 1970, p. 18; cited in Lohr et al., 2007). Catharsis is recognized in everyday speech when we say we should "let it out" or "get it off our chest." Although the immediate relief from anger and tension from such practices is familiar to many of us, it is unclear if this is necessarily effective long-term or even harmful treatment for aggression and violence.

Unfortunately, there is little evidence to support the effectiveness of catharsis in the treatment of aggression and violence which undermines the credibility of psychoanalytic theories of aggression. For example, in a review of primal scream therapy, a popular treatment fad from 1970s, and related therapies involving expressing anger, Lohr et al. concluded that catharsis was ineffective and sometimes positively harmful, not only by depriving the client of access to effective therapies, but also because of the possibility of client harms and increasing anger and violence in certain circumstances. In fact, several authors have concluded that such approaches may *increase* rather than decrease aggression (Bushman, 2002), perhaps through modeling, practicing aggression and violence, or increasing rather than abating anger. Lohr et al. concluded that catharsis was in effect a harmful and unethical practice. Thus, psychoanalytic theory is challenged by the failure of at least some of its predictions to be confirmed. (The outcomes of psychoanalytic treatments for aggression will be reviewed in Chap. 17, *Individual Therapies II for Violence and Aggression. Other Therapies*.)

9.2 The Frustration-Aggression Hypothesis

One of the earliest psychology/learning theories comes from Dollard et al. (1939), who boldly asserted that "aggression is always a consequence of frustration" (p. 1). This group of researchers had an intellectual agenda to translate the dubious and difficult-to-test psychoanalytic concepts to testable scientific concepts. Obviously, there are many everyday anecdotal examples of goal frustration sometimes resulting in aggressive behavior. In their analysis, however, Dollard et al. proposed that there is an *instigator*, an antecedent with a subsequent predicted response, such as an ice cream van and a bell that predict the availability of ice cream. The instigator was said to be broader than any immediate specific stimulus and referred to the broader context of the behavior. The *strength of the instigator* refers to the extent to which the instigated response competes with other ongoing responses. Repetitions of the response strengthens and reinforces the chain of behavior, leading up to the achievement of a goal. Frustration is said to exist when the organism could have performed certain acts and was prevented from doing so. A *substitute response* is one which reduces the strength of the goal-response, for example, having a cigarette while waiting for lunch to arrive when you are hungry. Aggression may occur when the goal-response is frustrated, when instigation is high and there are no substitute responses available, for example, when trying to wait for lunch when hungry and you are out of cigarettes. Aggression can include not only overt motor behavior, but also covert phantasies, dreams, or plans of revenge. Aggression may be directed at the person or object frustration the goal-response, but may also be directed at other people or objects, or the self, as in masochism, martyrdom, or suicide. Thus, Dollard et al. defined aggression as "that response which follows frustration, reduces only the secondary, frustration-produced instigation, and leaves the strength of the original instigation unaffected" (p. 11) and frustration as "condition which exists when a goal-response suffers interference. Aggression is independently defined as an act whose goal-response is injury to an organism (or organism-surrogate)" (p. 11). The authors give a wide range of everyday examples, which include broad examples, such as frustration building up over days and subsequently being redirected to other available people or objects. For example, according to Dollard et al., a built up frustration from work over several days may manifest itself by sudden bad temper with family members when the parent comes home.

Dollard et al. (1939) went on to identify variables that might influence this process. These variables included: (1) the amount of frustration, as measured by the strength of instigation, degree of interference, and number of frustrated response sequences; and (2) punishment of frustration-induced aggression. Aggression is most likely directed against the agent of the frustration, and, if this is blocked, aggression may be directed at substitute people and objects. Aggression toward the self is more likely if the person believes themselves to be responsible for the frustration. The authors also posited that acts of aggression were cathartic, that is, reduced the likelihood of future aggression. Finally, the notions of catharsis and displacement of aggression open the doors to considering a wide range of behavioral phenomena.

Dollard et al.'s (1939) *Frustration and Aggression* was a very important book because it opened up an alternative to psychoanalysis for explaining aggression and to potentially testable hypotheses. Today, its limits seem obvious. Namely, it expanded the definitions of both aggression and frustration considerably making it difficult to test the hypothesis. For example, in defining "aggression," they included humor and apparently limitless expressions of indirect aggression. Similarly, the expansion of displaced aggression makes it difficult to identify the frustrating event. Dollard et al. expansively attempted to link the rates of lynchings in southern American states with economic and other indices of apparent frustration. Others have criticized the frustration aggression hypothesis for not incorporating emotions and cognitions into an account of aggression (Berkowitz, 1990) (see below). The frustration-aggression hypothesis was and remains very influential, as it has been incorporated and modified into social learning theory (SLT) and the General Aggression Model (GAM), among other more recent accounts of aggression.

9.3 Social Learning Theory

SLT is one example of a broad-based learning explanation of aggression and violence which uses a wide range of learning concepts. Although many versions of SLT are not thoroughly behavior analytic, SLT is a useful vehicle to use to consider the acquisition of violence and aggression during the lifespan using learning concepts.

SLT refers to a broader learning approach to explain and treat aggression and violence which dates back to the 1940s (Miller & Dollard, 1941; Sears, 1941) and includes approaches characterized by rejection of psychoanalytic formulations of aggression, some rather general mention of learning as being important, or, in some cases, specific learning processes within families, educational settings, and elsewhere. Learning mechanisms might include classical behavioral mechanisms or cognitive theories of aggression. Building on and expanding the work of Bandura, SLT proposed that aggression is learned socially through siblings, parents, and peers and through modeling, positive reinforcement, such as social approval, and negative reinforcement, such as avoidance of chores or social interaction. These learning processes may interact with dispositions of the child, such as impulsivity, self-control, and sensitivity to reward and punishment, and dispositions of others, such as parents and other caretakers (Goodnight et al., 2017).

In a study using an SLT framework, Eron and Colleagues published a 22-year longitudinal study of the development of aggression in a cohort of participants in one semirural county of New York. The original sample consisted of approximately 870 children aged 7–9 years in 1960 (Eron, 1987). This group of authors was able to follow the sample up 10 and 22 years later. Like the other longitudinal studies of aggression such as that by Côté et al. (2006), Eron et al. found considerable stability of aggression over the lifespan but also lasting effects of environmental variables early during development on adult aggression and criminality. They found that at age 7–9 years, aggression was predicted by less nurturing and accepting parents, punishment in the home, and not identifying with their parents. They also found an interesting interaction between these variables in that children who identified with their parents were most sensitive to the suppressive effect of punishment on aggression and vice versa 10 years later. Eron et al. found that aggression at time 1 predicted persistence of aggression. Children who were aggressive at time 1 were three times more likely than the rest of the sample to have police records by the time they were 19 years old. Additionally, children who identified with their parents, who showed signs of remorse and guilt and who did not watch less violent TV than the rest of the sample at time 1 were less likely to be aggressive 10 years later. Behavior at time 1 *still* predicted aggression at 22-year follow-up! Aggression at time 1 continued to predict criminal behavior, aggression to spouses, and punishment of their own subsequent children. Additionally, measures of prosocial behavior, friendliness, and watching violent TV at time 1 also predicted aggression, seriousness of criminal behavior, and social and economic success 22 years later. The authors concluded that learning of aggressive and prosocial behavior early in childhood was a powerful influence on continued aggression over the lifespan.

9.3.1 The Coercive Family Processes Model

One of the most important statements of SLT comes from Gerald Patterson's (1982) Coercive Family Processes. In this model, the future aggressive child is both "architect and victim of a coercive system" (Patterson, 1976; Patterson et al., 1984).

Patterson and colleagues (1998) summarized a three-phase developmental model of antisocial behavior, including aggression. In phase 1, training and supervision in the home is deficient. Parents rely on discipline that is both harsh and inconsistent, they have little positive involvement with their children, and they fail to supervise and monitor their children. The children fail to identify with their parents and social norms for work and discipline, and fail to learn self-control. Parents reinforce deviant child behavior many times a day, sometimes through approval, attention, and laughing, but mostly through avoidance of aversive intrusions from other family members. Thus, children's antisocial behavior is highly functional in limiting aversive interactions with the family.

Over time the intensity of parental coercive practices and child antisocial behavior escalate. Further, parents ignore their children's appropriate behavior, and hence, children fail to learn prosocial behavior in the home, resulting in children who are antisocial, socially unskilled, and unable to manage their own frustrations. In the second phase, children are rejected in part because of their poor social skills and inability to tolerate daily frustrations, by their peers, and thus experience school failure. The children's uncontrolled problem behavior interferes with their learning of academic skills, they spend less time on task, and do not complete homework further weakening their academic and prevocational skills. Further, the children's antisocial behavior leads to rejection by many peers as the children fail to adopt behavioral norms of the group, and over-react to peer provocation. In the third phase, as a result of their antisocial behavior and lack of parental supervision, children become members of deviant groups where deviant peers train further antisocial behavior and drug use, punish typical prosocial behavior, and provide opportunities for delinquency and crime.

There is evidence of different developmental pathways in this learning model, some of which are less or more malignant than others. Patterson et al. (1998) pointed out that about half of antisocial children become delinquent teens. Of the delinquent teens, only about half or three-quarters go on to be adult criminals indicating that many adolescents drop out of this process. Teens with early onset criminal careers are at much greater risk than late onset teens continuing into adult criminal careers and other negative outcomes. Teens with late onset crime are more likely to commit only a single offense, again suggesting differentiation of developmental pathways toward delinquency, aggression, and violence.

Patterson et al. (1989) pointed out that the context of family disruption is also important in understanding the learning of aggression and delinquency. One facet of this problem is intergenerational transmission of aggression across multiple generations. Thus, grandparents of aggressive grandchildren are also more likely to be explosive. The grandparents teach their own adult children to be irritable, use ineffective parenting practices, and fail to model effective parenting practices. Hence, grandparents' behavior also impacts their grandchildren's aggression through the parenting practices of their adult children.

A second facet of family context is the myriad of problems associated with poverty, lack of parental education, and family stressors. Lower socioeconomic parents are more likely than middle class families to use physical management practices which are deficient that include less effective discipline, less effective monitoring, less positive reinforcement, and less involvement with their children. In contrast, middle income families are more likely to use verbal methods of discipline, offer children choice, use a lot of positive reinforcement, express positive affect, and support social and educational growth. Parents with marginal parenting skills may be especially vulnerable to family stressors, such as unemployment and poor parental relationships, which further amplify the effects of poor parenting.

An observational study by Wahler and Dumas (1986) illustrates these phenomena. They conducted extended naturalistic observations of social interactions between mothers and their children who had sought help with behavioral issues. All families had significant stressors: Two families were low income families with histories of child abuse, and economic and social problems. The third family was middle class and had chronic marital strife and were socially isolated. Wahler and Dumas described all three mothers as disinterested and detached from their children, rarely interfered with their children, had a sibling-like relationship with their children, and frequently used name calling and derogatory language with their children (e.g., "You jerk!").

The paper investigated two hypotheses. The first was the compliance hypothesis, that is, parental compliance with child demands was associated with escalating chains of increasingly aversive child behavior. The second hypothesis was the predictability hypothesis, that is, both the parties deliver aversive interactions unpredictably and that child problem behavior can make such interactions more predictable. The authors conducted in-home observations for 4–8 weeks. Data confirmed both the hypotheses. Over 90% of all aversive interactions were initiated by child demands for play and help, but these requests were largely ineffective in obtaining maternal compliance, supporting the compliance hypothesis. Second, all three mothers tended to punish appropriate child behavior quite often. Mothers were indiscriminate in their use of aversiveness on about 25% of opportunities, that is, often, the mothers were aversive in ways unrelated to the topography or intensity of their children's behavior. The mothers did, however, discriminate in their use of aversive interaction in that they were much more likely to deliver aversive interaction after their children escalated their own aversive behavior. Wahler and Dumas concluded that there was support for both the compliance and predictability hypotheses.

9.3.2 Summary

SLT has been a major advance in our understanding of how aggression can be learned across the lifespan and, indeed, across generations. The frustration-aggression hypothesis provided a framework for some aspects of aggression, Bandura's work on modeling provided a key foundation to this area, and finally, Patterson's elaboration of that work into understanding the mutually coercive family processes were important contributions. Further, SLT has considerable pragmatic implications for early intervention to prevent aggression and violence. If caretakers and children are involved in a mutual cycle of coercion, there may be ways to break the circle by changing each person's behavior. (See Chap. 16, *Individual Therapies for Violence and Aggression. I.: Cognitive and Behavioral Therapies*, Chap. 18. *Treatment of Violence and Aggression in Children, Adolescents, Families, and Schools* for reviews of various learning-based therapies that involve parent and child training.)

9.4 Behavior Analytic Theory

Behavior analysis views traditional psychological theory with suspicion, preferring to focus more on data collection and building an edifice of data to detect reliable generalizations concerning behavior and its relationship to environmental events (Skinner, 1950). Thus, there are few behavior analytics grand theories of aggression. However, one exception, building upon Skinner's earlier work, comes from Goldiamond (1974) and Layng (2006, 2017) who applied behavior analytic thinking to understanding and thinking about treatment of angry aggression. Layng (2006) wrote that "feelings are not the cause of action, nor are actions the cause of feelings. Instead what is felt, a private event, may describe a regnant contingency. For our purposes these private events will be designated as emotion. A feeling may be expressed publicly and may accompany other behaviors that together may be necessary to meet a contingency requirement. These behaviors will be designated as emotional behavior" (p. 157). Thus, an angry emotion is what the person feels, labels and describes, directly or indirectly. For example, they may say "My heart is racing and I am pumped. I am mad as Hell." The contingency that this labeling of an internal state refers to is that aggressive behavior will be reinforced by removal or the other person. This angry emotion may be accompanied by public correlates such as pacing, a flushed face, faster speech, and increased muscle tone. Finally, threatening and, if necessary, physical aggression to the other person may result in the other person suffering and moving away. Thus, emotions are tacts (labels) for contingencies. Layng suggested that once the overt angry behavior is evoked that it can then be subjected to contingencies that change its original function. Thus, angry-evoked aggression could come under the control of social contingencies and function as a generic mand (request) to request a variety of consequences unrelated to the original anger-based aggression.

Layng proposed that negative emotions such as fear and anger are similar in that they share a common consequence, namely removal of another person or situation. In the case of fear, the fearful person does so by moving away from the feared object and in anger the person does so by approaching the person to remove them through threats, aggression, etc. Layng characterized positive emotions, such as predation and longing, as those characterized by bringing people and objects toward oneself.

An important challenge with emotions and emotional behavior is that it can be difficult to identify the evoking conditions. Whereas observing deprivation of food and water and excessive stimulation such as too much noise may sometimes be readily observable, Layng proposed that the evocative conditions for emotions can be subtle and difficult to observe.

Using a case study of anger in Tim, a man living in a group home, Layng proposed that although traditional functional analysis-based contingency interventions might be helpful, the functional assessment might be more difficult. Although it was relatively clear that Tim obtained certain consequences delivered by staff through his aggression, Layng considered that identifying the more subtle evocative conditions might be helpful in developing a more nuanced and effective treatment plan. Noting that Tim aggressed inconsistently against peers, Layng suggested that Tim's aggression had come to function as a generic mand (request) to change something. Hence, staff should remind Tim to try to identify what it was that he wanted to change. A second suggested strand to the intervention plan, based on Goldiamond's constructional approach, was that although Tim had relatively strong receptive language skills, including following written and verbal instructions, his expressive-speaker verbal skills were relatively weak. Hence, making requests and engaging with staff were relatively difficult for him. Hence, Layng recommended general expansion training of his conversation skills. Layng recommended minimizing extrinsic, nonfunctional reinforcers as much as possible.

This understanding of the emotion of anger and the emotional behavior provides an interesting expansion and counterpoint to traditional behavioral accounts of aggression. It is also an effective response to critics that behaviorism does not consider emotions. As yet, there is little research on this approach to treating aggression and anger in this way.

9.5 Cognitive Neo-Associationist Theories

In contrast with behavioral models and in part due to dissatisfaction with the frustration-aggression hypothesis, Berkowitz (1989, 1990) proposed a model of anger and aggression that emphasized the roles of cognitive and affective processes in mediating people's responses to aversive stimuli. Bekowitz began with the notion that aggression, irritation, and annoyance are elicited by both social and nonsocial stimuli. Examples of social stimuli include mistreatment and frustration by others, and nonsocial stimuli include foul odors, high temperatures, extreme cold, and disgusting scenes. Berkowotz clarified here that by "anger" he means only the feeling or emotion of anger, which for him included irritation and annoyance. His theory is a psychological, not a neural or physiological one, and emphasizes the roles of automatic, nonthoughtful and effortful, longer-term cognitive processes, such as appraisals and attributions. The initial aversive stimuli activate associative networks that include thoughts, memories, expressive-motor behavior (e.g., facial expressions), and physiological responses with little initial thought. For example, following a social threat, the sight of a gun might initially activate a web of associations of words, emotions, and neural activities spreading activation throughout the semantic network. Sometimes, but not always, broadly defined negative affect elicits aggressive thoughts, feelings, and memories. Whether it does or not depends on the later activation of higher order cognitive processes such as attributions, in which individuals think about the elicited emotions and responses, makes attributions, and predicts future actions based on current or recent feelings such as fear or anger. Higher order cognitions may also include consideration of the causes of the aversive stimulation and arousal, available courses of action, goals that might be obtained, and what sensations and feelings have occurred.

Evidence for this model comes from a long series of experiments that Berkowitz and others have conducted. For example, in a typical experiment, a nonsocial aversive stimulus is presented, such as holding one's arm out for 3 min or placing one's hand in a bucket of ice water, in order to elicit negative feelings and activate the associationist network. Subsequently, some manipulation is made, for example, thinking about situations in which one would be frustrated, such as stuck in traffic, anxious, such as stuck in an elevator, or a neutral, nonfrustrating situation. The researchers assumed that thinking about a frustrating experience would induce a greater degree of anger than thinking about an anxiety-provoking situation and that thinking about the neutral scene would not produce anger. The experimenters then asked the participants to tell stories. The experimenters found that the highest number of angry references occurred in the high aversive situation and even more so in the group that thought about the frustrating situation. Berkowitz also noted that this model is similar to much research in clinical psychology adopting a cognitive-behavioral model in which intervention includes changing attributions and beliefs (Brewin, 1989).

9.6 General Aggression Model

Anderson and Bushman (2002) proposed the GAM which attempted to synthesize observations from multiple sources of data and concepts from individual psychological theories concerning aggression and violence including elements of Berkowitz's (1990) Neo-associationist theory (see above). Allen and Anderson (2017) described that the GAM "is an integrative bio-social-cognitive, developmental approach to understanding aggression that incorporates the best aspects of many domain-specific theories of aggression and takes into account a wide range of factors that affect aggression" (p. 1). They also proposed that this reorganization of existing knowledge can generate new interventions for aggression and violence.

The GAM focused on three phases: "the person in the situation"; cognitive, affective, and arousal routes; and outputs. Individual episodes of aggression are analyzed with respect to proximal and distal factors. Inputs include person factors, such as personality traits, attitudes, and genetics which are relatively stable over time; and situational factors such as immediate cues for aggression, provocation, frustration, pain, drugs, and incentives toward aggression. Routes include present internal states that mediate between inputs and outputs. These include cognitions, such as hostile thoughts and scripts; affect, such as mood and emotions; expressive motor behavior such as facial expression; and arousal. These three routes are said to be highly interconnected with one another. Finally, outcomes include thoughtful or effortful actions or impulsive, automatic, low effort actions.

The GAM also acknowledged other important factors. Among these are personality which reflects the lifetime of learning, aggressive attitudes, schemata and scripts, and desensitization to aggression. These factors result in personality traits of varying degrees of propensity toward aggression. Personality factors are then mediated by situational factors, such as social situations, and peer groups. Finally, aggression is also influenced by opportunity, inhibitions against anger, the effectiveness of aggression during interaction between big people (e.g., parents) and little people (e.g., children); preparedness to respond to pain with aggression, and varied roles for anger.

Noting the apparent failure of many interventions for aggression and violence, Anderson and Bushman (2002) identified multisystemic therapy as a promising approach. It appears that this is in part because MST adopts a broad analysis of family, cultural, and biological factors as the basis for its intervention.

GAM has been highly influential in the field of aggression and violence. For example, it has been extended to be applied to more extreme forms of violence (Anderson & Carnagey, 2004) and specific applications such as media violence (Anderson & Bushman, 2018), antigay aggression (Parrot, 2008), personality disorder (Gilbert & Daffern, 2011), offenders (Gilbert et al., 2013, 2017), and teacher aggression (Montuoro & Mainhard, 2017), etc. Thus, the GAM has become a commonly used framework to explain aggression and violence in several contexts and populations.

Despite or perhaps because of its popularity, GAM has also attracted critiques. For example, Ferguson and Dyck (2012) advocated that it was time to retire the GAM because of some of the following weaknesses. (a) Aggression is not clearly defined within the GAM and so it is not clear when GAM addresses instrumental, reactive, or other types of aggression. (b) Ferguson and Dyck claimed that the GAM is not really a theory at all but a moral crusade against violence masquerading as a psychological theory. (c) Evidence to support the GAM depends excessively on analog evidence rather than data from real-world violence. (d) The GAM assumed that all forms of aggression and violence are bad and hence should be reduced. Ferguson and Dyck argued that there are good forms of aggression and violence that should be distinguished from bad forms. (e) People can distinguish fantasy from fact from an early age. Hence, the citation of evidence based on violent cartoons rather than real-world violence is inappropriate. (f) The GAM depends excessively on recycled learning models of aggression and violence. Although it acknowledges biology, GAM does so in a cursory manner that is really only a fig leaf over the problem of the biological causes of aggression. (g) The evidence of short-term changes in reports of words in association tasks related to violence is not evidence of long-term formation of schemas. (h) The theory is not a testable, falsifiable theory.

9.7 Feminist Theories

Ali and Naylor (2013) defined feminism as "an ideology" based on the 1970s Women's liberation movement. There is no single feminist theory, rather there is a historically evolving political and intellectual movement. George and Stith (2014) distinguished three phases in the historical growth of feminism. The first was a nineteenth- and twentieth-century movement concerned with abolitionism and women's voting rights including the right to vote and inherit property. It was predominantly, but not exclusively, led by middle- and upper-class White women and some prominent African American women. The second phase was during the mid-twentieth century and focused on issues such as civil rights, elimination of sexism, changing women's traditional roles, and eliminating violence against women. This phase focused on heterosexual women as a homogenous group leading a significant number of women of color and lesbians to feel excluded. In this phase, some writers emphasized patriarchy—"the power of the fathers"—as the cause of male violence against women. Thus, the solution to intimate partner violence was to place violent men in batterers groups and women victims in support groups. Finally, the third phase attended to intersectionality of violence against women of all colors, classes, and sexual orientations and not to pigeon hole anyone into single, homogenous types of people. Third-wave feminists focus on anti-oppression and social justice and also tried to attend to men's issues, since men may also be subject to oppressive social forces.

Some forms of feminism propose that male on female violence is a mechanism to maintain male power in society in general and over women in particular. Examples run from the narrow actions, such as individual incidents of violence by a man against a woman when she threatens to leave a relationship, to broader actions, such as governments defunding women's groups when they become too troublesome to the government.

Some feminist perspectives that emphasize patriarchy as the cause of male violence against women are challenged by how to explain female on male violence in heterosexual relationships and women on women violence in lesbian relationships.

Feminist perspectives on therapy are quite varied. Some, such as the Duluth model, aspire to change men's behavior through education. Some feminists advocate that women must leave violent relationships and that attempts to support couples in violent relationships are immoral and should not be attempted. More recently, third-wave feminists have attempted to reduce violence against women by supporting both partners in a violent relationship by trying to understand each partner's perspectives from an intersectionalist perspective.

9.8 Social Role Theory

One theory which has a long and rich intellectual history that attempts to account for sex differences in aggression while also taking account of social, cultural, and biological causes of aggression is SRT. This theory is associated with the work of Eagly and Wood among others and has gradually evolved and expanded over some 40 or more years (Eagly & Wood, 2011).

SRT begins with the observation that there are marked differences in the behavior of men and women in many cultures; that is, men and women and boys and girls often behave differently in many circumstances and behave the same in other circumstances. Specifically, SRT posits that men are more likely than women to engage in task-oriented, instrumental, assertive, agentic behavior, and women are more likely than men to engage in socio-emotional, expressive, cooperative, helpful, and selfless behavior. Some have proposed that these differences evolve in part to ensure harmonious functioning of groups. SRT assumes that these sex differences originate in the physical differences between men and women. That is men's general greater upper body strength and endurance make men more suitable for physical labor (and indeed physical aggression). Women are adapted to gestation, nursing, and lactation and hence are more involved in child rearing and interaction with groups. Eagly and Wood concluded that many meta-analyses of sex differences and cross cultural studies of sex differences in such behavior confirm this generalization. This specialization of labor results in harmonious group living and deviations from it result in problems for society. If this is true, what explanations can be offered?

SRT proposes that when people see others behave they often infer that there is an inner cause to that observed behavior. That is, when men engage in stereotypical agentic behavior and women engage in stereotypical communal behavior, others are likely to attribute this to their inner traits such as competency and helpfulness. These gender stereotypes are often seen as stable and an inevitable consequence of an individual's biology; if individuals conform to these gender stereotypes, their behavior is likely to be reinforced and if their behavior is likely to deviate from these gender stereotypes, then their behavior is likely to be punished by others.

Hence, an assertive woman in a leadership role who shouts might be said to be strident or unable to cope, whereas a man engaging in the same behavior might be said to be a committed and assertive leader. Similarly, a man working as a classroom assistant who shows caring behavior toward children might be questioned as to his motives for engaging in such work, whereas a woman who engaged in the same behavior might be said to be nurturing. These cultural stereotypes of gender roles and behavior are said to influence evaluations of other people's behavior based on their gender conforming or nonconforming behavior and evaluations of one's own behavior. Thus, the proximal social causes of gender stereotypical behavior, including assertiveness, aggression, cooperation, and helping lie in people roles, their expectations of gender-role typical behavior. These proximal processes reflect more distal biological and cultural processes. For example, people may adapt their own behavior to the apparent gender of another hypothetical person. These ideas are said to be the core of SRT and are reflected in sex differences in aggression and related social behavior, such as helping (Eagly, 1987; Eagly & Wood, 2011).

Subsequently, these authors expanded the core of SRT theory in several ways. The first was to incorporate biological mechanism of sex differences including hormonal changes associated with parenting and cultural processes that influence gender roles. For example, Eagly and Wood (1999) examined mate preferences in a sample of 37 cultures. They found that in cultures that had the more distinct gender roles, the women had stronger preferences for older men with greater resources and men preferred women with mates who were better homemakers and providers. This was reflected in greater age differences between mates in cultures with more distinct gender roles.

9.9 Conclusions

As noted at the beginning of this chapter, there are numerous psychological and sociological theories of aggression and violence. They all differ in the variables and constructs that they find to be important and relevant; their testability; the extent to which they generate novel therapies; and the effectiveness or even harmful nature of these therapies. Psychotherapy offers the least satisfactory explanation of aggression and violence and generates treatments that are potentially harmful. The frustration-aggression hypothesis is also unsatisfactory due to offering an overgeneralized explanation of aggression and violence that is difficult to test. Social learning theory, ABA, and neurocognitive associationism are all potentially testable theories and have generated treatments that are potentially effective. The GAM is useful in that it attempts to integrate information from multiple theories into one comprehensive model and has been applied to many contexts; however, it is unclear whether it is a testable theory. Feminist theories are varied and offer very different perspectives on both the explanations for and treatments of aggression and violence. Some feminist-based therapies either seem to be ineffective or lack support. Finally, SRT appears to integrate evolutionary and social perspectives on gender-specific behavior.

References

Ali, P. A., & Naylor, P. B. (2013). Intimate partner violence: A narrative review of the feminist, social and ecological explanations for its causation. *Aggression and Violent Behavior, 18*(6), 611–619.

Allen, J. J., & Anderson, C. A. (2017). General aggression model. In P. Roessler, C. A. Hoffner, & L. van Zoonen (Eds.), *International encyclopedia of media effects*. Wiley-Blackwell. https://doi.org/10.1002/9781118783764.wbieme0078

Allen, J. J., Anderson, C. A., & Bushman, B. J. (2018). The general aggression model. *Current Opinion in Psychology, 19*, 75–80.

Anderson, C. A., & Bushman, B. J. (2002). Human aggression. *Annual Review of Psychology, 53*(1), 27–51.

Anderson, C. A., & Bushman, B. J. (2018). Media violence and the general aggression model. *Journal of Social Issues, 74*(2), 386–413.

Anderson, C. A., & Carnagey, N. L. (2004). Violent evil and the general aggression model. In A. G. Miller (Ed.), *The social psychology of good and evil* (pp. 168–192). Guildford.

Berkowitz, L. (1989). Frustration-aggression hypothesis: Examination and reformulation. *Psychological Bulletin, 106*(1), 59–73.

Berkowitz, L. (1990). On the formation and regulation of anger and aggression: A cognitive-neoassociationistic analysis. *American Psychologist, 45*(4), 494.

Bjørkly, S. (2006). Psychological theories of aggression: Principles and application to practice. In D. Richter & R. Whittington (Eds.), *Violence in mental health settings* (pp. 27–46). Springer.

Brewin, C. R. (1989). Cognitive change processes in psychotherapy. *Psychological Review, 96*, 379–394.

Bushman, B. J. (2002). Does venting anger feed or extinguish the flame? Catharsis, rumination, distraction, anger, and aggressive responding. *Personality and Social Psychology Bulletin, 28*(6), 724–731.

Côté, S., Vaillancourt, T., LeBlanc, J. C., Nagin, D. S., & Tremblay, R. E. (2006). The development of physical aggression from toddlerhood to pre-adolescence: A nation wide longitudinal study of Canadian children. *Journal of Abnormal Child Psychology, 34*(1), 68–82.

Dollard, J., Miller, N. E., Doob, L. W., Mowrer, O. H., & Sears, R. R. (1939). *Frustration and aggression*. Yale University Press.

Eagly, A. H. (1987). *Sex differences in social behavior: A social-role interpretation*. Lawrence Erlbaum.

Eagly, A. H., & Wood, W. (1999). The origins of sex differences in human behavior: Evolved dispositions versus social roles. *American Psychologist, 54*(6), 408.

Eagly, A. H., & Wood, W. (2011). Social role theory. In P. A. Van Lange, A. W. Kruglanski, & E. T. Higgins (Eds.), *Handbook of theories in social psychology* (Vol. 2, pp. 458–476). Psychology Press.

Eron, L. D. (1987). The development of aggressive behavior from the perspective of a developing behaviorism. *American Psychologist, 42*(5), 435.

Ferguson, C. J., & Dyck, D. (2012). Paradigm change in aggression research: The time has come to retire the general aggression model. *Aggression and Violent Behavior, 17*(3), 220–228.

Fonagy, P. (2003). Towards a developmental understanding of violence. *The British Journal of Psychiatry, 183*(3), 190–192.

Freud, A. (1947). Aggression in relation to emotional development; normal and pathological. *The Psychoanalytic Study of the Child, 3*(1), 37–42.

Freud, S. (1970). Why war? In E. I. Megargee & J. E. Hokanson (Eds.), *Dynamics of aggression: Individual, group, and international analyses* (pp. 10–21). Harper & Row.

George, J., & Stith, S. M. (2014). An updated feminist view of intimate partner violence. *Family Process, 53*(2), 179–193.

Gilbert, F., & Daffern, M. (2011). Illuminating the relationship between personality disorder and violence: Contributions of the general aggression model. *Psychology of Violence, 1*(3), 176–180.

Gilbert, F., Daffern, M., & Anderson, C. A. (2017). The general aggression model and its application to violent offender assessment and treatment. In P. Sturmey (Ed.), *The Wiley handbook of violence and aggression* (pp. 1–13). Wiley.

Gilbert, F., Daffern, M., Talevski, D., & Ogloff, J. R. (2013). The role of aggression-related cognition in the aggressive behavior of offenders: A general aggression model perspective. *Criminal Justice and Behavior, 40*(2), 119–138.

Goldiamond, I. (1974). Toward a constructional approach to social problems: Ethical and constitutional issues raised by applied behavior analysis. *Behaviorism, 2*(1), 1–84.

Goodnight, J. A., Wilhoit, S. A., & Receveur, A. (2017). Social learning theory and the development of aggression. In P. Sturmey (Ed.), *The Wiley handbook of violence and aggression* (pp. 1–13). Wiley.

Layng, T. V. J. (2006). Emotions and emotional behavior: A constructional approach to understanding some social benefits of aggression. *Brazilian Journal of Behavior Analysis, 2*(2), 155–170.

Layng, T. V. J. (2017). Private emotions as contingency descriptors: Emotions, emotional behavior, and their evolution. *European Journal of Behavior Analysis, 18*, 168–179. https://doi.org/10.1080/15021149.2017.1304875

Lohr, J. M., Olatunji, B. O., Baumeister, R. F., & Bushman, B. J. (2007). The psychology of anger venting and empirically supported alternatives that do no harm. *Scientific Review of Mental Health Practice, 5*(1), 53–64.

McKenna, J. J. (1983). Primate aggression and evolution: An overview of sociobiological and anthropological perspectives. *Journal of the American Academy of Psychiatry and the Law Online, 11*(2), 105–130.

McPhail, B. A. (2016). Feminist framework plus: Knitting feminist theories of rape etiology into a comprehensive model. *Trauma, Violence, & Abuse, 17*(3), 314–329.

Miller, N. E., & Dollard, J. (1941). *Social learning and imitation*. Yale University Press.

Montuoro, P., & Mainhard, T. (2017). An investigation of the mechanism underlying teacher aggression: Testing theory and the general aggression model. *British Journal of Educational Psychology, 87*(4), 497–517.

Parrot, D. J. (2008). A theoretical framework for antigay aggression: Review of established and hypothesized effects within the context of the general aggression model. *Clinical Psychology Review, 28*(6), 933–951.

Patterson, G. R. (1976). The aggressive child: Victim and architect of a coercive system. In E. J. Mash, E. L. Hamerlynck, & L. C. Handy (Eds.), *Behavior modification and families* (pp. 267–317). Bruner /Mazel.

Patterson, G. R. (1982). *A social learning approach, volume 3: Coercive family process*. Casalia.

Patterson, G. R., DeBaryshe, B. D., & Ramsey, E. (1989). A developmental perspective on antisocial behavior. *American Psychologist, 44*(2), 329–335.

Patterson, G. R., Dishion, T. J., & Bank, L. (1984). Family interaction: A process model of deviancy training. *Aggressive Behavior, 10*(3), 253–267.

Patterson, G. R., Forgatch, M. S., Yoerger, K. L., & Stoolmiller, M. (1998). Variables that initiate and maintain an early-onset trajectory for juvenile offending. *Development and Psychopathology, 10*(3), 531–547.

Sears, R. R. (1941). Non-aggressive reaction to frustration. *Psychological Review, 48*(4), 343–346.

Skinner, B. F. (1950). Are theories of learning necessary? *Psychological Review, 57*(4), 193–216.

Taubner, S., Rabung, S., Bateman, A., & Fonagy, P. (2017). Psychoanalytic concepts of violence and aggression. In P. Sturmey (Ed.), *The Wiley handbook of violence and aggression* (pp. 1–14). Wiley.

Wahler, R. G., & Dumas, J. E. (1986). Maintenance factors in coercive mother-child interactions: The compliance and predictability hypotheses. *Journal of Applied Behavior Analysis, 19*(1), 13–22.

Young-Bruehl, E. (2008). *Anna Freud*. Yale University Press.

Part III
Assessment, Treatment, and Prevention

The third part of this volume deals with assessment, treatment, and prevention of violence and aggression. The first task before starting any of these endeavors is assessment for treatment which is dealt with in Chap. 10. The subsequent chapters deal with three principal types of interventions: Individual biological interventions, cultural design, and individual therapies and interventions. The final chapter looks forward, somewhat optimistically, toward a future with less violence and aggression and how society can achieve this through the adoption of evidence-based practices and policies.

Chapter 10
Assessment for Treatment of Violence and Aggression

Assessment for treatment of aggression and violence is both an important practical and ethical question. Predicting the individual person's response to different treatments is important in order to find the best treatment for that specific person. Practitioners treating aggression and violence face a much more difficult problem than researchers evaluating and comparing treatments for aggression and violence. Researchers answer a rather general question, such as "Is this treatment better than no treatment or treatment as usual on average for this group of individuals?" Thus, researchers may be happy to know that: (a) this treatment does better than treatment as usual because seven out of ten participants in the treatment group showed less aggression than baseline compared to only three out of ten participants in the control group; and (b) this difference was not due to chance. Practitioners need to answer much more specific and demanding questions, namely "How can I produce and sustain large and meaningful reductions in aggression and violence for *this specific person* including low frequency but dangerous or socially significant incidents, and give them a better life to protect against future relapse?" Thus, practitioners should only be happy if there are: (a) a large, meaningful, and sustained reduction in aggression and violence for this specific person; (b) reductions in related behaviors, such as precursor behaviors that have led to aggression and violence in the past; (c) improvements in positive skills, such as communication, social behavior, stress management, problem-solving skills, etc.; and (d) change or management of relevant environmental factors, such as exposure to risk factors such as peer models, stressful relationships, etc. Whereas researchers are focused on questions of demonstrating causality, practitioners are ethically bound to produce not merely some change, but rather produce change and use methods that are socially valid (Wolf, 1978). Thus, the main aims of individual assessment for treatment of aggression and violence should be to: (a) identify and minimize harms to all concerned; (b) predict the most effective treatment that will be accepted and can be implemented with sufficient integrity to maximize client outcomes; (c) avoid harmful and potentially dangerous treatments; and (d) conduct the assessment in the

P. Sturmey, *Violence and Aggression*, Advances in Preventing and Treating Violence and Aggression, https://doi.org/10.1007/978-3-031-04386-4_10

most efficient manner possible to avoid the harms associated with delayed treatments and wasted time and resources. This chapter addresses these issues.

10.1 Functional Assessment and Analysis

Functional assessment and analysis of aggression has become a key aspect of good practice in treatment of aggression. Early research focused on the application of functional assessment and analysis of self-injury in individuals with developmental disabilities, but has since expanded to a very wide range of problems including aggression. As we saw in the preceding chapters that functional approaches to aggression and violence, while acknowledging the role of biological and cultural causes of aggression, focus on the current environmental causes of aggression, specifically the roles of establishing operations, discriminative stimuli, schedules of reinforcement, and rule-governed behavior for both aggression and other relevant behavior. Thus, the main purpose of functional assessments and analyses is to identify these controlling variables and their implications for idiographic treatment.

10.1.1 Methods

It is traditional to distinguish three broad methods of determining the functions of problem behavior. These are: (a) indirect methods that do not involve any direct observation such as questionnaires, interviews, reviews of records, case notes, etc.; (b) direct observational methods that do not involve any environmental manipulation such as antecedent–behavior–consequence (ABC) recordings and scatterplots; and (c) true experimental analyses in which the environment is systematically manipulated. Very often, true experimental analyses are preceded and informed by indirect and direct nonexperimental assessments as when practitioners and researchers interview family members and staff and conduct observations in the natural environment to identify potential target behaviors and their functions.

10.1.1.1 Indirect Methods

Almost all assessments of function begin by talking to someone, albeit a formal interview or phone conversation or chat over the water fountain, or a formal functional assessment interview. Thus, interviews, even when they receive little direct acknowledgement, are a key step in functional assessment and analysis of aggression. One approach has been to develop semistructured interviews such as the *Functional Analysis Interview* (FAI; O'Neill et al., 1990), modifications of the FAI such as the *Functional Assessment Checklist for Teachers and Staff* (McIntosh et al., 2008), the *Open-Ended Functional Assessment Interview* (Hanley, 2002/2009) among others. These interviews typically set out a number of predetermined questions covering skills and preferences, full range of potential target behaviors, prioritization of target behavior, operational definition of target behavior, environmental factors including establishing operations, discriminative stimuli, relevant situations and people, possible contingencies of reinforcement, current and past management strategies and treatments, and the informant's best guest as to the function (Hanley, 2002/2009). Semistructured interviews can be used flexibly by varying the wording, order of questions, and addition of other relevant questions. Informants are typically family members and staff, and students have only rarely been interviewed about their own behavior (Kern et al., 1994).

A second type of interview strategy has been unstructured hypothesis-driven interviews (Sturmey, 1991; Wolpe, 1970). This approach to interviewing is characterized by the interviewer developing hypotheses very early on or even before the interview begins based on referral and informal sources of information. Wolpe (1970) and Turkat (1985) characterized the clinicians as being a detective who generates hypotheses and makes predictions about the client's future behavior. These hypotheses and predictions guide the line of questioning. For example, if a child is referred for functional assessment due to aggression, he has been excluded from school and spends of lot of time on his own in his bedroom on the internet, then one might consider hypotheses such as: (a) he is avoiding activities and/or people at school or (b) staying at home is reinforced by lack of demands and access to preferred activities. To test this hypothesis, the interviewer might ask school staff (a) if he likes school work, has friends, and gets along well with others, and to test hypothesis, (b) if he spends time with preferred games at school and at home, etc. If hypothesis (a) is true, one would predict observing precursor or target behavior when presented with school tasks or people, and if hypothesis (b) is true, one would predict that he would be nonproblematic at school when with preferred items and problematic when the items were removed.

The second indirect assessment is performed by using psychometric instruments which typically consist of psychometric scales that correspond to behavioral functions with perhaps 8–20 items rated on Likert scales. These are typically completed by several family members or staff. There are several (Sturmey, 1994a) but some of the more commonly used ones include the *Motivation Assessment Scale* (Durand & Crimmins, 1992) and *Questions About Behavioral Function* (QABF, Matson & Vollmer, 1995). For example, Matson et al. (1999) administered the QABF to 398 persons with developmental disabilities and reported that functions could be determined in 84% of individuals. Further, in cases of aggression, the most common functions were attention (38.7%) and escape (35.3%) where in cases of stereotypy the most common function was nonsocial (93.3%), suggesting that aggression is primarily socially mediated unlike stereotypy.

The use of psychometric assessments alone is generally viewed as the beginning of a functional assessment due to numerous limitations in their use. Some limitations include poor or variable reliability (Conroy et al., 1996; Dracobly et al., 2018; Shogren & Rojahn, 2003; Zarcone et al., 1991), poor internal consistency of the scale (Newton & Sturmey, 1991), poor or unreliable factor structure (Singh et al., 1993), and poor convergent validity between different psychometric measures and between psychometric and other functional assessment/analysis methods (Fee et al., 2016). Thus, although easy, quick, and convenient to administer, psychometric assessments of the functions of aggression are not recommended, although they may be useful as part of the preliminary process of functional assessment/analysis.

10.1.1.2 Direct Observational Methods

Direct observation of aggression in the natural environment is the second, nonexperimental method of determining the function of aggression. It was first formally described in a classic paper by Bijou et al. (1968). They showed that objective narrative descriptions of behavior could be reorganized to identify the putative antecedents and consequences of behavior, so-called ABC analysis. In one of their case studies, Bijou et al. described the relationships between aggression in nursery children and environmental events such as ongoing activities and staff and peer behavior in response to the target child's behavior.

Touchette et al. (1985) developed a variation on this method which they called a scatterplot. In this study, data on the occurrences of aggression were plotted not simply as daily frequency, but rather as a grid of day of the week and time of day. Such a grid unpacked the daily totals to reveal temporal and spatial patterns of aggression. In a teen with autism, Joan, who lived in residential treatment, the data indicated that aggression tended to occur more frequently between 1 and 4 pm on Monday through Thursday, but did not occur in the morning and on Friday. The authors inferred that Joan's aggression functioned to escape specific group classes that were scheduled in the afternoon; she did not show aggression in group activities in the morning and during leisure and community activities. Based on this information, Touchette et al. designed an intervention which eliminated the provocative activities and replaced them with preferred nondemanding activities in the afternoon; activities at other times of the week were not altered. This resulted in near elimination of aggression over an approximately 1-year period. Kahng et al. (1998) found some evidence of temporal patterns of problem behavior including aggression in 12 of 15 data sets from scatterplots, suggesting their potential utility.

Direct observation has some potential advantages over indirect assessments of function. Namely, it brings the practitioner into direct contact with their client, their problem behavior, and environment. Through the use of quantitative ratio data such as frequency, rate, and duration, direct observation replaces informant's impressions and memories with objective data which is more likely to encourage the observer to detect functional relationships between the environment and aggression.

10.1.1.3 Functional Analysis

The final group of methods for determining the functions of aggressive behavior are collectively known as functional or experimental analysis in which the therapist or experimenter systematically manipulates some environmental variables in order to infer causality. This has the distinct potential advantage over nonexperimental methods because one can be more confident that treatment based on functional assessment will manipulate causal rather than correlational variables and because the results of the functional analysis will give their therapist some indication of the potential magnitude of the effect size this manipulation may have.

Following early demonstrations that aggression in nursery school children was responsive to attention extinction and differential reinforcement of incompatible behavior (Brown & Elliott, 1965), and the development of functional analysis methodologies applied to self-injury (Iwata et al., 1982/1994), researchers extended these methodologies to aggression. For example, Carr et al. (1980) conducted a functional analysis and treatment of aggression in two boys, Sam and Bob, aged 14- and 9-years, with severe cognitive impairments. Bob's aggression, which began at 5 years of age, included scratching, hitting, kicking, and biting mostly toward adults which was nonresponsive to multiple psychotropic medications. Sam's aggression began at 6 years of age and consisted of pinching, hair pulling, and scratching directed exclusively toward adults and have been nonresponsive to psychotropic medication, time out, and the Feingold diet. For both the children, teaching staff complained that aggression began when they presented any demand or attempts to teach. Staff who conducted functional analyses wore protective clothing such as thick coats and rubber gloves. Sessions were conducted in a small quiet room with only the boy and the experimenter being present.

The authors conducted several experiments. In Experiment 1, the authors evaluated the effects of demands, no demands, and demands plus a "safety signal," such as the experimenter removing their gloves and not presenting any further demands. The experimenters demonstrated clear experimental control of aggression in both boys showing that presentation of demands was strongly associated with aggression and signals associated with future removal of demands were strongly associated with little or no aggression. Sam participated in Experiment 2 which evaluated adding copious and varied reinforcers for following instruction to the demand condition. This experiment showed again that demands were associated with aggression and addition of reinforcement for following instructions greatly reduced aggression. Bob participated in Experiment 3. Since the authors did not find any tangible or edible reinforcers for Bob, they used escape as a negative reinforcer for a nonaggressive response, tapping on the back of his hand. This procedure resulted in a large increase in tapping and low rates of aggression. Finally, Bob participated in Experiment 4 which compared: (a) escape extinction, which consisted of preventing him from getting off his chair by placing a seat belt across his thighs; (b) negative reinforcement of aggression, which consisted of permitting 1 min of escape contingent upon aggression, and demand fading; and (c) gradual increase in demands which included removing the seat belt, the experimenter's coat and gloves, and fading in nonverbal imitation trials starting with one trial per session and increasing the number of trials until the experimenter presented 30 trials of imitation training on two responses. The initial extinction condition resulted in a large reduction in aggression within four sessions, negative reinforcement of aggression with escape resulted in high rates of aggression, and gradual increases in demands were associated with very low and often zero rates of aggression. The authors concluded that both the boys' aggression was maintained by negative reinforcement—escape from demands—and that interventions based on this hypothesis such as escape extinction, differential reinforcement of incompatible behavior with escape, and demand fading with were effective.

Research and practice on functional analysis have been ongoing since the late 1950s, and hence, the methods and range of applications of functional analysis have evolved and expanded in numerous ways. For example, Hanley et al. (2003) and Beavers et al. (2013) found that approximately 40–50% of published experimental articles on functional analysis addressed aggression. Early research focused predominantly on children with developmental disabilities, such as intellectual disabilities and autism spectrum disorders, and to a lesser extent on adults with developmental disabilities. Over time, however, functional approaches to aggression have been extended to a wide range of other populations including typically developing children (Gardner et al., 2012), children with attention deficit hyperactivity disorder (DuPaul et al., 2013), individuals with Alzheimer disease (Abraha et al., 2017; Cook et al., 2012), traumatic brain injury (Byrne & Coetzer, 2016; Verberne et al., 2019), Huntington disease (Lundervold & Jackson, 1992), and both youth (McDougall, 2000) and adults in forensic settings (Howells et al., 2008).

Many early functional analyses of aggression took the standard four conditions developed by Iwata et al. (1982/1994) to assess the functions of self-injury and applied them to aggression. This approach was often successful in identifying functions and effective function-based treatments; however, functional analysis does not require standard conditions (Jessel et al., 2020), and clinicians and researchers can be creative in designing appropriate test and control conditions. Several studies have identified idiosyncratic variables that control aggression. For example, Borrero et al. (2004) combined naturalistic observations and experimental analysis to develop a treatment plan for Tobias, a 13-year-old boy with moderate intellectual disabilities and a history of severe abuse who was admitted to a hospital inpatient unit. Interviews with Tobias' mother indicated that she and classroom staff often could not get Tobias to follow simple instructions and so, for example, when his mother reported becoming "frustrated" or "tired" and became abrasive with Tobias. Given Tobias' history and parent reports of physical abuse, the authors speculated that harsh instructions were discriminative stimuli for physical punishment. Using data taken during naturalistic observation in sessions conducted with the young man's mother, Borerro calculated the overall conditional probabilities of compliance or aggression and probabilities of compliance or aggression given pleasant or abrasive instructions. Pleasant instructions involved a "please," and abrasive instructions involved a loud exclamatory tone and a taught face. Borrero et al. found that the probability of compliance was 0.91 for pleasant instructions and 0.07 for abrasive instructions, and corresponding probabilities for aggression were 0.00 and 0.71. In a subsequent functional analysis using Iwata et al.'s (1982/1994) standard conditions using neutral instructions in the demand condition, they found little to no aggression. They then conducted a supplementary functional analysis comparing abrasive instructions in the demand condition and found high rates of aggression in that condition rather than a control condition.

Therefore, the function-based intervention, which was conducted during instructions to complete activities of daily living, such as brushing teeth, consisted of four components. These were: (a) praise contingent on compliance with instructions; (b) 30 s of noncontingent escape every 30 s with access to preferred toys during the break; (c) escape extinction consisted of prompts to follow instructions; and (d) pleasant instructions. This treatment package was effective in eliminating aggression and increasing compliance. A subsequent component analysis of the package (Ward-Horner & Sturmey, 2012) identified that differential reinforcement and noncontingent escape alone did not result in decreases in aggression; since no aggressive behavior occurred during implementation of the entire package, escape extinction was never implemented and so could not have been an effective component of the package. Hence, Borrero et al. concluded that changes in the topography of the instructions alone were the effective component of the treatment package. This study demonstrates the creative and individual manner in which functional analyses of aggression should be conducted.

Functional analyses of aggression have also been expanded to address the interactive effects of EOs and other independent variables. For example, in an experiment with one individual with disabilities, O'Reilly (1995) demonstrated that aggression occurred during demand conditions, but only during naturally occurring days after sleep deprivation. Others have expanded functional analyses of aggression to embed them within more naturalistic contexts by having natural change agents conduct them in typical settings. For example, Baker et al. (2006) conducted a functional analysis of aggression in a 96-year-old nonambulatory woman with Alzheimer disease who was referred because of aggression which had been treated ineffectively with non-function-based behavior management and Ativan. A 21-year-old direct care staff conducted all sessions in the bathroom, bedroom, and a common area. In the first setting, analysis observations revealed much higher rates of aggression in her bathroom routine than her recreational routine. The bathroom routine was the longest routine and involved the most extended period of time close to the care staff. In the functional analysis the authors compared contingent attention, contingent escape, and a control condition. They found much higher rates of aggression in the contingent escape condition, suggesting that her aggression was maintained by escape from the bathroom task. Thus, in the final treatment analysis, the experimenters compared noncontingent escape (frequent short breaks) versus contingent breaks (breaks contingent upon aggression) and found that noncontingent breaks were effective in greatly reducing aggression.

One concern about conducting functional analyses has been the time that they take to administer; hence, several researchers have developed faster and more economical methods of functional assessment that can be applied easily in routine settings. The first example comes from Northup et al. (1991) who developed *brief functional analyses*. This procedure was developed for use in an outpatient clinic with the aim of identifying the function of the target behavior and a functionally equivalent replacement behavior and treatment recommendations in one short appointment. There were three individuals aged between 13 and 24 years referred to a regional service to assess severe problem behavior. All three individuals displayed aggressive behavior. In each case, the children were exposed to two to four functional analysis conditions sequentially with only one data point per phase. The authors took data on both aggression and a relevant communication response. After implementing each phase once, they then implemented a brief contingency reversal based on the most likely hypothesis. For example, Curtis showed more aggression in the escape than the alone condition. The authors conducted a very brief reversal design experiment to confirm experimental control with a total of four short sessions. The authors then conducted a contingency reversal in which they made escape contingent upon the sign for escape or aggressive behavior. The authors replicated these results with two other participants in this study. Subsequently, Northup et al. (1991) found that in a series of 79 cases referred to the service over a 3-year period, that the target behavior was shown in 63% of clients and of those clients who showed the problem behavior a distinct function was identified in 74% of cases. Problem behavior decreased in 54% of cases and adaptive behavior (typically mands ["requests']) increased in 65% of cases. Thus, brief functional analyses are an economic method for identifying the function of aggression in about two thirds of cases which may be a good starting place for a functional analysis of aggression in many cases.

Another way to increase the efficiency of functional analyses is to use latency to the first instance of aggression as the dependent variable and only run sessions for a maximum of 5 min. For example, Roscoe et al. (2010) used the latency to first instance of aggression to evaluate the function of aggression in Carrie, a 13-year-old girl with autism. The authors used 5 min long assessment sessions. The authors used standard functional analysis conditions (Iwata et al., 1982/1994) and found no reliable differentiation between conditions, as aggression did not occur in most sessions. Based on the informal observation that Carrie often talked about specific topics a lot, they conducted a paired stimulus preference assessment of conversation topics and found that she chose certain topics such as zoos, dogs and cats over babies, bikes, and the weather. They then conducted a functional analysis comparing access to conversation about high versus low preferred topics and found reliably shorter latencies to aggression during access to high preferred topics compared to low preferred topics. Finally, in an extended functional analysis, they reran the functional analysis conditions but this time including highly preferred topics of conversation in the attention condition. This resulted in clear differentiation in the rates of aggression between conditions leading to the conclusion that aggression was maintained by access to conversation about highly preferred topics. This approach to functional analysis may be more efficient than standard functional analyses of aggression as the use of latency to first aggressive response may require fewer shorter sessions than standard experimental analyses.

In a third study using a brief functional analysis methodology, Sigafoos and Saggers (1995) embedded test and control trials for each function in the natural classroom environment over 5 days. This method is called trial-based functional analysis. They presented the EO and made the consequence available for 1 min and then ran a test trial for 1 min in which the reinforcer was continuously available. Trials were terminated when the participant displayed aggressive behavior. For example, to evaluate the escape function, demands were presented for 1 min and terminated if aggression was displayed, and in the subsequent control trial, no demands were presented. In both the participants, aggression was maintained by social-attention contingencies.

Subsequently, Bloom et al. (2011) compared trial based on typical functional analysis. Participants included ten children aged 6–18 years mostly with autism or other developmental disabilities, eight of whom displayed aggression. The two methods agreed on the function in six of ten comparisons, suggesting fair concordance between the two methods. Bloom et al. suggested that trial-based functional analysis might be a good enough first choice of functional analysis methods or an alternative when resources are unavailable for a more extensive experimental functional analysis.

Finally, a recent development in functional analyses has been the development of Interview Informed Synthesized Contingency Analysis (IISCAs). The aims of IISCAs are to make functional analysis both efficient (i.e., quickly produce accurate results that produce effective function-based treatment programs) and to enhance the ecological validity of functional analysis (i.e., emulate real-world controlling variables including both EOs, discriminative stimuli, and contingencies accurately). Thus, traditional functional analyses isolate and manipulate single contingencies—an approach which has been highly successful (Beavers et al., 2013)—it does so without incorporating an analysis of other potentially relevant contingencies.

To address this limitation, Hanley et al. developed an assessment procedure to identify synthesized contingencies, that is, the contingencies that occur both before and after the problem behavior. Thus, IISCAs emphasize the transition from one contingency to another. So, rather than emphasizing access to iPads from a classroom staff following aggression, IISCAs emphasize transition from an aversive task, such as a difficult academic problem, to the iPad and staff attention. Traditional functional analyses might (or might nor) identify the isolated contingencies with high rates of aggression in the escape and tangible conditions, whereas IISCAs identify and evaluate the synthesized contingencies of removal of task demands followed by access to iPad and staff attention.

The IISCA protocol includes a semistructured, open-ended interview and brief observation to identify potential synthesized contingencies, and a brief experimental analysis which compares the synthesized contingency with a control condition in which all putative reinforcers are made freely available noncontingently. The results of the IISCA analysis is then used as the basis of a function-based intervention which is often FCT for the reinforcers identified in the IISCA assessment protocol.

In an early evaluation of the IISCA protocol, Hanley et al. (2014) evaluated the IISCA protocol with three children aged 3–11 years diagnosed with pervasive developmental disorder or autism who displayed severe problem behavior. Gail was a 3-year old diagnosed with pervasive developmental disorder. When her mother asked her to clean up her toys or play independently, Gail engaged in multiple problem behavior, including hitting her mother. The IISCA interview and brief observation identified loss of adult attention and removal of preferred activities as potential antecedents and access to adult attention and preferred toys as relevant consequences. To evaluate the effect of this synthesized contingency on Gail's severe problem behavior, Hanley et al. designed and compared a test condition in which no attention was delivered and toys were removed and only provided continent on severe problem behavior with a control condition in which adult attention and preferred toys were freely and continuously available. Data from 14 brief sessions demonstrated that problem behavior occurred during the test condition but only when Gail's mother, but not the experimenter, conducted the sessions. Problem behavior never occurred during the control condition irrespective of who ran the session. A separate functional analysis of isolated attention and tangible conditions failed to evoke any problem behavior. Thus, the authors concluded that Gail's problem behavior was evoked by maternal attention and preferred toys and reinforced by the synthesized contingency of access to maternal attention and toys, but not isolated contingencies or attention from the experimenter.

Based on this hypothesis, Hanley et al. (2014) designed and evaluated an individual, function-based behavioral intervention for Gail. This consisted of teaching her the functionally equivalent communication response which was "Play with me" which at first immediately resulted in her mother's attention and preferred toys, which immediately resulted in elimination of the severe problem behavior. The experimenters then taught Gail to: (a) make more elaborate effortful responses (e.g., "excuse me … may I have my toys please"); (b) learn to tolerate delays and denial of access to preferred toys; and (c) to play with less preferred toys and complete non-preferred tasks. The treatment, including baseline, involved 22 sessions and took just under 3 months from beginning to end.

Hanley et al. (2014) replicated these results with two other children. Additionally, they found extremely high ratings of social validity (6.9 on a 7-point Likert scale). Further economic analysis of the assessment and treatment revealed that the direct costs were 8–14 weeks (M = 11.3) of treatment consisting of 22–32, 1-h visits (M = 27 h), with direct costs, including direct contact hours, supervision, report writing of $5900–8600 ($M$ = $6900). These costs did not include indirect costs, such as transport, service administration, costs of offices, etc. These costs should be evaluated against the presumably extremely large costs of years of ineffective and expensive early intervention, special education, professional services, costs of injuries to client and caregivers, and failure to act effectively that these children received.

These results have received multiple replications in several case series (Ghaemmaghami et al., 2016; Jessel et al., 2018). For example, Ghaemmaghami et al. (2016) replicated the effectiveness of the IISCA assessment procedure and demonstrated that the assessment time could be reduced from 25 to 3 min, suggesting that the IISCA procedure could be made even more efficient with little loss of accuracy.

10.1.1.4 Comment

This section has shown that several methods of functional assessment and analysis exist. Although most commonly applied to children with developmental disabilities, this approach has been replicated to varying extents in other populations with aggression (Sturmey, 2020). Although some reports of functional assessment have reported implementation by typical settings by typical caregivers over the long-term, as with other treatments, there are fewer such reports.

Functional assessment and analysis of aggression have at least two other potential limitations. First, in some cases, a true functional analysis of aggression may be undesirable out of concern for the safety of the client and therapists. To address this legitimate concern, functional analysis can be performed on precursor behaviors which may correctly identify the function of aggression without directly provoking it (Heath & Smith, 2019; Smith & Churchill, 2002). Second, sometimes functional assessments and analysis fail to identify relevant environmental variables (Dura, 1991) despite extensive assessments (Sturmey, 1994b). In such cases, clinicians may still implement nonfunction, evidence-based practices such as contingency-based interventions and environmental enrichment using consequences unrelated to the function of aggression.

10.2 Assessment of Aggression- and Anger-Related Cognition

The most widely spread model of treatment of aggression is cognitive behavior therapy (CBT). In this model, assessment of thinking and cognitive structures is a key component of assessment prior to treatment. Assessment of cognition for treatment of aggression and violence can includes identifying problematic thoughts that come and go and problematic core beliefs that are seen as relatively permanent. Cognitions are also sometimes treated as an outcome measure by themselves during treatment. Assessment of cognition can be done using psychometric instruments and clinical interviews. In both the cases, the information is necessarily self-report data.

An example of assessment of cognition in intimate partner violence comes from a study by Slaby and Guerra (1988) who evaluated two types of cognitions in three groups of adolescents. The two broad classes of cognitions were problem solving and beliefs supporting aggression. The three groups of adolescents who participated were adolescents incarcerated in a maximum security prison for aggression-related offenses such as rape and attempted murder, and high and low aggressive adolescents in high school who were assigned to these groups based on psychometric criteria using teacher ratings of aggression. The experimenters presented the participants with two vignettes of hypothetical situations with social problems to solve and the participants had to select from options that the experimenters provided. The groups were quite readily differentiated based on their problem-solving cognitions and beliefs. That is, generally, the incarcerated group was most likely to define the problem in terms of aggression, select hostile and aggressive solutions to problems, select fewer solutions, and fewer effective solutions. In terms of beliefs, individuals in the offender group were most likely to believe aggression as legitimate, believe that aggression would increase their self-esteem and avoid negative self-image, and believe that victims do not suffer. In addition, there were gender differences in that often males were more likely to report aggressive problem-solving skills and beliefs.

Studies such as these inform treatment by identifying cognitions that may be either targets for change during therapy to reduce aggression by changing thinking or as an outcome, for example, by reducing aggressive behavior and perhaps aggressive cognition through social skills training. Cognitive behavior therapists also use such assessments as part of case formulation and might include cognitions as part of behavior chains.

10.3 Case Formulation

Case formulation is an approach to understanding individual clinical cases to guide idiographic interventions (Eells, 2021; Sturmey, 2008, 2020). A case formulation abstracts out key features of a clinical case to predict effective and ineffective treatment and integrates client presentation, history, and previous treatment into a unified, integrated, and useful account.

There have been several definitions of a case formulation. For example, Eells (2007) wrote that a case formulation was "… a hypothesis about the causes, precipitants, and maintaining influences of a person's psychological, interpersonal and behavioral problems …" (p. 4). Kuyken et al. (2009) defined case formulation as "[a] process whereby therapist and client work collaboratively first to describe and then to explain the issues a client presents in therapy. Its primary function is to guide therapy in order to relieve client distress and build client resilience …" (Kuyken et al., 2009, p. 3). Tarrier and Calam (2002) defined it as "the elicitation of appropriate information and the application and integration of a body of theoretical psychological to a specific clinical problem in order to understand the origins, development and maintenance of that problem …" (p. 311). Hart et al. (2011) refined these ideas by distinguishing the *process* of case formulation—working with a client and relevant others, reviewing records, and developing the formulation—and the *product* of case formulation—often a written account, diagram, or map that summarizes the results of the process of the assessment and its implications for treatment. Case formulation is widely recognized as a core competency for many mental health professionals and in the last 10 years has received considerable attention in forensic contexts (Hart et al., 2011; Sturmey & McMurran, 2011).

Often treatment is said to be guided by psychiatric diagnosis. For example, psychiatrists and other prescribing professionals prescribe classes of medications, such as antidepressants, anxiolytic, and antipsychotic medication. (Recall that, other than impulse control disorder, really there is no DSM or ICD diagnosis directly related to aggression and violence and that many or any psychiatric disorder could be related to aggression and violence in various ways. Thus, some have argued that making an accurate diagnosis related to aggression and violence is essential to guiding treatment such as psychopharmacology and other diagnosis-based interventions.) Other mental health professionals often offer treatment based on diagnostic categories when they run psychotherapy groups for people with anxiety, depression, or Posttraumatic stress disorder. This diagnostic-based approach to treatment suggests that some treatments are more effective for some diagnoses than others and thus implying that matching diagnosis with treatment is an effective strategy for maximizing client outcomes and that mismatching diagnosis and treatment might be ineffective or even harmful.

Often, however, the rationale for case formulation is framed in terms of the limitations of psychiatric diagnosis and diagnostic-based treatment. Proponents of case formulation argue that psychiatric diagnosis cannot in fact guide treatment, at least in some circumstances. The first reason for this is that outcomes are highly variable from diagnosis-based treatment: Some people respond well, but a substantial minority respond minimally or do not respond at all, and some others are even harmed *because* of the treatment. Thus, there should be some account beyond diagnosis that should account for these varied outcomes. Additionally, clinicians must provide effective treatments when diagnostic-based treatments fail—case formulation might be one solution to this problem. Second, many individuals have multiple diagnoses, and it is unclear which of the diagnoses should guide treatment. For example, suppose someone is referred for treatment because of aggression to their partner and their diagnoses include major depression, generalized anxiety disorder, alcohol abuse, and impulse control disorder. Which diagnosis should guide treatment in such a case? Is there one treatment that might have an effect on all problems, or should the presenting diagnoses be dealt with sequentially, if so, what order might be best? A diagnostic approach alone cannot answer these questions. A third justification for case formulation is that there are many individuals with unique problems. Suppose an individual in an inpatient psychiatric unit with chronic aggression engages in licking walls constantly. What guides treatment for this problem? It is likely that there is little to no treatment literature on such a problem and diagnosis will not predict what to do. How can a clinician predict the best treatment for this individual absent guidance from diagnosis? Case formulation might provide a set of general principles to guide treatment design that can be applied flexibly to many different and novel clinical situations. Finally, many clinicians out of both pragmatic concerns and respect for individual differences and client autonomy wish to tailor treatment to individual interests, preferences, and circumstances, and involve the client in the assessment process as much as possible. Thus, there appears to be many good reasons to use case formulation.

Eells and Lombart (2011) identified five goals of case formulation. First, accurate formulations should predict effective and ineffective treatments, enabling clinicians to maximize benefits and avoid harms to the clients and others. Second, case formulation might improve treatment outcomes beyond what would have been done anyway, such as in a standard treatment group, such as anger management. A third goal of case formulation is to be both parsimonious and comprehensive. That is the case formulation should be short, concise and analytical, and account for everything about the presentation and the individual's history. (Some have observed that some novice or incompletely trained clinicians and practitioners have problems moving beyond listing all the client's historical events, recent events, and the myriad of presenting problems to abstracting out the core features of a case that will direct treatment to keystone interventions that produce the maximum benefit in the most efficient manner.) Fourth, there should be a balance between description and explanation of the problem. Finally, the intervention should be evidence-based.

There are many different approaches to case formulation (Eells, 2007, 2021; Sturmey, 2008). Some are atheoretical and make no explicit use of any particular theoretical framework. Some are eclectic and borrow freely and sometimes systematically from different theoretical frameworks (Weerasekera, 1996). Others use a specific theory such as psychodynamic case formulations (Messer & Wolitzky, 2007), cognitive behavioral approaches (Easden & Kazantzis, 2018), and behavior analytic approaches to case formulation (Skinner, 1953; Sturmey, 1996, 2008; Turkat, 1985; Wolpe & Turkat, 1985).

A commonly cited example of integrated or pantheoretical approach to case formulation comes from Weerasekera's (1996) "5 Ps" approach. The five Ps are as follows: (1) The *problem definition* and identification of the desired alternative to the problem, i.e., the therapy endpoint which includes both the absence of the problem (e.g., no aggression) and a positive outcome (e.g., obtaining and maintaining sustained and enjoyable employment). (2) *Predisposing factors* include the distal internal and external factors that increase the person's vulnerability to the problem, such as histories of failure to learn key self-management, anger management, problem-solving, and other skills related to aggression and violence; acquisition of problematic problem solving skills, such as alcohol and other substance use, to relax and regulate mood; acquisition of problematic reinforcers such as observing other people suffering or complying; and histories of trauma and abuse. (3) *Precipitating factors* include the proximal internal and external factors that trigger the problem. These might include recent changes at work such as changes in responsibilities or new supervisors, job losses, life transitions, acquisition of resources (such as money to enable drinking or drug use) that lead to aggression and violence. (4) *Perpetuating factors* refers to the internal and external factors that maintain the problem. An example might include core sustained beliefs, such as "Real men control women however they have to, because deep down women prefer that men are in charge," or a stable peer group of violent individuals that legitimizes violence. (5) Finally, *protective factors* which are the internal and external factors that help the person cope with or recover from the problem, such as religious beliefs and good peer models, social support and resources. This and other definitions of case formulation are useful in that they provide guidelines for practitioners to use when writing case formulations and for training.

An early theory-specific approach to case formulation comes from Skinner's work which proposed a model of psychopathology and an approach to its formulation and treatment (Skinner, 1953, 1971). While acknowledging the importance of biology, including genetics and physiology, and cultural evolution, Skinner proposed that an important individual source of psychopathology was society's strategies to control its members' behavior through the group, religion, parents, education, etc. relating to restricting behavior related to primary reinforcers, predominantly by positive punishment. Thus, we slap the greedy child's hand who selfishly reaches for the last cookie; roughly pull away the child who hits their sibling to obtain a toy; and scowl at and threaten the fidgeting child with trouble for fidgeting in class and place of worship. Such positive punishment strategies are of course quite effective, at least in the short term. Further, caretakers' and teachers' use of positive punishment is negatively reinforced when children keep their hands in the laps, ask for the last cookie, ask their sibling for a toy, and sit still so that classrooms and public places are quiet and orderly. Unfortunately, punishment also has undesirable side effects such as escape, revolt, and passive resistance, and emotional side effects which ironically sometimes but not always includes aggression (Fontes & Shahan, 2021)! The child learns to sneak a cookie when adults are not watching, pretends not to hear the instruction to put the cookie back, and, when finally called on their bad behavior, throws the cookie and the cookie jar at their parents and tells their parents that the cookies are horrible anyway. In addition, control of behavior by punishment also has undesirable emotional by-products including fear and anxiety; inhibition of adaptive operant behavior (Estes & Skinner, 1941); emotional side effects, such as anger and rage; broad response suppression that looks like depression; excessively vigorous behavior, such as agitation and excessive talking; excessively restrained behavior; and self-punishment, etc. to escape punishment (Skinner, 1953). Further some individuals learn to manage the aversive emotional side effects of punishment through problematic operant behavior, such as alcohol and drug addiction, which are negatively reinforced by reduction or removal of the aversive emotional side effects of punishment.

Skinner (1971) illustrated these processes with a hypothetical case formulation of "… A young man has … graduated from college and is going to work … Most of the behavior he has acquired … proves useless in this new environment …" Skinner then translated each vernacular description of the man's problems into behavioral explanations. For example, when we say, "He lacks assurance feels insecure [and] is unsure of himself," Skinner translated this as "He is seldom reinforced, his behavior undergoes extinction." When we say, "He feels frustrated," Skinner translated this as "Extinction is accompanied by emotional response." When we say, "he feels uneasy or anxious," Skinner translated this as "His behavior has unavoidable aversive consequences." When the man says that "He has no sense of accomplishment, craft, pride, purpose," Skinner translated as "He is rarely reinforced." Finally, when the man reports that "He feels guilty or ashamed," Skinner explained this as "He has previously been punished for idleness which now evokes emotional responses."

Thus, Skinner (1953, 1971) provided a narrative description of a common clinical case and used basic behavioral concepts to explain what is going on. Skinner hinted at interventions based on behavioral self-management but did prescribe specific interventions such as relaxation training, assertiveness training, etc. for the problems.

Subsequently, researchers extended behavioral models of psychopathology to cover a wide range of psychopathology including depression, anxiety, stuttering, psychiatric inpatients, autism spectrum disorders (Sturmey, 2020; Ullman & Krasner, 1965), and including true functional analyses of conditions such as phobias (Jones & Friman, 1999) and psychotic speech (Froján-Parga et al., 2019). Other broader behavioral and cognitive behavioral approaches have used multimethods of assessment but tend to rely on indirect methods of assessment. They are not typically not interested in experimental control of behavior. They also vary in terms of their implicit or explicit use of behavioral concepts (extinction, concurrent schedules of reinforcement, etc.), degree of acceptance of self-report, treating covert behavior as the cause of behavior and as the main target behavior, etc.

Behavioral approaches to case formulation have been extended to address the assessment and case formulation of conditions such as personality disorders which may be related to violence and aggression. Thus, although concepts such as personality disorders are diffuse and ill defined, they can be broken down into specific target behaviors, such as avoidance of criticism and evaluation by looking away, avoidance, making excuses, stomach pain, sweating, and thinking "Something terrible is going to happen," etc. Further, assessment can identify specific situations such as the presence of supervisors, exams, and public speaking. Thus, although personality disorders are diffuse, they may be broken down into multiple target behaviors, specific problematic situations, and recent temporally discrete events to permit analysis of the case and set priorities for behavior change. Although personality disorders may be lifelong, apparently rooted in early development, some authors have attempted to integrate that information into case formulations too (Kramer, 2019; Turkat, 1985).

Over the past 15 years or so, some work on case formulation has emphasized that the process of case formulation should be a collaborative one in order to be more respectful to the client, acknowledge their contribution, enhance collaboration, access information available elsewhere, build a better therapist–client alliance, etc. This is foreshadowed by Skinner (1953) who advised therapists against giving their opinions or telling patients what to do during psychotherapy for two reasons. A first pragmatic reason is that people will say "no" to the best of advice, but Skinner put forward a subtle reason. Skinner advised therapists to encourage clients to discover the relationship between the environment and their problems, for example, through self-recording, in effect learning to make their own case formulation. The reason he gave for this was that by doing so the therapist had already started the process of behavior change, thereby making change during intervention a little more likely.

Although making case formulation a joint enterprise may be more likely to engage the client, promote the beginning of change in client behavior, build therapeutic rapport, access personal information, improve case formulations, lead to valued treatment goals, and set the scene for engagement in therapy, the literature has acknowledged both advantages and disadvantages of coproduction. For example, some possible disadvantages may include clients being overwhelmed by the formulation and their history of their problems, inclusion of misleading information, may involve challenging the service user's opinions, and may take longer (Pain et al., 2008).

In the age of evidence-based practice, we should ask the question "But does case formulation make a difference to treatment outcome?" The answer appears to be "Sometimes, but not always!" (Ghaderi, 2011). There is a small literature on this question, but before considering the question, we should first consider the requirements to show effectiveness of case formulation. These include: (1) case formulation should produce better than treatment as usual; (2) case formulation must be done competently; and (3) the treatment that matches the case formulation must be the most effective treatment and treatments that do not must be less effective, ineffective, or harmful compared to nonmatched or evidence-based treatments. Although some evaluations of case formulation have produced outcomes similar to standard treatments for depression (Persons et al., 2006) and anxiety (Schulte et al., 1992), others have found that case formulation has produced outcomes superior to evidence-based practices such as Ghaderi's (2006) study evaluating standard cognitive behavior with individualized, function-based cognitive behavior therapy for eating disorders. Thus, there is some evidence that case formulation may sometimes but not always be superior to standardized, evidence-based treatment. It is notable that there is little evidence at this time specifically related to case formulation and aggression and violence.

10.3.1 Case Formulation and Aggression and Violence: Nonforensic Applications

Both clinicians and researchers have extended the use of case formulation to aggression and violence. In this section, we will consider case formulation in nonforensic contexts, and in the next section, we will consider forensic case formulation as applied to aggression and violence. In this section, we will consider two such contrasting case formulations. The first comes from McAndrew (1989) who presented a case formulation of obsessive-compulsive behavior that also involved aggression as part of the formulation. The second comes from Hunter et al. (2008) who described a multimodal approach to formulating aggression in a psychiatric inpatient with severe aggression.

In the first example, McAndrew (1989) reported an annotated behavioral interview, multiple psychometric assessments, and functional assessment of a 26-year-old woman with obsessive-compulsive disorder and emotional distress. She was disturbed by thoughts of harming others and became anxious when concerned that she might become aggressive. Her compulsions appeared to function to manage her experience of anxiety especially in situations where she believed she might lose control. That is, engaging in compulsive behavior was negatively reinforced by avoidance of more aversive thoughts, such as killing her children, and situations, such as guns and knives being left out in the home. She was very unassertive and very sensitive to negative evaluation, criticism, and rejection by others and again strove, in part though compulsive behavior, also to avoid these situations. She had a poor marital relationship in part because her husband did not like to talk about her intrusive thoughts and beliefs and was dismissive of her distress. Based on the formulation, the treatment plan involved 22 treatment sessions over 8 months. Treatment included: (a) in vivo exposure and response prevention while exposed to knives; (b) flooding to obsessions; (c) cognitive therapy; (d) parenting and assertiveness skills training; and (e) marital counseling. Immediately post treatment and at 6 months follow-up, she was almost symptom-free.

This case study is useful and instructive because it involves assessment through behavioral interviewing, which is the method that many clinicians use to conduct clinical case formulations, and it attempts to integrate history and presenting problems. It is also a notable example of the application of case formulation because it illustrates hypothesis-driven clinical interviews to develop an idiographic formulation-based treatment plan that involves aggression thoughts, impulses and fear of engaging in aggressive behavior, rather than overt aggressive behavior, as part of the formulation in an atypical presentation of aggression.

In a second example of case formulation and aggression, Hunter et al. (2008) described the application of the Multimodal Functional Model (MFM) to understanding and treating aggression and related behavioral issues in Gary, a 28-year-old African American man with multiple psychiatric and other diagnoses including: (a) Schizoaffective disorder, polysubstance abuse, and dependency involving cocaine, alcohol, and cannabis use; and (b) movement disorders characterized as choreiform movements and dystonia in all four limbs and other body parts. Gary had been resident in a psychiatric institution for 2 years, primarily due to behavioral issues including aggression managed ineffectively primarily through psychotropic medication and restrictive behavior management practices such as restraint. The primary target behavior was named behavioral dyscontrol which the authors defined as "loud moaning, yelling, crying, crawling and writhing on the floor, indiscriminate grabbing and hugging of anyone near him, removing clothing, and begging others to help him remove or adjust clothing" (p. 19). Behavioral dyscontrol was displayed for approximately 7 h on the first few days of admission to a new unit. It also resulted in restraints and seclusion used extensively on the first 2 days of admission. The target behavior was disruptive, frightening, and confusing to staff and other residents.

Early informal observation indicated that this behavior often resulted in others getting him water, adjusting his clothes for him, and reassuring and soothing him which he appeared to enjoy. Gary's clinical team completed a grid called the multimodal integrated intervention plan. This grid includes assessments of medical, psychiatric, psychological, and socio-environmental factors. For each of these four factors, the team evaluates causal hypotheses, causally based intervention plans, and staging plans, specifies the expected change, collects data, and specifies the responsible team members. Different team members complete different components of the assessment, and the entire assessment is then assembled into one concise document.

Some of the components of the multimodal integrated intervention plan for Gary included: (a) ruling out other potential causes of movement disorders and improved pharmacological management of movements disorders; (b) improved psychiatric management improved management of psychotic symptoms and by treatment of potential benzodiazepine withdrawal; (c) differential reinforcement of appropriate social behavior and communication, attention timeout, and gradual reduction in time allowed for behavioral dyscontrol; and (d) training peers to reduce attention for behavioral dyscontrol and use alternate patterns of interaction. Over a 43-day period, Gary's time engaged in behavioral dyscontrol reduced from 6 to 7 h/day on Days 1 and 2 to very low levels such that there was no target behavior during the last 7/11 days.

The authors indicated that this formulation was predominantly psychological/behavioral with medical, psychiatric, and socio-emotional factors being secondary in this case. Thus, in many ways, it has some plain parallels with behavior analytic case formulations described in the previous section. Some of the important differences are that this approach did not use preintervention functional analysis but mainly used indirect and direct nonexperimental assessments of function, did not use behavior analytic language as extensively as possible, did not show experimental control of the target behavior, and did not collect data on treatment integrity and social validity. Another feature that some readers would see as a strength is that it collected multimodal, multimodal assessment data which was integrated into one formulation and hence may have greater clinical validity and be close to typical practice.

10.3.2 Case Formulation and Aggression and Violence: Forensic Applications

Work on case formulation has been extended to forensic cases of aggression and violence and is sometimes termed "forensic case formulation" (FCF) which often addresses severe and chronic forms of violence. FCF is often more complex than regular case formulation. First, FCF is often used with individuals with long histories of deprivation and abuse, personality disorders, such as psychopathic personality disorder, substance abuse, and developmental disabilities, such as intellectual disabilities and autism spectrum disorders. Second, many individuals have extensive involvement with the law. Third, the quantity of the clinical material is often very large and complex spanning over a lifetime and multiple service settings. Fourth, because the target behavior is both low frequency and often highly dangerous, it cannot be evolved for the purpose of assessment. Fifth, errors in assessment, treatment, and decisions to discharge a violent offender may result in great harm to future victims which was potentially preventable. Finally, the clinical issues are often highly emotionally charged for both perpetrators and staff, due to the severe nature of the violent acts, the injuries to the victims and families, the motivations for perpetrators to lie and dissimulate, and the emotional and sometimes judgmental and revengeful reactions of staff and professionals (Hart et al., 2011; Sturmey & McMurran, 2011).

FCF often takes place in forensic services including outpatient and inpatient forensic services and specialized forensic psychiatric units and hospitals. In this context, FCF must not only fulfill the criteria for routine case formulation discussed above (Eells, 2021), but it must also address the issue of safety and risk assessment. Thus, a trend in FCF over the last 10 years has been to integrate risk assessment and FCF as the two may inform one another. For example, a good FCF might identify risk factors that can be avoided or modified as part of treatment or management of a violent offender.

Researchers have developed at least four methodologies specific to FCF. These include: (1) collection and analysis of data using various modifications of ABC recordings (Bijou et al., 1968) and taxonomies of functions of aggression (Daffern et al., 2007a, b); (2) multiple sequential functional analysis, which also uses an ABC framework, but which provides a developmental account of the presenting and current problem; (3) Vedel and Emmelkamp's (2011) function-based assessment protocol to address aggression and violence in the context of substance abuse; and, finally, (4) offense-paralleling behavior (OPB) is a relatively new methodology for understanding forensic cases including cases of aggression and violence, in which conducts an assessment nonoffense behavior which is believed to be the same function as offense-related behavior.

10.3.2.1 Daffern et al.'s Typology of Function

In the first method, Daffern et al. (2007a, b) used and modified ABC data to develop a "typology of functions" that is focused on proximal environmental variables. Their typology of functions could be grouped as: (a) operant aggression in which aggression is maintained by negative social reinforcement (e.g., demand avoidance, avoidance of interactions); (b) positive social reinforcement (e.g., access to, access to staff/other patients/specific persons; enhanced social status and social approval; sexual/sadistic gratification, or access to tangibles, such as money, cigarettes, and drugs; enforcing compliance with instruction; and observing suffering in others); (c) nonsocial positive reinforcement (e.g., sensation seeking) and nonsocial negative reinforcement (e.g., reduce tension or "catharsis"); and (d) elicited aggression, such as expressing anger following provocation.

To evaluate this approach, Daffern and Howells (2002) conducted a study in a high security forensic psychiatric unit in Australia using case notes. Staff were required to record incidents of aggression during a 12-month period using a modified version of the *Overt Aggression Scale* (Silver & Yudofsky, 1987) named the *Record of Aggressive Behavior*. Following completion of the *Record of Aggressive Behavior*, a psychologist conducted a semistructured behavioral interview if the individual was not in seclusion, and the staff determined it was appropriate to do so, to determine the function of aggression using the typology of functions described above. There were 502 aggressive incidents involving 105 of 232 individuals. Functions could be determined in approximately 95% of incidents analyzed. Incidents of aggression were much more likely to involve staff rather than other patients. The most common functions with staff were avoidance of staff demands, staff compliance, and to express anger and reduce tension, whereas incidents of aggression with peers were most likely to involve expression of anger, enhance social status, and reduce tension. There were relatively few examples of access to tangible items and observing suffering in both incidents involving staff or peers.

Daffern and Howell (2002) data illustrate the functional nature of severe aggression and violence in forensic inpatient settings and are conceptually similar to functional approaches to understanding aggression and violence in other populations. They are also very useful for the design of a unit-wide program. For example, based on these data, this unit could have a program with three elements including: (a) training staff on effective, nonprovocative presentation of requests and management of compliance; (b) teaching effective assertiveness and other social skills between peers; and (c) alternate ways to reduce tension, such as relaxation training or mindfulness practices.

10.3.2.2 Multiple Sequential Functional Analysis

The second approach to forensic case formulation is *Multiple Sequential Functional Analysis* (MSFA) (Gresswell & Hollin, 1992; Hollin, 1990) which is a developmentally oriented functional approach which goes beyond describing the history of the problem of aggression and violence to understand how and why the problem developed in the most parsimonious manner using a learning theory framework. MSFA identified the functional relationships and specific contexts during development that account for the acquisition of aggression and violence as well as a functional account of the presenting problems. This approach uses the ideas of ABCs to organize very large quantities of developmental material, offense histories, etc. sequentially over the lifespan in a series of interlocking ABC analyses that identify key learning experiences. In this way, MSFA offers an account of rare, apparently inexplicable terminal behavior, such as acts of extreme violence that have not occurred previously, and hence could not have a history of direct reinforcement (Hollin, 1990).

In the first paper using this method, for example, Gresswell and Hollin (1992) used MSFA to understand the development of a man referred to as "D", with an apparently motivationless plan to kill 20 people who had already attempted two murders on unknown victims. Following these two attempted murders, he was diagnosed with psychopathic disorder and detained in a maximum security (forensic) hospital. The analysis was based on 12 h of interviews with the perpetrator, 2 h of interviews with his father and sister, 2.5 h with his psychotherapist and family doctor, 300 pages of witness statements and professional reports. MSFA involves organizing material within an ABC framework and abstracting out and inferring key learning at each stage of development. Each individual ABC analysis interlocked with the subsequent analysis and often overlaped and forms the basis with the next

ABC analysis. This method demonstrated that a series of relatively small steps accumulated a lifetime of pathological learning and experiences that made sense of or perhaps explained an apparently senseless, horrendous violent crime.

In this section, we will consider some of the seven ABC analyses that Gresswell and Hollin (1992) reported. During stage 1, early development, antecedents included emotionally absent parents, medical problems, including surgery for an undescended testicle, and peer rejection and tormenting. His behavior included the covert behavior of feeling alienation, being unable to control others, fantasies of his real parents coming to save him, and miracle cures for his disfiguring acne; and overt behavior such as searching through his parents possessions, repeated face washing, self-isolation, attempted suicide, and visiting abandoned land where he fantasized about attacking women. The consequences included social isolation, tormenting from new workmates, poor relationships with his parents, and reduced feedback about his own behavior. Gresswell and Hollin summarized the key earning in stage 1 as "He is very different from others, unattractive, unlikable and helpless … [and] Revenge fantasies can make him feel good" (p. 332). In stage 2, early adult and work experience, antecedents included rejection by coworkers and peers making him be aware of his own isolation. Behaviors included covert behavior such as anxiety, feelings of rejection, being able to control customers, etc., and overt behaviors including changing jobs, lying at work to present himself as normal, putting up a good front, attempted suicide, and maintaining social isolation. The consequences include being friendly and superficial and being admitted with hospital following suicide but without follow-up. The key learning here included learning to protect himself by putting on a good front that he will never succeed in relationships and that he can only appear normal by lying, but others can discover this, resulting in further rejection.

The final two ABCs described the final learning that led to the attempted murders. In stage 6, the antecedent was a high profile, multiple murder, the Hungerford massacre. The behaviors were the covert behavior of fantasizing about exceeding the number of people killed at the Hungerford massacre, being excited while prowling around at night, feeling more confident, and eventually habituating to the excitement of prowling. He stopped talking about his fantasies and acquired more weapons. The consequences were that he nearly got caught and a worsening relationship with his therapist. The key learning included that prowling could be thrilling. Finally, in stage 7, the antecedent was a local attack on a woman near him and rejection of his paintings from a local art exhibition. The behaviors included the covert behavior of deciding that if others can attack, then so can he and the over behavior of the two attempted murders. The consequences included fear in his victims' eyes and being caught by the police. The key learning in stage 7 was the increased risk resulted in greater thrills.

Subsequently, Hart et al. (2011) reported the case of "Bob" who killed someone as an adult. During his early childhood and high school, there were distal antecedents, such as poor parenting and being sexually abused by his step-father and proximal antecedents such as being criticized by teachers and bullied by peers. These resulted in overt behavior, such as isolation and withdrawal, and, covert behavior, such as ambivalent feelings about his abuser and his own sexual orientation, violent fantasies, and social anxiety. The consequences included punishment of social interactions. This resulted in key learning that Bob could not gain peer recognition, that males are a threat, and that he was inadequate and undeserving of friendship.

Following a similar analysis of his childhood and high school development, Hart et al. then presented two offense sequences. In the first, the antecedent was that Bob meets acquaintances in pub with the presence of a gay man (Tom), his overt behavior was that he drank excessively, invited others back to his flat, and hid a knife, was isolated and withdrew. His covert behavior was that he thought that "I must become more sociable" and a sense of threat from Tom. The consequences were that the acquaintances came back to the flat, drank all his alcohol, and left him alone with Tom. The key learning here was that he was in danger, he had been set up by strangers, and that Tom meant to harm him. In the second offense sequence, the antecedent was that Tom touched him with the smell of alcohol on breath. His covert behavior was his memory of own abuse and immediate sense of physical threat, and his overt behavior was that he killed Tom, thereby removing the threat but which also resulted in his arrest. This resulted in the key learning that he was safe and the threat had been removed.

Others have extended MSFA in various ways. For example, Mappin et al. (2013) applied this approach to assess intimate partner violence perpetrated by three women. Finally, Gresswell and Dawson (2010) integrated both OPB and MSFA (see below).

10.3.2.3 Vedel and Emmelkamp's (2011) Method

These authors reported on a third functional assessment method in which they developed and evaluated an assessment package to assess criminal behavior including aggression and violence in individuals with substance abuse problems, interpersonal problems, anger, and legal problems in an outpatient setting. They developed an assessment protocol which included a balance sheet to evaluate the pros and cons of using and not using alcohol and other substances, and ABC's for cravings, etc. to develop case formulation of alcohol, substance abuse, and aggression and violence.

They described the application of this approach with "Bob," a 44-year-old man referred through the criminal justice system for mandatory treatment of substance abuse following an armed robbery. Tom has a long history of abuse of multiple substances and escalating violent crime. He also married a woman with a history of substance abuse. Following concern for the safety of their two children, family members took responsibility for their care, and the couple split. Tom then escalated his drug use, attempted suicide by driving his car off the road and was subsequently diagnosed with an antisocial personality disorder.

The balance sheet assessment indicated that alcohol use had short-term advantages of helping him to feel relaxed and feeling alive, but alcohol use had no obvious long-term benefits. The short-term disadvantages include hangovers and getting in trouble with his probation officers, and the long-term disadvantages include not being able to lead a normal life and health problems. The short-term advantages of change include keeping his probation officer off his back and saving money, and the long-term advantages include better health and being able to see his children. The short-term disadvantages include not being able to "blow off steam" and see his friends, and the long term disadvantages include life becoming boring.

Even this most simple assessment is very useful in terms of FCF-based idiographic intervention in that it specifies relevant reinforcers that treatment must supply or replace through more acceptable and safer behavior and suggests core values, such as being a father with associated short- and long-term goals. A similar sheet was completed for cocaine use.

The first part of therapy was to build a trusting relationship based in part on a functional view of Tom's antisocial personality trait. Vedel and Emmelkampt write that concerning many people with such antisocial personality traits, "Acting in a brutal, manipulative, or avoidant manner is often done in order to keep others at an emotionally safe distance … this implies that effort will be needed to build a good therapeutic relationship and a safe environment in which the patient with antisocial personality traits dares to be open about the difficulties he is facing … During the first phase of treatment, we tried to enhance the working alliance between Tom and his therapist by keeping sessions primarily focused on current substance use and associated problems. Tom was not prompted for emotional disclosure nor was he extensively questioned about his background …" (pp. 161–162). Thus, these authors used a good FCF to inform how to build a more effective therapeutic relationship and increase client motivation. The authors used behavioral self-monitoring to identify both external and internal cues and situations related to substance abuse and aggression. The authors negotiated goals for alcohol and substance use. Tom preferred limited use at weekends rather than abstinence, but this resulted in relapse after a couple of weeks and so he agreed to goals of abstinence.

In their summary functional analysis of Tom's alcohol use, Vedel and Emmelkamp (2011) identified external cues for drinking such as being home alone and contact with authority figures, evenings and weekends, preferred pubs, drinking pals, and specific TV ads. Internal cues included thoughts such as "They are out to make my life miserable" and feeling angry and tense. These resulted in cravings characterized by dry throat, thinking that he needed a drink immediately and feeling restless. The behavior of drinking 10–15 units of alcohol resulted in short-term consequences including decreased tension, feeling capable, and negative consequences, such as craving cocaine and loss of control, and long-term consequences, such as bad health and bad relationships with children. A similar analysis was done for cocaine use.

The FCF-based treatment plan included the following elements: (a) cognitive behavior therapy with an anger management module; (b) avoiding cues and managing trigger situations more effectively; (c) coping with cravings by strategies such as distraction an social support; (d) time out when feeling angry (Tom refused to use relaxation training); (e) relapse prevention such as strategies to use early in relapse to prevent a full relapse; and (f) role play addressing both verbal language, nonverbal skills and thoughts such as "They'll think I am a loser" to refuse alcohol and drugs when offered.

The outcomes for Tom's FCF treatment was quite impressive. Over a 10-month period, he had only one relapse into drinking and two episodes of verbal aggression associated with moving accommodation. By way of social validity, Tom appeared remorseful about his mistreatment of his children, his probation office began to question the diagnosis of antisocial personality disorder, and Tom reported much more confidence in trigger situations.

10.3.2.4 Offense Paralleling Behavior

Conducting a functional assessment or analysis of severe violence is both impractical because the target behavior occurs so infrequently, might be difficult to elicit, and almost certainly unethical because people would be exposed to severe injury. Thus, the final approach to FCF is OPB. Daffern et al. (2007a, b) described OPB as: "[A] behavioural sequence incorporating overt behaviours (that may be muted by environmental factors), appraisals, expectations, beliefs, affects, goals and behavioural scripts, all of which may be influenced by the patient's mental disorder, that is functionally similar to behavioral sequences that are functionally similar to previous criminal acts …" (p. 267). Examples of OPBs could be nondangerous precursor behavior in a stimulus–response behavior chain or functionally equivalent or functionally relevant behavior. For example, an offender with a history of violent rape that occurred following being humiliated by a woman might engage in frotteurism following a slight from a female staff member.

A second example of OPB applied to aggression and violence comes from Al-Attar (2010) who described a violent offender with temporal lobe epilepsy who received a custodial sentence following an incident at the age of 17 years in which he went into a violent rage and committed grievous bodily harm. Despite 3 years of apparently successful conventional therapy, such as cognitive behavior therapy, he continued to display violent outbursts resulting in restraint when other prisoners teased him. In this example, the OFB is similar to the index offense in that they share the similar antecedent stimuli and evoke similar behavior (extreme anger and violence), and both resulted in harm to the other person, presumably resulting in termination of the other person's aversive behavior be it threats or teasing.

In addition to assessing the OPB, this approach may also include assessment of two other classes of relevant behavior. These include prosocial, adaptive behavior, such as assertiveness or nondeviant sexual behavior, and detection evasion skills, such as avoidance, and lying, etc. Often information on these three classes of behavior are assembled from reviews of the index offense; offending history; developmental history; interviews with prison or other relevant staff, and the offender himself; and, psychometric instruments. As noted above, this is often done in tandem with risk assessment which often focuses on dynamic risk factors which could be changed through FCF based on OPB, removal or avoidance of risk factors, or individual intervention based on FCF (Daffern, 2010; Daffern et al., 2010).

One example of a specific and partially evaluated methodology to detect OPB comes from Akerman and Beech (2013) who developed a psychometric assessment and typology of OPBs. They developed a psychometric instrument, based on previous literature and piloting the measure with two groups of offenders. The literature review identified 15 categories of potential OPBs including sexual preoccupation, sexualized violence, paraphilias, lack of emotionally intimate relationships, lifestyle impulsiveness, poor problem solving, resistance to rules and supervision, grievance and hostility, hostile beliefs about women, lack of concern for others/callousness, lack of empathy, sexualized coping, social skills deficits, and lack of motivation for treatment. They administered a 55-item draft instrument to 30 men who had committed sexual offences and 20 violent offenders. The pool of items were then grouped into four domains: deviant sexual interest and preoccupation, distorted cognitions supporting offending, problematic socio-emotional functioning, and self-regulatory problems. A series of case studies revealed rather different patterns of responding and OPBs. For example, Mr. X had committed rape when angry and belittled and other violent sexual offences when he felt humiliated by a woman. The OPB was that when a female officer challenged him, he exhibited frotteurism against her and on other similar occasions when he felt slighted by a female staff, he apparently accidentally brushed against the female staff. Additionally, he used a confidential suicide line but staff suspected that he engaged in masturbation in a similar fashion to his history of obscene calls. His psychometric assessment revealed that his scores on measures of sexual interests and OPB were much higher than a matched nonsexual violent offender. Thus, this measure of OPB and other assessments identified ABCs that apparently parallels actual offending, and hence, this information could be used as the basis for function-based treatment, i.e., how to cope better with slights and offences from women and how to develop a healthier sexual and other relationship with women.

Interest in FCF has increased considerably over the last 15 years. It is an interesting expansion of the application of case formulation of aggression and violence to novel populations and contexts with additional challenges beyond those typically encountered in routine clinical work. Interestingly, it is one of the few areas where there is evidence from at least one true experimental study (Shaw et al., 2017) that case formulation results in significant positive outcomes (see below). Despite its popularity and potential utility, the quantity of empirical evidence for this approach to understanding violence remains relatively small although this literature is gradually growing over time.

10.3.3 Training Practitioners in Forensic Case Formulation

For FCF to be effective and impactful, routine clinicians and other practitioners must be competent to develop and use FCFs to design, implement, and evaluate FCF-based idiographic treatments that are superior to treatment as usual. Thus, recently, several studies have evaluated efforts to train staff such as probation officers to conduct FCFs. As a first step in this effort, Völlm (2014) conducted a Delphi survey with 31 experts with backgrounds in forensic psychiatry or psychology to determine what skill practitioners should learn related to FCF. There was broad consensus that the following elements should be in an FCF: (a) a comprehensive history, including offending history, childhood behavioral problems, and early trauma; (b) personality traits; (c) a comprehensive accounts of alcohol and drug abuse; (d) problem behaviors, including premeditated versus impulsive aggression, work, and leisure activities; (e) mental health issues; (f) risk assessment; (g) practical solutions to problems; (h) OPBs; (i) Wirasekeera's five Ps; and (j) the client's relationship style.

The findings of this Delphi study were useful to determine how to measure the adequacy of FCF. For example, McMurran and Bruford (2016) developed the *Case Formulation Quality Checklist* which raters can use to assess coherence, explanatory breadth, simplicity, predictive statements, and action-oriented components of an FCF. The *Case Formulation Quality Checklist* has adequate reliability and adequate validity, in that it distinguishes between case formulations from experts and novices (Minoudis et al., 2013) and is sensitive to the effects of training (Brown et al., 2018).

These developments have been useful in stimulating the beginning of efforts to train FCF skills in that they have defined the content of FCFs and developed preliminary measures that can be used to guide and evaluate training. Several studies have indicated that there is indeed a significant need for practitioner training in FCF. For example, Jeffcote et al. (2020) developed a checklist to evaluate six domains of FCFs which were problem behaviors, underlying thoughts and feelings, early experiences, particularly risks situations, self and staff support, and other support. The checklist was evaluated after initial pilot work with ten prisoners. The assessment used a card sort method in which the participant placed cards describing an element of the FCF applied to them next to five cards numbered one through five. Triads consisting of one prisoner, one clinician, and one key worker from 59 formulation teams each completed the FCF. The internal consistency of the six domains was adequate to good. Agreement across raters was poor (intraclass correlation coefficient [ICC] = 0.51) but slightly better between key worker and clinician (ICC = 0.65). Although agreement on some specifics of FCF was adequate, especially for overt problem behavior (ICC = 0.78), and history of early experiences (ICC = 0.64), agreement on other domains was poor, especially in areas of covert behavior such as underlying thoughts and feelings (ICC = 0.35), staff and self-support (ICC = 0.33), and particular risky situations (ICC = 0.48). Thus, these data and low baseline data on FCF skills in staff training studies point toward the need for training in FCF skills.

The outcomes of FCF training efforts have been mixed. For example, Brown et al. (2018) trained 18 probation officers over 5 days to make FCFs of offenders with personality disorders. They found that posttraining case formulations were generally better than pretraining vignettes especially in areas such as factual foundation, breadth of explanation, coherence of explanation over time, generativity, and overall quality, although there was no improvement in prioritizing and planning, treatment, consistency with psychological or criminological theory, and simplicity and freedom from unnecessary details. These probation officers tended to produce long lists of unprioritized problems rather than abstract out key treatment targets, did not link target problems to underlying causes, and did not identify future desirable behaviors or hypotheses to account for the presenting problems.

Subsequently, Shaw et al. (2012) evaluated a large-scale consultation and training program for 150 probation staff in four inner London boroughs. The authors administered a psychometric measure of self-evaluation of knowledge, attitudes, and skills before and 1-year after consultation had begun. They found that probation officers reported feeling more knowledgeable and capable regarding FCF, although the study did not include any objective measure of CFC skill.

Subsequently, Minoudis et al. (2013) evaluated training probation offices to make FCFs using Weerasekera's (1996) 5 P format. Unfortunately, they found little evidence of the effects of training suggesting that current efforts to train professionals in FCF need further development and evaluation. Probation officers completed two formulations before and after treatment, but there was no improvement in the content of the formulations. It is unclear why this training did not work. For example, it may not have been effective because participation and completion of exercises could have been monitored more rigorously, more training examples may have been required, or perhaps a more intense and/or greater quantity of training was required in order to achieve competency in case formulation. Shine and Cooper-Evans (2016) have extended training in FCF to address offenders with Autism, but data on outcomes are not yet available.

In a unique study providing validation evidence for the effectiveness of FCF, Shaw et al. (2017) reported an experimental posttest only design evaluation of training offender managers to write FCFs. The authors randomized 84 staff with direct face contact with high-risk offenders. All staff had already received a routine half-day of CFC training. Training consisted of constructing FCFs jointly and an associated risk management plan when working with the offender during routine supervision. Attrition was approximately 30% and 20% in the experimental and control groups, respectively. Offender managers in both the groups completed ratings of the quality of the relationship between themselves and the offender. Offender managers in the experimental group reported that they had overall better, more fair, and more trusting relationships with their offenders than offender managers in the control group. Effects were just statistically significant (p's between 0.01 and 0.05) and of small or moderate effect sizes. Data from a small subset of offenders themselves who consented reported broadly similar data.

10.3.4 Summary

This literature shows that FCF is possible, and there are models for practitioners to use to develop case formulations. The literature, however, is limited in quantity and quality, and the evidence that this approach makes a difference to aggression and violence is limited or absent. FCF comes with several special challenges that are less present than in typical case formulation. These include: (a) balancing the risks of assessment and treatment versus safety; (b) usually practitioners cannot observe the behavior of interest because it only occurs under special circumstances, there are no opportunities or it is too dangerous to permit (although offense paralleling behavior may partially circumvent this problem); (c) there are often years of data which may be unavailable, incomplete, or inaccurate; (d) it may take a long time because of the complexity and quantity of material; and (e) there may be motivations to lie, misrepresent, under-report, over-report information by client, family, victim, and staff and professionals.

10.4 Chapter Summary

One of the most important functions of assessment is to guide treatment toward doing something better than that which one would already have done without the assessment and the associated cost of delay in access to treatment. There are both ethical and pragmatic aspects to this. First, if one is going to consume resources for assessment and delay access to treatment during assessment, then there should be some commensurate benefit to the client, others, and society. Second, if therapists and services are going to invest limited resources into assessment, then the assessments should result in better treatment. For example, if there is already a fairly effective treatment available, there may be limited opportunities for pretreatment assessment to improve upon something that is already quite effective. If, however, there is no robustly effective treatment available or if standard treatment has failed, there may be a role for pretreatment assessment to result in more effective treatment.

The two approaches described here in this chapter are functional assessment and analysis and case formulation, including FCF. Both have similarities in goals—they are both forms of pretreatment assessment used to guide subsequent treatment—and sometimes they share the use of a similar functionalist framework and even methods. Disseminating effective pretreatment assessment practices is a challenge for both the approaches which research has begun to address; however, continued research is essential in this area to generate effective protocols to teach practitioners to be fluent and flexible in this area.

References

Abraha, I., Rimland, J. M., Trotta, F. M., Dell'Aquila, G., Cruz-Jentoft, A., Petrovic, M., Gudmundsson, A., Soiza, R., O'Mahony, D., Guaita, A., & Cherubini, A. (2017). Systematic review of systematic reviews of non-pharmacological interventions to treat behavioural disturbances in older patients with dementia. The SENATOR-OnTop series. *British Medical Journal Open, 7*(3), e012759.

Akerman, G., & Beech, A. R. (2013). Exploring offence paralleling behaviours in incarcerated offenders. *Forensic Update, 112*, 6–18.

Al-Attar, Z. (2010). The assessment and treatment of offence paralleling behaviours in young offenders: Added complications or greater opportunities for change? In M. Daffern, L. Jones, & J. Shine (Eds.), *Offence paralleling behaviour: A case formulation approach to offender assessment and intervention* (pp. 151–170). Wiley-Blackwell.

Baker, J. C., Hanley, G. P., & Mathews, R. M. (2006). Staff-administered functional analysis and treatment of aggression by an elder with dementia. *Journal of Applied Behavior Analysis, 39*(4), 469–474.

Beavers, G. A., Iwata, B. A., & Lerman, D. C. (2013). Thirty years of research on the functional analysis of problem behavior. *Journal of Applied Behavior Analysis, 46*(1), 1–21.

Bijou, S. W., Peterson, R. F., & Ault, M. H. (1968). A method to integrate descriptive and experimental field studies at the level of data and empirical concepts. *Journal of Applied Behavior Analysis, 1*, 175–191.

Bloom, S. E., Iwata, B. A., Fritz, J. N., Roscoe, E. M., & Carreau, A. B. (2011). Classroom application of a trial-based functional analysis. *Journal of Applied Behavior Analysis, 44*(1), 19–31.

Borrero, C. S., Vollmer, T. R., & Borrero, J. C. (2004). Combining descriptive and functional analysis logic to evaluate idiosyncratic variables maintaining aggression. *Behavioral Interventions: Theory & Practice in Residential & Community-Based Clinical Programs, 19*(4), 247–262.

Brown, P., & Elliott, R. (1965). Control of aggression in a nursery school class. *Journal of Experimental Child Psychology, 2*(2), 103–107.

Brown, S., Beeley, C., Patel, G., & Völlm, B. (2018). Training probation officers in case formulation for personality disordered offenders. *Criminal Behaviour and Mental Health, 28*(1), 50–60.

Byrne, C., & Coetzer, R. (2016). The effectiveness of psychological interventions for aggressive behavior following acquired brain injury: A meta-analysis and systematic review. *NeuroRehabilitation, 39*(2), 205–221.

Carr, E. G., Newsom, C. D., & Binkoff, J. A. (1980). Escape as a factor in the aggressive behavior of two retarded children. *Journal of Applied Behavior Analysis, 13*(1), 101–117.

Conroy, M. A., Fox, J. J., Bucklin, A., & Good, W. (1996). An analysis of the reliability and stability of the Motivation Assessment Scale in assessing the challenging behaviors of persons with developmental disabilities. *Education and Training in Mental Retardation and Developmental Disabilities, 31*, 243–250.

Cook, E. D. M., Swift, K., James, I., Malouf, R., De Vugt, M., & Verhey, F. (2012). Functional analysis-based interventions for challenging behaviour in dementia. *Cochrane Database of Systematic Reviews, 15*(2), CD006929. https://doi.org/10.1002/14651858.CD006929.pub2

Daffern, M. (Ed.). (2010). *Offence paralleling behaviour: A case formulation approach to offender assessment and intervention*. Wiley-Blackwell.

Daffern, M., & Howells, K. (2002). Psychiatric inpatient aggression: A review of structural and functional assessment approaches. *Aggression and Violent Behavior, 7*(5), 477–497.

Daffern, M., Howells, K., & Ogloff, J. (2007a). What's the point? Towards a methodology for assessing the function of psychiatric inpatient aggression. *Behaviour Research and Therapy, 45*(1), 101–111.

Daffern, M., Jones, L., Howells, K., Shine, J., Mikton, C., & Tunbridge, V. (2007b). Refining the definition of offence paralleling behaviour. *Criminal Behavior and Mental Health, 17*, 265–273.

Daffern, M., Jones, L., & Shine, J. (Eds.). (2010). *Offence paralleling behaviour*. Wiley.

Dracobly, J. D., Dozier, C. L., Briggs, A. M., & Juanico, J. F. (2018). Reliability and validity of indirect assessment outcomes: Experts versus caregivers. *Learning and Motivation, 62*, 77–90.

DuPaul, G. J., Kern, L., Volpe, R., Caskie, G. I., Sokol, N., Arbolino, L., Van Brakle, J., & Pipan, M. (2013). Comparison of parent education and functional assessment-based intervention across 24 months for young children with attention deficit hyperactivity disorder. *School Psychology Review, 42*(1), 56–75.

Dura, J. R. (1991). Controlling extremely dangerous aggressive outbursts when functional analysis fails. *Psychological Reports, 69*(2), 451–459.

Durand, V. M., & Crimmins, D. B. (1992). *The Motivation Assessment Scale (MAS) Administration guide*. Monaco and Associates.

Easden, M. H., & Kazantzis, N. (2018). Case conceptualization research in cognitive behavior therapy: A state of the science review. *Journal of Clinical Psychology, 74*(3), 356–384.

Eells, T. D. (2007). History and current status of psychotherapy case formulation. In T. D. Eells (Ed.), *Handbook of psychotherapy case formulation* (2nd ed., pp. 3–32). Guilford.

Eells, T. D. (2021). History and current status of psychotherapy case formulation. In *Handbook of psychotherapy case formulation* (pp. 5–32). Guilford.

Eells, T. D., & Lombart, K. G. (2011). Theoretical and evidence-based approaches to case formulation. In P. Sturmey & M. McMurran (Eds.), *Forensic case formulation* (pp. 1–32). John Wiley & Sons.

Estes, W. K., & Skinner, B. F. (1941). Some quantitative properties of anxiety. *Journal of Experimental Psychology, 29*(5), 390–400.

Fee, A., Schieber, E., Noble, N., & Valdovinos, M. G. (2016). Agreement between questions about behavior function, the motivation assessment scale, functional assessment interview, and brief functional analysis of children's challenging behaviors. *Behavior Analysis: Research and Practice, 16*(2), 94–102.

Fontes, R. M., & Shahan, T. A. (2021). Punishment and its putative fallout: A reappraisal. *Journal of the Experimental Analysis of Behavior, 115*(1), 185–203.

Froján-Parga, M. X., de Prado-Gordillo, M. N., Álvarez-Iglesias, A., & Alonso-Vega, J. (2019). Functional Behavioral Assessment-based interventions on adults' delusions, hallucinations and disorganized speech: A single case meta-analysis. *Behaviour Research and Therapy, 120*, 103444.

Gardner, A. W., Spencer, T. D., Boelter, E. W., DuBard, M., & Jennett, H. K. (2012). A systematic review of brief functional analysis methodology with typically developing children. *Education and Treatment of Children, 35*, 313–332.

Ghaderi, A. (2006). Does individualization matter? A randomized trial of standardized (focused) versus individualized (broad) cognitive behavior therapy for bulimia nervosa. *Behaviour Research and Therapy, 44*(2), 273–288.

Ghaderi, A. (2011). Does case formulation make a difference to treatment outcome? In P. Sturmey & M. McMurran (Eds.), *Forensic case formulation* (pp. 61–79). Wiley-Blackwell.

Ghaemmaghami, M., Hanley, G. P., Jin, S. C., & Vanselow, N. R. (2016). Affirming control by multiple reinforcers via progressive treatment analysis. *Behavioral Interventions, 31*(1), 70–86.

Gresswell, D. M., & Hollin, C. R. (1992). Towards a new methodology for making sense of case material: An illustrative case involving attempted multiple murder. *Criminal Behaviour and Mental Health, 2*(4), 329–341.

Gresswell, D. M., & Dawson, D. L. (2010). Offence paralleling behaviour and multiple sequential functional analysis. Offence paralleling behaviour: A case formulation approach to offender assessment and intervention. In P. Sturmey & M. McMurran (Eds.), *Forensic case formulation* (pp. 88–104). John Wiley & Sons.

Hanley, G. P. (2002/2009). *Practical functional assessment. Understanding problem behavior prior to its treatment*. Retrieved June 27, 2021, from https://practicalfunctionalassessment.com/implementation-materials/

Hanley, G. P., Iwata, B. A., & McCord, B. E. (2003). Functional analysis of problem behavior: A review. *Journal of Applied Behavior Analysis, 36*(2), 147–185.

Hanley, G. P., Jin, C. S., Vanselow, N. R., & Hanratty, L. A. (2014). Producing meaningful improvements in problem behavior of children with autism via synthesized analyses and treatments. *Journal of Applied Behavior Analysis, 47*(1), 16–36.

Hart, S., Sturmey, P., Logan, C., & McMurran, M. (2011). Forensic case formulation. *International Journal of Forensic Mental Health, 10*(2), 118–126.

Heath, H., Jr., & Smith, R. G. (2019). Precursor behavior and functional analysis: A brief review. *Journal of Applied Behavior Analysis, 52*(3), 804–810.

Howells, K., Daffern, M., & Day, A. (2008). Aggression and violence. In K. Soothill, M. Dolan, & P. Rogers (Eds.), *Handbook of forensic mental health* (pp. 379–402). Willan.

Hollin, C. R. (1990). *Cognitive-behavioral interventions with young offenders*. Pergamon Press.

Hunter, R. H., Wilkniss, S., Gardner, W. I., & Silverstein, S. M. (2008). The Multimodal Functional Model-- Advancing case formulation beyond the "diagnose and treat" paradigm: Improving outcomes and reducing aggression and the use of control procedures in psychiatric care. *Psychological Services, 5*(1), 11–25.

Iwata, B. A., Dorsey, M. F., Slifer, K. J., Bauman, K. E., & Richman, G. S. (1982/1994). Toward a functional analysis of self-injury. *Journal of Applied Behavior Analysis, 27(2)*, 197–209. Reprinted from Iwata, B. A., Dorsey, M. F., Slifer, K. J., Bauman, K. E., & Richman, G. S. (1982). Toward a functional analysis of self-injury. *Analysis and Intervention in Developmental Disabilities, 2(1)*, 3–20.

Jeffcote, N., Freestone, M., Heathcote, C., Shaw, J., & Kennedy, P. (2020). Agreement over content of case formulation in offender personality disorder pathway prison services. *Criminal Behaviour and Mental Health, 30*(2–3), 105–116.

Jones, K. M., & Friman, P. C. (1999). A case study of behavioral assessment and treatment of insect phobia. *Journal of Applied Behavior Analysis, 32*(1), 95–98.

Jessel, J., Ingvarsson, E. T., Metras, R., Kirk, H., & Whipple, R. (2018). Achieving socially significant reductions in problem behavior following the interview-informed synthesized contingency analysis: A summary of 25 outpatient applications. *Journal of Applied Behavior Analysis, 51*(1), 130–157.

Jessel, J., Metras, R., Hanley, G. P., Jessel, C., & Ingvarsson, E. T. (2020). Evaluating the boundaries of analytic efficiency and control: A consecutive controlled case series of 26 functional analyses. *Journal of Applied Behavior Analysis, 53*(1), 25–43.

Kahng, S., Iwata, B. A., Fischer, S. M., Page, T. J., Treadwell, K. R., Williams, D. E., & Smith, R. G. (1998). Temporal distributions of problem behavior based on scatter plot analysis. *Journal of Applied Behavior Analysis, 31*(4), 593–604.

Kern, L., Dunlap, G., Clarke, S., & Childs, K. E. (1994). Student-assisted functional assessment interview. *Diagnostique, 19*(2–3), 29–39. https://doi.org/10.1177/073724779401900203

Kramer, U. E. (2019). *Case formulation for personality disorders: Tailoring psychotherapy to the individual client*. Elsevier Academic Press.

Kuyken, W., Padesky, C., & Dudley, R. (2009). *Collaborative case conceptualization*. Guilford Press.

Lundervold, D. A., & Jackson, T. (1992). Use of applied behavior analysis in treating nursing home residents. *Hospital & Community Psychiatry, 43*(2), 171–173.

Mappin, L., Dawson, D. L., Gresswell, D. M., & Beckley, K. (2013). Female-perpetrated intimate partner violence: An examination of three cases using multiple sequential functional analysis. *Criminal Behaviour and Mental Health, 23*(4), 290–303.

Matson, J. L., & Vollmer, T. R. (1995). *The Questions About Behavioral Function (QABF) user's guide*. Scientific Publishers.

Matson, J. L., Bamburg, J. W., Cherry, K. E., & Paclawskyj, T. R. (1999). A validity study on the questions about behavioral function (QABF) scale: Predicting treatment success for self-injury, aggression, and stereotypies. *Research in Developmental Disabilities, 20*(2), 163–175.

McAndrew, J. F. (1989). Obsessive-compulsive disorder: A behavioral case formulation. *Journal of Behavior Therapy and Experimental Psychiatry, 20*(4), 311–318.

McDougall, T. (2000). Violent incidents in a forensic adolescent unit: A functional analysis. *NT Research, 5*(5), 346–362. https://doi.org/10.1177/136140960000500505

McIntosh, K., Borgmeier, C., Anderson, C. M., Horner, R. H., Rodriguez, B. J., & Tobin, T. J. (2008). Technical adequacy of the Functional Assessment Checklist: Teachers and Staff (FACTS) FBA interview measure. *Journal of Positive Behavior Interventions, 10*(1), 33–45.

McMurran, M., & Bruford, S. (2016). Case formulation quality checklist: A revision based upon clinicians' views. *Journal of Forensic Practice, 18*, 31–38.

Messer, S. B., & Wolitzky, D. L. (2007). The psychoanalytic approach to case formulation. In T. D. Eels (Ed.), *Handbook of psychotherapy case formulation* (pp. 67–104). Guilford.

Minoudis, P., Craissati, J., Shaw, J., McMurran, M., Freestone, M., Chuan, S. J., & Leonard, A. (2013). An evaluation of case formulation training and consultation with probation officers. *Criminal Behaviour and Mental Health, 23*(4), 252–262.

Northup, J., Wacker, D., Sasso, G., Steege, M., Cigrand, K., Cook, J., & DeRaad, A. (1991). A brief functional analysis of aggressive and alternative behavior in an outclinic setting. *Journal of Applied Behavior Analysis, 24*(3), 509–522.

O'Neill, R. E., Horner, R. H., Albin, R. W., Storey, K., & Sprague, J. R. (1990). *Functional analysis of problem behavior: A practical assessment guide*. Sycamore Publishing Company.

O'Reilly, M. F. (1995). Functional analysis and treatment of escape-maintained aggression correlated with sleep deprivation. *Journal of Applied Behavior Analysis, 28*(2), 225–226.

Pain, C. M., Chadwick, P., & Abba, N. (2008). Clients' experience of case formulation in cognitive behaviour therapy for psychosis. *British Journal of Clinical Psychology, 47*(2), 127–138.

Persons, J. B., Roberts, N. A., Zalecki, C. A., & Brechwald, W. A. G. (2006). Naturalistic outcome of case formulation-driven cognitive-behavior therapy for anxious depressed outpatients. *Behaviour Research and Therapy, 44*(7), 1041–1105.

Roscoe, E. M., Kindle, A. E., & Pence, S. T. (2010). Functional analysis and treatment of aggression maintained by preferred conversational topics. *Journal of Applied Behavior Analysis, 43*(4), 723–727.

Schulte, D., Künzel, R., Pepping, G., & Schulte-Bahrenberg, T. (1992). Tailor-made versus standardized therapy of phobic patients. *Advances in Behaviour Research and Therapy, 14*(2), 67–92.

Shaw, J., Minoudis, P., Craissati, J., & Bannerman, A. (2012). Developing probation staff competency for working with high risk of harm offenders with personality disorder: An evaluation of the Pathways Project. *Personality and Mental Health, 6*(2), 87–96.

Shaw, J., Higgins, C., & Quartey, C. (2017). The impact of collaborative case formulation with high risk offenders with personality disorder. *The Journal of Forensic Psychiatry & Psychology, 28*(6), 777–789.

Shogren, K. A., & Rojahn, J. (2003). Convergent reliability and validity of the Questions About Behavioral Function and the Motivation Assessment Scale: A replication study. *Journal of Developmental and Physical Disabilities, 15*, 367–375.

Sigafoos, J., & Saggers, E. (1995). A discrete-trial approach to the functional analysis of aggressive behavior in two boys with autism. *Australia & New Zealand Journal of Developmental Disabilities, 20*, 287–297.

Silver, J. M., & Yudofsky, S. C. (1987). Documentation of aggression in the assessment of the violent patient. *Psychiatric Annals, 17*(6), 367–370. https://doi.org/10.3928/0048-5713-19870601-06

Singh, N. N., Donatelli, L. S., Best, A., Williams, D. E., Barrera, F. J., Lenz, M. W., Landrum, T. J., Ellis, C. R., & Moe, T. L. (1993). Factor structure of the motivation assessment scale. *Journal of Intellectual Disability Research, 37*, 65–74.

Shine, J., & Cooper-Evans, S. (2016). Developing an autism specific framework for forensic case formulation. *Journal of Intellectual Disabilities and Offending Behaviour, 7*(3), 127–139.

Skinner, B. F. (1953). *Science and human behavior*. Simon and Schuster.

Skinner, B. F. (1971). *Beyond freedom and dignity*. Alfred A. Knopf.

Slaby, R. G., & Guerra, N. G. (1988). Cognitive mediators of aggression in adolescent offenders: I. Assessment. *Developmental Psychology, 24*(4), 580–588.

Smith, R. G., & Churchill, R. M. (2002). Identification of environmental determinants of behavior disorders through functional analysis of precursor behaviors. *Journal of Applied Behavior Analysis, 35*(2), 125–136.

Sturmey, P. (1991). Assessing challenging behaviour using a semi-structured behavioural interview: A case transcript. *Journal of the British Institute of Mental Handicap (APEX), 19*(2), 56–60.

Sturmey, P. (1994a). Assessing the functions of aberrant behaviors: A review of psychometric instruments. *Journal of Autism and Developmental Disorders, 24*, 293–304.

Sturmey, P. (1994b). Assessing the functions of self-injurious behavior: A case of assessment failure. *Journal of Behavior Therapy and Experimental Psychiatry, 25*(4), 331–336.

Sturmey, P. (1996). *Functional analysis in clinical psychology*. John Wiley & Sons.

Sturmey, P. (2008). *Behavioral case formulation and intervention: A functional analytic approach*. John Wiley & Sons.

Sturmey, P. (Ed.). (2020). *Functional analysis in clinical treatment*. Academic Press.

Sturmey, P., & McMurran, M. (Eds.). (2011). *Forensic case formulation* (Vol. 49). John Wiley & Sons.

Tarrier, N., & Calam, R. (2002). New developments in cognitive-behavioural case formulation. Epidemiological, systemic and social context: An integrative approach. *Behavioural and Cognitive Psychotherapy, 30*(3), 311–328.

Touchette, P. E., MacDonald, R. F., & Langer, S. N. (1985). A scatter plot for identifying stimulus control of problem behavior. *Journal of Applied Behavior Analysis, 18*(4), 343–351.

Turkat, I. D. (Ed.). (1985). *Behavioral case formulation*. Plenum Press.

Ullman, L. P., & Krasner, L. (1965). *Case studies in behaviour modification*. Holt, Rinehart and Winston.

Vedel, E., & Emmelkamp, P. M. (2011). Forensic case formulation, substance abuse disorders, and anger. In P. Sturmey & M. McMurran (Eds.), *Forensic case formulation* (pp. 61–79). Wiley-Blackwell.

Verberne, D. P., Spauwen, P. J., & van Heugten, C. M. (2019). Psychological interventions for treating neuropsychiatric consequences of acquired brain injury: A systematic review. *Neuropsychological Rehabilitation, 29*(10), 1509–1542.

Völlm, B. (2014). Case formulation in personality disordered offenders–A Delphi survey of professionals. *Criminal Behaviour and Mental Health, 24*(1), 60–80.

Ward-Horner, J., & Sturmey, P. (2012). Component analysis of behavior skills training in functional analysis. *Behavioral Interventions, 27*(2), 75–92.

Weerasekera, P. (1996). *Multiperspective case formulation: A step towards treatment integration*. Kriege.

Wolf, M. M. (1978). Social validity: The case for subjective measurement or how applied behavior analysis is finding its heart. *Journal of Applied Behavior Analysis, 11*(2), 203–214.

Wolpe, J. (1970). The instigation of assertive behavior: Transcripts from two cases. *Journal of Behavior Therapy and Experimental Psychiatry, 1*(2), 145–151.

Wolpe, J., & Turkat, I. D. (1985). Behavioral formulation of clinical cases. In I. D. Turkat (Ed.), *Behavioral case formulation* (pp. 5–36). Springer.

Zarcone, J. R., Rodgers, T. A., Iwata, B. A., Rourke, D. A., & Dorsey, M. F. (1991). Reliability analysis of the Motivation Assessment Scale: A failure to replicate. *Research in Developmental Disabilities, 12*(4), 349–360.

Chapter 11
Individual Biological Interventions for Violence and Aggression. I. Psychopharmacology and Hormonal Treatments

If biological determinism of aggression and violence is controversial, then biological treatment is sometimes equally if not more controversial. Yet, a wide range of biological treatments are commonly used to reduce aggression and violence. The most commonly used is psychopharmacology, but other biological interventions include hormonal treatments, psychosurgery, electrical stimulation of the brain, ECT, and dietary interventions. Most biological treatments are based on a traditional diagnostic approach to assessment of aggression and violence to predict the best treatment is the use of psychiatric diagnosis. Probably the most prototypical example of diagnosis-driven treatment is psychopharmacological treatment of aggression although some psychological treatments are also based on such diagnostic models as when there are treatment groups of anxiety disorders or prototypical case formulation for personality disorders, aggression, etc. Hence, we refer to antipsychotic, anxiolytic, and antidepressant psychotropic medications. This chapter will review two main classes of biological treatments—psychopharmacology and hormonal treatments for aggression and violence.

11.1 Psychopharmacological Intervention

In many service settings, the primary and most common biological treatment for aggression is psychopharmacology. Pharmacological treatment is now widespread for many mental health disorders in children and adolescents (Piovani et al., 2019), adults (Téllez-Lapeira et al., 2016), and older adults (Smeets et al., 2018). Studies report that their use continues to increase both nationally (Paulose-Ram et al., 2007; Seifert et al., 2021) and globally (Hálfdánarson et al., 2017) including young children and adolescents (Zhou et al., 2020) and adults (Paulose-Ram et al., 2007). Within specific populations and settings, the use of psychotropic medication is especially prevalent. For example, Lohr et al. (2018) found that around one third of

P. Sturmey, *Violence and Aggression*, Advances in Preventing and Treating Violence and Aggression, https://doi.org/10.1007/978-3-031-04386-4_11

children using Medicaid in Kentucky took at least one psychotropic medication for at least 90 days. Of those children taking psychotropic medication, around a third of them received polypharmacy for at least 90 days. Zhou et al. found that between 2011 and 2014, approximately a quarter of children and approximately one third of adolescents with attention deficit hyperactivity disorder (ADHD) took stimulants. Children and adults with developmental disabilities are especially likely to take psychotropic medication. A recent systematic review found that around a quarter of children and half of adults with autism take psychotropic medications (Jobski et al., 2017). Some studies have reported that over half of adults with intellectual disabilities take psychotropic medications (Koch et al., 2021; Tsiouris et al., 2013), and some studies report prevalence as high as around 80% within certain subpopulations such as adults in residential care and especially in institutional settings (Esbensen et al., 2009). Furthermore, some studies have reported that over half of older individuals in nursing homes takes psychotropic medications (Vasudev et al., 2015) and approximately half of individuals with dementias take psychotropic medication within 4 years of diagnosis with around one in five taking multiple psychotropic medications (Orsel et al., 2018). Thus, psychotropic medications are now widely used but particular populations and settings are much more likely to use them than the general population.

Prescription of psychotropic medication for acute mental health issues seen in outpatient settings is driven by the personal discomfort and suffering of the individual patient who is competent to report it and solicit help from professionals. Hence, the most common reasons for prescription of psychotropic medication in adult outpatients is depression, anxiety, stress-related disorders, and substance abuse (Benson et al., 2015). In children and youth, behavioral issues that are bothersome for caretakers such as parents and teachers and that many caretakers are often unskilled at dealing with in other ways drive the prescription of psychotropic medications (Lohr et al., 2018). In populations with chronic mental health needs and disabilities, prescription of psychotropic medications may sometimes be driven by behavioral issues rather than diagnoses, such as disruptive behavior, agitation, tantrums, hostility, self-harm, self-injury, and aggression and violence. For example, Tsakanikos et al. (2007) used multiple regression analysis to predict psychotropic medication use in two groups of adults one with ASD and ID and one matched group with ID alone. Aggressive behavior and externalizing behaviors, such as pestering staff, independently predicted use of psychotropic medication, and admission to inpatient units. Other studies have reported similar data. For example, Lake et al. (2012) found that aggression predicted both psychotropic medication use and polypharmacy in adults with autism in Canada. Similarly, Smeets et al. (2018) found that aggression and especially behavioral issues at night independently predicted prescription of antipsychotic medication in seniors with dementia in nursing homes. Further, admission to restrictive units where psychotropic medication is more likely to be used is also predicted by aggression (Tsakanikos et al., 2007). Thus, psychotropic medication is commonly used as a treatment for aggression and violence in many varied populations.

Another important predictor of the use of psychotropic medication for aggression is the absence of alternate effective therapies and services. For example, Smeets et al. (2018) found that prescription of antipsychotics is independently predicted by service-related factors (rather than resident-related factors) in nursing homes for seniors. These factors included nursing staff stress and lack of availability of physicians, although only the latter was statistically significant in multivariate analyses.

Further evidence that factors other than psychiatric diagnosis drives the use of psychotropic medication comes from some studies that have found that the use of behavioral interventions can reduce psychotropic medication as shown by studies of children in inpatient units (Carlson et al., 2020). Studies of PRN medication in adults with psychiatric disorders in inpatient units have also shown that function-based behavioral interventions can reduce dependence on PRN medication for behavior management (Donat, 1998, 2003, 2005). Multidisciplinary team reviews of psychotropic medication and standardized protocols for reduction in psychotropic medication can also result in reduction in psychotropic medication (Sheehan & Hassiotis, 2017), including on a state-wide basis (Carnahan et al., 2017) (see below for further discussion). Thus, quality of services—or the lack thereof—also predict use of psychotropic medications.

It is useful to distinguish several different populations and service contexts in which psychotropic medications are used to reduce violence and aggression. These include: (a) management of acute violence and aggression, such as upon admission to emergency rooms and inpatient psychiatric or forensic units; (b) ongoing use of emergency or "as needed" (*pro re nata* [PRN]) psychotropic medication; (c) routine prescription of psychotropic medication to treat ongoing aggression and violence; and finally, (d) management of aggression related to seizure disorders.

11.1.1 Acute Aggression and Violence

Psychotropic medication, alongside restrictive behavior management practices such as mechanical and personal restraints, is used quite commonly to manage aggression and violence to sedate individuals who are acutely violent in emergency rooms in general hospitals, inpatient psychiatric units, detox units for individuals with severe withdrawal reaction to alcohol or drugs, etc. (Sturmey, 2015). Sometimes these individuals may be delirious, agitated, physically ill, and/or intoxicated on alcohol or other substances or undergoing withdrawal from these substances. In these contexts, the immediate concern is the safety of the client and people around them who are placed in danger which sometimes might be imminent and severely threatening. Indeed, in such situations, the people themselves are at risk of harm both from harming themselves and from others. For example, intrusive behavior management practices such as restraint and seclusion and interventions from law enforcement and security staff which may be injurious and even lethal (Sturmey, 2015). Thus, one of the important ethical obligations for practitioners is client safety and the safety of others including members of the public and staff who are present.

In this situation, the paramount concern is usually the immediate safety of all concerned, rather than long-term management which can be dealt with later.

Thus, if the aggressive or violent person can be distracted, redirected, or calmed by nonpharmacological methods while keeping everyone safe, these methods are the first choice. Hence, many first responders, emergency room staff, and organizations serving people with mental health issues or developmental disabilities train their staff in crisis intervention methods and/or use crisis intervention teams to de-escalate aggression. More recently, educational and skills training has been made available to police, law enforcement, and other first responders to identify people with mental health needs and/or disabilities and use nonintrusive methods of dealing with aggression and violence (Booth et al., 2017). The extent of this training and its use in practice, however, are as yet unknown.

When these nonpharmacological methods have been tried and are unsuccessful or no effectively trained personnel are available, various pharmacological interventions may be used, especially in emergency rooms or inpatient units. In these situations, the immediate goal is to reduce aggression which is often done by suppressing most behavior including aggression and violence, i.e., to sedate the person safely.

There is a considerable evidence base for this practice which addresses both behavior change and medication safety. For example, Muir-Cochrane et al. (2020) reported a systematic review of chemical restraint for the management of agitation and aggression. They found 21 randomized controlled trials (RCTs) with 3,788 participants which took place in prehospital settings, general hospital, psychiatric emergency rooms, and inpatient psychiatric settings. The overall quality of the studies was moderately high; however, it was difficult for the authors to make broad conclusions concerning efficacy of specific agents or dosing strategies because studies differed substantially in terms of design. Some compared medication with placebo whereas others compared two or more medications with one another, different dosing schedules or routes of administration, and used different measures of aggression, such as time to calm or ratings of behavior intensity, etc. The most commonly evaluated medication was haloperidol, and intravenous administration of medication often resulted in calm behavior within 10 min of administration. Negative side effects occurred in 0–38% of participants depending upon the study. Some medications, such as the benzodiazepine and midazolam, were associated with faster more substantial calming than other medications. Unfortunately, they were effective for shorter periods of time than other medications and were also associated with more negative side effects such as sedation, reduced arousability, and respiratory suppression.

11.1.2 As Needed (PRN) Medications

A second somewhat similar but distinguishable situation is the use of psychotropic medication to reduce aggression and violence on an "as needed" basis for recurrent incidents. This is distinguishable from the acute situations discussed in the preceding paragraphs because the episodes are recurrent, may be predictable, and other less intrusive and less dangerous interventions may be available. A wide range of PRN agents are used including antipsychotics, benzodiazepines, combinations thereof (Patel et al., 2019), and antihistamines (Vitiello et al., 1991). Some services prefer to use antihistamines because the profile of negative side effects is better than for antipsychotics. In the past, chloral hydrate was used to sedate children and adolescents both for routine medical and dental treatments and for behavioral issues such as aggression and violence. Following reports of life-threatening and lethal effects, chloral hydrate is no longer recommended and largely unavailable (Institute for Safe Medication, 2016).

Although potentially effective in short-term reduction of aggression and violence through sedation, repeated PRNs for aggression and violence are ethically problematic. Management of aggression through repeated PRN medication demonstrates that they are not effective in long-term management of aggression, and their use may reflect programmatic failure or absence of effective treatment for aggression. For example, Carlson et al. (2020) reported that behavior modification programs resulted in both reduced aggression and reduced use of unnecessary PRN sedation of children in an inpatient unit. Similar results have also been found for behavioral consultation and intervention with psychiatric inpatients which may reduce not only PRN medication but also other restrictive behavior management practices such as restraint and seclusion (Donat, 1998, 2003, 2005). So, safe use of PRN on a short-term or very infrequent basis may be acceptable on occasions, but routine use of PRN to manage aggression and violence should be used as a red flag during quality assurance to trigger alternate, safer, and more dignified forms of behavioral intervention.

11.1.3 Routinely Prescribed Psychotropic Medications

Several classes of psychotropic medications have been evaluated as to their effects on aggression and violence. Sometimes, this has been done by matching the psychotropic medication with the alleged corresponding diagnoses. For example, matching of medication class and diagnosis could take place by treatment of ADHD with stimulants but not antipsychotic medications in order to reduce ADHD-related aggression or treatment of psychotic disorders with antipsychotics but not stimulants to reduce aggression related to schizophrenia. An alternate approach is to treat the symptoms rather than the underlying disorder; thus, one might prescribe psychotropic medication for a youth presenting with aggression either for the underlying ADHD or for the presenting problem of aggression (Gurnani et al., 2016). Several practice guidelines are available both generally on psychopharmacology for aggression in youth (Knapp et al., 2012) and specifically on antipsychotics (Pappadopulos et al., 2006). These guidelines often recommend implementing evidence-based psychosocial treatments such as evidence-based parent training before implementing psychopharmacology, and avoiding polypharmacy (Rosato et al., 2012).

There have been several expert panels, narrative reviews, systematic reviews, and meta-analyses of psychotropic medications to treat aggression in various populations. Gurnani et al. (2016), for example, reviewed a wide range of psychopharmacological studies of children and adolescents with aggression and psychiatric disorders. This narrative review was based upon papers published between 2000 and 2015 and also included earlier papers judged to be relevant. The review identified four classes of medications with evidence to support their use which were: (1) psychostimulants, (2) alpha-2 agonists for ADHD and disruptive behavior disorders, (3) mood stabilizers, and (4) atypical antipsychotics. Guralni et al. found the absence of evidence evaluating symptomatic treatment of aggression in the absence of diagnoses. Some papers have evaluated combined pharmacological treatment by adding on risperidone to stimulants (Linton et al., 2013) and other medication combinations (Blader et al., 2021) to treat aggression in youth with ADHD.

Pappadopulos et al. (2006) reported a meta-analysis of 45 RCTs of pharmacological treatment of aggression in children and adolescents. They found an overall effect size of $d = 0.56$ with somewhat larger effect sizes for methylphenidate for ADHD and comorbid aggression ($d = 0.9$, $n = 875$) and for risperidone for behavioral disturbance and conduct disorders in youth with below average IQ ($d = 0.9$, $n = 0.9$). These effect sizes would usually be interpreted as "medium" to "large."

11.1.3.1 Antipsychotics

One of the more commonly evaluated classes of psychotropic medications for the treatment of aggression and violence has been first- and second-wave antipsychotics. As their name suggests, this class of medications were first developed to treat psychotic disorders such as schizophrenia, but over time their use has broadened considerably so that those are used for other severe psychiatric disorders, such as bipolar disorder and autism spectrum disorder. They have also been used broadly for behavior management, including a wide range of disruptive behavior disorders in a wide range of populations, including children with emotional-behavioral disorders; ADHD; developmental disabilities, including intellectual disabilities and autism spectrum disorders; aggression and violence in adults in inpatient units and with traumatic brain injuries; and disruptive behavior and agitation in seniors.

First-wave antipsychotics were used extensively in the past both for psychotic disorders and for aggression. This was done despite their significant negative side effects, including concerns that they reduced aggression through sedation rather than specifically targeting aggression, perhaps because of the lack of alternative better medication at the time. They have largely been displaced by second-generation antipsychotics because when first marketed, second-generation antipsychotics appeared to have fewer negative side effects than first-generation antipsychotics. As time has gone on and more data has been collected, much more is known about the long-term negative side effects of second-generation antipsychotics clouding the relative merits of each class of medication (see Table 11.1).

Antipsychotic medications have been extensively evaluated in several different populations resulting in multiple systematic reviews and meta-analysis. See Table 11.2 for a brief summary of some of these meta-analyses. Generally, these meta-analyses agree that there appears to be some effect of both first- and second-generation antipsychotics on short-term reduction of aggression. For example, in a meta-analysis of RCTs for aggression in children and adolescents, Pappadopulos et al. (2006) reported effect sizes of 0.7 for first-generation and 0.90 for second-generation antipsychotics (all studies of risperidone). Tampi et al.'s review of 16 meta-analyses of antipsychotics found that two reported modest effects of first-generation antipsychotics in individuals with dementias. These observations must be tempered in a number of ways. First, there is a lack of long-term follow-ups in almost all reviews and meta-analyses. This is a significant limitation as aggression is often a chronic problem in many of these populations. Second, relatively little is known about either the relative efficacy of different medications within the classes and between the two classes of antipsychotics. Third, there are relatively few clear findings concerning the relationship between doses and treatment response.

As noted earlier, practice guidelines recommend first implementing psychosocial treatments before implementing antipsychotics. In practice, many individuals may experience both pharmacological and psychosocial treatments for aggression simultaneously, and these two treatments might act independently or interact to enhance or inhibit one another in various ways. The most common psychosocial treatments for aggression include some form of behavior therapy; thus, several studies have attempted to evaluate the combined and interactive effects of psychosocial treatments such as parent training in behavior management and antipsychotic

Table 11.1 Examples of first- and second-generation antipsychotic medications and their side effects

Medication class	Examples	Negative side effects
First-generation antipsychotics	Thioridazine Chlorpromazine Fluphenazine Pimozide	Dry mouth, stiff muscles, muscle cramping, and tremors Extrapyramidal side effects include dystonia, akathisia, and parkinsonism. Rigidity, bradykinesia (slowed movements) Tardive dyskinesia Cardiac side effects including arrhythmias and sudden death Neuroleptic malignant syndrome (potentially lethal).
Second-generation antipsychotics	Risperidone Olanzapine Aripiprazole Quetiapine Clozapine	Sedation, weight gain, metabolic syndrome, type II diabetes Increased mortality in older people

Table 11.2 A summary of meta-analytic studies of antipsychotic medications for aggression in children, youth, adults, and individuals with traumatic brain injury

Study/medication(s)/ population	Search and literature	Findings and conclusions
Pappadopulos et al. (2006) Pharmacotherapy generally Children and adolescents	Identified 45 RCTs from 1980 to 2005 including nine RCTs ($n = 8750$) of risperidone for aggression in children with conduct disorder, ADHD, low IQ, autism spectrum disorder, oppositional defiant disorder	The mean effect size was 0.90 (range 0.80–1.96 across 8 RCTs). Most trials of atypicals were of risperidone with a relatively large effect size ($g = 0.9$) but with negative side effects such as extrapyramidal symptoms, somnolence, headache, weight gain, and prolactin elevation. There were three RCTs of the older typical antipsychotics haloperidol ($k = 2$) and thioridazine ($k = 1$) (total $n = 136$, average age = 7.7 years, 85% boys) with an effect size of 0.8 for haloperidol and 0.35 for thioridazine. Again, negative side effects were more significant than alternate medications for treatment of aggression such as lithium and were sufficiently concerning that these medications may not be suitable for treatment of aggression in children
Loy et al. (2017) Atypical antipsychotics Children and youth aged 5–18 years with disruptive behavior disorders (8/10 trials in outpatient settings)	Ten studies published up to 2017 on risperidone ($k = 8$), one quetiapine ($k = 1$) and ziprasidone ($k = 1$) compared to placebo. Most were small scale (38 or fewer participants) and short term (nine were 4–10 weeks long, one was a 6-month trial)	There was low-quality evidence that risperidone reduced aggression and moderate quality evidence that it reduced conduct problems. There was short-term weight gain. Other side effects and long-term side effects have not been well studied. There were no studies with children aged under 5 years
Fallah et al. (2019) Atypical antipsychotics Pediatric autism	Double-blind RCTs using Aberrant Behavior Checklist-Irritability scale (Aman & Singh, 1986) published up to March 2018. There were eight trials on risperidone, aripiprazole, lurasidone, and placebo in $N = 878$ participants	Risperidone and aripiprazole were superior to placebo and equally effective in terms of reduction in ABC-I and side effects profile. Both were well tolerated
Van Schalkwyk et al. (2018) Antipsychotics Adults (including those with psychosis, bipolar disorder, dementia, and ID)	$k = 21$ RCTs with $N = 2610$ participants published up to August 2016	There was a small but statistically significant effect size (SMD = 0.29, $p < 0.0001$) comparable to nonpharmacological interventions. There was no difference between different medications and a small dose effect. There were no differences between diagnostic groups, except there was no significant effect on the study for adults with ID. Clinicians should consider the small effect size and negative side effects when using them for aggression in adults

(continued)

Table 11.2 (continued)

Study/medication(s)/ population	Search and literature	Findings and conclusions
Faay et al. (2018) Typical and atypical antipsychotics Psychosis spectrum disorders	$k = 18$ studies published up to 2014 which lasted at least 4 weeks, and compared typical and atypical antipsychotics on measures of hostility	There was a small effect size (0.26) favoring atypical antipsychotics over typical antipsychotics. There was a moderate effect size when comparing Clozapine with typical antipsychotics (0.415)
Williamson et al. (2019) Nine classes of pharmacological treatments generally including antipsychotics/ traumatic brain injury	21 RCTs for treatment of agitation, behavior and other targets with at least 50% of participants with TBI found in literature published up to December 2018 including one RCT on olanzapine and four nonexperimental studies on safety of antipsychotics (two on haloperidol)	Maturana and Maturana (2009) reported one RCT ($N = 38$) with reductions in irritability, aggressiveness, and insomnia ($p < 0.01$–0.0001), but did not report any safety data. Four nonexperimental studies reporting safety data found longer periods of posttraumatic amnesia and longer periods of hospitalization in one study

medications. For example, Van der Oord et al. (2008) conducted a meta-analysis of 26 RCTs conducted between 1985 and 2006 in children aged 6–12 years with ADHD and concurrent oppositional or conduct symptoms which compared methylphenidate, psychosocial treatments, such as behavior therapy or cognitive behavior therapy, and combined treatments. Several studies included specific psychometric measures of aggression and others measured aggression as part of composite measures of externalizing problem behavior. There were large effect sizes on measures of ADHD and oppositional and conduct symptoms for methylphenidate alone and combined treatments and moderate effect sizes for psychosocial treatments alone. For social behavior outcomes, there were moderate effect sizes for all treatments, and for academic outcomes, there were small effect sizes for all treatments. There were no differences between behavioral and cognitive behavioral treatments. The authors concluded that all treatment options were effective, but psychosocial treatments had smaller effects than methylphenidate. This conclusion was tempered by the observation that the effect sizes for psychosocial treatments might have been masked by treatment sequence effects. That is, several studies first implemented methylphenidate at doses likely to be maximally effective and then added on psychosocial treatments when there was a ceiling effect with little room left for improvement.

As well as evaluating medication effectiveness, medication trials are also conducted to evaluate the safety of the medications. Thus, some reviews have conducted systematic reviews and meta-analyses which have focused on the harms from antipsychotics in vulnerable children including both children and young adults who, since their nervous systems are still developing, may be especially vulnerable to negative side effects and long-term harms (Pillay et al., 2018), and seniors who may be especially vulnerable if they are frail (Corbett et al., 2014; Salzman et al., 2008). For example, Pillay et al.'s systematic review and meta-analysis of negative side effects of antipsychotics in children and young adults focused on potential harms based on 135 studies (95 experiments and 40 nonexperiments) published up to April 2016. One hundred twenty-six studies reported harms. They found that first-generation antipsychotics were associated with modestly less weight gain and more extrapyramidal symptoms than second-generation antipsychotics. They also found that second-generation antipsychotics were associated with a wide spectrum of negative side effects including weight gain; high triglycerides; extrapyramidal symptoms, such as uncontrollable muscle movements, tremors, and muscle contractions, which are stigmatizing and sometimes quite disabling; sedation and somnolence; and type II diabetes. Some of these negative side effects are particularly worrying. These studies were conducted on young people, who may carry a lifetime burden of preventable disease and disability related to these negative side effects. In addition, people with severe disabilities may have limited capacity to report such negative side effects early and receive appropriate dosage adjustment or treatment.

Some of these concerns are reflected in RCTs evaluating removing or reducing antipsychotic medications. For example, Tampi et al. (2016) reviewed two meta-analyses of trials reducing antipsychotics in individuals with dementia. They found that reduction in antipsychotic medication did not result in increases in aggression and other problem behavior in many individuals and that in one trial, drug reduction was associated with reduction in mortality. Studies with individuals with ID have reported similar effects (Ahmed et al., 2000; McNamara et al., 2017).

There are three final limitations in many of these studies. The first is that they did not include explicit controls to demonstrate that the effects of antipsychotics were specific to reducing aggression and related maladaptive behavior and did not broadly suppress adaptive behavior. The second is that, although many practice recommendations recommend that antipsychotics are used only after safe alternative effective treatments have been tried, there is little or no mention of whether this actually happened in almost all of the reviews. Third, there was no mention in any of the studies reviewed of functional analysis as the basis of psychosocial treatments, an approach which is considered to be standard treatment and which is likely to result in more effective psychosocial treatments in many situations (Sturmey, 2020).

Antipsychotics are widely used to control aggression and related behavioral phenomena in seniors, especially in nursing homes. Indeed, the intensity of interest and concern is attested to by the fact that by 2018 there were already at least 16 meta-analyses of this literature focusing on both the effectiveness of antipsychotics in treating aggression, but also the ongoing concern over risks, including risks of death (Tampi et al., 2016). These meta-analyses generally concur that some antipsychotics such as risperidone do result in short-term reduction of aggression and agitation (Corbett et al., 2014; Schneider et al., 2006), but this is based on trials lasting only 6–12 weeks and thus effectiveness in the longer term is unknown. In contrast, the harmful effects of long-term prescription of antipsychotics in seniors are well documented, extensive, and life-threatening. For example, Schneider et al. (2005) conducted a meta-analysis of mortality data from 15 trials (9 of which were unpublished) and found that antipsychotics increased mortality from 2.3% (placebo) to 3.5% (medication) during the course of the trial (odds ratio = 1.54, $p = 0.02$). Some longer-term trials have reported a 29% increased risk of death over 36 months (Schneider et al., 2006). Further, safe alternatives to antipsychotics exist in this population including better pain management, when pain presents as agitation and aggression, and improved psychosocial treatments (Corbett et al., 2014). These findings have resulted in both black box warnings from the Federal Drug Administration for the use of antipsychotics with seniors and recommendations that the use of antipsychotics should be restricted to only 6–12 weeks. Corbett et al. put it plainly: "Don't use antipsychotics routinely to treat agitation and aggression in people with dementia" (p. 1).

In summary, antipsychotics are widely used to treat aggression in clinical practice in several different contexts with quite different populations. Although most trials agree that short-term reductions in aggression do occur it is unclear if this is simply due to sedation or other broad behavioral effects rather than specific effects on aggression and violence, the long-term effectiveness is not known, and the long-term safety of these medications remains questionable.

11.1.4 Intermittent Explosive Disorder

The closest *DSM5* comes to explicitly defining a psychiatric disorder characterized by "abnormal" aggression is intermittent explosive disorder (IED). *DSM5* defined IED as multiple discrete episodes of aggression over at least 3 months that consist of either serious assaults or property destruction and are "grossly out of proportion to any precipitating psychosocial stressors". Verbal aggression must occur at least twice weekly for 3 months or physical aggression must occur for at least three times in 12 months resulting in property damage or injury to other people or animals. The exclusion criteria for IED include excluding other mental disorders that might explain the aggression, such as conduct, personality, psychotic disorders, a manic episode, ADHD, or the direct effects of prescribed or illegal drugs. Interestingly, when describing the aggressive behavior, *DSM5* uses the phrase "failure to resist aggressive impulses," suggesting an important element of failure of self-control as a cause of such aggression.

Given the potential severity and negative impact of IEP, it is surprising that most empirical research has focused on issues such as taxonomy, epidemiology, and assessment of psychosocial impairment, rather than treatment (Coccaro, 2012; Fanning et al., 2019; McCloskey et al., 2008). Thus, there are surprisingly few controlled studies of IED-related aggression, and they tend to be extensions of standard psychosocial or pharmacological treatment developed for other populations. For example, Coccaro et al. (2009) randomly assigned 100 individuals with IED and current aggression to fluoxetine or placebo. There were improvements in self-reports of aggression and irritability on standard psychometric measures which were unrelated to changes in anxiety or depression. Thus, fluoxetine may have a specific effect on reducing aggression in individuals with IED.

In one of the few papers evaluating psychosocial treatments for IED, McCloskey et al. (2008) compared the effectiveness of group and individual CBT and wait list control in 45 individuals with IED. They found that CBT resulted in large reductions in anger in individuals in both the individual and group CBT conditions compared to the wait-list control. The effect sizes were large for both individual ($d = 1.13$) and group ($d = 0.82$) CBT, and these effects were largely maintained at 3-month follow-up. There was a small difference between the individual and group formats ($d = 0.45$) favoring individual CBT, perhaps reflecting challenges in addressing each participant's individual cognitive distortions during the group format.

Thus, there is only a rather small literature on treatment of aggression in IED with either psychopharmacology or CBT. Interestingly, both the methods produce broadly similar effect sizes (Coccaro, 2012).

11.1.5 Epilepsy, Psychotropic Medications, and Aggression

Some individuals with seizure disorders demonstrate aggression either during a seizure or immediately afterwards (post-ictal aggression). Such aggression and violence are rare but can be severe and dangerous to all concerned and is primarily managed by treatment of the seizure disorders where possible (Marsh & Krauss, 2000). In some treatment refractory cases, pharmacological intervention in addition to management of the seizure disorder has been attempted but most of the literature on pharmacological management of post-ictal aggression tends to be case reports (e.g., Abd Wahab et al., 2019).

11.1.6 Conclusions

Psychopharmacology is widely used to control aggression and violence in children, youth, adults including older adults. It is probably the most widely used, default treatment. Yet, evaluating the effectiveness of psychotropic medications for the treatment of aggression and violence, even in research studies let alone routine clinical practice, is not straightforward. Apart from the usual issues of adequate experimental design, there are additional issues that are relatively specific for aggression and violence. For example, many RCTs of psychotropic medications are quite short-term, yet, like many other socially significant problems, many forms of aggression and violence are ongoing, long-term problems. This brings up two questions: (a) Do psychotropic medications work long-term? and (b) What are the long-term risks? Without this information, it is difficult for the client or their proxy to make a judgment about the risks and benefits of psychotropic medications as many of the long-term risks and benefits are unknown. It also places the practitioner in a difficult ethical position, in that if they are to be honest with their client, they should disclose that risks are unknown. In other situations, such as the use of antipsychotics with seniors with dementias, the short-term benefits are indeed known—they are modest—but the long-term benefits are unknown but the risks, including increased risk of mortality, are well documented.

A second issue that is common across populations and contexts is that there are often alternative psychosocial treatments or other safer medical interventions. In clinical practice, it is easy to assume that these alternatives have been implemented such as when clients or their caregivers say, "We tried that, but it did not work." In this situation, it is difficult to ascertain exactly whether or not the alternative treatment was appropriate and whether or not it was done adequately. Often the documentation of previous treatments is simply unavailable, incomplete, or too voluminous to review expeditiously in applied settings.

A third related issue is that sometimes alternate safer and potentially effective treatments are known but such treatments are not currently available. For example, a physically abusive partner might benefit from anger management but might need more sessions than insurance will cover and from a highly trained practitioner, but the insurance might cover routine psychotropic medication that may be equally or even less effective and more risky in some contexts. Finally, the outcomes of psychotropic medications are sometimes weak and disappointing. This may be because their use is not based on a clear understanding of the individual nature of violence and aggression and the use of psychotropic medication is not matched to a specific type of aggression. For example, if the binary distinction of emotionally hot/reactive/respondent aggression versus emotionally cold/planned/operant aggression is biologically real with different underlying biological mechanisms, then one might predict that there should be different forms of biological interventions for each type of aggression. Similarly, if one could reliably and accurately diagnose different forms of neurotransmitter deficits or excesses prior to prescription, then one might be able to match pharmacological interventions to the individual profiles of neurotransmitter functioning. Thus, the use of psychotropic medication remains controversial. Some believe that: (a) they are used as a default treatment when better, safer alternatives exist; (b) there are short- and long-term negative side effects which are underplayed or not revealed to patients when they are first prescribed; (c) they are unnecessarily expensive because they need monitoring for extensive periods of time when alternative cheaper and more effective alternatives exist; and (d) big pharma plays a dubious and often unethical role in promoting their use, often putting profits before patients.

11.2 Hormonal Treatments

Chapter 5 described the evidence for hormonal mechanisms for control of aggression and violence. This section will review evidence that there are individual treatments for aggression and violence including surgical and chemical castration, treatment of hyperthyroidism and aggression, and oxycontin- and vasopressin-based treatment of aggression. Hormonal treatments of aggression and violence have the possibility in some cases to offer treatment that, unlike neurosurgery, is potentially reversible and hence potentially safer.

11.2.1 Surgical and Chemical Castration

If testosterone causes or contributes to aggression and violence, then biological interventions to reduce the production of testosterone seem a rational intervention. Reviewing the role of testosterone in aggression in Chap. 5, revealed that there is no direct or simple relationship between testosterone and aggression. Rather, the effects of testosterone on aggression is mediated by baseline levels of testosterone and social mediation of the bidirectional effects between testosterone and behavior, such as competitive challenges (Carré & Robinson, 2020; Carré et al., 2017). Hence, it seems unlikely that decreasing testosterone in humans would have a linear effect of decreasing aggression. Despite the obvious conceptual, practical, and ethical difficulties in evaluating such treatments, there is surprisingly some evidence of the effectiveness of this approach to treatment, in both nonhumans and humans, including violent sexual offenders.

The effects of castration on nonhuman aggression and other behavior have been studied in a wide range of nonhuman animals. Indeed, castration has been widely studied as a treatment for behavioral issues including aggression in domestic dogs. For example, D'Onise et al. (2017) conducted a systematic review of desexing of dogs to reduce dog bites to humans. They identified a small sample of studies which, although they produced a consistent reduction in human bites, were associated with a very wide range of effect sizes. Thus, no clear conclusion as to how effective this approach was or could be made.

At least one systematic review of treatment of sexual offenders included data evaluating the effects of castration in humans and it found evidence of surprisingly large reductions in aggression-related outcomes. In a comprehensive review and meta-analysis of treatment of sexual offenders, Lösel and Schmucker (2005) included data outcomes of castration on aggression and violence. They identified eight studies of surgical castration with very large effect sizes (OR = 15.45) and six studies of hormonal medication (OR = 3.08). Both of these effect sizes were larger when compared to the overall effect size of ORs of 1.67–1.90 for outcomes such as sexual recidivism, violent recidivism, and general recidivism. This 2005 meta-analysis was relatively unselective in the studies it included as only six of 69 included studies used randomization. In a subsequent updated meta-analysis using more rigorous inclusion criteria, no organic treatments were included as no studies met inclusion criteria (Schmucker & Lösel, 2015). Thus, the evidence for the effectiveness of hormonal and surgical castration is weak and significantly inferior to psychosocial interventions since evaluations of psychosocial treatments meet these more rigorous standards of experimental design.

Although there is tentative evidence that castration results in some reduction in aggression in some people, this treatment is ethically dubious for many reasons. First, surgical castration is irreversible, and hence, once any harm is done, it cannot be repaired. Second, both surgical and chemical castration come with a range of negative side effects. For example, surgical castration may result in loss of sex drive and potency, hot flashes, loss of muscle tissue, increased fat deposits, perhaps depression, and loss of activity and vitality (Hedlund, 2000). Hormonal castration may also result in negative side effects in men including breast enlargement, cardiovascular, gastrointestinal side effects, and anemia among other negative side effects (Hedlund, 2000). Finally, there is a difficult question of informed consent for both treatment and research. There may be benefits to castration to the patient, in terms of increased independence and release from incarceration, and benefits to society in terms of protection from harm. It is unclear, however, that even if the personal costs and negative side effects of surgical or hormonal castration are fully explained, coercion to accept treatments that are harmful and sometimes irreversible cannot be avoided.

The choice of castration as a treatment option is especially challenging in light of the availability of relatively effective treatments with a better evidence base such as CBT (Schmucker & Lösel, 2015). Even when CBT has already been implemented, it is always unclear if it was done effectively, competently and with sufficient intensity and extensivity to be effective. Others (e.g., Douglas et al., 2013) have argued that it is almost impossible to have noncoercive consent for either surgical or chemical castration but that chemical castration is preferable as it enhances the person's autonomy. Thus, surgical and chemical castration remains an ethically highly controversial treatment.

11.2.2 Treatment of Hyperthyroidism and Aggression

The association between hyperthyroidism and irritability and anger suggest that it may be prudent for practitioners to screen individuals with aggression and violence for hyperthyroidism. Medical treatment of hyperthyroidism might help with the management of aggression and violence in some individuals with hyperthyroidism where their aggression and violence are related to hyperthyroidism, but there is little good empirical evidence to support this at this time. Where possible, the patient might benefit from a comprehensive case formulation that integrates biomedical with environmental factors that influence aggression and violence (see below).

11.2.3 Oxytocin- and Vasopressin-Based Treatment of Aggression

One interesting area of potential application is the use of oxytocin to treat aggression in individuals with ASD. There is a potentially good rationale for such an approach, namely that deficits in social behavior and social reinforcers are particularly marked in many individuals with ASD (Guastella & Hickie, 2016).

This has been an active area of recent research, indeed, there have been at least 12 RCTs published (Ooi et al., 2017). Although some trials have reported improvement in social behavior in young children (Yatawara et al., 2016) and adults (Guastella et al., 2010) with autism, meta-analyses of the effects on core symptoms of autism (Wang et al., 2019) suggest minimal effects. Further, there is little direct evidence of effective treatment of aggression per se, and the safety of these agents is not yet fully understood.

11.2.4 Hormones and Individual Case Formulation

Certain biomedical conditions, such as hyperthyroidism, might be important factors to consider when making an individual case formulation of aggression for some individuals. For example, poorly controlled thyroid levels might result in general irritability, overactivity, and over-responsiveness to stimulation and may sometimes present as anger. Poor control of thyroid levels might reflect compliance with medication and the availability and quality of medical and pharmacological services. These variables, which are potentially ones which can be modified, might enter along with other variables such as environmental antecedents and contingencies, into an individual case formulation of aggression. For example, in individuals with hyperthyroidism, poorly controlled levels of thyroxine might be considered establishing operations that potentiate the reinforcers maintaining aggressive behavior. Thus, formulation-based intervention could include one or more of the following: (a) improved compliance with taking thyroid medication, which would remove the establishing operation for aggression; (b) removing the antecedents for aggression, such as other people; and (c) increased function-based alternative behavior, such as removing oneself from other people's provocative social behavior.

Clinicians might also consider other examples of incorporating hormonal treatments for medical conditions into case formulations which might affect aggression and violence. These could include: (a) the use of corticosteroids prescribed for a range of inflammatory and autoimmune medical conditions such as asthma and eczema, painful inflammatory conditions such as rheumatoid arthritis and lupus; and (b) testosterone prescribed for erectile dysfunction, loss of sexual interest, increased muscle mass (as in some individuals who are HIV positive), loss of concentration and mood disorders. These substances can also be illegally obtained, may be abused, and, in the case of corticosteroids, may contribute to aggression in cases of "steroid rage." Finally, prescribed hormones can have negative side effects which might also be relevant to cases of aggression. For example, steroids can result in increased appetite, sleep loss, and mood changes, and stopping steroids can result in withdrawal effects. Again, these factors might be relevant in some individual cases of aggression and violence, and clinicians should consider them when formulating individual cases (White, 2015).

11.2.5 Summary

The endocrinology of aggression and violence has received less attention than other aspects of the biology of aggression and violence. This is especially true in the area of treatment. Although most physicians and other professionals will be aware of the need to screen for relevant hormonal conditions and to treat these conditions both to treat the associated medical conditions and to manage aggression and violence. This approach may have some pragmatic value where the underlying medical condition has a large effect on aggression and violence, but if other factors are also relevant and/or interact with the underlying medical condition, then this simple approach may be suboptimal or ineffective. At this time, there is little data to build on, but the general principles of case formulation may be applicable.

11.3 Chapter Summary

There are a wide range of biological treatments for aggression and violence. They have a long and checkered history. As a generalization, the evidence for the effectiveness of psychopharmacological and hormonal treatments is not compelling. Indeed, many biological treatments raise numerous ethical issues, especially regarding consent, safety, availability of alternate effective and safer treatments, and, in some contexts, coercion.

References

Abd Wahab, W., Collinson, K., & Gross, D. W. (2019). Successful management of postictal violence with pindolol in temporal lobe epilepsy. *Epilepsy & Behavior Reports, 12*, 100346.

Ahmed, Z., Fraser, W., Kerr, M. P., Kiernan, C., Emerson, E., Robertson, J., et al. (2000). Reducing antipsychotic medication in people with a learning disability. *The British Journal of Psychiatry, 176*(1), 42–46.

Aman, M. G., & Singh, N. N. (1986). *Aberrant behavior checklist*. Slosson.

Benson, T., O'Neill, S., Murphy, S., Ferry, F., & Bunting, B. (2015). Prevalence and predictors of psychotropic medication use: Results from the Northern Ireland Study of Health and Stress. *Epidemiology and Psychiatric Sciences, 24*(6), 542–552.

Blader, J. C., Pliszka, S. R., Kafantaris, V., Foley, C. A., Carlson, G. A., Crowell, J. A., ... & Margulies, D. M. (2021). Stepped treatment for attention-deficit/hyperactivity disorder and aggressive behavior: A randomized, controlled trial of adjunctive risperidone, divalproex sodium, or placebo after stimulant medication optimization. *Journal of the American Academy of Child & Adolescent Psychiatry, 60*(2), 236–251.

Booth, A., Scantlebury, A., Hughes-Morley, A., Mitchell, N., Wright, K., Scott, W., & McDaid, C. (2017). Mental health training programmes for non-mental health trained professionals coming into contact with people with mental ill health: A systematic review of effectiveness. *BMC Psychiatry, 17*(1), 196.

Carlson, G. A., Chua, J., Pan, K., Hasan, T., Bied, A., Martin, A., & Klein, D. N. (2020). Behavior modification is associated with reduced psychotropic medication use in children with aggression in inpatient treatment: A retrospective cohort study. *Journal of the American Academy of Child & Adolescent Psychiatry, 59*(5), 632–641.

Carnahan, R. M., Brown, G. D., Letuchy, E. M., Rubenstein, L. M., Gryzlak, B. M., Smith, M., et al. (2017). Impact of programs to reduce antipsychotic and anticholinergic use in nursing homes. *Alzheimer's & Dementia: Translational Research & Clinical Interventions, 3*(4), 553–561.

Carré, J. M., & Robinson, B. A. (2020). Testosterone administration in human social neuroendocrinology: Past, present, and future. *Hormones and Behavior, 122*, 104754.

Carré, J. M., Ruddick, E. L., Moreau, B. J., & Bird, B. M. (2017). Testosterone and human aggression. In P. Sturmey (Ed.), *The Wiley handbook of violence and aggression* (pp. 1–14). Wiley.

Coccaro, E. F. (2012). Intermittent explosive disorder as a disorder of impulsive aggression for DSM-5. *American Journal of Psychiatry, 169*(6), 577–588.

Coccaro, E. F., Lee, R. J., & Kavoussi, R. J. (2009). A double-blind, randomized, placebo-controlled trial of fluoxetine in patients with intermittent explosive disorder. *The Journal of Clinical Psychiatry, 70*(5), 653–662. https://doi.org/10.4088/JCP.08m04150

Corbett, A., Burns, A., & Ballard, C. (2014). Don't use antipsychotics routinely to treat agitation and aggression in people with dementia. *British Medical Journal, 349*, g6420. https://doi.org/10.1136/bmj.g6420

D'Onise, K., Hazel, S., & Caraguel, C. (2017). Mandatory desexing of dogs: One step in the right direction to reduce the risk of dog bite? A systematic review. *Injury Prevention, 23*(3), 212–218.

Donat, D. C. (1998). Impact of a mandatory behavioral consultation on seclusion/restraint utilization in a psychiatric hospital. *Journal of Behavior Therapy and Experimental Psychiatry, 29*(1), 13–19.

Donat, D. C. (2003). An analysis of successful efforts to reduce the use of seclusion and restraint at a public psychiatric hospital. *Psychiatric Services, 54*(8), 1119–1123.

Donat, D. C. (2005). Special section on seclusion and restraint: Encouraging alternatives to seclusion, restraint, and reliance on PRN drugs in a public psychiatric hospital. *Psychiatric Services, 56*(9), 1105–1108.

Douglas, T., Bonte, P., Focquaert, F., et al. (2013). Coercion, incarceration, and chemical castration: An argument from autonomy. *Bioethical Inquiry, 10*, 393–405. https://doi.org/10.1007/s11673-013-9465-4

Esbensen, A. J., Greenberg, J. S., Seltzer, M. M., & Aman, M. G. (2009). A longitudinal investigation of psychotropic and non-psychotropic medication use among adolescents and adults with autism spectrum disorders. *Journal of Autism and Developmental Disorders, 39*(9), 1339–1349.

Fallah, M. S., Shaikh, M. R., Neupane, B., Rusiecki, D., Bennett, T. A., & Beyene, J. (2019). Atypical antipsychotics for irritability in pediatric autism: A systematic review and network meta-analysis. *Journal of Child and Adolescent Psychopharmacology, 29*(3), 168–180.

Fanning, J. R., Coleman, M., Lee, R., & Coccaro, E. F. (2019). Subtypes of aggression in intermittent explosive disorder. *Journal of Psychiatric Research, 109*, 164–172.

Faay, M. D., Czobor, P., & Sommer, I. E. (2018). Efficacy of typical and atypical antipsychotic medication on hostility in patients with psychosis-spectrum disorders: a review and meta-analysis. *Neuropsychopharmacology, 43*(12), 2340–2349.

Guastella, A. J., & Hickie, I. B. (2016). Oxytocin treatment, circuitry, and autism: A critical review of the literature placing oxytocin into the autism context. *Biological Psychiatry, 79*(3), 234–242.

Guastella, A. J., Einfeld, S. L., Gray, K. M., Rinehart, N. J., Tonge, B. J., Lambert, T. J., et al. (2010). Intranasal oxytocin improves emotion recognition for youth with autism spectrum disorders. *Biological Psychiatry, 67*, 692–694.

Gurnani, T., Ivanov, I., & Newcorn, J. H. (2016). Pharmacotherapy of aggression in child and adolescent psychiatric disorders. *Journal of Child and Adolescent Psychopharmacology, 26*(1), 65–73.

Hálfdánarson, Ó., Zoëga, H., Aagaard, L., Bernardo, M., Brandt, L., Fusté, A. C., et al. (2017). International trends in antipsychotic use: A study in 16 countries, 2005–2014. *European Neuropsychopharmacology, 27*(10), 1064–1076.

Hedlund, P. O. (2000). Side effects of endocrine treatment and their mechanisms: Castration, antiandrogens, and estrogens. *The Prostate, 45*(S10), 32–37.

Institute for Safe Medication. (2016). *Chloral hydrate: Is it still being used? Are there safer alternatives?* Author. Retrieved December 29, 2021, from https://www.ismp.org/resources/chloral-hydrate-it-still-being-used-are-there-safer-alternatives

Jobski, K., Höfer, J., Hoffmann, F., & Bachmann, C. (2017). Use of psychotropic drugs in patients with autism spectrum disorders: A systematic review. *Acta Psychiatrica Scandinavica, 135*(1), 8–28.

Knapp, P., Chait, A., Pappadopulos, E., Crystal, S., Jensen, P. S., & T-May Steering Group. (2012). Treatment of maladaptive aggression in youth: CERT guidelines I. Engagement, assessment, and management. *Pediatrics, 129*(6), e1562–e1576.

Koch, A. D., Dobrindt, J., & Schützwohl, M. (2021). Prevalence of psychotropic medication and factors associated with antipsychotic treatment in adults with intellectual disabilities: A cross-sectional, epidemiological study in Germany. *Journal of Intellectual Disability Research, 65*(2), 186–198.

Lake, J. K., Balogh, R., & Lunsky, Y. (2012). Polypharmacy profiles and predictors among adults with autism spectrum disorders. *Research in Autism Spectrum Disorders, 6*(3), 1142–1149.

Linton, D., Barr, A. M., Honer, W. G., & Procyshyn, R. M. (2013). Antipsychotic and psychostimulant drug combination therapy in attention deficit/hyperactivity and disruptive behavior disorders: A systematic review of efficacy and tolerability. *Current Psychiatry Reports, 15*(5), 1–11.

Lohr, W. D., Creel, L., Feygin, Y., Stevenson, M., Smith, M. J., Myers, J., et al. (2018). Psychotropic polypharmacy among children and youth receiving Medicaid, 2012-2015. *Journal of Managed Care & Specialty Pharmacy, 24*(8), 736–744.

Lösel, F., & Schmucker, M. (2005). The effectiveness of treatment for sexual offenders: A comprehensive meta-analysis. *Journal of Experimental Criminology, 1*(1), 117–146.

Loy, J. H., Merry, S. N., Hetrick, S. E., & Stasiak, K. (2017). Atypical antipsychotics for disruptive behaviour disorders in children and youths. *Cochrane Database of Systematic Reviews, 8*, CD008559.

Marsh, L., & Krauss, G. L. (2000). Aggression and violence in patients with epilepsy. *Epilepsy & Behavior, 1*(3), 160–168.

Maturana, W. R., & Maturana, R. R. (2009). Control de la agresividad con olanzapina en pacientes post tec. *Ciencia & Trabajo, 11*, 22–24.

McCloskey, M. S., Noblett, K. L., Deffenbacher, J. L., Gollan, J. K., & Coccaro, E. F. (2008). Cognitive-behavioral therapy for intermittent explosive disorder: A pilot randomized clinical trial. *Journal of Consulting and Clinical Psychology, 76*(5), 876.

McNamara, R., Randell, E., Gillespie, D., Wood, F., Felce, D., Romeo, R., et al. (2017). A pilot randomised controlled trial of community-led ANtipsychotic Drug REduction for Adults with Learning Disabilities. *Health Technology Assessment, 21*(47), 1.

Muir-Cochrane, E., Oster, C., Gerace, A., Dawson, S., Damarell, R., & Grimmer, K. (2020). The effectiveness of chemical restraint in managing acute agitation and aggression: A systematic review of randomized controlled trials. *International Journal of Mental Health Nursing, 29*(2), 110–126.

Ooi, Y. P., Weng, S.-J., Kossowsky, J., Gerger, H., & Sungl, M. (2017). Oxytocin and autism spectrum disorders: A systematic review and meta-analysis of randomized controlled trials. *Pharmacopsychiatry, 50*, 5–13.

Orsel, K., Taipale, H., Tolppanen, A. M., Koponen, M., Tanskanen, A., Tiihonen, J., et al. (2018). Psychotropic drugs use and psychotropic polypharmacy among persons with Alzheimer's disease. *European Neuropsychopharmacology, 28*(11), 1260–1269.

Pappadopulos, E., Woolston, S., Chait, A., Perkins, M., Connor, D. F., & Jensen, P. S. (2006). Pharmacotherapy of aggression in children and adolescents: Efficacy and effect size. *Journal of the Canadian Academy of Child and Adolescent Psychiatry/Journal de l'Académie Canadienne de Psychiatrie de l'Enfant et de l'Adolescent, 15*(1), 27–39.

Patel, J., Frankel, S., & Tampi, R. R. (2019). Evidence for using PRN pharmacotherapy to treat undifferentiated acute agitation or aggression. *Annals of Clinical Psychiatry: Official Journal of the American Academy of Clinical Psychiatrists, 31*(1), 54–69.

Paulose-Ram, R., Safran, M. A., Jonas, B. S., Gu, Q., & Orwig, D. (2007). Trends in psychotropic medication use among US adults. *Pharmacoepidemiology and Drug Safety, 16*(5), 560–570.

Pillay, J., Boylan, K., Newton, A., Hartling, L., Vandermeer, B., Nuspl, M., et al. (2018). Harms of antipsychotics in children and young adults: A systematic review update. *The Canadian Journal of Psychiatry, 63*(10), 661–678.

Piovani, D., Clavenna, A., & Bonati, M. (2019). Prescription prevalence of psychotropic drugs in children and adolescents: An analysis of international data. *European Journal of Clinical Pharmacology, 75*, 1333–1346. https://doi.org/10.1007/s00228-019-02711-3

Rosato, N. S., Correll, C. U., Pappadopulos, E., Chait, A., Crystal, S., Jensen, P. S., & Treatment of Maladaptive Aggressive in Youth Steering Committee. (2012). Treatment of maladaptive aggression in youth: CERT guidelines II. Treatments and ongoing management. *Pediatrics, 129*(6), e1577–e1586.

Salzman, C., Jeste, D., Meyer, R. E., Cohen-Mansfield, J., Cummings, J., Grossberg, G., et al. (2008). Elderly patients with dementia-related symptoms of severe agitation and aggression: Consensus statement on treatment options, clinical trials methodology, and policy. *The Journal of Clinical Psychiatry, 69*(6), 889.

van Schalkwyk, G. I., Beyer, C., Johnson, J., Deal, M., & Bloch, M. H. (2018). Antipsychotics for aggression in adults: A meta-analysis. *Progress in Neuro-Psychopharmacology & Biological Psychiatry, 81*, 452–458.

Schmucker, M., & Lösel, F. (2015). The effects of sexual offender treatment on recidivism: An international meta-analysis of sound quality evaluations. *Journal of Experimental Criminology, 11*(4), 597–630.

Schneider, L. S., Dagerman, K. S., & Insel, P. (2005). Risk of death with atypical antipsychotic drug treatment for dementia: Meta-analysis of randomized placebo-controlled trials. Jama, 294(15), 1934–1943.

Schneider, L. S., Tariot, P. N., Dagerman, K. S., Davis, S. M., Hsiao, J. K., Ismail, M. S., ... & Lieberman, J. A. (2006). Effectiveness of atypical antipsychotic drugs in patients with Alzheimer's disease. *New England Journal of Medicine, 355*(15), 1525–1538.

Seifert, J., Engel, R. R., Bernegger, X., Führmann, F., Bleich, S., Stübner, S., et al. (2021). Time trends in pharmacological treatment of major depressive disorder: Results from the AMSP Pharmacovigilance Program from 2001–2017. *Journal of Affective Disorders, 281*, 547–556.

Sheehan, R., & Hassiotis, A. (2017). Reduction or discontinuation of antipsychotics for challenging behaviour in adults with intellectual disability: A systematic review. *The Lancet Psychiatry, 4*(3), 238–256.

Smeets, C. H., Gerritsen, D. L., Zuidema, S. U., Teerenstra, S., van der Spek, K., Smalbrugge, M., & Koopmans, R. T. (2018). Psychotropic drug prescription for nursing home residents with dementia: Prevalence and associations with non-resident-related factors. *Aging & Mental Health, 22*(9), 1244–1252.

Sturmey, P. (2015). *Reducing restraint and restrictive behavior management practices*. Springer.

Sturmey, P. (Ed.). (2020). *Functional analysis in clinical treatment*. Academic.

Tampi, R. R., Tampi, D. J., Balachandran, S., & Srinivasan, S. (2016). Antipsychotic use in dementia: A systematic review of benefits and risks from meta-analyses. *Therapeutic Advances in Chronic Disease, 7*(5), 229–245.

Téllez-Lapeira, J., López-Torres Hidalgo, J., García-Agua Soler, N., Gálvez-Alcaraz, L., Escobar-Rabadán, F., & García-Ruiz, A. (2016). Prevalence of psychotropic medication use and associated factors in the elderly. *The European Journal of Psychiatry, 30*(3), 183–194.

Tsakanikos, E., Costello, H., Holt, G., Sturmey, P., & Bouras, N. (2007). Behaviour management problems as predictors of psychotropic medication and use of psychiatric services in adults with autism. *Journal of Autism and Developmental Disorders, 37*(6), 1080–1085.

Tsiouris, J. A., Kim, S. Y., Brown, W. T., Pettinger, J., & Cohen, I. L. (2013). Prevalence of psychotropic drug use in adults with intellectual disability: Positive and negative findings from a large scale study. *Journal of Autism and Developmental Disorders, 43*(3), 719–731.

Vasudev, A., Shariff, S. Z., Liu, K., Burhan, A. M., Herrmann, N., Leonard, S., & Mamdani, M. (2015). Trends in psychotropic dispensing among older adults with dementia living in long-term care facilities: 2004–2013. *The American Journal of Geriatric Psychiatry, 23*(12), 1259–1269.

Van der Oord, S., Prins, P. J., Oosterlaan, J., & Emmelkamp, P. M. (2008). Efficacy of methylphenidate, psychosocial treatments and their combination in school-aged children with ADHD: a meta-analysis. *Clinical Psychology Review, 28*(5), 783–800.

Vitiello, B., Hill, J. L., Elia, J., Cunningham, E., McLeer, S. V., & Behar, D. (1991). P.R.N. medications in child psychiatric patients: A pilot placebo-controlled study. *The Journal of Clinical Psychiatry, 52*(12), 499–501.

Wang, Y., Wang, M. J., Rong, Y., He, H. Z., & Yang, C. J. (2019). Oxytocin therapy for core symptoms in autism spectrum disorder: An updated meta-analysis of randomized controlled trials. *Research in Autism Spectrum Disorders, 64*, 63–75.

White, C. A. (2015). Physical health problems: A framework and checklist for case formulation. In N. Tarrier & J. Johnson (Eds.), *Case formulation in cognitive behaviour therapy: The treatment of challenging and complex cases* (pp. 302–321). Routledge.

Williamson, D., Frenette, A. J., Burry, L. D., Perreault, M., Charbonney, E., Lamontagne, F., ... & Bernard, F. (2019). Pharmacological interventions for agitated behaviours in patients with traumatic brain injury: A systematic review. *BMJ open, 9*(7), e029604.

Yatawara, C., Einfeld, S., Hickie, I., et al. (2016). The effect of oxytocin nasal spray on social interaction deficits observed in young children with autism: A randomized clinical crossover trial. *Molecular Psychiatry, 21*, 1225–1231. https://doi.org/10.1038/mp.2015.162

Zhou, Z., Betts, K. A., Bocharova, I., Kinrich, D., & Spalding, W. M. (2020). Concomitant use of psychotropic medication with stimulants for the treatment of ADHD in children and adolescents: A retrospective insurance claims study in the United States. *Journal of Attention Disorders, 24*(2), 336–347.

Chapter 12
Individual Biological Interventions for Violence and Aggression. II. Other Biological Treatments

As well as pharmacological and hormonal treatments reviewed in the preceding chapters, there are also numerous other biological treatments. This chapter reviews a sample of the more common ones. The first section considers neurosurgery, including lobotomy, amygdalotomy, hypothalamic surgery, and cingulotomy. Neurosurgery has a long and notorious history and was more widely used in the past and largely faded but not completely disappeared. To some extent, it has been replaced with less invasive procedures involving electrical and magnetic stimulation of the brain; these procedures have become widely used for many problems, including aggression and violence. Electroconvulsive therapy (ECT) is another controversial treatment with limited data supporting its use for aggression and violence. One way to potentially manipulate neurotransmitters might be to change one's diet. The most common dietary intervention is serotonergic diets. Like many new treatments, initial optimism has been replaced by caution as to their effectiveness. Other diets related to aggression and violence include omega-3 diets and regulating caffeine consumption.

12.1 Neurosurgery

12.1.1 History of Neurosurgery

Neurosurgery for mental illness generally, and pathological aggression specifically, has a long cultural history. Trephination or trepanation involves opening of the human skull and was observed in Neolithic (8000–800 BC) and bronze age (3000–1200 BC) skulls in Europe and in Central and South America in pre-Columbian times as evidenced by skulls with holes often indicating recovery from surgery and ancient tools used to bore holes in skulls (Faria, 2013a). Similar

P. Sturmey, *Violence and Aggression*, Advances in Preventing and Treating Violence and Aggression, https://doi.org/10.1007/978-3-031-04386-4_12

instruments and written accounts of trephination exist from classic Greek and Roman times, including in Africa (El Khamlichi, 1998; Rawlings & Rossitch, 1994).

Such practices continued during the Renaissance in Europe. There is evidence of this from the existence of surgical instruments used to elevate depressed skull fractures, suggesting continuation of such practices. Many authors have speculated that early use of trephination must have been used for headaches and mental illness, perhaps based on the idea of releasing demons and spirits. By Medieval times, there were explicit accounts of the use of trephination for treatment of mental illness (Faria, 2013a, b, c) in order to remove the so-called stones of madness or lithotomy, which might lurk in patients' brains and cause mental illness. Images of psychosurgery in art occurred in paintings by Hieronymus Bosch's painting *The Extraction of the Stone of Madness* and Peter Bruegel the Elder's *Cutting out the Stone of Madness, or an Operation on the Head*.

By the nineteenth century, people recognized that there were cerebral tumors, and surgeries could be performed to remove them, making people aware of cerebral localization of function. The well-known case of Phineas Gage, a man who was injured when a 3-ft 7-in. iron rod was blown and lodged into his head, confirmed these findings. His physician documented the case carefully noting the changes in personality and behavior in what was later known as frontal lobe syndrome. Phineas Gage changed from an energetic, shrewd, and persistent business man to being an obstinate, irreverent, and disinhibited one (Faria, 2013a).

Following some case studies and studies on nonhuman animals which appeared to show that sectioning of the prefrontal cortex reduced emotionality, response to frustration, and aggression generally, and a preliminary series of six case studies of psychosurgery was published by Gottlieb Burkhardt in 1891. Around 1935, Monitz and colleagues began to perform bilateral frontal lobotomies, using either alcohol injections or surgery to destroy tissue including the pathways between the frontal lobes and thalamus, to treat a wide range of mental health problems including aggression and violence. These procedures were also used on psychiatric inpatients for whom few or no alternative effective treatments were available. Thus, questionable treatments, such as insulin shock therapy, barbiturates, bromides, morphine, and ECT, and management procedures, such as restraint, seclusion and locked wards, were available but were of limited effectiveness.

There were several variations in lobotomy procedures. They became very widely used when they could be performed in the outpatient clinic using a modified ice pick tool! Tens of thousands of procedures were performed in the USA alone (Faria, 2013a), and around 2,000 were performed on Veterans of World War II in the USA, mostly for trauma-related psychopathology (Komesaroff & Rosenfeld, 2020). In 1949, Moitz was awarded the Nobel Prize for his work. It was argued that although the science was incomplete, the urgency of the problem outweighed the lack of robust evidence from well-controlled trials, thereby justifying its continued use (Komesaroff & Rosenfeld, 2020). After widespread use, concerns were raised concerning many failed cases and safety concerns indexed by mortality rates as high as nearly 20% (Ögren & Sandlund, 2007).

Although controversial, work on lobotomies gave rise to a more detailed knowledge of the neuroanatomy of aggression, in particular the role of the prefrontal cortex and its connections to the thalamus as was the role of the limbic system were recognized as important in the control of aggression and autonomic functions. This and the development of more precise stereotactic surgical methods, resulted in alternatives to lobotomies, such as cingulotomy, which were apparently safer than lobotomies (see below). Additionally, other methods of tissue damage were developed such as thermocoagulation.

Concern over the safety and ethics of neurosurgery in the 1980s resulted in greater professional regulation of neurosurgery. Controversial applications included its use with prisoners who were violent sex offenders, individuals with intellectual disabilities, individuals with chronic mental illness, and children with behavior disorders including aggression, where obtaining meaningful consent is dubious. Thus, many practitioners gave up neurosurgery in response to social pressure to do so (Faria, 2013b).

Another line of treatments came from studies of electrical stimulation of the brain (ESB) much of which was conducted by Delgado. Studies of nonhuman animals and humans have shown that electrical stimulation to the brain elicits various forms of aggression in a variety of organisms, including cats, monkeys, and humans. Examples include: (a) stimulation of the anterior hypothalamus elicits "sham rage" in cats in which they displayed generalized offensive behavior but they would not retaliate against conspecifics of they attacked; and (b) stimulation of the lateral hypothalamus elicited "true rage" in which the animal hissed, prowled, and also directly searched for and counter-attacked conspecifics and friendly experimenters. This aggression was sometimes complex, responsive to environmental factors, such as the selection of a suitable victim, and ceased when ESB ceased (Faria, 2013b). These studies suggested the possibility of treatment of aggression and violence using various forms of ESB including various forms of brain stimulation that are sometimes used for aggression and violence (Faria, 2013b) (see below for further discussion of the evidence on brain stimulation therapies for aggression and violence).

12.1.2 Lobotomy

As noted above, the lobotomy was the first neurosurgical procedure used for treatment of aggression and violence, among other behavioral and psychiatric problems. Although today lobotomy is almost completely rejected as a treatment for aggression and violence, it is worthwhile reviewing the evidence for its effectiveness and safety and how it fell out of favor as important lessons can be learned that could be applied to contemporary procedures.

Many reports of lobotomies for aggression tended to be retrospective case series rather than true experiments. A true experiment would require an adequate control procedure, such as a sham operation, competent alternative treatment, randomized assignment, double-blind assessment using reliable and valid measures, and meaningful consent, among other things. This is a tall order under any circumstance, but is especially challenging in the populations for which lobotomies and other forms of neurosurgery are used (Komesaroff & Rosenfeld, 2020).

One example of such studies comes from Ögren and Sandlund (2007) who reported a series of 771 lobotomies performed in a Swedish psychiatric hospital between 1947 and 1958. Four hundred seventy-two lobotomies were performed on females and 299 on males. The most common psychiatric diagnosis was schizophrenia (about 80% of patients), and the most common behavioral issues were quite nonspecific such as "disturbing behavior," "fluctuations of mood," and "violence." The main outcome reported was mortality and general impressions of resulting passivity and docility (see above).

A second example comes from Braslow (1999) who conducted a retrospective chart review of lobotomies performed in a California state hospital. He found that lobotomies had been performed on 232 patients 245 times, primarily of behavior that was disruptive to the ward routine and/or resulted in restraint. Braslow concluded that many lobotomies were performed based on implicit judgments concerning appropriate behavior for women—85% ($N = 205$) of lobotomies were performed on women rather than men. Although it was not possible to provide conclusive proof that this was the reason for the larger number of lobotomies on women, Braslow commented that problems were often defined in terms of women not wishing to raise children at home or working outside the home, and recovery was defined in terms of not worrying and doing housework contentedly.

Finally, in a disturbing review of the literature, Breggin (1975) recounted how he had reviewed hundreds of studies, which he characterized as "few if any that qualify as 'scientific,'" that is, almost all studies were without experimental designs and used general impressions of outcomes such as clinicians' impression of patients being "blunted" or "subdued." Only three studies were true experiments, and they reported negative outcomes and poor functioning at follow-up. He recounted use of lobotomies on children as young as 6 years old and on individuals with intellectual disabilities. Finally, he described an ideology whereby neurosurgeons attributed the causes of drunk driving, wife beating, and opposition to the war in Vietnam to a minority of individuals who must be brain damaged. The phrase "physical control of the mind" became a byword to explain and treat violence.

It might seem that this section is flogging a dead horse since lobotomies may largely be a thing of the past, but the issues that the use of lobotomies raises are not. The literature on lobotomies lacked scientific methods, and hence, conclusions were readily open to misinterpretation and misrepresentation, and given the severity of the problems lobotomies and contemporary psychosurgery are used for, it is tempting to sacrifice science on the alter of social need. These issues are around today, but in the new context of DBS, vagal nerve stimulation, and more precise surgeries and imaging methods (Komesaroff & Rosenfeld, 2020).

12.1.3 Amygdalotomy

Based on the notion that the amygdala is the so-called aggression center of the brain, several empirical studies have examined whether removal or destruction of part or all of the amygdala results in reduced aggression. For example, Heimburger et al. (1966) reported a case series of 25 individuals with "behavioral abnormalities, epilepsy and depression" (p. 741) treated with amygdalotomies. They reported that behavioral abnormalities were eliminated in seven and improved in nine of 20 patients and that there were overall improvements in 23 of 25 patients. They concluded that amygdalotomy was safe and could be considered for a subgroup of aggressive patients who were treatment nonresponsive.

In a subsequent systematic review 20 years later, Mpakopoulou et al. (2008) identified 13 studies with widely varying outcomes. Positive outcomes carried between 33% and 100% and complications from 0% to 42% with mortality of nearly 4%. They did not find evidence of cognitive impairment but did note that there was a lack of long-term follow-up. They noted that these studies were difficult to evaluate as a group as they involved very different types of participants, different surgical procedures, and different measurements of outcomes.

More recently, Gouveia et al. (2019) conducted a systematic review which identified 27 studies published over a 60-year period with 1217 patients who had undergone amygdalotomy. Nearly 70% of patients were reported to have "good" or "excellent" improvement including reductions in severity and frequency of aggression. Several studies reported long-term maintenance of treatment effects over several years. When psychiatric status was reported many patients had cerebral insults, severe intellectual disabilities, or schizophrenia, and many had severe cognitive impairments and were nonverbal prior to the surgery. A proportion also had previous psychosurgery.

These studies were quite limited in a number of ways. For example, almost all studies were case series or case studies, used relatively informal, nonstandardized assessment of behavior change, failed to assess changes in adaptive behavior and cognitive functioning, and were nonblind, leading to an overall low quality of evidence.

12.1.4 Hypothalamic Surgery

In their systematic review of surgical treatment of aggression, Gouveia et al. (2019) also reported data on the effects of hypothalamic surgery on human aggression. Their systematic search identified 243 cases of aggression treated with hypothalamic surgery, and these articles claimed there was an approximately 80% improvement in aggression. The cautions regarding the poor quality of the studies noted with amygdalotomy are also applicable here. In addition, it should be noted that psychosurgery of the hypothalamus comes with additional risks including problems with control of basic physiology, including problems with homeostasis, such as problems with thermoregulation, and water and electrolyte balance, and a 4% risk of mortality.

12.1.5 Cingulotomy

Cingulotomy has been used and evaluated for a wide range of treatment-resistant psychiatric disorders (Ballantine et al., 1987), including several case series specifically for treatment of aggression. For example, Jiménez et al. (2012) reported treatment of 23 consecutive cases of aggression: six were treated with adjustment of psychopharmacology, seven refused the procedures, and ten underwent cingulotomy. Of those ten individuals, eight were diagnosed with intellectual disabilities and two with schizophrenia. They had an average history of 9 years of aggressive behavior. The authors reported large reductions in aggression on a standard psychometric measure of aggressive behavior and improvements in ratings of *Global Assessment of Functioning* at 6 months and 4 years post surgery. Negative side effects were reported in five participants including transient hyperphagia (40%), somnolence (30%), disinhibition (20%), hypersexuality (10%), and infection (10%); one person died.

12.1.6 Summary and Conclusion

To date there have been several neurosurgical procedures for aggression and violence. Overall there is little evidence of their effectiveness other than from poorly controlled case studies and case series; indeed, several of these procedures are sometimes harmful and sometimes even lethal. Although this literature typically describes these procedures as procedures of last resort for individuals who are treatment nonresponsive, it is often unclear how these patients were selected and what their history of alternate treatments was prior to surgery; hence, it is in fact unclear if these procedures are actually used as a procedure of last resort. It is notable that none of the papers reviewed here mentioned functional assessment and analysis or applied behavior analysis–based treatment. Only one mentioned changes in psychopharmacology prior to surgery (Jiménez et al., 2012), but this protocol was not fully described.

12.2 Electrical and Magnetic Brain Stimulation

Since electrical stimulation of the brain in nonhuman animals can function as powerful reinforcement and punishment of behavior (Delgado et al., 1954) and can elicit relatively specific emotional behaviors such as aggression (Panksepp, 1971) and fear, it seems possible that electrical stimulation of the brain in various ways could inhibit aggression, stimulate the expression of behavior that is incompatible with aggression, or change neurotransmission to modify aggression.

Denson (2021) made an appeal for the use of "neuronormalization" procedures to break the cycle of violent crime, punishment, and recidivism. Citing the apparent ineffectiveness or even harmfulness of public policy, nonbiological interventions, such as the reliance on punishment and incarceration, Denson advocated for the use of "moral neuroenhancement" based on "neuroprediction" of violence. That is in this model, measures of brain function would be used to identify structural and functional brain abnormalities causing violence and would also be used to predict violence. A range of biological methods, including neurostimulation, would then be used to remedy these brain dysfunctions. These methods of neurostimulation are potentially appealing since some are noninvasive and, unlike psychosurgery, are potentially reversible or perhaps nonharmful and, thus, may have a better safety profile than other biological interventions. There have been several methods used to this including transcranial direct current stimulation (tDCS), repetitive transcranial magnetic stimulation (rTMS), and cranial electrotherapy. The use of ECT to treat aggression and violence will be discussed below.

12.2.1 Transcranial Direct Current Stimulation

Recently, tDCS has been developed and used to treat a wide range of physical and psychiatric disorders including aggression. tDCS involves passing a small direct current via electrodes on the scalp. tDCS has been used for a wide range of psychiatric and behavioral conditions such as schizophrenia. Recently, Sergiou et al. (2020) reviewed evidence for its use in increasing empathy and decreasing violence. They found several such studies including studies showing tDCS can increase empathy, moral judgment, and norm compliance, and that tDCS can decrease aggression and risk taking. The authors concluded that tDCS was a promising treatment, especially when targeting the prefrontal ventromedial cortex. They also noted numerous procedural differences between studies and the participants suggesting the need for systematic studies evaluating these parameters. It is also important to note that most of these studies used self-report and laboratory analog measures of aggression, rather than objective measurement. There were no RCTs in clinical populations with severe aggression.

The need for such systematic studies was highlighted by two meta-analyses that have not found any effect of tDCS on aggression (Bell & DeWall, 2018; Ling et al., 2020). Bell and DeWaal, for example, reported a meta-analysis of 48 studies with 2,816 participants and found a modest effect size on all outcome measures ($d = -0.20$), but no effect on measures of aggression, impulsivity, and dishonesty. Further, a more recent relatively well-controlled RCT using students and an analog laboratory task found no effects of tDCS on how many pins participants stuck in a voodoo doll while thinking about someone who had recently angered them (Ling et al., 2020). Thus, at this point, there is little evidence for any effects of tDCS on aggression in either nonclinical or clinical/forensic populations.

12.2.2 Repetitive Transcranial Magnetic Stimulation

rTMS is the positioning of an electromagnetic coil over the brain region of interest which delivers electromagnetic pulses and is intended to modify the electrical activity of the brain region of interest. The procedure is noninvasive. It has been used for a wide range of psychiatric and neurological conditions including depression, pain, stroke, and schizophrenia (Lefaucheur et al., 2014). It is possible to implement well-conducted trials of rTMS using sham control double-blind conditions.

There have been several experimental studies of the effects of rTMS in analog situations showing that rTMS may reduce aggression (Dambacher et al., 2015; Riva et al., 2015). Riva et al. (2015), for example, had 80 healthy undergraduates play a computer game simulating basketball in which they were either included (other players passed the ball to them) or excluded (other players did not pass the ball to them). They were randomized to either rTMS or a sham control procedure applied to the right ventrolateral prefrontal cortex. Aggression was measured as the amount of hot sauce they delivered to a hypothetical other person. As expected, the authors found that social exclusion precipitated more aggression than inclusion. More importantly, they also found that rTMS mitigated the effects of aggression-inducing social exclusion such that the individuals who were socially excluded but also received tCMS were no more aggressive than nonexcluded participants. These analog experiments are interesting and point toward the possibility of the applied use of tCMS in applied settings with clinically significant aggression.

Although there are case reports and uncontrolled studies of the treatment of aggression and violence using rTMS (Jardri et al., 2007), and limited reported of changes in aggression-related constructs such as anger in two small-scale RCTs for borderline personality disorder (Cailhol et al., 2014; Reyes-López et al., 2018), no RCT or systematic review of rTMS for aggression or violence could be located. Yang et al. (2018), however, published a systematic review and meta-analysis of the effects of rTMS on impulsivity in the general population, which might be related to problematic aggression in clinical or forensic populations. They located 27 papers with 50 trials of rTMS and found an effect size of $g = 0.30$ for motor impulsivity (ability to inhibit a motor response) and $g = 0.59$ on temporal impulsivity (ability to delay accessing reinforcers or delay discounting). Thus, at this time, there is only very limited evidence for the effectiveness of rTMS for the treatment of aggression and violence.

12.2.3 Deep Brain Stimulation

Although DBS has been used widely for many psychiatric and behavioral conditions such as Parkinson's disease, obsessive compulsive disorder, depression, essential tremor, etc., its application to aggression and violence has been limited to case reports (e.g., Harat et al., 2015) and case series (Franzini et al., 2013) rather than controlled trials. For example, Franzini et al. reported a case series of DBS around the posterior hypothalamic regions in seven adults with severe or profound intellectual disabilities aged 21–64 years with chronic histories of aggression and multiple failed psychotropic medications. The procedure involved general anesthesia to place the DBS device. They reported that six individuals responded favorably.

12.2.4 *Vagal Nerve Stimulation*

A more recent biological treatment that has been tried for aggression is vagal nerve stimulation (VNS). There are two vagal nerves connecting the lower brain with larynx, diaphragm, heart, chest, and abdomen. VNS involves implantation of an electrical stimulating device to modulate electrical impulses in one of the vagal nerves.

The Federal Drug Administration has approved VNS as a second choice treatment for refractory seizure disorders and depression after primary treatments have failed or have been insufficiently effective. The rationale for the use of VNS for aggression and violence is unclear, and the evidence is very limited. Van Hoorn et al. (2019) conducted a systematic review of VNS and autism spectrum disorders. They found only 11 empirical papers which consisted of eight case series and three case studies with mostly anecdotal reports and weak data collection. There were no experimental evaluations. There are also a few uncontrolled case studies in other populations which sometimes anecdotally report reductions in aggression sometimes in the context of improved management of seizure disorders (e.g., Ramnarayan, 2021.)

12.2.5 *Summary*

Studies of various methods of electromagnetic stimulation of the brain on aggression are intriguing and especially interesting because of the possibility that they are safer than many invasive and potentially dangerous biological interventions. Although there are analog studies and clinical case studies indicating their potential, evidence in clinical and forensic populations is lacking at this time. Indeed, recently, Yang et al. (2020) extended their earlier review of noninvasive brain stimulation including both tCMS and tDCS, on impulsivity in people with mental disorders. They located 11 tDCS and 18 rTMS studies and found significant but rather small effect sizes for tDCS ($g = 0.29$, $p = 0.004$) but not for rTMS ($g = -0.08$, NS). These results led them to conclude pessimistically that "there is insufficient evidence to support the use of rTMS or tDCS in clinical practice to reduce impulsivity in patients with mental health disorders" (p. 1). Thus at this time, the clinical and forensic application to the treatment of aggression and violence does not have sufficient support.

12.3 Electro-Convulsive Therapy

ECT is a medical treatment which refers to passing a small electrical current through the brain using electrodes on the scalp while the patient is under general anesthetic which induces a seizure. Under general anesthesia, the patient does not show a full blown seizure. This procedure alters brain chemistry. ECT has been and continues to be used widely for problems such as depression, bipolar disorder, catatonia, other mental health disorders, and aggression and violence in some circumstances.

Meta-analyses reveal robust effects on depression including comparisons against placebo conditions and antidepressant medications (Pagnin et al., 2008), and some practitioners prefer ECT over antidepressants in some circumstances, such as depression and suicidality, because of the apparent faster response to ECT than antidepressants. Its use, however, remains controversial, due to negative side effects such as memory loss (Calev et al., 1991) and long-standing and inaccurate, melodramatic portraits of ECT in the media, such as its application without general anesthesia. Either way, clients and practitioners are faced with difficult decisions regarding the effectiveness, speed of response, negative side effects, and capacity to consent when selecting ECT from other treatment options.

ECT has been used to treat aggression and violence in older people with schizophrenia (Gu et al., 2017), dementias (van den Berg et al., 2018), and individuals with intellectual disabilities or autism spectrum disorders (Bird et al., 2021; Consoli et al., 2013; Haq & Ghaziuddin, 2014; Guinchat et al., 2015; Luiselli et al., 2021). ECT has also been incorporated into treatment algorithms for aggression and violence in autism (Guinchat et al., 2015) and seniors (van den Berg et al., 2018). Research with individuals with dementias and developmental disabilities has exclusively used case studies, retrospective chart reviews, and case series with the likelihood of reporting and selection bias in publication (van den Berg et al., 2018). In the case of research with individuals with schizophrenia, there are indeed several RCTs that address aggression or related constructs.

Gu et al. conducted a systematic review and meta-analysis of RCTs in the international literature and located only seven RCTs published in China. These RCTs evaluated ECT alone (4 RCTS, 5 arms, $n = 240$), and combined ECT and antipsychotics (three RCTs, four arms, $n = 240$). Most participants were young adults (mean age = approximately 34 years), and the trials were about 4 weeks long. All RCTs were of low or very low quality, being nonblinded, and without clear descriptions of randomization procedures. Concerning changes in the agitation score of the *Positive and Negative Syndrome Scale* (Kay et al., 1987), there was no difference in agitation between ECT alone and ECT combined with antipsychotics compared to antipsychotic monotherapy. Negative side effects, when reported, were described as mild and transient. Thus, there was no evidence from a small number of low to very low quality evidence that ECT improved agitation in young adults with schizophrenia taking antipsychotic medications.

The literature on ECT and aggression and violence is very limited in size and quality. The only area with experimental studies—schizophrenia—had a small number of low quality, short-term studies. In addition, although some studies listed previous alternative treatments (Gu et al., 2017), there was not much known about the quality and integrity of those alternative treatments. Hence, there is little good quality, experimental evidence supporting the use of ECT for the treatment of aggression and violence at this time.

12.4 Dietary Treatments

Dietary treatments offer an interesting option for treating aggression. They are potentially quite attractive to many individuals, because people often change their diets for many reasons, and hence, it is a relatively acceptable intervention option. Further, dietary interventions may be reversible and so, unlike some other biological treatments, any harm done may be potentially reversible.

A review by Du (2019) suggested that there may be several possible links between diet and violence reflected in popular beliefs that contemporary high sugar, high fat, junk foods, and sensitivities to foods and additives cause a range of behavioral problems and unrefined, natural foods are good for us in many ways, including behavioral health. An important distinction is that diets can influence behavior in two ways. First, there may be long-term effects due to diet during fetal development, infancy, and childhood. Second, diets may influence our behavior due to recent or intake throughout our lifespan.

A frequently cited study that illustrates the promise of dietary interventions comes from Gesch et al. (2002) who evaluated the effects of a multicomponent dietary supplement on aggressive and violent behavior in prisoners using an RCT. They allocated 231 young adult prisoners to a dietary supplement of placebo. The dietary supplement was a multicomponent supplement containing supplementary multiple vitamins, multiple minerals, and essential fatty acids including omega-3 and omega-6 supplementation. The authors found that those prisoners taking active treatment reduced offenses from 16 to 10.4 per 1,000 person days. The prisoners in the active treatment group had an average of 26% fewer offenses than those on placebo ($p < 0.03$) and 35% fewer offenses than baseline ($p < 0.0001$). These effects were especially marked for serious offenses (37% versus 10% reduction) compared to minor offenses (33% versus 7%). This study has been much cited and has resulted in considerable interest in the potential for diets to reduce aggression and violence but requires independent replication.

A recent systematic review by Qureshi et al. (2021) portrayed a more cautious picture than the initial study. These authors conducted a systematic search of the literature up to March 2020. They were unable to conduct a meta-analysis due to excessive variability in reporting in these studies. They identified 14 studies investigating nutritional supplementation to treat aggression and violence of which 12 were RCTs. The studies investigated the effects of a wide range of different supplements including micro and macronutrients. Five studies reported positive results when evaluating omega-3 fatty acids, vitamins and minerals,

S-adenosyl-L-methionine, or tryptophan, and five studies reported no benefits of omega-3 fatty acids, folic acid, tryptophan, a broad range of vitamins and fatty acids, and fatty acids and valproic acid. The authors concluded that fatty acids were the most promising nutritional supplements for treatment of aggression, but noted numerous limitations in this literature. Some of these limitations included a lack of replication, a lack of systematic evaluation of dose–response relationships, too much emphasis on statistical significance over socially meaningful outcomes and effect size, and not accounting for current dietary practices and related nutritional deficiencies where relevant.

The list of potential dietary influences on aggression and violence is endless. So, in this section, we will consider three examples of dietary influence on behavior that illustrate some of the interests in this field. These include serotonergic diets, omega-3 supplementation, and caffeine.

12.4.1 Serotonergic Diets

Some researchers have used the hypothesis that if aggression is controlled by neurotransmitters such as serotonin, then changing serotonin levels through diet might result in changes in aggression. Two uncontrolled case studies by Gedye illustrate this approach. Gedye (1991a) presented a case study of treatment of aggression in an adult with Down syndrome and apparently unprovoked aggression during the onset of dementia. Low-dose trazodone and a serotonergic diet result in a 96% reduction in aggression over a 4-month period. Gedye (1991b) presented a second case of aggression in an adult with intellectual disabilities treated with buspirone, a serotonergic psychotropic medication, and a serotonergic diet. The combined treatment results in a 67% reduction in aggression and a decrease in self-injury.

Another approach to increasing serotonin to treat aggression is to use serotonin reuptake inhibitors to increase the level of serotonin. In a systematic review and meta-analysis of preclinical studies in nonhuman animals, Carrillo et al. (2009) found a modest effect size ($r = 0.3$), but this effect was mediated by strain of animal used, types of aggression, methods used to increase 5-hydroxytryptophan, etc. among other factors. At this time, meta-analysis of human data has revealed an even more modest and tenuous relationship (Duke et al., 2013), suggesting that serotonergic diets are unlikely to be a robustly effective treatment for human aggression.

12.4.2 Omega-3 Fatty Acid Diets

Omega-3 fatty acids are a group of fatty acids that are part of cell membranes, synapses, and retinal tissue. They can influence neurotransmission by altering the structure and function of ionic channels and cell membranes. Omega-3 fatty acids are not produced naturally, rather they are obtained only through one's diet either as food in fish, eggs, vegetables, and nuts, or dietary supplements. Hence, they can potentially be manipulated quite readily. Omega-3 fatty acids have been implicated in fetal and early childhood development and have also been linked to a wide range of mental health issues including cognition, mood, and behavioral issues including aggression. Some research has supported the hypothesis that consumption of insufficient omega-3 fatty acids may increase the risk of aggressive behavior (Hibbeln & Gow, 2014). Hence, several studies including experimental studies have evaluated the impact of dietary supplementation with omega-3 fatty acids on aggression and violence.

In a systematic review and meta-analysis of these studies, Gajos and Beaver (2016) identified 40 studies from 17 countries, with 73 effect sizes and 7,773 participants including children and adolescents with and without developmental disabilities and adults. Studies were published between 1987 and 2016. There were both pre- and poststudies including RCTs and correlational studies. The effect size for group studies was moderate ($r = 0.62$, $p < 0.001$) and effect sizes from correlational studies was significant but rather small ($r = -0.06$, $p < 0.001$). Finally, the authors calculated the probability of so many consistent positive effects across studies and found it was highly probable that the effects of omega-3 acid on aggression were real. Thus, the authors concluded that omega-3 had an effect on aggression. They also recommended that omega-3 supplementation should be used as an economic approach to managing aggression in a wide range of populations, including prisoners.

12.4.3 Caffeine

Another dietary agent that might be related to aggression is the consumption of caffeine (Du, 2019). Dietary caffeine can come from tea, coffee, chocolate, energy drinks, and as an over-the-counter preparation, thus consumption of large quantities of caffeine is relatively easy. Caffeine consumption increases alertness and can drive the autonomic nervous system resulting in increased heart rate, blood pressure, etc.

Some studies have found correlations between self-reports of violence and of conduct disorders in adolescents independently of other likely predictors (Kristjansson et al., 2013). Zaslove et al. (1991), for example, reported that rates of aggression decreased in a large psychiatric hospital following a ban on caffeinated drinks, although this was a nonexperimental paper and thus confounds cannot be excluded. Hence, although caffeine consumption might be an important factor in aggression in some individual cases, little is known about this potential relationship. It might be useful to inquire concerning caffeine consumption in individual cases and where relevant to include in clinical case formulations, for example, when excess caffeine consumption increases physiological responses that might be labeled as anxiety or anger.

12.4.4 *Summary*

The literature on dietary interventions for aggression and violence is relatively weak (Qureshi et al., 2021). A common issue in research on dietary supplements is treatment integrity. Whereas it is easier to control consumption of dietary supplements using prepared capsules, changing people's consumption through eating food is more difficult as often consumption of nutrients is measured by self- or other reports of food consumption. Many people have problems adhering to diets, and many treatment studies have problems demonstrating the integrity of the independent variables. Treatment integrity can mean both consumption of certain food but can also mean establishing certain levels of metabolites in the body for some dietary substances. Future research should address this important aspect of treatment integrity.

Issues in the quality of RCTs for dietary supplementation studies comes from a review of drop out, reporting practices and drop out in studies of omega-3 supplementation in children and adolescents by Van der Wurff et al. (2017). They found that the majority of studies were significantly flawed in that they failed to report sufficient details to judge the quality of the studies. For example, most studies did not follow CONSORT standard and did not report a flowchart or study dates. Of 75 studies, only 25 reported treatment adherence which on average was 85%, and 65/75 studies reported drop-out which on average was 17%.

An important potential improvement in this area of research might be to select clients who have relevant dietary deficiencies prior to intervention. Thus, a strong test of the effectiveness of dietary supplementation to treat aggression would be to cross type of deficiency with type of supplementation in a double-blind crossover trial with convincing data on accurate intake of all foods and supplements.

12.5 Chapter Summary

There are a wide range of biological treatments for aggression and violence. They have a long and checkered history. As a generalization, the evidence for the effectiveness of biological treatments is not compelling. Indeed, many biological treatments raise numerous ethical issues, especially regarding consent, safety, availability of alternate effective and safer treatments, and, in some contexts, coercion.

References

Ballantine, H. T., Jr., Bouckoms, A. J., Thomas, E. K., & Giriunas, I. E. (1987). Treatment of psychiatric illness by stereotactic cingulotomy. *Biological Psychiatry, 22*(7), 807–819.

Bell, S. B., & DeWall, N. (2018). Does transcranial direct current stimulation to the prefrontal cortex affect social behavior? A meta-analysis. *Social Cognitive and Affective Neuroscience, 13*(9), 899–906.

van den Berg, J. F., Kruithof, H. C., Kok, R. M., Verwijk, E., & Spaans, H. P. (2018). Electroconvulsive therapy for agitation and aggression in dementia: A systematic review. *The American Journal of Geriatric Psychiatry, 26*(4), 419–434.

Bird, F., Wachtel, L. E., Henry, M., Gold, J., Fernandez-Robles, C., Orchanian, S., Shlesinger, A., & Luiselli, J. K. (2021). Treatment and maintenance effects of behavioral intervention and Electroconvulsive Therapy (ECT) in a man with catatonia, life-threatening self-injury, and Autism Spectrum Disorder. *Advances in Neurodevelopmental Disorders, 5*(2), 135–143.

Braslow, J. (1999). Therapeutic effectiveness and social context: The case of lobotomy in a California state hospital, 1947-1954. *The Western Journal of Medicine, 170*(5), 293–296.

Breggin, P. (1975). Psychosurgery for the control of violence: A critical review. In W. S. Fields & W. H. Sweet (Eds.), *Neural bases of violence and aggression* (pp. 350–378). Warren H. Green Inc.

Cailhol, L., Roussignol, B., Klein, R., Bousquet, B., Simonetta-Moreau, M., Schmitt, L., Thalamas, C., Tap, G., & Birmes, P. (2014). Borderline personality disorder and rTMS: A pilot trial. *Psychiatry Research, 216*(1), 155–157.

Calev, A., Nigal, D., Shapira, B., Tubi, N., Chazan, S., Ben-Yehuda, Y., Kugelmass, S., & Lerer, B. (1991). Early and long-term effects of electroconvulsive therapy and depression on memory and other cognitive functions. *Journal of Nervous and Mental Disease, 179*(9), 526–533. https://doi.org/10.1097/00005053-199109000-00002

Carrillo, M., Ricci, L. A., Coppersmith, G. A., & Melloni, R. H. (2009). The effect of increased serotonergic neurotransmission on aggression: A critical meta-analytical review of preclinical studies. *Psychopharmacology, 205*(3), 349–368.

Consoli, A., Cohen, J., Bodeau, N., Guinchat, V., Wachtel, L., & Cohen, D. (2013). Electroconvulsive therapy in adolescents with intellectual disability and severe self-injurious behavior and aggression: A retrospective study. *European Child and Adolescent Psychiatry, 22*(1), 55–62. https://doi.org/10.1007/s00787-012-0320-7. PMID: 22923049.

Dambacher, F., Schuhmann, T., Lobbestael, J., Arntz, A., Brugman, S., & Sack, A. T. (2015). Reducing proactive aggression through non-invasive brain stimulation. *Social Cognitive and Affective Neuroscience, 10*(10), 1303–1309.

Denson, T. F. (2021). Breaking the cycle of violent crime and punishment: The promise of neuronormalization. *Social Issues and Policy Review, 15*(1), 237–276.

Delgado, J. M., Roberts, W. W., & Miller, N. E. (1954). Learning motivated by electrical stimulation of the brain. *American Journal of Physiology, 179*(3), 587–593.

Du, Y. (2019). You may be what you eat, so can you be violent due to your food? *European Journal of Biomedical and Pharmaceutical Sciences, 6*(7), 20–28.

Duke, A. A., Bègue, L., Bell, R., & Eisenlohr-Moul, T. (2013). Revisiting the serotonin–aggression relation in humans: A meta-analysis. *Psychological Bulletin, 139*(5), 1148–1172.

El Khamlichi, A. (1998). African neurosurgery. Part I: Historical outline. *Surgical Neurology, 49*(2), 222–227. https://doi.org/10.1016/s0090-3019(96)00422-3.2340-2349

Faria, M. A., Jr. (2013a). Violence, mental illness, and the brain - A brief history of psychosurgery: Part 1 - From trephination to lobotomy. *Surgical Neurology International, 4*, 49. https://doi.org/10.4103/2152-7806.110146

Faria, M. A., Jr. (2013b). Violence, mental illness, and the brain–A brief history of psychosurgery: Part 2–From the limbic system and cingulotomy to deep brain stimulation. *Surgical Neurology International, 4*, 75.

Faria, M. A., Jr. (2013c). Violence, mental illness, and the brain–A brief history of psychosurgery: Part 3–From deep brain stimulation to amygdalotomy for violent behavior, seizures, and pathological aggression in humans. *Surgical Neurology International, 4*, 91.

Franzini, A., Broggi, G., Cordella, R., Dones, I., & Messina, G. (2013). Deep-brain stimulation for aggressive and disruptive behavior. *World Neurosurgery, 80*(3–4), S29–e11.

Gajos, J. M., & Beaver, K. M. (2016). The effect of omega-3 fatty acids on aggression: A meta-analysis. *Neuroscience & Biobehavioral Reviews, 69*, 147–158.

Gedye, A. (1991a). Serotonergic treatment for aggression in a Down's syndrome adult showing signs of Alzheimer's disease. *Journal of Intellectual Disability Research, 35*(3), 247–258.

Gedye, A. (1991b). Buspirone alone or with serotonergic diet reduced aggression in a developmentally disabled adult. *Biological Psychiatry, 30*(1), 88–91.

Gesch, C. B., Hammond, S. M., Hampson, S. E., Eves, A., & Crowder, M. J. (2002). Influence of supplementary vitamins, minerals and essential fatty acids on the antisocial behaviour of young adult prisoners: Randomized placebo-controlled trial. *British Journal of Psychiatry, 181*, 22–28.

Gouveia, F. V., Hamani, C., Fonoff, E. T., Brentani, H., Alho, E. J. L., De Morais, R. M. C. B., de Souza, A. L., Rigonatti, S. P., & Martinez, R. C. (2019). Amygdala and hypothalamus: Historical overview with focus on aggression. *Neurosurgery, 85*(1), 11–30.

Gu, X., Zheng, W., Guo, T., Ungvari, G. S., Chiu, H., Cao, X., D'Arcy, C., Meng, X., Ning, Y., & Xiang, Y. (2017). Electroconvulsive therapy for agitation in Schizophrenia: Metaanalysis of randomized controlled trials. *Shanghai Archives of Psychiatry, 29*(1), 1–14. https://doi.org/10.11919/j.issn.1002-0829.217003

Guinchat, V., Cravero, C., Diaz, L., Perisse, D., Xavier, J., Amiet, C., Gourfinkel-An, I., Bodeau, N., Wachtel, L., Cohen, D., & Consoli, A. (2015). Acute behavioral crises in psychiatric inpatients with autism spectrum disorder (ASD): Recognition of concomitant medical or non-ASD psychiatric conditions predicts enhanced improvement. *Research in Developmental Disabilities, 38*, 242–255.

Harat, M., Rudaś, M., Zieliński, P., Birska, J., & Sokal, P. (2015). Deep brain stimulation in pathological aggression. *Stereotactic and Functional Neurosurgery, 93*(5), 310–315.

Haq, A. U., & Ghaziuddin, N. (2014). Maintenance electroconvulsive therapy for aggression and self-injurious behavior in two adolescents with autism and catatonia. *The Journal of Neuropsychiatry and Clinical Neurosciences, 26*(1), 64–72.

Heimburger, R. F., Whitlock, C. C., & Kalsbeck, J. E. (1966). Stereotaxic amygdalotomy for epilepsy with aggressive behavior. *Journal of the American Medical Association, 198*(7), 741–745.

Hibbeln, J. R., & Gow, R. V. (2014). The potential for military diets to reduce depression, suicide, and impulsive aggression: A review of current evidence for omega-3 and omega-6 fatty acids. *Military Medicine, 179*, 117–128.

van Hoorn, A., Carpenter, T., Oak, K., Laugharne, R., Ring, H., & Shankar, R. (2019). Neuromodulation of autism spectrum disorders using vagal nerve stimulation. *Journal of Clinical Neuroscience, 63*, 8–12.

Jardri, R., Lucas, B., Delevoye-Turrell, Y., Delmaire, C., Delion, P., Thomas, P., & Goeb, J. L. (2007). An 11-year-old boy with drug-resistant schizophrenia treated with temporo-parietal rTMS. *Molecular Psychiatry, 2*(4), 320. https://doi.org/10.1038/sj.mp.4001968. PMID: 17389898.

Jiménez, F., Soto, J. E., Velasco, F., Andrade, P., Bustamante, J. J., Gómez, P., Ramírez, Y., & Carrillo-Ruiz, J. D. (2012). Bilateral cingulotomy and anterior capsulotomy applied to patients with aggressiveness. *Stereotactic and Functional Neurosurgery, 90*(3), 151–160.

Kay, S. R., Fiszbein, A., & Opler, L. A. (1987). The positive and negative syndrome scale (PANSS) for schizophrenia. *Schizophrenia Bulletin, 13*(2), 261.

Komesaroff, P. A., & Rosenfeld, J. V. (2020). Ethical challenges in psychosurgery: A new start or more of the same. In S. Honeybull (Ed.), *Ethics in neurosurgical practice* (pp. 175–192). Cambridge University Press.

Kristjansson, A. L., Sigfusdottir, I. D., Frost, S. S., & James, J. E. (2013). Adolescent caffeine consumption and self-reported violence and conduct disorder. *Journal of Youth and Adolescence, 42*(7), 1053–1062.

Lefaucheur, J. P., André-Obadia, N., Antal, A., Ayache, S. S., Baeken, C., Benninger, D. H., ... & Garcia-Larrea, L. (2014). Evidence-based guidelines on the therapeutic use of repetitive transcranial magnetic stimulation (rTMS). *Clinical Neurophysiology, 125*(11), 2150–2206.

Ling, S., Raine, A., Choy, O., & Hamilton, R. (2020). Effects of prefrontal cortical stimulation on aggressive and antisocial behavior: A double-blind, stratified, randomized, sham-controlled, parallel-group trial. *Journal of Experimental Criminology, 16*(3), 367–387.

Luiselli, J. K., Bird, F., & Wachtel, L. E. (2021). Electroconvulsive Therapy (ECT) for autism spectrum disorder associated with catatonia and self-injury: A clinical review. *Advances in Neurodevelopmental Disorders, 5*, 1–9.

Mpakopoulou, M., Gatos, H., Brotis, A., Paterakis, K. N., & Fountas, K. N. (2008). Stereotactic amygdalotomy in the management of severe aggressive behavioral disorders. *Neurosurgical Focus, 25*(1), E6.5–E6.13.

Ögren, K., & Sandlund, M. (2007). Lobotomy at a state mental hospital in Sweden. A survey of patients operated on during the period 1947–1958. *Nordic Journal of Psychiatry, 61*(5), 355–362. https://doi.org/10.1080/08039480701643498

Pagnin, D., de Queiroz, V., Pini, S., & Cassano, G. B. (2008). Efficacy of ECT in depression: A meta-analytic review. *Focus, 6*(1), 155–162.

Panksepp, J. (1971). Effects of hypothalamic lesions on mouse-killing and shock-induced fighting in rats. *Physiology & Behavior, 6*(4), 311–316.

Qureshi, F. M., Kunaratnam, N., Kolla, N. J., & Konkoly Thege, B. (2021). Nutritional supplementation in the treatment of violent and aggressive behavior: A systematic review. *Aggressive Behavior, 47*, 296–309. https://doi.org/10.1002/ab.21953

Ramnarayan, R. (2021). Vagal Nerve Stimulation (VNS) for depression and uncontrolled seizures: Report of a case. *Journal of Medical and Research Neurology, 3*(1), 1–4.

Rawlings, C. E., III, & Rossitch, E., Jr. (1994). The history of trephination in Africa with a discussion of its current status and continuing practice. *Surgical Neurology, 41*(6), 507–513.

Reyes-López, J., Ricardo-Garcell, J., Armas-Castañeda, G., García-Anaya, M., Montis, A. D., González-Olvera, J. J., & Pellicer, F. (2018). Clinical improvement in patients with borderline personality disorder after treatment with repetitive transcranial magnetic stimulation: Preliminary results. *Brazilian Journal of Psychiatry, 40*(1), 97–104.

Riva, P., Romero Lauro, L. J., DeWall, C. N., Chester, D. S., & Bushman, B. J. (2015). Reducing aggressive responses to social exclusion using transcranial direct current stimulation. *Social Cognitive and Affective Neuroscience, 10*(3), 352–356.7.

Sergiou, C. S., Santarnecchi, E., Franken, I. H., & van Dongen, J. D. (2020). The effectiveness of Transcranial Direct Current Stimulation as an intervention to improve empathic abilities and reduce violent behavior: A literature review. *Aggression and Violent Behavior, 55*, 101463.

Van der Wurff, I. S., Meyer, B. J., & De Groot, R. H. (2017). A review of recruitment, adherence and drop-out rates in omega-3 polyunsaturated fatty acid supplementation trials in children and adolescents. *Nutrients, 9*(5), 474.

Yang, C. C., Völlm, B., & Khalifa, N. (2018). The effects of rTMS on impulsivity in normal adults: A systematic review and meta-analysis. *Neuropsychological Review, 28*, 377–392. https://doi.org/10.1007/s11065-018-9376-6

Yang, C. C., Mauer, L., Völlm, B., & Khalifa, N. (2020). The effects of non-invasive brain stimulation on impulsivity in people with mental disorders: A systematic review and explanatory meta-analysis. *Neuropsychology Review, 30*, 1–22.

Zaslove, M. O., Beal, M., & McKinney, R. E. (1991). Changes in behaviors of inpatients after a ban on the sale of caffeinated drinks. *Hospital & Community Psychiatry, 42*(1), 84–85.

Chapter 13
Cultural Design to Change Violence and Aggression. I. Radical Behavioral Approaches

The next three chapters deal with different aspects of designing cultures to prevent and treat violence and aggression. This chapter reviews behavior analytic approaches to cultural design based largely on Skinner's writing and attempts to implement a real-world version of his vision of a better society described in his utopian novel, *Walden Two*. Chapter 14 describes some examples of intuitive approaches to designing cultures with the goal of broad behavioral change. These include religious and other utopias and attempts to heal the wounds left by wars and civil unrest using procedures that fall broadly under the umbrella of restorative justice. Finally, Chap. 15 describes cultural interventions that take place within the existing cultures, but use more circumscribed interventions to attempt to change more focused aspects of behavior. Examples include parent training, street lighting, and other environmental interventions to reduce violent crime, and interventions to adopt evidence-based policing practices to reduce violence, such as intimate partner violence.

13.1 Skinner and Cultural Design

Skinner wrote extensively about cultural control of behavior, evolution of cultural practices as one of the three causes of behavior—the other two being biological evolution and evolution of operant behavior during the lifespan—and explicit cultural design (Skinner, 1953, 1961). So, how do cultures come to control the behavior of its members?

Skinner noted that there were many behavioral similarities between members of the same culture; language is the most obvious one, but the way we eat, drink, smile, laugh, walk, and travel are often similar to other members of our own culture. Some of these commonalities reflect the shared physical, nonsocial environment, such as the way weather and geography may determine some of the characteristics of

P. Sturmey, *Violence and Aggression*, Advances in Preventing and Treating Violence and Aggression, https://doi.org/10.1007/978-3-031-04386-4_13

clothes, footwear, and transport. More important than the shared physical environment are community standards and the related contingencies maintaining "proper," conventional behavior: Whether, when, where and how we should cover or uncover our head is strongly controlled by the contingencies of reinforcement and punishment from fellow community members. For Skinner, the existence of such contingencies is obvious, and their existence is a trivial point. If a community member deviates too far from the standards of conventional behavior, we quickly observe contingencies of positive punishment (scowls and head shakes from others) and negative punishment (being asked to leave because you are improperly dressed for the context).

The important question is why such contingencies are there. Some answers are obvious. Using the right, conventional effective verbal behavior to obtain reinforcers is maintained by obtaining those reinforcers. For example, we ask for something we are deprived of and ask for the removal of aversive stimulation in a manner that others will respond to using the right language, voice volume, tone, and degree of politeness. When we speak too quietly, mumble, ask rudely, use contextually inappropriate language, or use the wrong language, those responses may be extinguished such as when you do not get what you asked for or they may elicit aversive stimulation from others when they ask loudly with a frustrated facial expression, "Speak up I can't hear you!" "What did you say you want?" or "Is that the way we ask here?" Our history of speaking and the responses of others among other things determine our current speech.

Conventional, cultural-specific behavior however is much broader than conventional verbal behavior. What explains why there are contingencies in place to maintain conventional patterns of grammatical speech when nongrammatical speech would be readily understood or to maintain conventional patterns of eating or dressing when nonconventional behavior could still get the food in our mouth or keep us warm and dry? Some culture-specific patterns of behavior may have functional historical roots, such as not eating dangerous foods. But why do cultures continue to maintain this behavior in its members when the food is no longer dangerous? Skinner stated that engaging in conventional behavior is often categorized as prestigious by members of the culture. To speak well, eat correctly, and dress well are behaviors esteemed by members of the culture and hence are differentially reinforced by members of the culture. People receive compliments, prizes, and degrees for speaking in culturally specific valued ways; people get asked back for dinner if they eat nicely and are polite and entertaining; nicely dressed people are asked to sit at the front and on stage. It seems that behavior that fails to follow the culture's rules is aversive to other members of the culture and hence they punish nonconventional behavior. Vulgar speech, crude table manners, and sloppy dress is often followed by rude looks and rude comments from others. According to Skinner, by suppressing other people's aversive behavior, such as stealing and assaulting, broader classes of responses in community members are established that we might call "good" and "bad" behavior or "right" and "wrong." Through differential reinforcement, bad, nonconventional behavior that is aversive to other members of the culture may be suppressed, but other nonconventional behavior that is not aversive to other members of the culture may also come to be suppressed as with sufficient stimulus pairings all behavior labeled as nonconventional becomes treated as aversive.

As well as the actions of other community members, society has agencies of institutions that control our behavior. In *Science and Human Behavior*, Skinner (1953) identified five important agencies which controlled human behavior: government and the law, religion, psychotherapy, economics, and education. Skinner analyzed each of these agencies in terms of the contingencies on the person's behavior that is controlled, the contingencies on the person's behavior that control, and the possibilities of counter-control, such as lying, cheating, passivity, and rebellion. Although these agencies often use some forms of positive reinforcement, they often do so suboptimally. Hence, many of these controlling agencies often resort to excessive use of punishment with short-term benefits for the person punishing and the culture but often long-term costs for all concerned. Governments control behavior using negative punishment procedures by fining citizens, suspending licenses of all kinds, and imposing token and monetary fines. If that does not work, society requires members to engage in unpaid or minimal labor and sometimes applies painful stimuli of many kinds as positive punishment procedures and even death for some illegal behavior. Religions use negative punishment procedures by excluding members from the mass, loss of salvation, or opportunity for marriage until the members of the congregation repent and show sufficient contrition. If that does not work, they use positive punishment procedures to coerce their members to make penance, by threatening to send them to Hell, beating or flogging them, or induce their members to self-punish by doing painful penances like walking on their knees or flogging themselves. Psychotherapists often use milder punishments, such as asking their clients if there was a better way to handle something or telling their clients that "That is not appropriate" or reminding their clients of the negative outcomes of not changing. Economists and employers set the price of labor at the minimum to ensure large profits for the employer and refuse to pay their employees under certain circumstances. If these procedures are ineffective, they fine their employees, give them bad working conditions until they change their behavior, demote them, reduce their wages, and, eventually, fire their employees. Schools and universities give their students bad grades and require their students to repeat assignments, shame their students, and suspend them from programs delaying or denying them their graduation, certificates, degrees, and licenses.

Use of punishment seems to be a common and somewhat effective strategy to control behavior at least in the short-term. Unfortunately, it comes with many costs. Some of the most obvious ones are that punishment teaches individuals to avoid, lie, cheat, and dissimulate. Unfortunately, punishment also is accompanied by anxiety (Estes & Skinner, 1941) which inhibits adaptive operant behavior in turn sometimes results in psychopathology managed by further problematic behavior such as avoidance, alcoholism, drug abuse, etc. In turn, societies pay enormous costs in human suffering, lost productivity, coercive control, and treatment.

13.2 Designing Cultures

Cultures often undertake interventions to maintain conventional behavior and to suppress problematic behavior, but they often do so intuitively and unsystematically with varying degrees of success. If cultures control its members' behavior in ways that are relatively unplanned and unsystematic, why not design the culture explicitly? Skinner asked what virtue could there be in leaving things to chance? Could there not be considerable harm that could occur when we do so? In any case, cultures are often explicit in the mechanisms of control and the goals of doing so. Religious texts and exhortations, laws, educational programs, and new psychotherapies are all mechanisms to control the members of a culture, and setting up new industries or agencies is often examples of Utopian thinking. Skinner asked why individuals engage in cultural intervention, and, unsurprisingly, he stated that it is due to their previous learning history for engaging in that or similar behavior.

Skinner's ideas on explicit cultural design were influential in several ways (see below). Some groups of people took up the challenge of *Walden Two* and set up several behavioral utopias with differing degrees of success and permanency (Kuhlman, 2010). Others have more modestly attempted to intervene in existing organizations to undertake interventions to establish a "culture of safety" in businesses using Skinner's ideas on cultural design (Geller, 1991).

Skinner began in his mid-career as an optimist about cultural design. Living in a world which had almost destroyed itself in World War II and had the capacity to destroy itself many times over with atomic bombs, Skinner was aware of the terrible effects that human behaviors, such as overconsumption, pollution, stockpiling atomic weapons, etc., were having and might have. He seems to have believed that answers were available: "The methods of science have been enormously successful wherever they have been tried. Let us then apply them to human affairs," he wrote blithely in *Science and Human Behavior* (1953, p. 5). Having written a piece of fiction on cultural design, *Walden Two* (Skinner, 1948), he still had grounds for optimism. Later in his career, Skinner became quietly pessimistic about cultural design. By 1982 Skinner could ask the question "Why are we not acting to save the world?" (Skinner, 1982) and privately said "I used to believe that a science of behavior could show us how to solve the problems confronting us—pollution, overpopulation, poverty, the threat of nuclear war. But I am forced to conclude that what the science of behavior shows us is that we can't solve these problems" (private conversation, 1989; reported in Chance & Harris, 1990, p. 10). Why is it so difficult to change the world, if we know what works and how might this apply to prevention and treatment of aggression and violence?

Let us now turn to the younger, optimistic Skinner began on cultural design with *Walden Two*.

13.3 *Walden Two*

Skinner's (1948) *Walden Two* described a fictional behavioral utopia, a purposefully designed community, based on behavioral principles. He did so in response to the opportunity and challenge of designing a better society for American veterans returning to the USA after World War II. Skinner (1979) wondered if "... they would abandon their crusading spirit and come back only to fall into the old lock-step of American life—getting a job, marrying, renting an apartment, making a down payment on a car, having a child or two" (p. 292).

Prior to writing *Walden Two*, Skinner was aware of several utopian or perfectionist communities, including the Oneida community (Olin, 1980) located close to his alma mater, Shakers who were present near where he was raised, and Mormons who were also connected historically to the part of Pennsylvania where also he was raised (Altus, 2013). He was probably also inspired by the Victorian utopian novel *Looking Backwards*, published by Edward Bellamy in 1888, which was an all-time best seller in the nineteenth century.

Although the key emphasis in *Walden Two* was to experiment rather than specific practices, it contained many examples of behavioral technology to design a better culture. Two of these were: (1) a labor-credit system, essentially a token economy, to maintain members completing the necessary and desirable tasks to maintain the community; and (2) the planner-manager system, a revolving committee responsible for oversight of the community, managing the behavior of its members and coordinating the maintenance and evolution of its cultural practices. But there were many other examples from preventative dental practices, shared child rearing, sheep eating the grass inside a movable electric fence rather than humans cutting the grass, and careful design of dishes and utensils to make eating and washing dishes easy for everyone. In Skinner's fictional Utopian community members worked happily 4 h a day on their preferred tasks and had ample leisure time: No one worried where the money for all this came from! (see below).

After the publication of *Walden Two*, some attempted to take Skinner's ideas and translate them into deliberately planned cultural interventions to address many of society's evils, including aggression and violence. Thus, starting in the 1970s, there were several attempts to implement *Walden Two*-inspired designed behavioral communities in the USA and Mexico. Like many other earlier utopias, most fizzled out fairly quickly. Sometimes, this was because the communities failed to attract and maintain members; overrelied on founder charismatic leaders that could not be maintained; and had problems related to disagreement and drift in both program goals and methods. Two or three communities do seem to have survived a relatively long time and one or two may still have survived into the present time.

13.4 Planned Behavioral Communities

The publication of *Walden Two* inspired several attempts to plan communities using behavioral principles. Some examples included Los Horcones, Sunflower House, Twin Oaks, Lake Village, Sydelle House, Dandelion, East Wind, Aloe, and North Mountain among others (Altus et al., 1999; Graham, 1982, 1984; Kinkade, 1999; Kuhlman, 1999, 2005). Many of these communities had explicit goals related directly or indirectly to the control of aggression and violence including promoting prosocial behavior, noncompetitiveness, egalitarianism, nonsexism, nonviolence, reduction of consumption, economic self-support, and shared parenting. The methods different communities used varied in detail. These behavioral communities emulated either specific procedures described in *Walden Two*, such as the labor-credit system, or modifications of it. Other *Walden Two*-inspired communities took Skinner's message of continual experimental and evaluation of cultural practices to heart and used *Walden Two*'s specific procedures as starting points but continuously modified and developed them.

The first two of these communities, Twin Oaks and Los Horcones, may have survived for some extended time. Others closed quickly because of lack of commitment to a behavioral model, lack of skilled behaviorists to help design the culture, inability to attract members, programmatic drift in terms of the goals and methods used, disadvantages of charismatic founders, economic problems, and a poor standard of living resulting in members leaving. The histories of each community and the procedures used in each community are documented in different degrees of detail. Here, we will consider three examples which have better public documentation.

13.4.1 Comunidad Los Horcones

One well-published *Walden Two*-inspired community that has adhered consistently to a behavioral model of community design from the beginning is Comunidad Los Horcones. Founded in 1973, Los Horcones is located in the Mexican Sonoran desert. Unlike other *Walden Two*-inspired communities, Los Horcones was systematically, explicitly, and consistently based on behavioral philosophy, analytic principles, and technology. It has published conceptual papers on behavior analytic issues such as analysis of the concept of consequences (Comunidad Los Horcones, 1987), types of reinforcers (Comunidad Los Horcones, 1992), relationship between behavior analysis and psychology (Comunidad Los Horcones, 1986a, b), and how external contingencies limit the dissemination and adoption of behavior analysis (Comunidad Los horcones, 2002). It also published descriptions of its original program (Comunidad Los Horcones, 1982, 1986a, b, 1989) and experimentation in evolving its government (Comunidad Los Horcones, C., 1989/2016). There is also a sunny personal account of living and growing up in Los Horcones by the founder's son (Altus, 1999).

Comunidad Los Horcones (1982) described its principal objectives, programs, and philosophy in designing a new community using procedures that were relatively close to *Walden Two*. In general, it emphasizes: (a) experimentation (see below); (b) action from within the community, rather than seeking political power; (c) teaching social alternatives to reduce problem behavior including aggression; (d) equality among members so that everyone can participate in governing, if they wish; (e) an active form of pacifism that avoids aggression by managing one's own community and avoiding politics, but that does involve actively changing one's own community; (f) acting as models for a new society by being happy, reinforced models that outsiders will attend to; (g) creating conditions in which members do not support and reinforce the inappropriate behavior of other members; and (h) an extensive commitment to educating one another and the outside society about behavioral design of communities. Los Horcones is also concerned with creating natural conditioned reinforcers, such as clean dishes rather than praise or points for cleaning dishes, so that appropriate behavior is maintained naturally.

Until 1979, there were between six and 22 adults including both permanent and provisional members. Similar to Skinner's planner-manager system, Los Horcones originally had two elected planners serving for 18 months who named up to 22 managers with specific responsibilities such as agriculture, etc. There is a code of behavioral standard of communitarian and noncommunitarian behaviors. Community members work with behavior managers to obtain information about current behavior and develop personal behavior management programs apparently mostly using behavioral self-management, such as self-recording, review meetings, and positive reinforcement from other members of the community. Behavioral objectives for members are generated both by the members themselves (i.e., a form of behavioral self-management rather than explicit control from others) and other community members.

Considerable emphasis is placed on a characteristic approach to rearing and educating children. This is done by both biological parents and "behavioral parents," i.e., other specific trained community members. All community members are educated as to what children should do and how to teach them. There are also people who are specialists in child behavior called "metas" who provide specialized training for all community members regarding child behavior and education. Children learn basic self-care and social behavior, academic skills, and self-management skills, such as self-recording from the age of around 3 years. Child education takes place in all environments, not such classrooms.

A good example of applying Skinner's principle of continuous experimentation comes from the evolution of Los Horcones' system of government. Los Horcones (1989/2016) described the gradual evolution of their system of government. After four years of implementing the planner-manager system from *Walden Two*'s original model, they noted that the authority of the system had become somewhat more democratic with planners having less authority than had originally, but that they were dissatisfied with it because this system was insufficiently democratic. So, they adopted a "direct democratic" form of government, meaning they eliminated planners and required 51% approval votes during weekly community meetings. This form of government was also subsequently abandoned because of negative competitive statements from members; encouragement of adversarial behavior among apparent winners and losers; passive reliance on certain members' opinions rather than true evaluation of proposal resulting in few members effectively taking most decisions; and dissenting members not working toward agreed goals. Other forms of government that were tried were described as totalitarian, gerontocratic, and technocratic.

The best solution as of 1982–1988 was called the Personalized System of Government (PSG) designed to support cooperation, sharing, nonviolence, and equality and active participation in government from all community members. This included allowing more than one individual to be a manager, now called coordinators, for a program area, increasing the number of program areas from 12 under the planner manager system to 61 under the PSG, having coordinators educate each other about skills needed for coordination and management, as another form of sharing. PSG resulted in a very large increase in decisive participation in government from nearly all community members sustained over a 7-year period. Anecdotally, the authors also reported that the PSG system results in increases in responsible participation and extinction of submissive and dependent participation, increases in productivity and better equality (i.e., participation by nearly everyone) while becoming more economically self-sufficient and maintaining balance with their environment.

One of the goals of Los Horcones was nonviolence. How did they attempt to achieve this? Los Horcones addressed violence by numerous direct and indirect methods. Some direct methods include stating clear goals to foster cooperation, sharing, noncompetition, and active pacifism. Another direct approach was to use skill training including behavioral self-management from an early age and the surrounding infrastructure to support doing this. More indirect interventions included safe, shared parenting to reduce parental workload and reduce possessiveness of one another, shared materials and wealth, to reduce competition and jealousy, and designing a society where people are comfortable, have their needs met, and are happy. (Some of these practices are similar to those adopted by some religious utopias, such as Hutterites.)

To date there are no public data on aggression and violence at Los Horcones. After a period of active publication, nothing has been published by this intentional community over the last 10 years or so.

13.4.2 Sunflower House

One designed *Walden Two*-inspired community was Sunflower House, a student housing cooperative, associated with the University of Kansas. It used the work-sharing system to maintain equitable participation in household chores and maintained 95% of chores completed over several years using point systems, contingent reduction in rents for residents, public governance, and researchers who collected and evaluated data on program performance (Feallock & Miller, 1976). This community is distinguished from the others reported here because of a series of data-based experiments evaluating their community practices primarily related to the community's governance and decision-making mechanisms and the procedures to maintain them over extended periods of time as individuals join and leave the community.

Feallock and Miller (1976) reported three experiments evaluating the work-credit system. In the first experiment, these authors showed that work credits were effective in maintaining completion of household tasks. When points were available for household chores members completed a high proportion of chores in a timely fashion, but when points were available for outside painting but not household cleaning, residents only completed outside painting and stopped completing household chores. This experiment showed that contingencies were effective in maintaining community members' behavior. The subsequent experiment demonstrated that points had to be contingent upon completing household chores, merely providing points independently of task completion resulted in fewer household chores being completed. The third experiment demonstrated that the backup reinforcer of contingent rent reduction was also an essential and effective essential component of the system.

In a subsequent paper, Altus et al. (1991) demonstrated that the original system was originally dependent upon the presence of researchers implementing the system. With systematic training, however, the system was handed over to community members, and this maintained resident task completion for over a five-year follow-up period. In addition, although not targeted directly, the number of fines and resident complaints about the system systematically reduced and was maintained over time, suggesting the evolution of a less coercive community. Further, the procedure was easy and cheap to implement.

Similar studies from Sunflower House have demonstrated that it is possible to maintain effective management and cultural practices, such as data collection performance related to maintaining key community components such as coordinating chores, maintaining the points system, maintaining member participation in education programs (Altus et al., 1993; Johnson et al., 1991), and running meetings more effectively and efficiently (Welsh et al., 1989). Further, this was done over very long periods of time up to 37 years (Miller, et al., 2006, cited in Altus, 2013).

These empirical studies are impressive. They demonstrated that practices within a designed culture can be evaluated experimentally; handed over to members of the culture and run independently by the original researchers or original leaders; were robust across changes in people, including both residents and residents in program administrative roles; and were maintained over extensive periods of time. Over time these cultural practices did indeed involve. For example, members of Sunflower decided to replace independent data collection with self-recording without loss of program outcomes (Altus, 2013). These studies helped community members understand which components of the cultural practices were effective and thus do indeed exemplify the implementation of Skinner's prime mandate that cultures should experiment on their own practices. The current status of Sunflower House is unknown but publications ceased around 2006, the time of Miller's retirement.

13.4.3 Twin Oaks

One planned behavioral community that survived a long time but that has changed a great deal is Twin Oaks whose early history was mostly documented by Kinkade (1999, 1973). In the 1960s, there were groups of people speculating about setting up *Walden Two*-inspired communities resulting in loosely affiliated groups and some national conferences on this question addressing what such communities might look like, how to plan and to fund them. After failed attempts to set up a *Walden Two*-inspired housing cooperative in an urban area, Twin Oaks began in 1967 in rural Virginia with only eight adults.

Some 6 years later, there were about 55 adults, including a smaller spin off community of eight individuals. Using a version of the labor-credit system, members worked 45.5 h/week (rather than *Walden Two*'s overly optimistic 4-h work days) and got to choose their jobs. About 40% of the community's income came from making and selling hammocks for Pier One Imports, and about 30% from other external jobs. They had no nuclear families, and many married couples chose to live separately. The three-member Board of Planners managed much of the community's activities. Personal possessions were minimized, and members were asked to turn over their personal assets to the community over time. The population was unstable and at one point Hippies arrived as new communards. The community was still going quite strongly in 2009 and had around 100 members and reportedly continued to use the planner-manager labor-credit systems. Twin Oaks was substantially economically self-supporting and grew around three quarters of its own food (Renwick, 2009). There were also two daughter communities East Wind and Acorn. The community's current web page (Twin Oaks Community, 2021) reported that as of 2010 they continued to use the labor-credit system and planner-manager decision-making system., proudly referring to itself as "Walden Two's bastard child"!

A key person in the history of Twin Oaks was Kat Kinkade, the community's visionary and founder who read *Walden Two* and, "struck by lightning," ran with it (Altus, 2009). Not a psychologist or formally trained behavior analyst, Kinkade did the daily hard work for years of translating Skinner's utopian novel into a practical and sustainable community over decades. She variously pointed out the importance of making the community economically self-sustaining. Living cheaply costs a lot of money when you have to raise $5000 a year per person to support a hundred communards—cheap at half a million dollars or so a year! It also involves extensive work to manage, evaluate, and develop the labor-credit system to ensure that community members are sufficiently productive to maintain the community economically, in addition to other work needed to manage an intentional community with occasional hippie invasions (Kinkade, 1999).

Twin Oaks appears to have kept going for at least 50 years (Butcher, 2017). The current Twin Oaks website (as of March 2021) documents that it is no longer a behavioral community, although it is rooted in *Walden Two*. It has experienced waves of hippies, primal scream, and other therapy groups along the way. It appears to remain an intentional community committed to values of peaceful behavior, environmentalism, self-sufficiency, and feminism.

13.4.4 Comment

Designing and implementing cultures based on Skinner's ideas has quite a checkered history. Most *Walden Two*-inspired communities, like other utopias and perfectionistic communities, failed quickly; others drifted and slowly withered away. Some never really got off the ground because of lack of resources, economic problems, inability to attract and retain sufficient members, founders leaving early on, lack of shared vision, lack of behavioral expertise, failure to adopt the principle of experimentation, program drift over the years—sometimes early on—over-reliance on founder member leaders, and hippie invasions. Not so many people wanted to give up traditional family life and hand over their children to other people even if the other people are trained metas.

Even utopias have many temptations. Temptations from the outside world included a life that might have been easier, might have given a more immediate higher standard of living, and other conditioned reinforcers. Temptations within the community included interesting, addictive, and fun programs within the alleged *Walden Two* inspired community, such as drugs, free love, and not working if you did not want to. Despite all this work, there is little data to work with as of today other than the experiments published by Miller and colleagues from Sunflower.

What accounted for the differential survival of these *Walden Two* Communities? Table 13.1 summarizes some features that may be relevant. Two factors may be key. The first is a key leader willing to put in consistent work over many years, as was the case in Los Horcones, Sunflower, and Twin Oaks. It is notable that Sunflower may have withered away after a key researcher retired. Second, it may be that a clear commitment to conceptual consistency and behavioral technology prior to starting the community, having a group of behavioral experts with a high level of conceptual sophistication, and a vision that includes both a breadth of the project and the long-term visions may be helpful. This requires commitment to engaging in practices that are proactive and preventative and avoidance of temptations of all kinds. Perhaps this distinguished Comunidad Los Horcones and Sunflower from Twin Oaks.

It is rare for people designing cultures to be able to start from scratch. Indeed, perhaps no-one ever does as new cultures are placed within the physical and social milieu of existing cultures. The next section goes on to deal with the more typical situation of addressing violence and aggression through behavior analytic cultural design, that is addressing violence and aggression in an ongoing cultural context.

Table 13.1 A comparison of some features of Los Horcones, Sunflower, and Twin Oaks communities

Community	Behavioral expertise	Breadth of visions and long-term vision	Adoption of nonbehavioral practices, including over-reliance on specific leaders
Los Horcones	Rigorous and explicit commitment to radical behavioral conception of all aspects of the program prior to and after conception; contributes to conceptual basis of behavior analysis through publications; limited publication of empirical studies	Clear vision of a comprehensive program designing all aspects of the environment, extensive training of membership, and development of teaching materials; committed to be a very cultural laboratory and model for other program in the future	None clearly documented other than possible over-reliance on a few leaders for their technical expertise, at least early on in the program
Sunflower	Program added to an existing student residence using expertise from University of Kansas behavioral program, especially Keith Miller; produced a substantial cadre of data-based papers on their practices until Miller retired	Focused on completing chores, training members, and management practices to support those behaviors; less explicit commitment to environmental design than Los Horcones	Difficult to know what practices were ongoing before the program was initiated; although programs were handed over successfully to residents, Miller was probably a key figure and technical resource
Twin Oaks	Inspired by *Walden Two*, but little conceptual or practice background in behavior analysis; produced a newsletter and several descriptive articles but probably no data-based publications	Unclear, perhaps the initial enthusiasm for the specific practices of *Walden Two* blinded founder members as to the central importance of the principle of experimentation and enormous amount of work needed to develop a *Walden Two*-inspired community	Readily adopted a wide range of nonbehavioral practices early on; Kat Kinkade a key figure in managing the economics and other aspects of the community

13.5 Behavioral Cultural Interventions for Violence

Walden Two-inspired communities attempt to start designing communities from scratch. Of course, they cannot truly start from scratch; prospective members come with a learning history from their community of origin, and the new community is physically and functionally part of the larger community, but these planned communities often start in a new physical setting and a new group of communards. A second strand of interest in behavioral approaches to cultural design has been to modify cultural practices within existing cultures. Behaviorists have in fact attempted to do exactly this related to both violence in several contexts including gang violence, cultures with high rates of violence, such as La Violencia in Medellin, Colombia, and police violence. These are specific examples of a broad, behavioral-ecological approach to understanding and changing violence at the community, rather than individual level (Mattaini, 2001, 2003, 2019).

13.5.1 Gang Violence

One set of cultures with high rates of violence is gangs. Mattaini and Aspholm (2016) and Aspholm and Mattaini (2017) used the notion of cultural evolution and maintenance of gang-related violence to describe and explain why this might be the case. Working in Southside Chicago, USA, and Medellin, Colombia, they noted that in neighborhoods with gang-related violence, there is extensive modeling of violence and a lack of models of peaceful behavior. They also noted that high levels of community violence weakens conventional informal and formal mechanisms of social control and results in lack of protection from violence and ineffective opportunities for social mobility and escape from such neighborhoods. Further, gang-related violence results in increased healthcare costs and reduced tax base further weakening conventional nonviolence behavior.

These authors also noted the advantages of violence. For example, violence has acquisitive functions, both in obtaining material benefits, justice in the absence of effective conventional institutions, and increased self-worth and self-protection from violent others, including members of other gangs and one's own gang. This often takes place in the absence of protection from violence from conventional social agencies, such as the police, neighbors, and nongang member peers. In the context of ineffective conventional sources of social and material support, such as families, schools, and other social institutions, gangs also provide social, material, and affective support and physical protection to its members from other gangs and the police. When legal oversight is largely ineffective, or even harmful or dangerous, engaging in gang-related violence appears more like a rational response to a pathological community than pathological behavior in a healthy community. The contingencies of punishment that conventional social agencies apply may leave the person inclined to behave in the way that resulted in punishment (Skinner, 1953) because it often results in avoidance only of the terminal aggressive behavior, but may leave all the precursor behavior intact and may not teach alternative behavior. Thus, adolescents who are repeatedly the recipient of sarcasm, humiliation, and other forms of punishment at the hands of their schools, and who are told—directly or indirectly—to stay away, are less likely to acquire and lack reinforcement for conventional prosocial behavior. Hence, when they encounter the contingencies related to gang membership, they may quasi rationally spend time with the gang rather than mainstream society and thereby learn the life of the gang and its associated aggression and violence. These authors described the putative current contingencies that fail to support conventional, nondangerous behavior and maintain gang-related violence. Thus, these authors go at least part of the way to explain how an individual can learn to engage in gang-related violence and go from a life of nonviolence to a life of violence.

The implication is that interventions to reduce gang-related violence should be based on this understanding of the rational basis of gang violence and give community members alternate, effective repertoires of prosocial behavior. Mattaini and McGuire (2006) reviewed the literature on the question of constructing nonviolent cultures with youth and concluded that two strategies were likely to be effective—universal skills training and ecological strategies. Universal skills are evidence-based practices, such as attending school regularly, academic skills, and social behaviors, such as parent training, and aggression replacement training. These two strategies are likely to result in prosocial repertoires and avoid gang membership. They should be delivered through positive reinforcement rather than punishment methods, behavioral skills training rather than verbal methods of delivery, and use universal classroom preventative strategies. These evidence-based universal skill training approaches are not only likely to be effective in reducing aggression and violence, but are also likely to result in a more pleasant school environment and less stress for all concerned.

Ecological interventions include a wide range of strategies, but include practices such as the good behavior game, etc. among a host of other strategies, which are likely to not only solve immediate academic and behavioral problems, but are also likely to have positive ripple effects across the entire ecosystem. One important ecological intervention which has the potential to reduce gang violence is youth activism which can be construed as a functionally equivalent repertoire that can sometimes address issues of social injustice and violence in a constructive, peaceful way (Aspholm & Mattaini, 2017). These authors noted that there are many young people, and even gang members, who are sometimes highly engaged in youth activism. Youth activism may be either supported or opposed by various practices from news media, police, local government, and nongovernmental organizations. For example, police might reach out to youth to collaborate on violence prevention or police might primarily function to surveille the community and use punishment including excessive use of force.

13.5.2 La Violencia

A second example of Mattaini's work comes from La Violencia in Colombia. La Violencia refers to 60 years of violence. La Violencia displaced millions of people from the dangerous countryside to major cities resulting in enormous problems in urban areas due to the influx of people from the countryside. (This problem continues today in 2021 in a less dramatic form, despite a signed peace accord between militias and the Colombian government.)

Mattaini (2010) reported on intervention in Medellin where violence was especially frequent. The chief of police was a key figure. He constituted a committee of experts from Colombia, Chile, and USA who wrote papers on the interlocking contingencies of gang behavior and mainstream behavior especially related to the best approaches to *bandas delincuentes* ("criminal gangs"). The paper focused on the behavior of gang members who were alienated from mainstream society, often at an early age, and who were deeply embedded in drug trafficking culture and identified the interlocking contingencies of reinforcement supporting gang-related behavior and preventing participation in mainstream society. The work focused on constructionist approaches to working with gang members and retraining of police to use "policing with love" and engaging with community members.

This work is interesting in that its ideas are clearly based on applied behavior analysis. A limitation is the absence of any outcome data.

13.5.3 Police Violence

Recently, Mattaini and Rehfeldt (2020) provided an interesting editorial and review of police violence from a behavior analytic perspective using a verbal behavior framework as part of redesigning cultures, including the police culture. They begin by noting the widespread violence and injustice against African American men and other groups, many recent calls for police reform in the context of the Black Lives Matter movement, and similar contemporary movements across the world, and more than endorsement of statements of support is needed for behavior change. They proposed that Skinnerean approaches to cultural design face enormous challenges (see discussion of Chance, 2007 below) and that consideration of the verbal behavior of community members was a useful addition to understanding and perhaps also changing racist and violent behavior through verbal behavior interventions such as Acceptance and Commitment Therapy. They noted that Relational Frame Theory could be the basis for racist beliefs and overt behavior. When a poor hungry person hears from fellow community members that poverty and hunger (A) are caused by Americans (B), he may directly learn an equivalence relationship. If he then learns that terrorist groups promise a better tomorrow, he may learn a relationship in opposition indirectly, perhaps resulting in learning rules such as "If I join this group, I won't be hungry and Americans won't ruin our future" (p. 3).

When exploring police culture, they note that the history of policing is one of 200 years of prevention of disorder, often in the context of controlling the behavior of members of minority and other disadvantaged groups through aversive control, for example, through slave patrols. Contemporaneously, this should also be understood in the context of the extremely aversive nature of police work which includes: (a) shift work with excessive hours of work; (b) multiple jobs and excessive paperwork; (c) exposure to trauma, such as failed resuscitation, death, and violence; (d) lack of appreciation from supervisors and members of the public; (e) absent, inadequate, and ineffective training; and (f) discrimination at work. These factors result in a high stress, aversive job with impossible requirements where members of traditional police culture support each other to do their best to avoid the very worst of their job and changes in police policy and simplistic single factor solutions to complex problems have little impact on the daily functioning of the police or reductions in crime.

What solutions did Mattaini and Rehfeldt (2020) propose? Noting the ineffectiveness of nonspecific "community policing," they gave examples of several promising practices. These included: (a) Camden New Jersey abolishing a corrupt police service and replacing it with a new service using direct community engagement practices which, it was claimed, reduced homicide by 50%; (b) use of community practices that can reduce crime, such as local nonprofit agencies to promote community collaboration; (c) youth activism and justice-seeking as an alternative to violence and coercion; (d) working alongside community members using accompaniment over the long-term, rather than leading; (e) various forms of nonviolent action; (f) public policy advocacy such as making specific "asks," rather than requests for nonspecific endorsements or expressions of support; and (g) use of ACT as an intervention to change relational frames related to racism and violence.

13.5.4 Comment

Behavioral approaches to design and intervention in existing cultures to reduce violence are an interesting extension of traditional behavioral work which, rather than focusing on the behavior of individuals and experimental control, extends the use of behavioral principles to analyzing the present culture, how violence is maintained and prosocial behavior is not supported, and to the implications for prevention. Unlike traditional behavioral research and practice, this literature lacks data and experimental control making clear conclusions concerning treatment implementation and outcome difficult to ascertain. Although this is sometimes possible, this is unusual due to the practical difficulties and cost of doing evaluative research on cultural design.

13.6 Challenges in Explicit Cultural Design

Designing cultures is not easy for many reasons, but an important reason is that it is difficult to foresee the long-term outcomes of a new cultural practice. For example, Skinner (1961) noted that sanitation and medical science are cultural practices that are designed in part to escape the costs of illness and the aversive stimulation of illness and death. While the immediate benefits to the survival of the culture may sometimes be readily apparent—namely maintaining health, avoiding suffering, and increasing productivity—the long-term benefits are not so obvious. Presciently, Skinner noted that in the future a new virus might wipe out clean, healthy people who have not had the immunological benefit of exposure to filth and disease. Only the future will tell if a contemporary cultural practice will contribute to long-term cultural survival in some unknown future environment!

Chance (2007) listed nine challenges to designing cultures. First, the reinforcing consequences for doing the right thing—whether increasing taxes on gasoline or teaching young children self-management skills—are delayed and uncertain. Indeed, engaging in the prevention strategies are effortful and take us away from more immediately reinforcing and easier activities. Second, problematic behavior often comes with good consequences for the individual and bad consequences for the group. The angry child does get to eat the snack that she steals but her siblings lose out; later society has to lock up snacks, supervise their consumption, and deal with an adult who has become a lifelong bully and thief. Third, we are susceptible to coincidental reinforcement. Praying for a violent partner to change their ways or policing broken windows is sometimes followed by reduction in violence even though they may be unconnected; in the future people may then repeat these ineffective behaviors superstitiously due to adventitious reinforcement. Fourth, humans are susceptible to powerful positive primary reinforcers, sometimes referred to as "dark reinforcers," such as sugar, salt, and drugs of addiction, which lead to many inappropriate and dangerous behaviors. Changing cultural practices to remove or defer these powerful reinforcers or to develop new conditioned reinforcers to compete with these primary reinforcers in order to invest in an uncertain future is very challenging, especially when punishment procedures are sometimes effective quite quickly in the short-term. It is much easier to lock up the cookie jar and smack a child hard when he or she reach for the cookie than to teach the child to put the cookies away and eat a banana. Fifth, people trust familiar ideas even if they are ineffective, but do not trust novel ideas. Thus, people may believe that counseling and education may help intimate partner violence even if the evidence is that such approaches do not work, but novel solutions may not be heeded, even when they do indeed work. The failure of society in general and therapists in particular to adopt evidence-based practices such as evidence-based psychotherapy and education are good examples of this challenge. Sixth, social reinforcement of extremist views can be very powerful, especially when people are not exposed to contrary views. Hence, gang and cult members, terrorists, and members of other outcaste groups such as incel may all hold and maintain views and engage in cultural practices that are supportive of aggression and violence when gang members are not exposed to contrary views. Seventh, punishment is highly addictive to the punisher since it may be effective in the short-term. Thus, ever more aversive solutions to aggression and violence, such as more and more severe punishment, death penalty, boot camps, and the like, are all highly addictive to the punisher as sometimes they work and may remove the offender for some time. Unfortunately, they also teach people to avoid the punisher and related situations and some with further emotional side effects which predispose the person to aggression and violence even more. Eighth, when progressing from mild to intense punishment is much less effective than starting with intense punishment. Many intuitive punishment systems, such as those used in schools and the justice systems, often consist of progressing from mild to intense punishment, such as a progression from reprimands to detention, bad grades, short suspensions, and finally expulsion from school. Further, dealing with aggression and violence in this way often means that we can avoid the effort to be effective in reducing aggression and violence, for example, teaching alternate behavior that takes longer than use of primary punishers and dealing with the effort necessary to administer more intense punishment such as the administrative effort to suspend a student from school and deal with the parent's responses. Finally, many people believe in the supernatural and continue to deny science. Many people still do not believe in biological evolution, but do believe in ghosts and spirits and praying COVID away. Thus, when it comes to aggression and violence, many people believe that victims are justly punished for their sins or that people who commit violence must be mentally ill or possessed by evil spirits. The problem is not that positive changes do not occur, but they occur too infrequently and too late.

Chance (2007) described many of the issues that *Walden Two*-inspired communities and that interventions in current cultures have faced and sometimes, perhaps, have solved at least in part. Many utopias—behavioral and nonbehavioral—have faced considerable resource issues especially early on. The dire descriptions of life at the short-lived Brook Farm experimental community in the nineteenth century are an example of this. Unable to put their own turkey on the table for Thanksgiving (a local farmer gave it to them), once smallpox swept through the community, the community folded in a manner reminiscent of Diamond's (2013) account of the end of Viking Greenland. Thus, the early days of planning for Twin Oaks were plagued with the question of how to get funding to buy land and Kincaid wrote extensively about the practical economics required to sustain such a community, despite remarkably low personal consumption. Remarkably, Twin Oaks appears to have largely solved these problems over several decades. Without solving these problems communards are faced with an easy choice between a high effort lifestyle of poverty (lack of reinforcement) which sets the occasion for coercive practices and the mainstream community with an easier life with more resources perhaps for less work.

13.7 Conclusions

In the tripartite approach to the evolution of aggression and violence, cultural evolution of violence has received less attention than biological evolution and evolution of operant behavior throughout the lifespan. Yet, cultural evolution is a core component of functional approaches to behavior. This chapter illustrates that designing behavioral utopias and cultural interventions based on behavior analysis are possible albeit very challenging.

An interesting lacuna in this literature has been the studies of planned communities that use coercion as their main behavioral strategy. These communities are often referred to as "cults" with the negative connotations of coercion and various kinds of behavioral practices often called abuse. Again, as Skinner pointed out, understanding the contingencies is straightforward. Coercive cultural practices ensure members' compliance with the rules, people hand over property to leaders, sometimes engage in sex with leaders, engage in practices that ensure the short-term survival of the community such as proselyting—but understanding how these contingencies came to be selected and are maintained is the real and more difficult question. Future research might address this question as it might help us avoid such coercive practices in the future.

References

Altus, D. (1999). Growing up at Los Horcones. *Communities, 103*, 53–57.

Altus, D. (2009). Visionary and activist: Kathleen Kinkade (1930–2008). *The Behavior Analyst, 32*, 255–258.

Altus, D. (2013). Revisiting Walden two: Sustainability from a natural science perspective. *Social Sciences Directory, 2*(4), 2–8.

Altus, D. E., Welsh, T. M., & Miller, L. K. (1991). A technology for program maintenance: Programming key researcher behaviors in a student housing cooperative. *Journal of Applied Behavior Analysis, 24*(4), 667–675.

Altus, D. E., Welsh, T. M., Miller, L. K., & Merrill, M. H. (1993). Efficacy and maintenance of an education program for a consumer cooperative. *Journal of Applied Behavior Analysis, 26*(3), 403–404.

Altus, D. E., Kuhlmann, H., & Welsh, T. (1999). Walden two communities: Where are they now? *Communities, 103*, 27–28.

Aspholm, R. R., & Mattaini, M. A. (2017). Youth activism as violence prevention. In P. Sturmey (Ed.), *The Wiley handbook of violence and aggression* (pp. 1–12).

Butcher, A. A. (2017). Fifty years of Utopian intentioneering at Twin Oaks Community. *Communities, 176*, 27.

Chance, P. (2007). The ultimate challenge: Prove B. F. Skinner wrong. *The Behavior Analyst, 30*(2), 153–160. https://doi.org/10.1007/BF03392152

Chance, P., & Harris, T. G. (1990). The best of psychology today. In *Humanities, social sciences & world languages*. McGraw-Hill.

Comunidad Los Horcones. (1982). Pilot Walden Two experiments: Beginning of a planned society. *Behaviorists for Social Action Journal, 3*, 25–29.

Comunidad Los Horcones. (1986a). Behaviorology: An integrative denomination. *The Behavior Analyst, 9*(2), 227–228.

Comunidad Los Horcones. (1986b). News from now-here, 1986: A response to "News from nowhere, 1984". *The Behavior Analyst, 9*(1), 129.

Comunidad Los Horcones. (1989). Personalized Government: A governmental system based on behavior analysis. *Behavior Analysis and Social Action, 7*, 42–47.

Comunidad Los Horcones. (1992). Natural reinforcement: A way to improve education. *Journal of Applied Behavior Analysis, 25*(1), 71.

Comunidad Los Horcones. (2002). Western cultural influences in behavior analysis as seen from a Walden two. *Behavior and Social Issues, 11*, 204–212.

Comunidad Los Horcones, C. (1987). The concept of consequences in the analysis of behavior. *The Behavior Analyst, 10*(2), 291–294.

Comunidad Los Horcones, C. (1989/2016). Walden Two and social change: The application of behavior analysis to cultural design. *Behavior Analysis and Social Action, 7*(1–2), 35–41.

Diamond, J. (2013). *Guns, germs and steel: A short history of everybody for the last 13,000 years*. Random House.

Estes, W. K., & Skinner, B. F. (1941). Some quantitative properties of anxiety. *Journal of Experimental Psychology, 29*(5), 390.

Feallock, R., & Miller, L. K. (1976). The design and evaluation of a worksharing system for experimental group living. *Journal of Applied Behavior Analysis, 9*(3), 277–288.

Geller, S. E. (1991). A total safety culture: From corporate achievement to a global vision. *Behavioral and Social Issues, 11*, 18–19.

Graham, R. (1982). Path with a behaviorist heart. *Communities, 103*, 42–44.

Graham, R. (1984). The dandelion community. *Behaviorists for Social Action Journal, 4*(2), 11–14.

Johnson, S. P., Welsh, T. M., Miller, L. K., & Altus, D. E. (1991). Participatory management: Maintaining staff performance in a university housing cooperative. *Journal of Applied Behavior Analysis, 24*(1), 119–127.

Kinkade, K. (1973). *A Walden two experiment: The first five years of Twin Oaks Community*. Twin Oaks.

Kinkade, K. (1999). But can he design a community? *Communities, 103*, 49–52.

Kuhlman, H. (1999). Walden two communities. What were they all about? *Communities, 103*, 35–41.

Kuhlman, H. (2005). *Living Walden two: B. F. Skinner's behaviorist utopia and experimental communities*. University of Illinois press.

Kuhlman, H. (2010). *Living Walden two: BF Skinner's behaviorist Utopia and experimental communities*. University of Illinois Press.

Mattaini, M. L. (2001). Constructing cultures of non-violence: The peace power! strategy. *Education and Treatment of Children, 24*, 430–447.

Mattaini, M. L. (2003). Understanding and reducing collective violence. *Behavioral and Social Issues, 12*, 90–108.

Mattaini, M. A. (2010). Cultural analysis and social change in Medellín. *Behavior and Social Issues, 19*(1), 1–5.

Mattaini, M. A. (2019). Out of the lab: Shaping an ecological and constructional cultural systems science. *Perspectives on Behavior Science, 42*(4), 713–731.

Mattaini, M. A., & Aspholm, R. (2016). Contributions of behavioral systems science to leadership for a new progressive movement. *The Behavior Analyst, 39*(1), 109–121.

Mattaini, M. A., & McGuire, M. S. (2006). Behavioral strategies for constructing nonviolent cultures with youth: A review. *Behavior Modification, 30*(2), 184–224.

Mattaini, M. A., & Rehfeldt, R. A. (2020). Editorial rendezvous with truth and discovery. *Behavior and Social Issues, 29*, 1–14.

Miller, L. K., Welsh, T. W., Altus, D. E., & Zwicker, T. (2006, May). A 37-year case study in the design and analysis of a program that survives post-research. Symposium presented at the annual conference of the Association for Behavior Analysis International, Atlanta, GA.

Olin, S. C., Jr. (1980). The Oneida Community and the instability of charismatic authority. *The Journal of American History, 67*, 285–300.

Renwick, V. (2009). Kathleen Kinkade. *The Behavior Analyst, 32*, 337. https://doi.org/10.1007/BF03392196

Skinner, B. F. (1948). *Walden two*. Hackett Publishing.

Skinner, B. F. (1953). *Science and human behavior*. Simon & Schuster.

Skinner, B. F. (1961). The design of cultures. *Dadelus, 90*(3), 534–546.

Skinner, B. F. (1979). *The Shaping of a behaviorist: Part two of an autobiography*. Knopf.

Skinner, B. F. (1982). *Why we are not acting to save the world*. Paper presented: American Psychological Association.

Twin Oaks Community. (2021). Retrieved April 5, 2021, from https://www.twinoaks.org/culture-government-65/walden-two-community

Welsh, T. M., Johnson, S. P., Miller, L. K., Merrill, M. H., & Altus, D. E. (1989). A practical procedure for training meeting chairpersons. *Journal of Organizational Behavior Management, 10*(1), 151–166.

Chapter 14
Cultural Interventions to Change Violence and Aggression. II. Broad-Based Intuitive Approaches

Cultures engage in numerous practices to change their members' behavior, and many of these practices address aggression and violence directly and indirectly. Some of the most common cultural practices that do so include: (a) utopian and perfectionist communities, many but not all of which are religious; (b) nonviolent approaches to atrocities and wars such as Truth and Reconciliation commissions; and (c) attempts to design more violent societies by inciting aggression and violence. This chapter will review these three examples of cultural design and their approach to controlling aggression and violence.

14.1 Utopias

There have been and continue to be many attempts to design utopian and perfectionist cultures intuitively. In a sense, every family, neighborhood, subdivision, village, town, and state is engaged in cultural design when they agree on common ground rules, pass laws, set curfews, change speeding limits for cars, and raise taxes. Some of the more interesting examples of cultural intuitive design are deliberately and explicitly designed religious utopias. Some examples include Shakers, Mennonites, Hutterites, Amish, German-Texan utopias, and the Oneida Community in upstate New York. In recent history, there have been more than a hundred of such communities in North American alone (Friesen, 2016; Hayden, 1976). There have also been nonreligious, ethical, and political communities. Some of these communities have had radical ideas on personal and sexual relationships, work, personal possessions, perhaps designed in order to address the potential problems of overly possessive relationships including overly possessive sexual relationships and parenting, jealousy, and other factors that might be related to aggression and violence. These communities also often have procedures to maintain their continued existence, such as implicit and explicit rules and expectations concerning the number of children, age

P. Sturmey, *Violence and Aggression*, Advances in Preventing and Treating Violence and Aggression, https://doi.org/10.1007/978-3-031-04386-4_14

of marriage, and when communities should be split to found daughter communities.

These designed utopian communities have many different goals and different procedures to design their cultures to move toward achieving and maintaining these goals. For example, these communities differ considerably from one another in terms of interaction with the outside world, economic activity and adoption of modern technologies. Some religious utopias, such as Shakers, Quakers, Hutterites were and are economically quite successful by producing crops, herbs, quality crafts and furniture, and commercial goods to fund their communities. Other utopian communities shun the outside secular world, self-support and try to be as independent as possible.

The control of aggression and violence is often an important part of cultural design in utopian communities. Failure to address aggression and violence will at least result in some costs such as time lost from other activities and perhaps costs of avoidable injuries, disability, and death. The utopias sometimes initiate setting up a new community in response to violence in their former host community. Perhaps the cultural and personal memories of their historical experiences of past violence and persecution may never be forgotten and hence sensitize community members to the issue of control of aggression and violence within the community and with the outside world. Alternatively, they may simply recognize that control of aggression and violence is desirable in any community.

Most intuitively designed communities fail. They voluntarily disband or more often simply die out. Their initial charismatic and committed leaders, on whom they may depend initially, may die, and no effective replacement may be found. The seeds of their destruction may be sown early on as they adopt unsustainable practices and destroy their environment (Diamond, 2011) or they fail to adopt practices to ensure maintaining their numbers over long periods of time. Let us consider one prototypical utopia that, after decades, failed, The Oneida Community, and one quite successful group, the Hutterites, which is still very much alive, but experiencing challenges.

14.1.1 The Oneida Community

The Oneida Community was a religious Utopia in upstate New York that lasted from approximately March 1848 to September 1880. It was founded by a charismatic founder John Noyes. He was recognized early in life as having a special talent as a religious and charismatic leader. After studying marriage and an earlier experiment in a perfectionist utopia in Putney, Vermont, Noyes moved to Oneida, New York to set up a new perfectionist community based on his own ideas.

The Oneida community has several characteristic social practices. They practiced shared possessions. They also practiced a kind of semi-formal, public mutual criticism to improve everyone's behavior and moral character. The Oneida

Community had a variety of successful businesses. Even after the community dissolved, Oneida silverware remained an American standard product that outlived the original utopia by more than a 100 years. Thus, the Oneida Community interacted economically with the outside world considerably (Olin, 1980).

The Oneida Community also had some quite characteristic social practices. Long before sexually liberated hippie communities of the 1960s and Twin Oaks, and similar to other nineteenth-century "free love" communities, the Oneida Community espoused communal marriage in which everyone was married to each other simultaneously. Part of the rationale for so doing was to minimize jealousy and selfishness. Thus, a distinction was made between amative and procreative intercourse. The community arranged preapproved sexual encounters for amative intercourse through intermediaries, and required men to be continent, i.e., refrain from orgasm during sex. The Oneida community had preapproved sexual procreative encounters designed to produce children with the positive eugenic overtones in that prospective parents were carefully matched up with one another. This practice was called "stirpiculture." These cultural practices were highly effective in controlling the birth rate and resulted in a stable low birth rate. After children were about 1 year of age, they were raised by a children's department, a separate wing of the main building. Parents were permitted to visit their children, but not have the final say in how their children were raised. These latter shared child rearing practices are echoed in Skinner's *Walden Two* and Comunidad Los Horcones (see Chap. 13). Outsiders could join the community, but the process was onerous and carefully controlled. The Oneida Community was economically successful and largely stable for over 30 years. The Oneida Community was relatively self-sufficient in terms of raising their own food, making their own clothes, etc.

The Oneida Community broke up voluntarily after its leader fled New York for Canada before he was prosecuted for statutory rape. In addition, the community had changed, become less religious, and had less faith in Noyes as a leader. No effective and acceptable replacement leader could be found, despite attempts by his son to lead the community. A new generation of community members who had been raised in the community preferred traditional marriage and sexual relationships and practices such as mutual criticism waned. So, they decided voluntarily to abandon the community. The members continued to associate with one another in part through economic ventures. Upon dissolution, the members all held shares in a limited company involved in various businesses such as the manufacture of silverware (Olin, 1980) which lasted in various guises until the early 2000s. Its historic buildings are now a state park.

Opinions are divided as to the causes of the extinction of the Oneida Community. Some suggest conflict with the outside world as being a key factor. Others have suggested that internal factors were more important. These included growing dissent, loss of authority by a theocratic authoritarian leader, the influence of outsiders joining, and the failure to grow nonfamily members as future leaders were the real causes (Kephart, 1963; Olin, 1980).

14.1.2 Hutterites

The Hutterites are a German-speaking, Anabaptist group. Anabaptists believe in adult Baptism, following one's own conscience, and pacifism, including not following the laws of the country if necessary. There are currently around 50,000 Hutterites living in colonies in the northern USA and Canada. Murdered and persecuted in Germany and Moravia for their religious beliefs in the sixteenth century, they moved to Ukraine. When persecuted by Russia in Ukraine in the nineteenth century, they moved en masse to Dakota territory in the USA, leaving no Hutterite colonies behind in Europe. Following persecution of four Hutterite men for pacifism and refusal to participate in the US army in World War II, two of whom died in custody, Hutterites moved en masse again to Canada. In Canada too, they experienced economic suppression by the Canadian government for some time in the form of restricting their ability to purchase land. Some Hutterites have since moved back to the USA, and there are now a fair number of Hutterite colonies in the Dakotas and surrounding northern US states. Thus, Hutterites have an interesting cultural history of violent and murderous persecution by the dominant cultural group to which they have adapted by moving en masse to other countries and continents.

The Hutterites are a religious group who attempt to live in a peaceful manner in relatively isolated colonies of about 70–150 individuals. Many colony members are related to one another. Marriages can only be to other Hutterites and so are commonly between members of different colonies as a form of prohibition against consanguinity. (There is a literature on genetic disorders in Hutterite communities.) They have a characteristic set of cultural practices. The first is that colony members live a collective life in which they look after one another at all times from little everyday things to larger things, such as looking after colony members who are older or with disabilities. They also look over one another to ensure that other colony members are not beginning to succumb to temptations or ways of the world. This is based on Jesus' commandment "Love one another. As I have loved you, so you must love one another. By this everyone will know that you are my disciples, if you love one another."

Second, there are strong traditional gender roles. Men are baptized at around 25 years, have dominant social roles, engage in agriculture, and control the colony's money. Women have largely traditional stereotypical gender roles involving cooking, cleaning, and gardening and are not involved in decision-making or financial decisions directly. Third, Hutterite families until recently were often larger than other cultural groups. Fourth, some interesting cultural practices include shared child rearing, where much child care is given over to or shared with teachers, kindergarten, German school, and religious classes. Child rearing avoids punishment. The caretakers and the community generally discourage individualism of any kind, fighting, and possessiveness of things and people. Rather, they emphasize doing things for one another, community singing and prayer, pacifism, and a simple, low consumption life. They eschew personal photographs as much as possible to discourage vanity and rivalry over appearance (Hostetler, 1997, 1983; Schludermann & Schludermann, 1971a, b, c, d).

Hutterites have minimal personal possessions, no personal money other than perhaps a personal allowance from the colony manager of a few dollars each month, and shared property and finances. Much of the finances are handled by a male colony manager who deals with colony financial matters. The community is primarily engaged in farming, which is mechanized with modern machinery, and all colony members are involved in hard manual work from an early age. Hutterite colonies tend to be located on land which is marginal and of less interest to other people. Hutterite society, unlike some other religious groups, is flexible in adopting new and modern technology to sustain their communities and engage with the outside world economically. They avoid alcohol consumption, but some members may break these rules by trading for alcohol from outside the community.

Hutterites believe that it is possible to obtain Heaven through living well in the colony. Individuals may leave the colony, and the colony will not stop them doing so. They may also return if they wish, but only on probation until the colony elders consider them members in good standing. Members who leave the colony are excluded from important cultural events, such as weddings and Christmas, and may only visit their family with permission from the colony elders. Colony members believe that former members of Hutterite colonies are going to Hell.

Daily decisions to manage the colonies are made by a colony manager and a small group of colony leaders, all of whom are men. Colony members need permission to leave the colony for any purpose, such as doctors' appointments, from the elders. If colony members need additional items, they may ask the colony financial manager who may buy these things if they deem it permissible.

Colony life seems quite regimented. Mealtimes are at set times. Colony members sit separated by gender and age. Church attendance is daily and twice on Sunday. There are few explicit contingencies of punishment for not following the implicit rules, but various forms of disapproval from colony members and elders might be quite powerful mechanisms of influencing behavior.

Unlike some other religious communities, Hutterites do not completely eschew interaction with the outside world. Rather, they engage in commerce related to agriculture. They also actively pursue expansion of the number of their members, the number of their colonies requiring purchasing of land, etc. Once the number of individuals in a colony exceeds about a hundred or so individuals, the community buys new land and splits off a daughter community (Diener, 1974). Thus, they have a fairly explicit and effective mechanism for sustaining and propagating their culture, perhaps despite its pacifist practices and the periodic loss of often younger colony members to the world.

Hutterite cultural pacifist practices may have cost many lives in the past, yet has both survived persecution and maintained a peaceful society by locating colonies on the edge of international capitalist society, adapting modern agricultural practices, having large families, and splitting colonies. They engage with the world to keep colonies financially viable and minimize personal consumption by devaluing possessiveness and emphasizing sharing and community life. It may be important that the mechanism of cultural evolution and replication does not depend on any individual charismatic leader, but rather on a set of cultural practices that are stable and survive from generation to generation, at least so far.

The current status of Hutterite communities is difficult to determine. Katz and Lehr (2019) recently interviewed and observed Hutterite colonies informally. The authors reported that there were over 520 colonies almost all in Montana, Minnesota, and the Dakotas in the USA, and in Alberta, Saskatchewan, and Manitoba in Canada with perhaps around 50,000 community members. They found that little by little Hutterites communities, although originally opposing such new technologies with easy access to the outside world, had adopted personal use of telephones, cell phones, emails, and internet, despite formal prohibitions from doing so in most circumstances. Although the advantages of the use of the internet for schools and farming became apparent, there was a formal ban on its use in 2002, but many younger people obtained laptops and the personal use of the internet continued to expand, including by community managers. Today, some Hutterites blog and use smartphones.

In the past, Hutterite communities were geographically isolated; but today geographical isolation is no longer effective in keeping the secular world at bay. The adoption of this new technology presents Hutterite communities with two problems that may threaten their cultural survival. The first is the much greater exposure to the outside world beyond trade. Today, Hutterites have much greater knowledge of alternative lifestyles and the temptations of secular society. The second threat is the need for individuals to pay for new technology when they have almost no personal income forcing members to engage in clandestine personal business to pay for this technology. It is unclear what effect this greater exposure to the world will have on the evolution of Hutterite culture (Hostetler, 1997).

14.1.3 Comment

There were hundreds of utopias in the nineteenth century and most failed quickly. It is interesting to contrast the Oneida Community, Hutterites, and other such communities to consider what factors might contribute to success and cultural replication or cultural extinction.

If too many members decide to leave or the community may fail to produce enough children, the community may no longer be variable. In the nineteenth century, there were around 4000 Shakers. Shakers were celibate and so depended upon new members joining the group. There are now no more Shaker communities left. The Shakers failed to address one key aspect of sustainability—maintaining membership. Although celibacy could have been a viable community practice, the failure to attract members to the community long-term was fatal to their survival.

Both the Oneida Community and Hutterites appear to be economically quite successful. Both produce products that generate sufficient money to maintain themselves. The Oneida Community seems to have failed for at least two reasons. First, the community did not survive the loss of their founding member because no effective mechanism was in place to hand this over to others. Recall from Chap. 13, that Sunflower specifically addressed one aspect of this problem by defining clear roles for each community member, training many people to fulfill these roles, and having a mechanism to maintain this training, although Sunflower may no longer be operating today as a designed community. Second, the second-generation community members chose not to adopt stirpiculture and preferred traditional nuclear families and typical exclusive child rearing practices. As noted earlier, the future of the Hutterite communities is uncertain as adoption of forbidden technology and consumption beyond the prescribed low levels both challenges the communities economic viability and produces tensions between community leaders exhorting members to refrain from highly reinforcing activities, resulting in behavior that is nonconsonant with community values, namely unchristian lying and dishonesty.

It is easy to romanticize such religious groups. Indeed, periodically apparently peaceful groups have problems with violence, murder, and abuse including Hutterites. Nevertheless, several of these groups provide an interesting and peaceful counterpoint to mainstream modern, relatively violent, life.

14.2 Cultural Evolution of Peace: Truth and Reconciliation

It is tempting to focus only on the cultural evolution of aggression and violence, but the other side of this coin is the cultural evolution of peaceful practices which is just as important to consider and explain. Indeed, given the very large reduction in murder described in earlier chapters, it seems a useful thing to consider how this happened. Cultural evolution of peaceful behavior may occur in an unplanned and haphazard way, but may also occur as an example of explicit cultural design. In this section, we will present an example of attempts to design cultures to promote peace in circumstances in which aggression and violence might be expected.

14.2.1 Rough Justice After World War II

Immediately after many major conflicts and dictatorships, both military and civilians often take revenge on perpetrators, collaborators, and even the victims of war and occupation. Temporary removal of state controls permits aggression and violence including civil rough justice which previously would have been punished but immediately after the breakdown of a violent government was permitted or even encouraged. For example, immediately after the liberation of Vichy France in World War II and before a formal civil government could be constituted effectively, French citizens engaged in retribution against perhaps 9,000 or so Nazi collaborators and profiteers who were rounded up and summarily murdered without trial. In addition, French women prostitutes who served occupying Germans, and women who had romantic relationships with Germans were publicly humiliated by shaving their heads and worse. This time was known as “the wild purge” (*epuration suavage*) to distinguish it from the more formal legalized forms of dealing with French Nazi government members and collaborators, a small number of whom were executed, and many others were subject to various legal forms of punishment (the *epuration legale*). Some have estimated that around 80,000–100,000 people were killed between the wild and civil phases of the purge.

Another example of post-World War II atrocities is the murder of Jews returning to Poland after being released from German concentration camps immediately after World War II. Approximately 2,000 Jews were murdered in Poland immediately after World War II. These included 42 Jews murdered in the Kielce pogrom on July 4, 1946, by Polish militia and Polish citizens who even tried to remove wounded Jews from the hospital in order to kill them. Polish civilians committed murders for several months by pulling Jews off trains passing through Kielce and murdering them.

Both of these examples of aggression and violence in France and Poland at the end of World War II occurred in the absence of formal legal controls of behavior or, in the case of Kielce, with the collaboration of the state. They only stopped upon later reinstitution of effective civil legal controls, such as state control of violence.

14.2.1.1 Nuremberg

After World War II, the Allies set up the Nuremberg trials to fully document the Nazi atrocities and bring some public justice to the Nazi perpetrators. Karstedt (2010) noted that victims were largely absent from the Nuremberg trials which were dominated by state and international prosecutors. The evidence presented at trial was primarily impersonal documents rather than personal statements from living witnesses and the focus was on fairly identifying and punishing perpetrators, rather than repairing harm to victims.

The Nuremberg trials also did little to inspire faith in the legal process as so many perpetrators got away with mass murder with no punishment or received relatively minor punishment, given their crimes. Subsequently, individual Nazi hunters, including organizations such as the Simon Wiesenthal Center and the Mossad continued to track down former Nazis in an attempt to bring them to justice before they died. For example, Adolf Eichmann, the architect and major organizer of the holocaust, escaped from arrest by American forces and after hiding in Germany and the middle east, set up life under an assumed name Ricardo Klement, in Buenos Aires, Argentina with the collusion of pro-Nazi Catholic clergy who arranged his passage to Argentina. He worked for several years and eventually was the head of a department for German car manufacturer Daimler Benz. After a tip to Simon Wiesenthal, the Mossad abducted Eichmann and, after keeping him prisoner for 9 days, flew him to Jerusalem for a public trial. Eichmann’s location in Buenos Aires was already known to the CIA and West German government who decided not to act on that knowledge. In contrast to the Nuremberg trials, Eichmann’s trial included eye witness testimony and documentation of his participation in the events of the holocaust. He was found guilty of 15 counts and hung (Arendt, 2006).

After these and similar atrocities, many governments deny events or twist and manipulate the truth to minimize events or blame others. For example, the current Turkish government continues to deny the Armenian holocaust and the Japanese government minimizes or denies atrocities in China during World War II, such as the Rape of Nanking, when thousands were murdered, and many women, children, and the elderly were murdered after being raped and sexually brutalized. After these and many other examples of large-scale murders and related aggression and violence, many people asked can something better be done?

14.2.2 Restorative Justice

Sometimes after crimes, wars, or following large-scale outbreaks of violence, societies have attempted to take actions which, at least in part, are motivated to promote peace and reduce violence. One approach to this problem is called restorative justice (Braithwaite, 2003; Marshall, 1999; Ness, 2004). Restorative justice is motivated in part by the apparent limitations of traditional justice. Traditional justice practices are seen as bureaucratic, slow, punitive, focused on the perpetrator, do not involve the victim or their community, and do not involve restoration of the harm done. Many believe that traditional justice practices are largely ineffective or even harmful. They seem to result in *more* rather than less violence and foster a life of crime, resentments by all parties, and unresolved harms to the victims and communities. Thus, rather than preventing violence, traditional justice practices seem to foster further harm to the victim, perpetrator, and community by dehumanizing all concerned.

Marshall (1999) stated that restorative justice did not have specific prescribed practices, but rather had four main principles which were "making room for the personal involvement of those mainly concerned (particularly the offender and the victim, but also their families and communities) … seeing crime problems in their social context … a forward-looking (or preventative) problem-solving orientation … [and] flexibility of practice (creativity)" (p. 4). In addition, the five objectives of restorative justice were "to attend fully to victims' needs—material, financial, emotional and social (including those personally close to the victim who may be similarly affected) … to prevent re–offending by reintegrating offenders into the community … to enable offenders to assume active responsibility for their actions … to recreate a working community that supports the rehabilitation of offenders and victims and is active in preventing crime … [and] to provide a means of avoiding escalation of legal justice and the associated costs and delays" (p. 6).

In contrast to traditional approaches to justice, restorative justice includes: (a) participation of all concerned, including victims, family members, perpetrators, and community members; (b) encounters between victims and perpetrators, including indirect encounters, such as letters and direct face to face encounters; (c) acknowledging the crimes committed and the effects on the victims; (d) making amends for the crime and the damage done to everyone by paying for the damages done, community work, and other restorative acts; and (e) reintegration of both victim and perpetrator into society. Restorative justice is personal, emphasizes healing for all involved, and involves active participation of all involved. Advocates of restorative justice sometimes use non-European models of community justice as models for this approach.

Restorative justice has no specific methods but often involves face to face encounters of perpetrators, victims, and mediators through meditation and individual and group conferences in which all actors get to speak, the harms are acknowledged by perpetrators and victim, apologies made and, if possible, harms repaired in some manner (Braithwaite, 2003; Marshall, 1999). For example, Marshall (1999) placed restorative justice practices at the apex of a pyramid between victim, perpetrator and community. (See Table 14.1 for examples of restorative justice actions, processes, and potential benefits.) Some researchers have attempted to evaluate the effectiveness of restorative justice practices, investigating the potential benefits in terms of satisfaction to actors, reduction in recidivism, and economic benefits in terms of avoided costs.

Restorative justice practices can be seen as an example of evolution of cultural practices in response to the limitations of traditional justice. Restorative justice has been used with both the individuals' crimes, such as victims of robbery, rape and murder, and wider community violence. Restorative justice has also been attempted after major violent conflicts in countries such as South Africa (Gibson, 2005, 2006), Northern Ireland, Serbia, Timor Leste, Rwanda (Eriksson, 2015), Colombia (Bueno & Rozas, 2013), Peru (Heilman, 2018), and elsewhere with varying degrees of community acceptance and success.

14.2.2.1 South Africa and the End of Apartheid

South Africa is a former British colony with a complex and unique ethnic and cultural history. Originally a Dutch colony settled in 1652, South Africa has an ethnically and culturally diverse Black African indigenous majority, and Afrikaans-speaking, White minority descended from Dutch settlers, a large South Asian minority from India descended from slaves imported from India by the Dutch, and a British, White minority. During the early twentieth century, South Africa was a British Dominion with racist laws that disenfranchised Black South Africans and apportioned 90% of the land to the 20% White South Africans. In 1948 this system was legally formalized as *apartheid* ("apartness" in Afrikaans) which included segregation of non-Whites into geographically segregated homelands. Apartheid eventually resulted in militarization of South African society involving hundreds of extrajudicial killings by the South African state in both South Africa and surrounding countries.

There was some nonviolent opposition to apartheid post World War II which was ineffective in resulting in change. In 1966, however, apartheid was condemned as a

Table 14.1 Summary of some of the actions and processes and potential outcome of restorative justice for each of the various relationships between victims, offenders, community, and the justice system based on Marshall (1999)

Relationship	Actions and processes	Potential benefits
Victim–offender	Mediated victim–offender meetings by skilled facilitators between all parties and chosen others to permit the offender reparations to the victim and allow the victim to understand the offender. Offenders may make financial payment, do work for the victim or a community organization of the victims choice, etc. This may take place with other agencies including police and voluntary agencies	Victim's needs may be met and feel consulted; both victim and offender may empathize with each other as individuals, rather than stereotypes; and the offender may be more personally involved in reparation and feel that society may re-accept them
Victim–community	Support comes from victim's family and friends and victim support groups	Improved victim trust and social functioning and reduced victim alienation
Offender–community	Community agencies that offer services for offenders from literacy, substance abuse, job finding, etc. and improving family functioning of offenders' families. School programs for at-risk individuals and offenders. Offenders making reparation through community volunteering	Reduced recidivism, improved community functioning such as better safety and improved offender mental health
Victim–offender–community	Community takes part in mediation and takes on some responsibility for social justice	Reduced recidivism and more peaceful society
Justice agency and victims	Assigning criminal responsibility through criminal trials and financial compensation and protection of the victim; sometimes rehabilitation programs for long-term offenders	Some partial justice and safety for victims and potential prevention of recidivism
Justice agency and offender	Assigning criminal responsibility to offenders, rehabilitation efforts, and inter-agency collaboration	Reduction in recidivism, improved offender functioning and reduced crime
Justice agency and community	Police take preventative actions to establish good community relationships before violence occurs. Probation services and inter-agency collaboration may undertake preventative work	Prevention of crime and recidivism; a more peaceful less violent community

crime against humanity by the United Nations, but this condemnation resulted in little immediate change. Following international economic and arms boycotts and the Soweto massacre, in which 170–700 people were killed, including unarmed children as young as 13 years, the system began to unravel so that apartheid ended in 1994. During the process, several thousands of South Africans were killed by the state, by White rightwing groups, and by Black African political parties involved in political struggles for power between different Black political groups. A heinous example of this violence was the practice of so-called necklacing in which tires soaked with gasoline were hung around the Black collaborators' necks and set alight by fellow Black South Africans.

As part of the transition to a peaceful, integrated South Africa, the post-apartheid government set up the Truth and Reconciliation Commission (TRC) chaired by Archbishop Desmond Tutu. The TRC heard from witnesses from all parts of the community, some of whom told their stories of violence and abuse publicly and some of whom confronted their perpetrators directly face to face. These hearings were held publicly and televised and broadcast around the world (Skelton, 2002).

Following the end of apartheid, South Africans faced numerous challenges. Some of these included continued poverty for many Black South Africans, a large national debt, severe health problems related to HIV, financial corruption in new state leaders described by Desmond Tutu as a "gravy train," a large brain drain involving emigration of many White South Africans, continued violence, and a failure of the state to follow through on promised reparations for apartheid.

The outcomes of the TRC were seen as mixed by many. Many South Africans felt that the truth had largely been told but that there was little justice and little reconciliation. For example, of over 300 names that the TRC forwarded to the South African judiciary fewer than six prosecutions were made. Many perpetrators received pardons without penalty or being required to engage in reparation for their actions. Thus, the TRC appeared to be an incomplete implementation of restorative justice which involved airing of a lot of the truth, some face-to-face confrontation of victims and perpetrators but little healing of the harms done.

14.2.2.2 Northern Ireland and the Troubles

The second example of restorative justice comes from Northern Ireland and the Troubles. Anti-Catholic discrimination when the island of Ireland was part of the United Kingdom dates back at least over 300 years, but the seeds of discontent which resulted in the more recent Troubles were sown when the Republic of Ireland separated from the United Kingdom in 1922 (see Chap. 7, for a brief description of the Belfast pogrom of 1920–1922 and Lynch, 2008).

Northern Ireland, which remained part of the United Kingdom, had a large Catholic minority, many of whom wanted Northern Ireland to unite with the Republic of Ireland, and a Protestant majority, many of whom were vociferously loyal to the United Kingdom and opposed any form of union or political power sharing with the Republic of Ireland. In the years prior to the Troubles, both the groups lived mostly in segregated communities. Individual's religious affiliations were readily apparent based on where they lived, their names, and the schools they went to (schools were largely segregated by religion). Hence, it was easy and quick to identify individuals as Catholic or Protestant. The Catholic minority were subject to numerous forms of discrimination for many years. There was gerrymandering of local electoral districts such that Catholic majorities were excluded from local power and Protestant minorities elected representatives to local government, thereby ensuring continued anti-Catholic discrimination. Catholics were excluded from well-paying jobs, which were often advertised in newspapers with the tag "Catholic need not apply." Catholics were also denied union protections, and unions were also sometimes openly anti-Catholic. Public services in local Catholic areas were also systematically underfunded by the Protestant minority who held political power.

During the Troubles, a civil war between the Protestants and Catholic communities broke out in which the two communities separated even further than previously. Following civil rights marches in the 1960s, which paralleled civil rights marches in the USA at that time but proved ineffective in obtaining political goals, violence broke out in 1969. Both Catholic and Protestant civilians and paramilitary groups engaged in deliberately provocative public marches, burning out one another's houses, bombings of public buildings, and punitive acts, such as "knee capping" (shooting individuals in the knee) of alleged traitors. The Irish Republican Army (IRA) instigated a campaign of bombing public places including downtown areas. This eventually included bombings on mainland Britain including bombing of pubs and the Tory annual conference in which the IRA nearly killed the then Prime Minister, Margret Thatcher, and killed five British politicians and family members and injured some 30 or so others. During the Troubles, over 3500 people died of whom about half were civilians, a third were British armed forces and the remainder were members of Protestant and Catholic paramilitary groups. Eventually, British troops were stationed in Northern Ireland and were eventually regarded by Catholics as pro-Protestant. There were British internment camps for suspects without trial, torture of suspects by the British government, and flawed trials in which innocent people were jailed and only finally released after many years.

It was a dirty war. The British government had spies in Catholic paramilitary organizations, and there is evidence that the British government colluded with the Protestant community by leaking information about planned IRA bombings and potential suspects resulting in extra-judicial murders of Catholic paramilitaries. In addition, the British government undermined its own public enquiries into these events and probably orchestrated break-ins into the offices of its own investigators in attempts to destroy evidence of its own wrong doings. Within the paramilitary organizations, individuals deemed to be disloyal or British spies were maimed or shot.

Following the 1998 Good Friday Peace agreement, the outcome was access to power for members of both Catholic and Protestant communities and eventual peace, although not complete cessation of all violence, for over 30 years. Like some other civil wars, violence appears to have succeeded in achieving political objectives where peaceful marches and protests did not. After the Good Friday Agreement, a surprising sight was the Reverend Ian Paisley, Protestant leader and possible funder of anti-Catholic bombings, and Gerry Adams, who had probably been involved in anti-Protestant bombings and murders, sitting around the table together, smiling and laughing; their pictures face next to each other at Stormont, the seat of political power in Northern Ireland.

Eriksson (2008) and Eriksson (2015) give an account of restorative justice in Northern Ireland. The Good Friday Agreement included several significant elements including the decommissioning of weapons in paramilitary groups which largely occurred and a recognition of civil rights for both groups. In order to replace local, informal violence as a way of settling injustice—in part because of frustration at the lack of effective state justice—two separate restorative justice projects, one for Protestant and one for Catholic neighborhoods, were initiated. Participants included former prisoners, and members of paramilitary organizations, community members including religious leaders and representatives of the state. In addition, there were projects addressing restorative justice for youth. These organizations dealt with several thousand individual cases over many years.

At first there were many problems, including yet further separation of communities with so-called peace walls (walls that were painted declaring certain neighborhoods Protestant or Catholic and other structures built to separate the two communities) and periodic increases in violence. Over about 10 years, these restorative justice projects achieved some degree of cooperation between Protestant and Catholic restorative justice organizations that had not been present at first. Eriksson attributed this progress in part to decreasing social distances between communities and communities and the state, increased local community responsibility for conflict resolution and eventual reduction in physical and symbolic barriers between communities.

As of 2022, there has been a great reduction in violence on all parts and much greater peace between the two communities which has largely been sustained. It is unclear how stable these changes are. The two communities remain geographically and functionally isolated—schools are still segregated by religion, for example. The demographics of Northern Ireland have changed substantially since the Good Friday Agreement, such that Catholics will become a majority shortly. Further, BREXIT has raised the specter of power sharing with the Republic of Ireland, which remains an anathema to some loyalist Protestants. The post-BREXIT arrangements treat Northern Ireland differently from other parts of the United Kingdom, in terms of Northern Ireland having to follow European rules and standards regarding food and soil sanitation standards and practices, which the rest of the United Kingdom does not follow. This too is highly provocative to loyalist Protestants. There was a modest uptick in violence in Northern Ireland in 2021.

14.2.2.3 Canada and First Nations

In Canada, a TRC was established to investigate abuses in the system of residential schools run predominantly by religious groups for First Nation indigenous children who were separated from their families with the goal of eradicating indigenous languages, identity, and cultures. Cruel punishments and beatings were common, and children were forbidden to engage in their own cultural practices including speaking their own native languages on threat of painful punishment.

The report was published in 2015 (The Truth and Reconciliation Reports of Canada, 2015). Although the report cost millions of Canadian dollars, the report itself focused on obtaining the truth as a foundation for future action and reconciliation. In 2021, horror stories of hidden unmarked mass graves of Indigenous Canadian children hit the media worldwide, and there will probably continue to be more revelations as more truth is uncovered. It is unclear at this time whether the truth will result in any justice.

14.2.2.4 Comment

These three examples of restorative justice seem to give rather contrasting impressions of outcomes: There were relatively negative outcomes in South Africa, somewhat positive outcomes in Northern Ireland as of 2021, and an incomplete ongoing process in Canada. Evaluation of such practices are of course difficult because the practices themselves are hard to define, and many other factors come into play at the time that restorative justice efforts take place. Some outcomes, such as the number of killings, are immediate and concrete. Other outcomes may be more ephemeral and may only occur in the distant future. The reports of outcomes are largely not data-based, and the reports of their effects are colored by contemporary political considerations and the motivations of participants including the authors of the reports. Finally, the outcomes are sometimes delicate, and it is difficult to know if they will be sustained over time.

In some cases, there has been little adoption of restorative justice practices for violence both for individuals and for communities. For example, in the cases of the Armenian genocide, Japanese atrocities in China and elsewhere during World War II, state violence in parts of South and Central America, and numerous examples of state-sponsored mass violence during decolonization of European former colonies, there have been little or no efforts at restorative justice in part because those countries that engaged in past violence deny or minimize their acts, perpetrators are still in power or their allies continue to support them. Likewise, although many Romanies, people with intellectual disabilities, and homosexuals were murdered during World War II, it took many years for this to be acknowledged, but there has been little effort at restorative justice for these groups. It is interesting to consider what factors promote the adoption of restorative justice, promote its complete and full implementation, and what other factors make it effective or ineffective.

14.3 Building a More Violent Society

Although most of the literature on changing aggression and violence at a social level addresses making society less violent, it should be acknowledged that from time to time societies do attempt to increase violent behavior, for example, at times of war and national self-defense. Thus, this should be considered both as a naturally occurring phenomenon that is worthy of study and, for those who wish to increase aggression, might inform how to do so.

In a sense the proximal mechanisms to increase violence are apparent for all to see. People are instructed to be violent; violence models are presented and extolled; imitating the violent model and engaging in modeled violent behavior is reinforced and made easy; alternate nonviolent behavior is punished. Thus, at an individual level, the parent of a bullied child may tell them to stand up for themself, punch the bully on the nose first before they get punched on the nose, sign their child up for boxing classes, practice fighting at home, and give their child money when they come home victorious with a black eye (see Llewellyn's 1939 *How Green Was My Valley* for a sentimental fictional example).

At a societal level, groups of people are induced to violence with broadly similar mechanisms. Unemployed and low income people are paid more to join the army and other parts of society are paid to take over their civilian work. Future murderers (or heroes) are trained to follow instructions without hesitation. Over time, through modeling, practicing on dummies how to kill with the gun and bayonet, and threats of punishment and violence, they are finally encouraged to shoot real living people. Heroic, patriotic, and brave models are presented and the violent behavior is seen to be reinforced in the media. When individuals or groups kill, their violent behavior is reinforced with money and medals. If they hesitate or refuse to kill, fake injuries or disability, or rebel, then they are sent to the front line as cannon fodder or taken out and shot for cowardice as a warning to others.

14.3.1 Hate Schools

A curious example of an intuitive approach to this is the so-called British hate schools used in World War II to increase effective fighting in members of the military. Bourke (2018) provided an interesting history of this project, principally involving the participation of psychiatry and the other then new and dubious "psych" professions, to prepare civilians for military service. The problem was framed as how to replace old fashioned and ineffective drilling and parading with something more effective for modern war. Preparation of new recruits had to take peaceful British shopkeepers and clerks and turn them into lethal fighters who would kill other people upon command. Further, the draft called up large swaths of the population without much concern as to their interest and skills for killing other people. Generally, only executioners could legitimately kill people, but during world war, recruits' histories of punishment and feeling guilty for being rude, let alone being violent, had to be overcome quickly and efficiently. Preparation addressed several different aspects of preparing civilians to combat. Psychiatry was involved generally in selection of recruits—mostly limited to four or sometimes even only one screening question ("Do you like girls?")—preparation for combat, treatment of war neuroses, and return of combatants to fighting and transition back into civilian life. This increased participation in the military by the fledgling mental health professionals reflected a gradual rise in the role of psychiatry, psychoanalysis, and psychology in many aspects of society since the 1930s.

Hate schools addressed preparation for combat to address several aspects of participation in the military. These included habituating combatants to the noise and horror of combat, being able to kill efficiently, and inciting hatred to increase motivation to kill. Habituation to the noise and horror of combat included *in vivo* exposure to live ammunition, explosions, and grueling combat exercises. Presumably the rationale was that if combatants were not exposed to this during their preparation that they may be unable to function during battle. Hate schools addressed efficient killing by training soldiers to practice killing during bayonet practice. Finally, hate schools attempted to provoke anger and hatred in combatants by showing them movies of civilian casualties and atrocities and instructing the combatants that they were defending the honor of their women and daughters from their enemies who were beastial rapists.

Bourke (2018) concluded that hate schools were both not well received and ineffective. There were many public critics. Within the military senior officials were divided. Many were skeptical of the new "psych professionals," who may never had even seen combat and were seen by many as a soft touch for malingerers, rather than a true and respected medical profession. Military officials and psychiatrists were divided across a spectrum from liberals, advocating sympathetic treatment of war neuroses, to conservatives, some advocating sterilization of malingerers to protect the race from degeneration. Some critics took an essentialist view that English soldiers could not be trained to be violent, but that Scottish soldiers needed no such training. Some from the Church of England viewed such a project as unchristian and immoral. The national and regional press were sometimes skeptical about the effectiveness of such an endeavor and echoed the moral objections from the military and Church of England. Finally, some critics were concerned that, if the training was indeed effective, the British society would face an unleashed cohort of violent killers who could not be controlled upon return to society. Bourke (2018) concluded that Hate Schools were both ineffective and controversial. Hence, interest in them fizzled out and other governments, such as in the USA, did not adopt such practices.

This vivid historical vignette illustrates some interesting conceptual, technological, and ethical challenges for the study of aggression and violence. Bourke's (2018) article does analyze parts of the problem of preparation of military recruits for combat which can be parsed into apparently respondent problems (habituation of new recruits to intense stimuli), operant processes (skilled killing and stimulus control of operant behavior in the form of following instructions, etc.), and treatment of war neuroses (through respondent extinction, etc.). It does also raise the issue of how to do this most efficiently and the ethical question of whether or not we should do it under what circumstances.

14.4 Conclusions

Societies take a broad range of actions to change the overall level of aggression and violence in a number of intuitive ways, some of which may sometimes be quite effective. Principally, these efforts are to reduce aggression and violence and deliver justice to the victims of widespread aggression and violence by designing a better society. Occasionally, societies promote aggression and violence at times of war and other times; perhaps such practices promote survival of societies over time.

References

Arendt, H. (2006). *Eichmann in Jerusalem*. Penguin.

Bourke, J. (2018). Psychiatry, hate training, and the Second World War. *Journal of Social History, 52*(1), 101–120.

Braithwaite, J. (2003). Principles of restorative justice. In A. von Hirsch, J. V. Roberts, A. E. Bottoms, K. Roach, & M. Schiff (Eds.), *Restorative justice and criminal justice: Competing or reconcilable paradigms?* (pp. 1–20). Hart Publishing.

Bueno, I., & Rozas, A. D. (2013). Which approach to justice in Colombia under the era of the ICC. *International Criminal Law Review, 13*(1), 211–247.

Chenoweth, E., & Cunningham, K. G. (2013). Understanding nonviolent resistance: An introduction. *Journal of Peace Research, 50*(3), 271–276.

Diamond, J. (2011). *Collapse: How societies choose to fail or succeed*. Penguin.

Diener, P. (1974). Ecology or evolution? The Hutterite case. *American Ethnologist, 1*(4), 601–618.

Eriksson, A. (2008). Challenging cultures of violence through community restorative justice in Northern Ireland. *Sociology of Crime, Law and Deviance, 11*(2), 231–260.

Eriksson, A. (2015). Restorative justice in the Northern Ireland transition. In A. M. McAlinden & C. Dwyer (Eds.), *Criminal justice transition: The Northern Ireland context* (pp. 341–361). Hart Publishing, Bloomsbury Publishing.

Friesen, J. (2016). *The Palgrave companion to North American Utopias*. Springer.

Gibson, J. L. (2005). The truth about truth and reconciliation in South Africa. *International Political Science Review, 26*(4), 341–361.

Gibson, J. L. (2006). The contributions of truth to reconciliation: Lessons from South Africa. *Journal of Conflict Resolution, 50*(3), 409–432.

Hayden, D. (1976). *Seven American Utopias: The architecture of communitarian socialism, 1790-1975*. The MIT Press.

Heilman, J. (2018). Truth and Reconciliation Commission of Peru. In *Oxford Research Encyclopedia of Latin American History*. Retrieved August 18, 2021, from https://oxfordre.com/latinamericanhistory

Hostetler, J. A. (1983). *Amish life Mennonite life Hutterite life*. Herald Press.

Hostetler, J. A. (1997). *Hutterite society*. Johns Hopkins University Press.

Karstedt, S. (2010). From absence to presence, From silence to voice: Victims in international and transitional justice since the Nuremberg trials. *International Review of Victimology, 17*(1), 9–30. https://doi.org/10.1177/026975801001700102

Katz, J., & Lehr, J. (2019). The digital revolution and the Hutterite community: The rules and reality. *Prairie Perspectives: Geographical Essays, 21*, 9–15.

Kephart, W. M. (1963). Experimental family organization: An historico-cultural report on the Oneida community. *Marriage and Family Living, 25*(3), 261–271.

Llewellyn, R. (1939). *How green was my valley*. M. Joseph.

Lynch, R. (2008). The people's protectors? The Irish Republican Army and the "Belfast Pogrom," 1920–1922. *Journal of British Studies, 47*(2), 375–391. https://doi.org/10.1086/526757

Marshall, T. F. (1999). *Restorative justice: An overview*. Home Office.

Ness, D. W. V. (2004). Justice that restores: From impersonal to personal justice. *Journal of Religion & Spirituality in Social Work: Social Thought, 23*(1–2), 93–109.

Olin, S. C., Jr. (1980). The Oneida Community and the instability of charismatic authority. *The Journal of American History, 67*, 285–300.

Schludermann, S., & Schludermann, E. (1971a). Adolescent perception of parent behavior (CRPBI) in Hutterite communal society. *The Journal of Psychology, 79*(1), 29–39.

Schludermann, S., & Schludermann, E. (1971b). Maternal child rearing attitudes in Hutterite communal society. *The Journal of Psychology, 79*(2), 169–177.

Schludermann, S., & Schludermann, E. (1971c). Adolescents' perception of themselves and adults in Hutterite communal society. *The Journal of Psychology, 78*(1), 39–48.

Schludermann, S., & Schludermann, E. (1971d). Paternal attitudes in Hutterite communal society. *The Journal of Psychology, 79*(1), 41–48.

Skelton, A. (2002). Restorative justice as a framework for juvenile justice reform: A South African perspective. *British Journal of Criminology, 42*(3), 496–513.

Stephan, M. J., & Chenoweth, E. (2008). Why civil resistance works: The strategic logic of non-violent conflict. *International Security, 33*(1), 7–44.

The Truth and Reconciliation Reports of Canada. (2015). Retrieved January 4, 2022, from https://nctr.ca/records/reports/

Chapter 15
Cultural Interventions to Change Violence and Aggression. III. Intuitive Focused Approaches

Unlike the preceding two chapters on behavioral approaches to cultural design, this chapter will review some examples of more typical approaches to cultural intervention. There are many such interventions so this chapter does not pretend to be comprehensive, rather it will select some of the more commonly used and researched interventions. Reflecting the possibility that starting early to prevent or treat aggression and violence, many programs attempt to divert the developmental course toward aggression and violence by intervening very early through various forms of parental support, including parent training, and through intervention in schools. Prevention and reduction of violent crime is another common approach. Since we know that violent crime is specific to time and place, it is interesting to ask what it is about these contexts that contribute to aggression and violence and if something about the environment can be modified to reduce the risk of violence. Thus, crime prevention through environmental design attempts to reduce violent crime through environmental modification, such as improving lighting and public cameras. A second approach to reducing violent crime is the use of evidence-based policing. Governments have also attempted to reduce aggression and violence through procedures such as modifying media, including pornography, bystander training, and restorative justice. Finally, governments often implement national policies to address common forms of violence such as intimate partner violence (IPV) and bullying.

15.1 Child Rearing Practices

Support for families raising children is a universal concern for many child outcomes. With respect to children becoming aggressive, there are many parenting variables which correlate with subsequent child aggression. Thus, interventions to prevent or reduce current and future child aggression and violence is a major

P. Sturmey, *Violence and Aggression*, Advances in Preventing and Treating Violence and Aggression, https://doi.org/10.1007/978-3-031-04386-4_15

endeavor for all societies which is reflected in a very large parent training literature on this topic. Studies have addressed both parental violence to their children and the development of child aggression and violence and some universal parent training programs may address one or both of these outcomes. When considering this literature it is important to note that these studies take place in widely differing contexts including: (a) primary, secondary, and tertiary prevention; (b) many different kinds of families with different personal, social, and economic circumstances; and (c) different cultural contexts such as high versus low/middle income countries, etc. Thus, it is important to consider whether the conclusions of research on parent training in one context will generalize to other contexts. A further important consideration is that many universal parent training programs may address child aggression and violence indirectly, for example, by promoting prosocial behavior or broadly defined child development rather than reduction of child aggression and violence specifically.

One example of a systematic review of parent training interventions relating to aggression and violence comes from Altafim and Linhares (2016) who searched the parent training literature published between 2008 and 2014. They found 23 empirical studies of 16 programs, 91% of which were conducted in high income countries, 52% in the USA, and only two from middle/low income countries. The most commonly evaluated programs were *Adults and Children Together—Raising Safe Kids* (7 studies) and Positive Parent Program (4 studies). There were six randomized controlled trials (RCTs) and only one study compared more than one program. Commonly addressed parental outcomes included parenting practices, stress, beliefs, behavior, and mental health, including anger; and commonly addressed child outcomes typically addressed child behavior. An example of typical outcomes comes from evaluations of *Adults and Children Together*, the program with the most research. Altafim et al. reported that studies of *Adults and Children Together* had found reductions in parental harsh and physical discipline, spanking, using objects against children, aggression towards children. Studies of *Adults and Children Together* also reported reduction in child problem behavior. Interestingly, one *Adults and Children Together* study found *Adults and Children Together* resulted in *increased* parental stress, perhaps because of attending to the challenges of parenting.

The importance of parent training was underlined by an extensive meta-analysis of child maltreatment interventions by van der Put et al. (2018). Based on 121 independent studies with 352 effect sizes, they found a moderate effect size of $d = 0.26$ and $d = 0.36$ (p's <0.001) for preventative and curative interventions, respectively. Importantly, parent training consistently had somewhat larger effect sizes. In multivariate analyses, parent training, along with substance abuse interventions and increasing parental confidence predicted outcomes for preventative strategies and for curative strategies, parenting strategies, and emotional support predicted outcome. Thus, parent skills training is one of the more robust approaches to reducing both parental use of aggressive disciplinary practices and reducing child behavior problems including aggression.

Until recently, most studies of parent training to reduce violence in the family were from high income countries. Reflecting the global nature of family violence, this has now begun to change (Shenderovich et al., 2021) with increasing numbers of studies from middle and low income countries, especially in sub-Saharan Africa (Pundir et al., 2020). For example, McCoy et al. (2020) conducted a meta-analysis of 11 studies of parent training in East and South-east Asian countries and found moderate evidence of reduction of abusive, harsh, and negative parenting ($d = -0.42$, $p < 0.01$).

Parent training is an extensively evaluated intervention for many family problems including aggression and violence in both parents and children. It is also quite flexible in that it has been applied to many different contexts and, with appropriate local modifications, is likely to be an effective intervention in many contexts.

15.2 Schools and Other Educational Settings

Schools and the education system generally are not only places where children learn academic skills. Rather, they should be considered a social agency with a broad range of goals and programs. They are often the place where non-educational programs such as nutritional programs, programs for parent support, surveillance and prevention of child abuse and neglect, mental health services for students, and a wide range of broader community interventions are delivered.

Aggression and violence in schools is relatively common and a widely recognized problem that many feel is amenable to change with potentially broad, long-term benefits to both individuals and society. Common school-based forms of aggression and violence include bullying, cyberbullying, verbal and other threats and physical violence, dating violence among teens, and non-compliance, whereas gun violence is relatively rare. Many authors now also include corporal punishment as a form of teacher-to-student violence.

Violence, however, is usually directed from students to other students. Less frequently violence can occur from students and family members to teachers and other staff. Violence from teachers and staff to students and other staff also occurs, but less frequently than student-to-student violence. Yet, too commonly, schools' response to violence is reactive, punitive, and tends to blame the bad students and their bad family rather than consider the school context of the student's aggressive and violent behavior. Thus, schools often use threats, punishment, and exclusion of individual students after aggression has occurred in the absence of effective prevention and training. Although prevention is commonly acknowledged, adoption of effective preventative strategies is less common and hence there are many missed opportunities with potential benefits for all.

Violence in schools is common, costly, and disruptive to the educational goals of schools and students; it inhibits attainment of important educational and social goals. It may also have long-term negative student consequences and considerable direct and indirect economic costs. Polanin et al. (2021), in a systematic review and meta-analysis of 114 independent studies investigated the effects of school violence on long-term student outcomes. They found that, after controlling for other factors, school violence still predicted long-term outcomes aggregated across student mental health, school performance, and involvement in crime and delinquency. Although the effect appeared to be of quite small magnitude (partial correlation = 0.06), the authors concluded that was still meaningful. In a systematic review and meta-analysis of international studies, Fry et al. (2018) found that violence, including school violence, results in dropout and failure to graduate from school which themselves will also have long-term negative impact on personal functioning. Thus, a common aspect of designing a better world is promoting non-violence in schools and reducing aggression and violence.

15.2.1 Meta-Analyses of School-Based Interventions

Given the pervasiveness of school violence there are many general approaches and specific programs available to address this issue some of which have received considerable attention from research. Thus, there have been a large number of RCTs (Fraguas et al., 2021; Jiménez-Barbero et al., 2016; Mytton et al., 2002; Ng et al., 2022; Park-Higgerson et al., 2008; Scheckner et al., 2002; Ting, 2009; Wilson et al., 2001, 2003). Indeed, there have been over 35 meta-analyses of this literature between 2005 and 2015 alone (Lester et al., 2017)! A review of systematic reviews of violence in the schools by Lester et al. (2017) identified 36 systematic reviews of four types of school-based violence: (a) peer violence; (b) corporal punishment; (c) student-on-teacher violence; and (d) teacher-on student violence. Lester et al. found systematic reviews for only two of these domains, mostly peer to peer violence and IPV. They located five systematic reviews of IPV and identified 11 programs that had been evaluated. Ten had been evaluated in the USA and one in South Africa. Four programs were deemed promising in that they had one RCT supporting their use. These programs were typically short-term with approximately 10–21 sessions being delivered over around 7 weeks. When follow-up data were available they were generally effective up to 3 years. No IPV programs, however, had been shown to be robustly effective due to the lack of replication. Two programs were potentially harmful in that they were associated with *increased* in violence! There was rather more research on treatment of peer-to-peer violence with 31 reviews available. Most studies were conducted in the USA on universal strategies. Most were conducted in primary schools and used around 16 sessions. Most interventions were judged to be effective. In terms of the types of interventions the authors noted that "Socio-emotional programmes have been found to be one of the most promising approaches, while cognitive behavioral and peer mentoring/mediation interventions have also fairly consistently demonstrated positive results" (p. 206). Finally, the authors noted that there were no systematic reviews on teach-on-student violence and student-on teacher violence including corporal punishment.

Subsequently, Heekes et al. (2022) reported a systematic review of corporal punishment. They found that corporal punishment was commonly used in schools across all regions of the world, especially with non-White students and boys. Corporal punishment was associated with many student harms including physical injuries, such as bruising, poor academic performance and emotional and behavioral problems, although causal relationships for the latter were difficult to establish. There were surprisingly few prevention and interventions studies and those that were identified tended to focus on teacher training in alternative strategies to corporal punishment. For example, Devries et al. (2017) evaluated a package of training teachers in positive behavioral interventions in a RCT conducted in Uganda with 42 schools. They found large reductions in teacher corporal punishment as reported by both students and teachers. In another large-scale RCT, Nkuba et al. (2018) reported similar large reductions in teacher violence toward students using a teacher training approach.

15.2.2 School-Wide Positive Behavior Support

A recently popular approach to school-wide prevention of violence is school-wide positive behavior support (SWPBS). The goals of positive behavior support are to prevent future violence and reduce existing violence. This approach to reducing school violence is very widespread. Gage et al. (2018), for example, reported that approximately 23,000 schools implemented SWPBS nationally and internationally. SWPBS interventions include clear expectations, rules, skills training, and interventions based on functional assessment. There are three levels of interventions. Tier one interventions are universal interventions for all children. Tier two interventions are skills training interventions targeting some children. Tier three interventions are specific intensive interventions for only a few students. The approach is data-driven and uses a team approach. SWPBS involves behavior theory and applied behavior analysis, a focus on prevention, an instructional focus in which teachers apply behavioral prevention and reactive strategies, evidence-based practices, and a systems approach (Sugai & Horner, 2009). SWPBS is quite popular as it contains several elements that are very acceptable such as focus on prevention, positive reinforcement, etc.

There have been at least two meta-analyses of SWPBS. In a recent meta-analysis, Solomon et al. (2012) identified 16 studies published over a 20 year period. They reported a moderate, positive effect size. Effect sizes were somewhat larger in non-classroom than classroom settings, quite varied in magnitude from study to study, and low-average for problem behavior. Hence, the authors described these outcomes as "promising."

There are few true experimental studies of SWPBS. Fortunately, Waasdorp et al. (2012) reported a large-scale RCT in which 37 schools in Maryland with 12,334 children (53% male, 45% African American) participated. They randomized schools to SWPBS or no SWPBS over a 4-year period. The outcome measures were teacher-reported bullying-related behavior and social rejection. There were statistically significant improvements in both measures and some evidence that implementing SWPBS earlier was associated with larger effect sizes. Such large-scale RCTs have limitations such as lack of data on treatment integrity and use of teacher reports as the dependent variable.

Recently, Gage et al. (2018) conducted a systematic review and meta-analysis of group designs evaluating the effects of SWPBS on school exclusion, a student outcome that is often related to school violence. They found four studies, two of which were RCTs and two quasi-experimental designs. There were 90 schools. Two studies implemented only tier one interventions and two studies implemented all three tiers. The dependent variable was combined disciplinary exclusions and suspensions. Although the overall effect size was $g = 0.61$ for disciplinary exclusions it was not statistically significant, but there was a large and statistically significant effect on school suspensions ($g = 0.86$) which was equivalent to a 16% reduction in school suspensions.

SWPBS appears to be a promising and acceptable approach to reducing violence school-wide. The outcomes, especially from the meta-analysis by Gage et al. (2018) are promising. The literature is limited due to the insufficient number of true experimental evaluations, the use of indirect measures of school violence such as school suspensions, and the common issues such as clearly defining what constitutes SWPBS and the related issue of treatment integrity and program maintenance.

15.2.3 Student Violence Against Teachers

Student-on-teacher violence ranges from bullying, threats, property destruction, such as vandalism against teacher property, and assault, including sexual assault. Student-on-teacher violence may be related to teacher mental health, injury, calling in sick, reduced participation, and leaving the profession (Espelage et al., 2013). In a systematic review of the epidemiological literature, Longobardi et al. (2019) reported that approximately 53% of teachers (range 20–75%) report violence from students over the preceding 2 years. The most common forms of violence were as follows: "obscene gestures (44%), offensive/obscene remarks (29%), verbal violence (29%), damage or theft of personal property (17%), intimidation (10%), physical attacks (3%), and sexual violence (3%)" (p. 606).

Despite student-on-teacher violence being a well-recognized problem, surprisingly little is known about its prevention or treatment. A good step in this direction was a couple of studies of functional assessment of violence toward teachers. In the first, Wehby et al. (1995) conducted an observational study of antecedents and consequences of aggression in a sample of 23 elementary school children with emotional and behavioral difficulties including aggression toward teachers. They found that teacher social commands, teacher removal of children's property, and teachers' instructional commands preceded around 40% of incidents of student aggression toward teachers. Common teacher delivered consequences for student aggression toward teachers included interaction with teacher, teacher commands, and negative physical interaction from teacher to child (e.g., grabbing student materials).

In the second study, McMahon et al. (2020) reported a larger scale study of both primary and secondary teachers using an anonymous online survey of 193 teachers. The authors used open-ended questions and a standardized coding protocol to conduct a functional assessment of violence toward teachers. They identified seven common patterns of antecedents, behavior, and consequences. In the first three examples, the antecedent was breaking up a fight and the consequences included police/legal involvement, involvement of other school staff, and student removal. The next three common examples included the antecedent of disciplining a student which was followed by inaction, student removal, or a positive outcome. Finally, the seventh common antecedent was a teacher directive which was followed by inaction.

These results are useful in that they point toward two important areas of intervention to prevent student violence toward teachers. First, there should be better and more effective training in better prevention and interaction skills related to breaking up fights, discipline, and directives. Second, teachers often need better support from school administrators, for example, in the consistent and fair implementation of school disciplinary problems when students are violent toward teachers and in related policies such as cell phone use, etc. These last two studies are a good start to understanding violence toward teachers in mainstream education and can form the basis for future interventions to prevent and reduce this serious problem.

15.2.4 Summary

Violence in schools is quite common and an important part of the process by which people become delinquents and violent criminals take place in schools, hence, there is an enormous literature on prevention and treatment of violence in school settings. Prevention and intervention in schools also have the potential to impact the problem of violence at an early age and so have the potential for long lasting effects on both prosocial behavior and violence itself. Fortunately, there is now a large evidence-base that has identified interventions that are most likely to work for common forms of violence and to guide schools as to what to do. In addition, there are specific programs and triage-type frameworks such as SWPBS that are palatable, efficient, and likely to be helpful.

15.3 Evidence-Based Policing to Reduce Violence

Governments spend billions of dollars in attempts to reduce violent crime, but often such initiatives are of unknown effectiveness and influenced by vivid but unsystematic anecdotal evidence of unknown quality. They are often based on intuitively designed programs that appeal to "common sense." The choice of programs is influenced by political and financial considerations and vested interests, rather than consideration of program effectiveness.

This section looks at empirical approaches to evaluating policing practices that might be effective in reducing violent crime and keeping the public safe and studies that have addressed reducing police violence. Evidence-based policing is an ever growing field. For example, McGuire et al. (2021) presented multiple reviews of evidence-based policing and violence including managing violence in the night-time economy (see below), managing sex offenders in community settings, police actions relating to IPV, prevention of gang membership, etc.

15.3.1 Overviews

Attempts to reduce gun violence include educational approaches, such as information, training, and storage campaigns; gun buyback programs; gun laws; and law enforcement campaigns. Makarios and Pratt (2012) conducted a systematic review and meta-analysis of such programs. They found 29 studies with an overall very weak effect size ($r = -0.144$, $p < 0.05$) with little evidence that gun buyback programs or gun laws were effective. A recent systematic review of legislative efforts to reduce gun crime in Australia (McPhedran, 2016) also found no evidence that legislation reduced gun-related homicide. The only types of intervention that Makarios and Pratt (2012) found were effective were various systematic law enforcement campaigns and programs. These programs included hotspot policing; some gun court programs that speeded up processing of offenders; multi-component packages that included education, enhanced supervision with professionals having the power to tighten or relax probation conditions such as curfews, and area restrictions; and multi-component community partnership programs.

15.3.2 Hotspot Policing and Violent Crime

The field of situational crime prevention (Clarke, 1995, 2009) focuses on the environment rather than the criminal to make crime less attractive and more difficult and to make prosocial behavior easier to engage in. Rather than focusing on punishing criminals in the criminal justice system after crimes have been committed, it focuses on the community before crimes occur and involves many other formal and informal agencies of change. Clarke identified 25 techniques organized in a matrix. The main five strategies to prevent crime include increasing effort, increasing risks, reducing rewards, reducing provocations, and removing excuses for criminal behavior. Examples of particular strategies include: surveillance cameras; making public spaces defensible, such as limiting access and egress to public housing and requiring check-ins; electronic access to cars and phones; street closure and other modifications of traffic patterns; control of access to alcohol; and training in conflict management for a variety of formal and informal monitors of public behavior, including publicans and bouncers, etc. Thus, the focus is on controlling crime and environmentally designing out opportunities for criminal behavior, rather than controlling the criminal, and the type of intervention is often highly specific to the type of crime and location.

Braga et al. (2012) reported a meta-analysis of hot spot policing. They located 19 studies of which 10 were RCTs and nine were quasi-experiments. There was a small, statistically significant reduction in community members calling the police when hot spot policing was used. An updated report by the same group yielded a similar conclusion (Braga et al., 2014).

These effects have also been observed for violent crime. For example, Rosenfeld et al. (2014) conducted a RCT of targeted policing on violent firearm offenses in 32 hotspots over a 9-month period. The authors randomized the hotspots to two treatment conditions and a control condition. In one treatment condition the officers patrolled the hot spots but limited their self-initiated activities, whereas in the other experimental condition they were encouraged to initiate activities. Self-initiated activities included checks on pedestrians, vehicles, and buildings; foot patrols; and problems solving. The number of firearm incidents decreased by about 45% in both the control and low self-initiated condition, whereas firearm incidents decreased by 64% in the self-initiated condition. The authors found little evidence of displacement of violent crime in terms of location, time of day, or type of crime; in fact, they found evidence of diffusion of benefits from experimental locations to neighboring locations.

15.3.3 Reducing Police Violence

Research has identified police violence as a common public health problem. For example, DeVylder et al. (2018) interviewed 1000 adults in Baltimore and New York concerning their experience of police violence over the preceding year. Using self-reports, they found that around one in ten had been exposed to police violence without a weapon and around one in five had been exposed to police violence with a weapon over the last year. The prevalence of violence ranged from 3% for sexual violence to 15% for neglect. Demographic risk factors of police violence were being male, a person of color, and homosexual or transgender. Police violence was associated with suicidality and psychosis in members of the public and was more likely when police violence was involved a weapon of sexual violence. Following the deaths of Travon Martin (2012), Michael Brown (2014), and Freddie Gray (2015) and many others in the USA and the rise of the Black Lives Matter movement, there has been considerable public interest in holding police accountable for violence and police reform and reallocation of police finding to other agencies (Lum et al., 2020).

Complete, accurate, and meaningful data on police homicides are difficult to come by. One attempt to do so comes from Schwartz and Jahn (2020) who analyzed data from the US Fatal Encounters database, a citizen database that aggregates and verifies police homicides reported in the media. They analyzed data from 2013 to 2017 broken down by Metropolitan Statistical Areas (MSAs), which roughly correspond to individual US urban areas. To estimate police homicides they excluded deaths due to suicides, drug overdoses, drownings, falling from a height, vehicular collisions, etc. This resulted in a database of 5,494 police homicides. The main causes were gunshots (94%); tasers (3.4%); 1.7% asphyxiation, pepper spray or bludgeoning; and 0.7% from other causes. The overall rate was 0.39/100,000. When calculating the rates of homicide they found a ninefold difference between geographical areas with rates ranging from 0.13 to 1.117/100,000 with highest rates of homicides generally being located in the Southwest and Central regions of the USA and lowest rates in the Northeast. The authors then calculated the racial inequalities overall and by regions. The MSA-specific rates for Black Americans ranged from 0.43 to 2.10/100,000 (a ratio of maximum to minimum of 4.9)—MSA-specific rates for White people ranged from 0.08 to 1.10/100,000 (a maximum/minimum ratio of nearly 14). Finally, for Latinos the MSA-specific rates ranged from 0.11 to 1.27 (a maximum/minimum ratio of more than 12). Thus, Black Americans were much more likely to be killed by police than Latinos and Whites. In contrast with ratio data, racial disparities for Black Americans were largest in the north and northeast and in the west and for Latinos in the West and Northeastern regions of the USA. Schwartz and Jahn conclude that there were large differences in geographical and ethnicity but the explanations for these differences were unclear and could be answered by these data.

This paper provoked a response from Nix (2020). While acknowledging the noteworthy findings, Nix made three critiques of Schwatz and Jahn's study (2020). First, Nix noted that the database only reported shootings resulting in fatalities and did not distinguish between justified, legal homicide, and illegal police homicide. Second, the homicide data was biased in that it did not report near-death events from police shootings, some of which may have included attempted murder by the police or the effects of the availability of emergency health services in different geographical locations. Finally, Nix noted that the data are difficult to interpret because the differences in exposure to police are unknown and not accounted for in this analysis. For example, different geographical areas may have the same rate of police homicides, but if the same geographical areas differed in the number of incidents, the rates of homicide per police encounter might be quite different. Nix concluded that if the geographical and racial differences are real—which he seems to acknowledge since they are so large—the reasons behind them are not. These differences might reflect many factors, including: (a) differences in police behavior, trainings, degrees of exposure of member of the public to the police, and policies between MSAs; (b) differences in citizen behavior between MSAs; (c) different access to emergency services; and (d) regional differences in laws related to firearms, access to firearms, degrees of segregation, poverty, and levels of income disparity. At this time we do not have good data to explain these large regional and racial differences in police homicide.

15.3.3.1 Dashboard and Other Cameras

Dashboard cameras to record police–public encounters during traffic stops have been long used in the USA. One recent approach to reducing police violence, however, has been to require the police to wear body-worn cameras (BWCs). This approach may have several potential benefits including reliable collection of evidence; increased rates of arrests, citations, and convictions; reducing police violence; increased police civility; and protecting the police themselves from false accusations of violence.

The intense interest in BWC has been driven by public concern in reducing police violence, homicide, and murder, especially against people of color and thus BWCs have become widely implemented. Nevertheless, questions concerning effectiveness and costs of this approach remain only partially answered despite widespread implementation before these methods were evaluated. Public concern has also focused on police turning off BWCs, reluctance to release video recordings, and police destroying evidence.

There have in fact been several true RCTs and quasi-experimental evaluations of BWCs such that Lum et al. (2020) reported a Campbell systematic review and synthesis of the evidence of the effects of BWCs. Their search located 30 empirical studies, mostly from the USA and UK, which generated 116 effect sizes on 12 dependent variables. Lum et al. grouped them into effects on officers' behavior, such as reduced violence, increased arrests, complaints against officers, officer use of violence, and on citizens' behavior, including citizen assaults on officers.

Results were mixed and effect sizes were modest. There were only two statistically significant effects. First, BWCs resulted in an approximately 17% reduction in citizen complaints against officers. Second, if officers had less discretion in turning BWCs off there was less officer use of force. These outcomes, although statistically significant, were of quite varied magnitude. Other potential effects of BWCs, such as increased officer civility, reduction in citizen violence to officers, reductions in officer preventative work were all non-significant.

In contrast to other recent reviews of BWCs, Lum et al. (2020) were quite cautious in their conclusions and emphasized that the outcomes were modest, varied, and in many cases absent. They did conclude, however, that police agencies should temper their expectations of the impact of BWCs on the behavior of police and citizens but should consider restricting officer discretion in turning cameras off.

An interesting and unanswered question is if BWCs do work, how do they work? BWCs may function as discriminative stimuli that both promote prosocial behavior in both police officers and citizens alike and may also signal the likelihood of punishment for officer violent behavior. Lum et al. hinted at this when they described how fellow officers and citizens might remind officers with BWCs of the possible consequences of violent behavior. Equally possible is that BWCs might inhibit provocative and aggressive citizen behavior and increase compliance with officer commands. These latter possibilities raise measurement issues in evaluating BWCs. Namely, their effects can only be meaningfully evaluated by considering the probabilities of both parties' behaviors and the contingencies that exist between officer and citizen behavior. For example, BWCs might work by both increasing citizen compliance and increasing officer civility and compliance with established non-violent procedures while the probability of police violence, given citizen's aggression remains the same. Finally, BWCs provide excellent opportunities for officer training both real world models of how to respond to citizens in various contexts (complaint, drunk, high, threatening, aggressive, etc.) and as a potential feedback mechanism to train officers to be more effective and less violent.

15.3.3.2 Replacing Police Services

As part of the response to recent police violence in the US, there has been a movement to defund some or all of police services, such as responding to mental health crises and community youth services. The rationale given for this course of action is that many deaths and violent incidents at the hands of the police are related to individuals in mental health and behavioral crises and that the police are poorly trained or untrained to recognize and manage these situations effectively and non-violently (Pollack & Humphreys, 2020).

Obviously such proposals are highly politically charged. There are, however, some examples of discussion papers and reviews of relevant evidence-based practices (Pollack & Humphreys, 2020). Unfortunately, there are few empirical reports demonstrating reduction in deaths that might support and direct such a public policy. A much cited report concerning the Crisis Assistance Helping Out on the Streets (CAHOOTS) service in Eugene, Oregon was founded in 1989. This report (Dyches et al., 2002) claimed to divert around 17% of police calls to its services, thereby saving up to $14 M per year from not using ambulance and emergency room services, etc. The authors did not report any direct data on reduction in homicides by police, but it seems to be implied that reducing contact between the police and the public during mental health crises may do so, thereby reducing the likelihood of police violence and homicide. The report also claimed that several other police forces have adopted this model but as of this time, this author was unable to locate peer-reviewed journal articles on this program.

15.3.4 Comment

There is an extensive and growing literature on evidence-based policing (McGuire et al., 2021). Some of this work has given empirical support to certain police practices that usually use data to identify the physical and temporal distribution of violence crime and the use of experimental methods to evaluate whether a program has been implemented as planned and its immediate and long-term effects including possible spatial and temporal displacement of violent crime.

15.4 National Prevention Programs

Many countries have national violence prevention programs which address common forms of violence such as IPV, alcohol-related violence, violence during teenage years, adoption of evidence-based policing, etc. These programs often include elements such as large-scale community education, promoting positive and respectful relationships, involvement and mobilization of community organizations, and government funding for research and development. In this section we will illustrate this approach to prevention and treatment of aggression and violence on the national scale with one example, Finland's KiVa anti-bullying program. Anti-bullying programs have an extensive evidence base allowing us to evaluate their effectiveness. For example, Fraguas et al. (2021) conducted a systematic review and meta-analysis and found 77 RCTs. Based on 69 independent RCTS, the effect sizes were −0.150 (95% CI, −0.191 to −0.109) for bullying and −0.205 (95% CI, −0.277 to −0.133) for improving mental health. Thus, anti-bullying programs seem a good choice for dissemination, as they have a large evidence base.

The average effect sizes reported by Fraguas et al. (2021) are relatively modest. The widespread nature of bullying and its high personal and economic costs indicate the possibility of widespread benefits and cost savings. The KiVa program was selected to illustrate national policy since it is relatively well documented, has some evaluation data from both the original authors and authors from other countries to support its effectiveness, and has reported data on cost-effectiveness.

Noting that Finland is a wealthy country with good academic outcomes from schools, but with relatively high student dissatisfaction with school, Finland passed a law in 1998 requiring stating that all students had a right to a safe school. In 2003, the law was amended to require all education providers to have a plan to reduce school bullying, implement the plan, and ensure its adherence. In subsequent years school bullying remained unchanged, despite these legislative efforts to reduce it. In response, the University of Turku developed the KiVa anti-bullying program, based on the notion that peer behavior was key in maintaining bullying. That is, when bullying occurs, peers passively or actively encourage it by saying nothing, sniggering, cheering, perhaps encouraging it, and even participating in bullying. Importantly, victims and peers and others do not provide sufficient effective negative consequences for the bully's behavior. Thus, the KiVa program included universal actions for all students and indicated actions for both the victim and bully. The program was delivered by lessons from teachers during regular classroom time using a manualized program. The program included: (a) skills practice; (b) school-wide components such as parents' guide, posters, visible reminders (such as staff vests), and an annual survey; and (c) reviews of bullying incidents including peer support from a high status peer (Herkama et al., 2017; Sainio et al., 2019). KiVa also addresses cyberbullying (Herkama & Salmivalli, 2018).

In order to evaluate KiVa, Kärnä et al. (2011a) conducted an RCT in which 78 schools with 8,237 children from grades 4 to 6 (ages 10–12 years) were randomized to the KiVa program or a no treatment control condition. Data were collected at the middle and the end of the school year. Dependent variables were self- and other-report of victimization, self-report of bullying, and attitudinal measures. There were positive effects on seven of 11 outcomes. For example, the rates of self-reported victimization fell by 46% and the rate of self-reported bullying fell by 61% in the KiVa schools. The odds ratios of victimization and bullying were approximately 1.5–1.8 and 1.2–1.3 in the control group compared to the KiVa group. The authors concluded that "On the whole, the results give clear support to the effectiveness of the KiVa program" (p. 323).

In a subsequent evaluation of a large-scale rollout in 888 schools with approximately 150,000 students, Kärnä et al. (2011b) found odds ratios of 1.22 and 1.18 for reduction of victimization and bullying, respectively. They estimated the program resulted in 7,500 fewer bullies and 12,500 fewer victims in Finnish schools. Similar positive findings have been reported in a replication RCT study in the Netherlands (Huitsing et al., 2020) and a pilot study in Wales (Hutchings et al., 2015), although the full scale Welsh study (Axford et al., 2020) did not find significant effects of KiVa (see below for further discussion of these findings). A cost–benefit analysis from the Dutch group using conservative estimates (Huitsing et al., 2019) found that the KiVa program returned approximately 4–7 times the financial investment to society.

The Finnish national program is impressive in several ways. The KiVa program addresses a common and expensive form of violence that is preventable, it is based on a plausible model of bullying, and the program is operationalized and rationally linked to the model it is founded on. Further, the Finnish government has been able to implement it on a national scale for many years and there are student outcome and economic data to evaluate its effectiveness and value.

There are four main weaknesses of the literature on KiVA. First, the use of self-report data without confirming its accuracy is a potential problem as it is possible that KiVa might only change the way people talk, rather than how they behave with one another. Second, is the issue of maintaining treatment integrity during large-scale, long-term implementation. For example, in a Welsh replication RCT, KiVa was not found to be effective, but the quantity of lessons was less than the original study and implementation was highly variable across schools (Axford et al., 2020). Similarly, Johander et al. (2021) reported that in KiVA programs in Finland from 2009 to 2015, program drift, which occurred when some schools began to add their own non-evidence-based practices to the core KiVA program, resulted in poorer outcomes than schools using the KiVa program with integrity as evaluated. Third, some have hypothesized that the KiVa program might have unintended negative effects on children, such as the use of peers to defend victims night result in disempowerment of victims, might reinforce bullying, and might erode broader peer support for victims (Healy, 2020), although these possibilities await empirical investigation. Finally, the program model, although broadly environmental, does not explicitly consider the functions of bullying and the antecedents and contingencies of reinforcement for the bully, victim, and peers, and other relevant parties such as school staff and family members. Future iterations of KiVA might benefit from incorporating explicit consideration of functional assessment and analysis of incidents of bullying.

15.5 Crime Prevention Through Environmental Design

Partly in response to the observation that crime is so localized in time and space, environmental planners and government agencies have attempted to discourage crime by redesigning Crime Prevention through Environmental Design (CPED; Jeffrey, 1975). There are a wide range of applications of CPED ranging from redesigning environments in parks (McCormick & Holland, 2015), railway stations (Cozens et al., 2003), and neighborhoods. CPED has many approaches including using surveillance of the public using street cameras; improved street lighting; placing street barriers to restrict movement and making it more difficult to engage in crime; increasing visible surveillance, such as security guards, having place managers, such as bus guards and attendants present; adding fencing and barriers, and pruning trees, etc. to make spaces visible, etc.

People have speculated that such interventions might work either because they make crime more difficult to commit or because they enhance informal social control and a "sense of community cohesion" (Welsh et al., 2015). These practices are now commonplace and widely accepted in many cities. So, do these strategies reduce violence crime, which ones work best, and are they worth the effort and money?

15.5.1 Street Lighting

Several studies and systematic reviews have evaluated the effects of increasing lighting in public places on aggression and violence. Fewer studies have also shown the effects of turning public street lights *off* on aggression and violence with sometimes counterintuitive results.

In an early review, Farrington and Welsh (2002) conducted a systematic review for the British Home Office of American and British studies on the effects of improving street lighting on crime. The authors noted two theories that might explain why increased street lighting might reduce crime. The first is a simple two-part explanation, which is that increased street lighting permits increased effective surveillance by increasing visibility of criminals and crime and increasing the numbers of people on the street which deters crime. Hence, this strategy should result in reductions in crime primarily at night. The second explanation is that increased public street lights is a signal of increased community investment, pride, and cohesion and an increase in informal surveillance. If this is so, increased street lighting should be associated with decreases in crime during both night and day.

The effects of street lighting in eight American and five British studies were associated with an approximately 20% reduction in crime with the effect size being smaller in the US studies (7% reduction) than British studies (approximately 30% reduction in crime). The effects were about the same for both day- and night-time crime. Effects were found for both violent and non-violent crimes. Some of the studies reviewed conducted economic evaluations of the treatment and concluded that street lighting was also worthwhile economically with the modest costs of street lighting being readily offset by avoidance of the much larger costs of crime. The authors recommended the use of street lighting as an economically and relatively easy approach to crime reduction and as one element of a more comprehensive community crime prevention strategy.

A unique study of the effects of street lighting on crime comes from Chalfin et al. (2021); what makes this study unique is that it was a true experiment. Chalfin et al. randomized subsample of 40 public housing developments from a total population of 340 public housing sites to a lighting intervention or routine lighting. The intervention consisted of two mobile, prominent diesel-powered light towers which emitted a slight diesel odor. For pragmatic reasons, two towers were assigned to each location regardless of the size of the housing project resulting in somewhat different lighting densities. Felony assaults and robbery at night were selected and the primary outcomes variable because they were the most serious crimes. Lighting resulted in an approximately 12% reduction in serious night-time outdoor violent crimes. The authors noted that the light towers used were very bright and focused and unlike typical street lighting and hence these results may not be generalizable to other lighting systems.

While most studies have evaluated the effects of turning street lights on, some have evaluated the effects of turning street lights *off*. Davies and Farrington (2020) reported an empirical study of a naturally occurring change when lights were switched off in streets in Madon, Essex and data from a contrast town, Braintree, Essex. Using police recorded crime, they found that following turning the lights off in Braintree burglary and vehicular crime increased, but criminal damage did not change. Surprisingly, violence in Madon decreased, perhaps because fewer people went out at night when it was dark.

Taken together, these studies suggest that increasing public street lighting is a potentially quite effective treatment to reduce public violence at night and may have other benefits too. It is relatively easy to implement and may be cheaper than other options such as public cameras (see below). Future research should evaluate the possibility of displacement of crime, the practical dissemination and widespread implementation of this approach, diffusion of benefits, and further economic analyses of street lighting. Other potentially important questions include features of “street lighting,” such as the brightness and intensity of the lights, the density of lighting devices, and other features of lighting and the locations in which they are stationed that may impact their effectiveness.

15.5.2 Public Cameras

In recent years public places have become routinely surveilled with cameras, including monitoring cars and drivers on streets, camera monitoring pedestrians, cameras in schools, hospitals, public housing, and other workplaces, and private gated communities and other public settings. For example, in the UK, the number of public cameras has increased from 100 to four million between 1990 and 2007 and more than half of US police departments and almost all large police departments were using cameras as of 2015 (Piza et al., 2019).

Public cameras could decrease crime in several ways, such as deterring crime directly, facilitating detection of crime, enhancing arrest of criminals, releasing police from surveillance and reallocating resources to prevention, facilitating obtaining evidence, and enabling earlier guilty pleas from suspects (Robin et al., 2020). Public cameras can be used in various ways to prevent violence including passive surveillance, active surveillance, and combining public cameras with other environmental design interventions, such as lighting, warning signs, publicly visible monitors, etc. (Piza et al., 2019). This environmental approach to reducing crime is based on the situational crime prevention model in which criminal behavior is seen as a quasi-rational choice between alternatives including when and where to commit a crime. As such, it appears similar to behavioral models of choice, namely that responses are distributed predictable depending on the availability of reinforcement, response effort, and reinforcer quality between two or more choices (Herrnstein, 1974). Hence, environmental interventions to reduce crime attempt to reduce the reinforcing value of criminal behavior by reducing access to the reinforcer, make criminal behavior more effortful, and make the alternatives more attractive and easier.

Some time ago, Welsh and Farrington (2008) conducted a meta-analysis of the effectiveness of street cameras on crime reduction. They included 13 studies and found that on average street lighting reduced crime by about 21%, although the magnitude of effects was quite variable between studies with some reporting large effects and nearly half of the studies reporting no effects.

More recently, in a comprehensive update by this group, Piza et al. (2019) identified 76 studies with an average of 18 months of follow-up data. Using both overall, best case, and conservative estimates of effect sizes, public cameras were effective and resulted in an average of 13% reduction in all kinds of crime (overall OR = 1.141, $p < 0.001$). Public cameras were most effective in car parks, when used in combination with other interventions such as signage, patrols, and lighting, and for non-violent crime. Unfortunately, public cameras were ineffective in reducing violent crime. Piza et al. recommended that the police should use active surveillance combined with multiple methods and focus on prevention strategies for violent crime, since some behavioral precursors to violence, such as disorder, or drug transactions are observable and provide the opportunity for police intervention prior to violence occurring.

Today, many private businesses and residences employ private security to monitor cameras and take other actions to prevent violent crime. In an interesting analysis of a global database Welsh et al. (2015) analyzed data from 165 evaluations of public versus private security. They found that private security (OR = 1.225, 18% reduction in crime) was rather more effective than police surveillance (OR = 1.081, 7% reduction in crime). Separate data on violent crime was not reported.

This literature is now sufficiently developed that we can go beyond questions of mere effectiveness to comparisons of the costs of different approaches to environmental design to prevent crime. For example, Lawson et al. (2018) reported a simulation of the cost-effectiveness of public cameras versus street lighting. They used simulations based on data from Glasgow, a city with relatively high crime rates, using data on the costs of installation of cameras and light and running both interventions, the costs of crime, and the savings from crime prevention. Their simulations showed that, generally, street lighting was more cost-effective than public cameras when improved street lighting was based on existing structures. Further, both interventions were most cost-effective when placed in hot spots. They recommended the initial use of street lighting initially in hot spots and subsequently its use spread out to progressively larger areas. They did not recommend the use of public cameras as a cost-effective option in the control of crime.

Public cameras, especially when used actively and in combination with multiple other environmental interventions, seem effective in reducing crime, but not violent crime. Future research should evaluate ways of supplementing public cameras and alternative approaches to reducing violence.

15.5.3 Preventing Access to Hotspot Locations

Some forms of environmental crime prevention have been evaluated but have not specifically reported data on violence. Thus, although some cities and neighborhoods have used physical barriers to reduce crime by preventing access to locations where crimes are likely to be committed, such as alleys.

One example comes from Sidebottom et al. (2018) who identified six quasi-experimental studies of the effectiveness of barriers with pre-post data in both an intervention and a control location. They observed a modest reduction in crime associated with placing alley gates to prevent access to high crime locations. The studies they reviewed mostly measured burglary rather than violent crime per se and did not segregate out data on violent crimes. Future research should expand on this promising literature and evaluate the effects of this and similar interventions on violence.

15.5.4 Conclusions

These studies on environmental design and reduction in aggression and violence give cause for modest optimism. That is, some but not all forms of environmental redesigns are associated with modest reductions in aggression and violence. Even when reliable effects are found in meta-analyses, however, the effects may not be robust across locations. Additionally, there is some evidence that when CPED interventions are part of a larger package of interventions to reduce crime, that these approaches may be economically viable (Sidebottom et al., 2018). Perhaps more importantly, these literatures demonstrate that empirical approaches to evaluate the effects of environmental interventions to reduce crime are possible and hence provide models for how to do this in the future.

15.6 Media Interventions

Another aspect of our environment is the media we are exposed to and consume including the web, online games, print media, TV and radio, children's cartoons, various forms of sometimes violent pornography, and written literature. It is plausible, given the research literature on modeling and aggressive behavior (Bandura et al., 1961) that such a causal relationship might exist. Thus, it is often assumed that exposure to violent media violence increases violence, and, conversely, that exposure to prosocial models in the media might both increase prosocial behavior and attenuate aggression and violence. Thus, popular children's media such as *Sesame Street* engages in social engineering when it incorporates characters with various disabilities and promotes play between White children and non-White characters (Goldberg & Gorn, 1979), etc. by instructing and modeling prosocial behavior in situations where society might be concerned about violence.

Of course, some such as psychoanalysts have held just the opposite belief, namely that exposure to violence is cathartic and thereby a form of prevention. Even if there is an association between violent media and subsequent violent behavior, in order to conclude that violent media causes violent behavior, several alternative explanations for such an association must be ruled out. For example, first, research must show that this association occurs prospectively rather cross-sectionally. That is, research must show that current relatively non-violent individuals become more violent in the future after exposure to violent media. Second, if such an association exists, research must show that the association is not due to selection effects. That is, the association must not be due to individuals who are already prone to violence seeking out violent media. Third, much of this research relies on self-report data for both the independent and dependent variables. For example, a study might show that self-reported exposure to violent media is associated with self-reported violent behavior. To be sure about such an association, research must show that the self-reports are accurate. Finally, each of these studies takes place in a specific context, including year, culture, and population. Research needs to demonstrate convincingly that findings in one context apply to another. For example, a study in the 1970s of association with consumption of violent movies, the contents of which might be relatively mild by today's standards, with violent behavior in teenage boys may have little to do with an association between consumption of sexually explicit violent pornography in prisoners convicted of violent rape in the 2020s. Finally, when designing experimental studies the procedure in the control group must be an appropriate one, so that the control group received everything, including novelty, expectations, etc. other than the exposure to violent models. That said and with all these caveats, a number of studies, including experimental studies, have attempted to redesign public media to reduce violence.

This topic, like many others related to aggression and violence, is complicated by the politics and passionately held beliefs concerning degrading images of children and women in violent pornography and lurid examples of violence apparently directly inspired and influenced by use of violent media and copycat incidents of violence portrayed on the media.

Nevertheless, there are programs to modify media content to reduce subsequent violence. For example, Christakis et al. (2013) recruited 565 parents of children aged 3–5 years from community pediatric practices. The intervention consisted of assisting parents to substitute prosocial and educational programming for aggressive content media without reducing overall screen time. They did this by having a home visitor assess initial media use, give handouts, and set goals with the parents, and mailings and follow-up phone calls containing a program guide, DVDs sampling appropriate materials, and reiterating program goals and messages to parents. The control group received a novel intervention related to nutrition with similar follow-up and newsletter procedures. The authors reported that, based on parent reports, exposure to violent media decreased in the experimental group. Using standard psychometric instruments to measure child behavior, parents reported decreases in externalizing behavior problems, increased prosocial behavior, and non-significant decreases in internalizing behavior in the experimental group. Parents were generally positive about the program and its outcomes. There are several similar RCTs which have evaluated manipulating the content of media with generally positive results (e.g., Grossman et al., 1997; Robinson et al., 2001).

In response to the problem of violent media potentially causing increases in violence, the American Psychological Association set up a task force on media and violence to update earlier reviews and APA statements on this issue (Calvert et al., 2017). The authors conducted a systematic review and meta-analysis. They reported that 12 of 14 studies published found a relationship between exposure to video violence and aggression with effect sizes similar to previous meta-analyses. The authors concluded that "there is an effect of violent video game use on aggression." (p. 141), but they did not find a link between media violence and violent crime.

The findings of the task force were controversial with numerous criticisms. Ferguson et al. (2020) wrote that "These criticisms include but are not limited to the composition of the members of the task force, methodological issues, and conclusions drawn from the meta-analysis" (pp. 1423–1424). In response to the APAs report, Ferguson et al. (2020) attempted to replicate the APA's task force's procedures and findings. Unfortunately, this new group of authors was unable to reproduce the original analyses because of incomplete data in the studies included in the original meta-analysis, inclusion of studies that did not meet the original authors' own inclusion criteria, and exclusion of appropriate studies that met the inclusion criteria. Ferguson et al. then conducted a new meta-analysis which attempted to correct for these defects and found almost *no* relationship between content of videogames and numerous measures of aggression and violence. Ferguson and Beresin (2017) comment that despite 50 years of research on the possible link between media violence and subsequent violent behavior, and despite the position statements from several professional guilds and associations, that scholars are in fact far from unanimous on the relationship between media and violence. Indeed, they recommended that professional guilds retire their position statements and replace them with more modest practice guidelines until better research is available.

There are numerous limitations to this literature. One is the lack of convincing demonstration of the manipulation of the independent variable; in many cases the independent variable is not measured or is measured only by self-report. A second is that these studies tend to refer rather broadly to "special cognitive theory" as the basis for their intervention and often reference Bandura's early work. It might be more fruitful to take a behavior analytic perspective on this problem and construe choice of violent versus non-violent media as a concurrent schedule of reinforcement (Herrnstein, 1974). Today, where individuals can access media through multiple devices (their own phones, iPads, TV, and the devices that their peers and friends have access to, etc.), an important consideration is that reducing access to violent content on one medium may result in increased consumption through other media. In addition, if modeling is supposed to be a mechanism responsible for violent pornography to increase subsequent violence, future research might consider the parameters related to modeling, such as model similarity, reinforcement and punishment of the model, etc. The current landscape of media is shifting rapidly. Thus, this literature needs frequent updating to be relevant to the contemporary context.

15.7 Bystander Training

It is often perplexing that when violence occurs that many people nearby do nothing, watch, or withdraw from the situation. Today, bystanders may videotape the violent incident on their phones and post it online but not intervene. When the public sees this they often ask why bystanders did not intervene and what could and should be done to facilitate them to intervene effectively to prevent violence. Bystander training programs have a surprisingly long history. For example, Darley and Latané (1968) listed five actions that bystanders could take to prevent violence. These included "(1) notice a situation as potentially harmful, (2) recognize the need to take action, (3) take responsibility for helping someone vulnerable to harm, (4) know how to help, and (5) take action to interrupt (e.g., distract, step in to prevent) abusive behavior" (Coker et al., 2020, p. 434).

Coker et al. (2020) reported a large-scale cluster RCT conducted over a 5-year period in 26 high schools with data on 74,836 participants. The dependent variable was annual self-reported surveys of perpetrating violence and victimization. The program was a two-component program consisting of an educational component for all students and intensive bystander training for students identified as popular opinion leader students. The intervention resulted in reductions in perpetration and victimization but primarily among sexual majority students. Sexual minority students only reported reduction in stalking victimization. Thus, this bystander intervention program required further revision to be effective in reducing violence for all students.

There have been at least two meta-analyses of this literature (DeGue et al., 2014; Mujal et al., 2021). Mujal et al. (2021), for example, reported an increasing number of RCTs with nearly 40% of 44 studies using RCTs to change attitudes, beliefs, and behavior related to intervening safely in risky or violent situations. Commonly evaluated procedures have included educational approaches, vignettes, and skills training incorporated into training packages and workshops such as *Bringing in the Bystander*. These approaches have been evaluated in quite varied settings such as high schools, college, and the military. Thus, it seems that bystander training is a promising approach to promote a more peaceful society and to prevent aggression and violence in many different organizations.

15.8 Arms Reduction Programs and Safe Gun Use

Another approach to reduce aggression and violence has been to reduce or remove the opportunities for gun violence by reducing access to firearms. This can be done by limiting access to firearms generally as well as limiting access to firearms for specific groups such as Felons in the USA and individuals with histories of violence and/or mental health issues. In the USA this can be done by both Federal and state laws limiting access to guns. Federal legislature has stalled for many years and states vary considerably in implementation of laws that sometimes restrict but more often increases access to firearms. This has led some to consider that waiting for effective legislative action is not the best course of action when effective immediate alternatives are available that can be implemented without state or federal legislation on firearms (Branas et al., 2021).

In any case, the evidence that arms reduction programs are effective in reducing violence is mixed and mostly of poor quality. In a critical review of scientific evaluations of gun policy, Morral et al. (2018) used 13 classes of gun policies (e.g., background checks, stand-your-ground, etc.), eight outcomes (e.g., violent crime, suicide, mass shootings), and five levels of evidence (no evidence, inconclusive, limited, moderate and supportive evidence) to evaluate scientific research into gun policies and legislation since 2003. The most glaring finding was the general absence of evidence for almost all combinations for policies and outcomes. Despite this overall lack of evidence, the authors did make eight main conclusions based on the available evidence. First, child access and safe storage laws appeared to be effective in reducing self-inflicted injuries and death and moderately effective in reducing suicide in children and teenagers. Second, these laws also appear effective in reducing accidental injuries to children and youth and there is limited evidence that they are effective in reducing accidental injuries to adults. Third, there was moderate evidence that background checks could reduce suicide and homicide. Fourth, there was moderate evidence that stand-your-ground laws *increased* homicide. Fifth, there was moderate evidence that laws limiting purchase of firearms reduces suicide, violent crime, and homicide. Sixth, bans on sales of assault weapons and high capacity magazines did not appear to increase sales or prices prior to implementation. Seventh, there was limited evidence that restricting gun purchase to people over 21 reduced suicides. Finally, there was no evidence that reporting stolen firearms, or recording gun sales reduced violence or suicide.

Thus, the overall state of the evidence for the effectiveness of gun policies is quite weak, other than for the impact of child access and safe storage laws on violence involving children and youth. These findings make sense in that these laws increase response effort in accessing guns and, if implemented correctly on a family-by-family basis, may reduce the access to guns when impulses toward violence run high such as when people are drunk, high, or angry. Branas et al. (2021) pointed out, however, that relatively small effect sizes, such as comparatively small reductions in gun access, might result in large and socially significant outcomes because the scale of the problem is so large and so costly.

15.8.1 Teaching Gun Safety Skills

Restricting access to firearms through legislation appears to have some benefits in reducing gun-related violence. Of course, in some countries, there is considerable access to firearms that comes with the danger that minors may access them resulting in accidental and intentional injuries. One approach to this aspect of reducing violence is to teach children how to behave safely around guns. In an early study (Himle et al., 2004b), these authors used behavioral skills training and, if needed, in situ training to teach eight children aged 4–5 years to behave safely when they found a gun. The authors used a disabled firearm left in a room and observed the child using a hidden camera. The authors defined behaving safely as not touching the gun, leaving the room, and telling an adult. The training package was highly effective in teaching the children to behave safely around guns. Subsequently, Miltenberger et al. (2004) systematically replicated these findings on children aged 6–7 years. Impressively, in two comparisons of this approach to the National Rifleman's Association's standard program to teach children gun safety skills, this behavioral package was superior on both occasions (Himle et al., 2004a; Kelso et al., 2007). Subsequently, there have been multiple replications and extensions of this procedure and variations of it including independent replications and extensions (Lee et al., 2019).

This approach to controlling children's access to guns is impressive due to the large number of studies, and comparisons showing it to be superior to industry standard approaches. At this time the most important area for future research on this program would be to provide follow-up data on child safety outcomes.

15.9 Alcohol Reduction Programs

Since alcohol is a widely available substance that facilitates violence, several programs have evaluated the effectiveness in reducing access to alcohol on violence. Research has evaluated strategies such as controlling the age of public legal drinking licensing bars; limiting hours that bars can operate and who can legally operate a bar or club; controlling the location and density of legal outlets; and taxation on alcohol (Carpenter & Dobkin, 2010). Such restrictions on alcohol access are often motivated by highly dangerous behavior, such as alcohol-related violence and drunk driving, but may be limited by the possibility that restrictions and punitive approaches to behavior control result in avoidance, black markets and illegal covert alcohol consumption.

Neighborhoods of downtown with late night drinking outlets and venues, sometimes described as being part of the "night-time economy" (McGuire et al., 2021), may contain multiple factors that facilitate violence. In such neighborhoods groups of people congregate, many of whom are disinhibited by alcohol, have close contact with other similar individuals, have expectations of fighting, are exposed to models of violence and its precursors, and may have active encouragement from bystanders to engage in violence. Other factors that are sometimes present can include factors related to sports, such as soccer team loyalties or national, cultural, class, and ethnic identities. Thus, such environments may be hotspots for alcohol-related violence. Nevertheless, such environments have numerous variables that can be modified that may inhibit or facilitate violence.

An interesting study that addresses the effects and interactions of licensing hours and outlet density comes from Taylor et al. (2021). They studied police records of the number of violent incidents between 2010 and 2018 in "safe night precincts" (SNPs), typically small areas of the city of approximately 1 Km squared. Each SNP varied in terms of size, number of outlets, number of high risk venues and density of high risk venues and whether alcohol was sold until midnight, 3 am or 5 am. Using multivariate statistics, Taylor et al. found that the effects of the density of high risk venues on violence were mediated by hours of alcohol availability. That is, availability of alcohol after 5 am resulted in greater risk of violence in high density SNPs than when alcohol was available only until 4 am. Importantly, variables such as density of outlets selling alcohol and hours of licensed operation are readily controllable by society. Perhaps combined with other evidence-based policing practices, such as identifying hot spots within each SNP and other environmental modifications such as lighting, cameras, signage and publicly visible monitors, alcohol-related violence can be reduced in high risk neighborhoods.

Building upon earlier meta-analytic research which has showed that a 10% increase in alcohol prices results in a 5% reduction in alcohol consumption, Wagenaar et al. (2010) reported a meta-analysis of the effects of alcohol tax prices and policy on morbidity and mortality including violence-related outcomes. The authors found that increasing alcohol prices resulted in decreases in violence (9 studies; $r = -0.022$; $Z = -3.579$; $p < 0.001$) and crime and misbehavior (5 studies; partial $r = -0.014$), which was significant ($Z = -2.943$; $p = 0.003$). The authors concluded that alcohol pricing was inversely related to a wide range of behavioral outcomes. They noted that these average effect sizes, statistically significant and on average rather small, hid large differences between locations and contexts. Thus, increasing alcohol pricing might produce no decrease in violence in one location and large decreases in violence in other locations.

15.10 Restorative Justice for Individuals

As we saw in previous chapters, after wars or following large-scale outbreaks of violence, societies have attempted to use restorative justice practices to take actions which, at least in part, are motivated to promote peace and reduce violence. In addition to use at a regional, national, or international level, people have used restorative justice practices at an individual level.

In a comprehensive systematic review of individual restorative justice practices, Sherman et al. (2015b) analyzed the results of 10 RCTs that involved face-to-face meetings between victims, offenders, kin, and community members. Retained studies had to have all eight of the following features: "1. Facilitators conduct a pre-conference screening discussion one-on-one with offenders and victims … 2. Scheduling of a conference at the victims' convenience. 3. Seating all participants in a circle in a private space … 4. Introducing all participants in terms of how they are emotionally connected to the crime … 5 … asking offenders to describe the crime they committed. 6. Inviting … all participants to describe the harm 7. … inviting all participants … to suggest how the harm might be repaired, usually reaching a consensus … that is written up … and signed by the offenders … [and] 8. Filing the agreement with a court, [etc.] … for monitoring and encouraging compliance by the offender with the agreement" (pp. 2–3).

In Sherman et al.'s (2015b) meta-analysis there were 1,880 participants with at least 2 years of data on reconviction. Most crimes included assault and violence, although some RCTs only reported data on crimes against property. There were overall rather small effect sizes suggesting that individual restorative justice practices were effective and economic data suggested that individual restorative justice practices were highly worthwhile. The mean effect size was $d = -0.155$ ($p < 0.001$) but with only two of ten individual RCTs reporting statistically significant reductions in reoffending. This represented a reduction in reoffending of between 7% and 45%. Moderator analysis revealed that: (a) effect sizes were somewhat larger for violent crimes ($d = 0.2$) which was 28% large than overall, although this difference was not statistically significant; (b) effect sizes were smaller for juveniles ($d = 0.199$) than adults ($d = 0.150$), but again this difference was not statistically significant; and (c) the effects were somewhat larger when individual restorative justice practices were a supplement to conventional justice. Cost–benefit analysis revealed that each dollar invested in individual restorative justice practices paid society $3.7–8.1 back primarily through the avoided costs of processing, convicting, and incarcerating future offenders. Thus, individual restorative justice practices have a decent evidence-based in reducing violent crime. The effect sizes are quite small, but, because of the high cost of reoffending, may be economically very worthwhile.

Individual restorative justice practices are an interesting and promising treatment for violent crime, but future research has much to do. The rather varied effect sizes from individual RCTs suggest that procedural variations, such as the content, intensity, and quantity of sessions may be important in determining outcome. Further, the mechanisms behind the effectiveness are unclear. For example, the content of the pre-conference preparation, content of the agreement, and what happens during enforcement and monitoring could contain many different procedures which might determine different outcomes. Future research should address this.

15.11 Summary

Cultures attempt to reduce (and sometimes increase) aggression and violence across all or many members of the culture using a wide variety of interventions using cultural interventions. Many of these interventions have focused on the family and educational settings using strategies such as parent and teacher training, increasing prosocial behavior in children and caretakers. These approaches have the potential to have a large and perhaps lifelong effect and hence be associated with the greatest personal, economic, and society-wide benefits. In addition, society-wide interventions for reduction of aggression and violence have included better and more effective, rational, and evidence-based policing, designing violence crime out of the environment through environmental design, modifying the media, bystander training, controlling access to alcohol and firearms and individual restorative justice interventions. As individuals attempt to design a more peaceful society, they may combine evidence-based, effective strategies into multi-element policies that address the highest need areas in the most economic and effective manner.

References

Altafim, E. R. P., & Linhares, M. B. M. (2016). Universal violence and child maltreatment prevention programs for parents: A systematic review. *Psychosocial Intervention, 25*(1), 27–38.

Axford, N., Bjornstad, G., Clarkson, S., Ukoumunne, O. C., Wrigley, Z., Matthews, J., et al. (2020). The effectiveness of the KiVa bullying prevention program in Wales, UK: Results from a pragmatic cluster randomized controlled trial. *Prevention Science, 21*(5), 615–626.

Bandura, A., Ross, D., & Ross, S. A. (1961). Transmission of aggression through imitation of aggressive models. *The Journal of Abnormal and Social Psychology, 63*(3), 575.

Braga, A., Papachristos, A., & Hureau, D. (2012). Hot spots policing effects on crime. *Campbell Systematic Reviews, 8*(1), 1–96.

Braga, A. A., Papachristos, A. V., & Hureau, D. M. (2014). The effects of hot spots policing on crime: An updated systematic review and meta-analysis. *Justice Quarterly, 31*(4), 633–663.

Branas, C. C., Reeping, P. M., & Rudolph, K. E. (2021). Beyond gun laws—Innovative interventions to reduce gun violence in the United States. *JAMA Psychiatry, 78*(3), 243–244. Retrieved from https://www.ncbi.nlm.nih.gov/pmc/articles/PMC6075800/

Calvert, S. L., Appelbaum, M., Dodge, K. A., Graham, S., Nagayama Hall, G. C., Hamby, S., et al. (2017). The American Psychological Association Task Force assessment of violent video games: Science in the service of public interest. *American Psychologist, 72*(2), 126–143. https://doi.org/10.1037/a0040413

Carpenter, C., & Dobkin, C. (2010). Alcohol regulation and crime. In P. Cook, J. Ludwig, & J. McCrary (Eds.), *Controlling crime: Strategies and tradeoffs* (pp. 291–329). University of Chicago Press.

Chalfin, A., Hansen, B., Lerner, J., & Parker, L. (2021). Reducing crime through environmental design: Evidence from a randomized experiment of street lighting in New York City. *Journal of Quantitative Criminology, 38*, 1–31.

Christakis, D. A., Garrison, M. M., Herrenkohl, T., Haggerty, K., Rivara, F. P., Zhou, C., & Liekweg, K. (2013). Modifying media content for preschool children: A randomized controlled trial. *Pediatrics, 131*(3), 431–438.

Clarke, R. V. (1995). Situational crime prevention. *Crime and Justice, 19*, 91–150.

Clarke, R. (2009). The theory of crime prevention through environmental design. *Police Management Studies Quarterly (PMSQ), 3*(3), 344–356. Retrieved from https://www.sid.ir/en/journal/ViewPaper.aspx?id=200562

Coker, A. L., Bush, H. M., Clear, E. R., Brancato, C. J., & McCauley, H. L. (2020). Bystander program effectiveness to reduce violence and violence acceptance within sexual minority male and female high school students using a cluster RCT. *Prevention Science, 21*, 434–444. https://doi.org/10.1007/s11121-019-01073-7

Cozens, P., Neale, R., Whitaker, J., & Hillier, D. (2003). Managing crime and the fear of crime at railway stations—A case study in South Wales (UK). *International Journal of Transport Management, 1*(3), 121–132.

Darley, J. M., & Latané, B. (1968). Bystander intervention in emergencies: Diffusion of responsibility. *Journal of Personality and Social Psychology, 8*(4), 377–383.

Davies, M. W., & Farrington, D. P. (2020). An examination of the effects on crime of switching off street lighting. *Criminology & Criminal Justice, 20*(3), 339–357. https://doi.org/10.1177/1748895818818861

DeGue, S., Valle, L. A., Holt, M. K., Massetti, G. M., Matjasko, J. L., & Tharp, A. T. (2014). A systematic review of primary prevention strategies for sexual violence perpetration. *Aggression and Violent Behavior, 19*(4), 346–362.

Devries, K. M., Knight, L., Child, J. C., Kyegombe, N., Hossain, M., Lees, S., et al. (2017). Witnessing intimate partner violence and child maltreatment in Ugandan children: A cross-sectional survey. *BMJ Open, 7*(2), e013583.

DeVylder, J. E., Jun, H., Fedina, L., et al. (2018). Association of exposure to police violence with prevalence of mental health symptoms among urban residents in the United States. *JAMA Network Open, 1*(7), e184945. https://doi.org/10.1001/jamanetworkopen.2018.4945

Dyches, H., Biegel, D. E., Johnsen, J. A., Guo, S., & Min, M. O. (2002). The impact of mobile crisis services on the use of community-based mental health services. *Research on Social Work Practice, 12*(6), 731–751. https://doi.org/10.1177/104973102237470

Espelage, D., Anderman, E. M., Brown, V. E., Jones, A., Lane, K. L., McMahon, S. D., et al. (2013). Understanding and preventing violence directed against teachers: Recommendations for a national research, practice, and policy agenda. *American Psychologist, 68*(2), 75–87.

Farrington, D. P., & Welsh, B. C. (2002). Improved street lighting and crime prevention. *Justice Quarterly, 19*(2), 313–342.

Ferguson, C. J., & Beresin, E. (2017). Social science's curious war with pop culture and how it was lost: The media violence debate and the risks it holds for social science. *Preventive Medicine, 99*, 69–76.

Ferguson, C. J., Copenhaver, A., & Markey, P. (2020). Reexamining the findings of the American Psychological Association's 2015 task force on violent media: A meta-analysis. *Perspectives on Psychological Science, 15*(6), 1423–1443.

Fraguas, D., Díaz-Caneja, C. M., Ayora, M., Durán-Cutilla, M., Abregú-Crespo, R., Ezquiaga-Bravo, I., et al. (2021). Assessment of school anti-bullying interventions: A meta-analysis of randomized clinical trials. *JAMA Pediatrics, 175*(1), 44–55.

Fry, D., Fang, X., Elliott, S., Casey, T., Zheng, X., Li, J., et al. (2018). The relationships between violence in childhood and educational outcomes: A global systematic review and meta-analysis. *Child Abuse & Neglect, 75*, 6–28.

Gage, N. A., Whitford, D. K., & Katsiyannis, A. (2018). A review of schoolwide positive behavior interventions and supports as a framework for reducing disciplinary exclusions. *The Journal of Special Education, 52*(3), 142–151.

Goldberg, M. E., & Gorn, G. J. (1979). Television's impact on preferences for non-white playmates: Canadian Sesame Street inserts. *Journal of Broadcasting, 23*, 27–32.

Grossman, D. C., Neckerman, H. J., Koepsell, T. D., Liu, P. Y., Asher, K. N., Beland, K., et al. (1997). Effectiveness of a violence prevention curriculum among children in elementary school: A randomized controlled trial. *Journal of the American Medical Association, 277*(20), 1605–1611.

Healy, K. L. (2020). Hypotheses for possible iatrogenic impacts of school bullying prevention programs. *Child Development Perspectives, 14*(4), 221–228.

Heekes, S. L., Kruger, C. B., Lester, S. N., & Ward, C. L. (2022). A systematic review of corporal punishment in schools: Global prevalence and correlates. *Trauma, Violence & Abuse, 23*, 52.

Herkama, S., & Salmivalli, C. (2018). KiVa antibullying program. In M. Campbell & S. Bauman (Eds.), *Reducing cyberbullying in schools* (pp. 125–134). Academic Press.

Herkama, S., Saarento, S., & Salmivalli, C. (2017). The KiVa antibullying program: Lessons learned and future directions. In P. Sturmey (Ed.), *The Wiley handbook of violence and aggression* (pp. 1–12). Wiley.

Herrnstein, R. J. (1974). Formal properties of the matching law. *Journal of the Experimental Analysis of Behavior, 21*(1), 159–164.

Himle, M. B., Miltenberger, R. G., Gatheridge, B. J., & Flessner, C. A. (2004a). An evaluation of two procedures for training skills to prevent gun play in children. *Pediatrics, 113*(1), 70–77.

Himle, M. B., Miltenberger, R. G., Flessner, C., & Gatheridge, B. (2004b). Teaching safety skills to children to prevent gun play. *Journal of Applied Behavior Analysis, 37*(1), 1–9.

Huitsing, G., Barends, S. I., & Lokkerbol, J. (2019). Cost-benefit analysis of the KiVa anti-bullying program in the Netherlands. *International Journal of Bullying Prevention, 2*, 1–10.

Huitsing, G., Lodder, G. M., Browne, W. J., Oldenburg, B., Van der Ploeg, R., & Veenstra, R. (2020). A large-scale replication of the effectiveness of the KiVa antibullying program: A randomized controlled trial in the Netherlands. *Prevention Science, 21*, 627–638.

Hutchings, J., Clarkson, S., Hutchings, J., & Clarkson, S. (2015). Introducing and piloting the KiVa bullying prevention programme in the UK. *Educational and Child Psychology, 32*(1), 49–61.

Jeffrey, C. R. (1975). Behavior control techniques and criminology: 1975–2075. In *Ecology youth development workshop*. University of Hawaii School of Social Work.

Jiménez-Barbero, J. A., Ruiz-Hernández, J. A., Llor-Zaragoza, L., Pérez-García, M., & Llor-Esteban, B. (2016). Effectiveness of anti-bullying school programs: A meta-analysis. *Children and Youth Services Review, 61*, 165–175.

Johander, E., Turunen, T., Garandeau, C. F., & Salmivalli, C. (2021). Different approaches to address bullying in KiVa schools: Adherence to guidelines, strategies implemented, and outcomes obtained. *Prevention Science, 22*(3), 299–310.

Kärnä, A., Voeten, M., Little, T. D., Poskiparta, E., Kaljonen, A., & Salmivalli, C. (2011a). A large-scale evaluation of the KiVa antibullying program: Grades 4–6. *Child Development, 82*(1), 311–330.

Kärnä, A., Voeten, M., Little, T. D., Poskiparta, E., Alanen, E., & Salmivalli, C. (2011b). Going to scale: A nonrandomized nationwide trial of the KiVa antibullying program for grades 1–9. *Journal of Consulting and Clinical Psychology, 79*, 796–805.

Kelso, P. D., Miltenberger, R. G., Waters, M. A., Egemo-Helm, K., & Bagne, A. G. (2007). Teaching skills to second and third grade children to prevent gun play: A comparison of procedures. *Education and Treatment of Children, 30*(3), 29–48.

Lawson, T., Rogerson, R., & Barnacle, M. (2018). A comparison between the cost effectiveness of CCTV and improved street lighting as a means of crime reduction. *Computers, Environment and Urban Systems, 68*, 17–25.

Lee, N., Vladescu, J. C., Reeve, K. F., Peterson, K. M., & Giannakakos, A. R. (2019). Effects of behavioral skills training on the stimulus control of gun safety responding. *Journal of Behavioral Education, 28*(2), 187–203.

Lester, S., Lawrence, C., & Ward, C. L. (2017). What do we know about preventing school violence? A systematic review of systematic reviews. *Psychology, Health & Medicine, 22*(Suppl 1), 187–223.

Longobardi, C., Badenes-Ribera, L., Fabris, M. A., Martinez, A., & McMahon, S. D. (2019). Prevalence of student violence against teachers: A meta-analysis. *Psychology of Violence, 9*, 596–610. https://doi.org/10.1037/vio0000202

Lum, C., Koper, C. S., Wilson, D. B., Stoltz, M., Goodier, M., Eggins, E., et al. (2020). Body-worn cameras' effects on police officers and citizen behavior: A systematic review. *Campbell Systematic Reviews, 16*, e1112.

Makarios, M. D., & Pratt, T. C. (2012). The effectiveness of policies and programs that attempt to reduce firearm violence: A meta-analysis. *Crime & Delinquency, 58*(2), 222–244.

McCormick, J. G., & Holland, S. M. (2015). Strategies in use to reduce incivilities, provide security and reduce crime in urban parks. *Security Journal, 28*(4), 374–391.

McCoy, A., Melendez-Torres, G. J., & Gardner, F. (2020). Parenting interventions to prevent violence against children in low-and middle-income countries in East and Southeast Asia: A systematic review and multi-level meta-analysis. *Child Abuse & Neglect, 103*, 104444.

McGuire, J., Evans, E., & Kane, E. (2021). *Evidence-based policing and community crime prevention*. Springer.

McMahon, S. D., Peist, E., Davis, J. O., Bare, K., Martinez, A., Reddy, L. A., et al. (2020). Physical aggression toward teachers: Antecedents, behaviors, and consequences. *Aggressive Behavior, 46*(1), 116–126.

McPhedran, S. (2016). A systematic review of quantitative evidence about the impacts of Australian legislative reform on firearm homicide. *Aggression and Violent Behavior, 28*, 64–72.

Miltenberger, R. G., Flessner, C., Gatheridge, B., Johnson, B., Satterlund, M., & Egemo, K. (2004). Evaluation of behavioral skills training to prevent gun play in children. *Journal of Applied Behavior Analysis, 37*(4), 513–516.

Morral, A. R., Ramchand, R., & Smart R. (2018). *The science of gun policy: A critical synthesis of research evidence on the effects of gun policies in the United States*. RAND. Retrieved from https://www.rand.org/pubs/research_reports/RR2088.html

Mujal, G. N., Taylor, M. E., Fry, J. L., Gochez-Kerr, T. H., & Weaver, N. L. (2021). A systematic review of bystander interventions for the prevention of sexual violence. *Trauma, Violence & Abuse, 22*, 381.

Mytton, J. A., DiGuiseppi, C., Gough, D. A., Taylor, R. S., & Logan, S. (2002). School-based violence prevention programs: Systematic review of secondary prevention trials. *Archives of Pediatrics & Adolescent Medicine, 156*(8), 752–762.

Ng, E. D., Chua, J. Y. X., & Shorey, S. (2022). The effectiveness of educational interventions on traditional bullying and cyberbullying among adolescents: A systematic review and meta-analysis. *Trauma, Violence & Abuse, 23*, 132.

Nix, J. (2020). On the challenges associated with the study of police use of deadly force in the United States: A response to Schwartz & Jahn. *PLoS one, 15*(7), e0236158.

Nkuba, M., Hermenau, K., Goessmann, K., & Hecker, T. (2018). Reducing violence by teachers using the preventative intervention Interaction Competencies with Children for Teachers (ICC-T): A cluster randomized controlled trial at public secondary schools in Tanzania. *PLoS One, 13*(8), e0201362. https://doi.org/10.1371/journal.pone.0201362

Park-Higgerson, H. K., Perumean-Chaney, S. E., Bartolucci, A. A., Grimley, D. M., & Singh, K. P. (2008). The evaluation of school-based violence prevention programs: A meta-analysis. *Journal of School Health, 78*(9), 465–479.

Piza, E. L., Welsh, B. C., Farrington, D. P., & Thomas, A. L. (2019). CCTV surveillance for crime prevention: A 40-year systematic review with meta-analysis. *Criminology & Public Policy, 18*(1), 135–159.

Polanin, J. R., Espelage, D. L., Grotpeter, J. K., Spinney, E., Ingram, K. M., Valido, A., El Sheikh, A., Torgal, C., & Robinson, L. (2021). A meta-analysis of longitudinal partial correlations between school violence and mental health, school performance, and criminal or delinquent acts. *Psychological Bulletin, 147*(2), 115–133. https://doi.org/10.1037/bul0000314

Pollack, H. A., & Humphreys, K. (2020). Reducing violent Incidents between police officers and people with psychiatric or substance use disorders. *The Annals of the American Academy of Political and Social Science, 687*(1), 166–184. https://doi.org/10.1177/0002716219897057

Pundir, P., Saran, A., White, H., Subrahmanian, R., & Adona, J. (2020). Interventions for reducing violence against children in low-and middle-income countries: An evidence and gap map. *Campbell Systematic Reviews, 16*(4), e1120.

van der Put, C. E., Assink, M., Gubbels, J., & van Solinge, N. F. B. (2018). Identifying effective components of child maltreatment interventions: A meta-analysis. *Clinical Child and Family Psychology Review, 21*(2), 171–202.

Robinson, T. N., Wilde, M. L., Navracruz, L. C., Haydel, K. F., & Varady, A. (2001). Effects of reducing children's television and video game use on aggressive behavior: A randomized controlled trial. *Archives of Pediatrics & Adolescent Medicine, 155*(1), 17–23.

Rosenfeld, R., Deckard, M. J., & Blackburn, E. (2014). The effects of directed patrol and self-initiated enforcement on firearm violence: A randomized controlled study of hot spot policing. *Criminology, 52*(3), 428–449.

Robin, L., Peterson, B., & Lawrence, D. (2020). *Public surveillance cameras and crime: The impact of different camera types on crimes and clearances*.

Sainio, M., Herkama, S., Kontio, M., & Salmivalli, C. (2019). KiVa anti-bullying programme. In P. K. Smith (Ed.), *Routledge psychological impacts. Making an impact on school bullying: Interventions and recommendations* (pp. 45–66). Routledge/Taylor & Francis Group. https://doi.org/10.4324/9781351201957-3

Scheckner, S., Rollin, S. A., Kaiser-Ulrey, C., & Wagner, R. (2002). School violence in children and adolescents: A meta-analysis of the effectiveness of current interventions. *Journal of School Violence, 1*(2), 5–32.

Schwartz, G. L., & Jahn, J. L. (2020). Mapping fatal police violence across US metropolitan areas: Overall rates and racial/ethnic inequities, 2013–2017. *PloS one, 15*(6), e0229686.

Shenderovich, Y., Lachman, J. M., Ward, C. L., Wessels, I., Gardner, F., Tomlinson, M., et al. (2021). The science of scale for violence prevention: A new agenda for family strengthening in low-and middle-income countries. *Frontiers in Public Health, 9*, 199.

Sherman, L. W., Strang, H., Mayo-Wilson, E., Woods, D. J., & Ariel, B. (2015b). Are restorative justice conferences effective in reducing repeat offending? Findings from a Campbell systematic review. *Journal of Quantitative Criminology, 31*(1), 1–24.

Sidebottom, A., Tompson, L., Thornton, A., Bullock, K., Tilley, N., Bowers, K., & Johnson, S. D. (2018). Gating alleys to reduce crime: A meta-analysis and realist synthesis. *Justice Quarterly, 35*(1), 55–86.

Solomon, B. G., Klein, S. A., Hintze, J. M., Cressey, J. M., & Peller, S. L. (2012). A meta-analysis of school-wide positive behavior support: An exploratory study using single-case synthesis. *Psychology in the Schools, 49*(2), 105–121.

Sugai, G., & Horner, R. H. (2009). Defining and describing schoolwide positive behavior support. In W. Sailor, W. Dunlap, & G. Sugai (Eds.), *Handbook of positive behavior support* (pp. 307–326). Springer.

Taylor, N., Livingston, M., Coomber, K., Mayshak, R., Zahnow, R., Ferris, J., et al. (2021). The combined impact of higher-risk on-license venue outlet density and trading hours on serious assaults in night-time entertainment precincts. *Drug and Alcohol Dependence, 223*, 108720.

Ting, S. M. R. (2009). Meta-analysis on dating violence prevention among middle and high schools. *Journal of School Violence, 8*(4), 328–337.

Waasdorp, T. E., Bradshaw, C. P., & Leaf, P. J. (2012). The impact of schoolwide positive behavioral interventions and supports on bullying and peer rejection: A randomized controlled effectiveness trial. *Archives of Pediatrics & Adolescent Medicine, 166*(2), 149–156.

Wagenaar, A. C., Tobler, A. L., & Komro, K. A. (2010). Effects of alcohol tax and price policies on morbidity and mortality: A systematic review. *American Journal of Public Health, 100*(11), 2270–2278.

Wehby, J. H., Symons, F. J., & Shores, R. E. (1995). A descriptive analysis of aggressive behavior in classrooms for children with emotional and behavioral disorders. *Behavioral Disorders, 20*(2), 87–105.

Welsh, B. C., & Farrington, D. P. (2008). Effects of closed circuit television surveillance on crime. *Campbell Systematic Reviews, 4*(1), 1–73.

Welsh, B. C., Farrington, D. P., & Taheri, S. A. (2015). Effectiveness and social costs of public area surveillance for crime prevention. *Annual Review of Law & Social Science, 11*, 111–130.

Wilson, D. B., Gottfredson, D. C., & Najaka, S. S. (2001). School-based prevention of problem behaviors: A meta-analysis. *Journal of Quantitative Criminology, 17*(3), 247–272.

Wilson, S. J., Lipsey, M. W., & Derzon, J. H. (2003). The effects of school-based intervention programs on aggressive behavior: A meta-analysis. *Journal of Consulting and Clinical Psychology, 71*(1), 136.

Chapter 16
Individual Therapies for Violence and Aggression: I. Cognitive and Behavioral Therapies

This chapter reviews psychological treatments for aggression based mostly on treatment of the aggressive individual. There is a very extensive literature on this topic. So this chapter focuses on mainstream approaches with the largest corpus of research, namely CBT, and a variety of behavioral interventions. These approaches have three shared characteristics. First, more than many other approaches to therapy, they espouse a scientific and empirical approach to understanding the mechanisms for change during individual therapy. Second, while acknowledging the importance of biology, they see some form of learning as the basis of acquisition and maintenance of aggressive and violent behavior. Thus, third, interventions involve some form of educational approach in which the client learns to think and behave differently in a non-problematic way. Thus, among other things, the therapist is a kind of teacher, educationalist, or mentor to help the client learn to solve life's problems more effectively.

16.1 Cognitive Behavior Therapies

16.1.1 General Approach

One of the most common forms of psychotherapy today is CBT. Emerging from the earlier behavior therapy in the 1970s, CBT assumes that the cause of psychopathology is various forms of faulty thinking and that by changing the way people think their behavior will subsequently change. Examples of faulty thinking can include: (a) making incorrect attributions of other people's motives, such as assuming that when another person is looking at you or they scowl, they must be thinking critical thoughts about you; (b) ruminating on past injustices and planning future retributions for those injustices; and (c) negative automatic thoughts, such as "he thinks I am weak," or "he is disrespecting me when he smiles." CBT also involves a

P. Sturmey, *Violence and Aggression*, Advances in Preventing and Treating Violence and Aggression, https://doi.org/10.1007/978-3-031-04386-4_16

substantial behavioral component including goal setting and self-monitoring of behavior; education and skills training, such as relaxation training, social skills and assertiveness training, and exposure-like procedures, such as progressive in vivo exposure to anger- or fear-provoking situations in order to practice alternate behavior. More recently, the so-called third wave behavior therapies based on and extending Skinner's (1957) *Verbal Behavior*, such as acceptance and commitment therapy (ACT), dialectical behavior therapy (DBT), and functional analytic psychotherapy (FAP), and mindfulness have emerged. These third wave therapies have also have been used to treat aggression and violence.

Assessment of aggression and violence within a CBT framework includes assessment of private behavior such as thoughts, feelings, and core beliefs and environmental factors, such as triggers for aggression and violence, and alternate skills such as assertiveness, relaxation, problem solving, and social skills. Since thoughts, feelings, and core beliefs are private events they can only be accessed through client self-report through interviews and questionnaires, or inferred through observation of client behavior or third party opinions about the client's beliefs. Some authors have proposed combining CBT methods with functional analytic principles to consider the functional nature of the client's thoughts, feelings, and behavior (Drossel & Rummel, 2009).

As with earlier behavior therapy, CBT has relatively strong empirical and scientific roots and, thus, many applications have an extensive evidence-based, especially in areas such as depression, anxiety, addictions, and eating disorders and this is true in areas such as anger, aggression, and violence. It is not possible to review all CBT programs for aggression and violence here. This section will describe general features of this approach and other sections below will illustrate some well-known programs.

16.1.2 Cognitive Behavior Therapy Outcome Research

There have now been numerous systematic reviews and meta-analyses of CBT for aggression, violence, and anger (Lee & DiGiuseppe, 2018). An early meta-analysis by Beck and Fernandez (1998) reviewed studies published between 1970 and 1995. They identified 50 studies with 1640 participants, 40 of which had control groups. This meta-analysis documented that anger management had been applied to a wide range of populations, such as abusive spouses, school children, adolescents, juvenile delinquents, and forensic patients. The grand unweighted d was 0.81 (range −0.32 to 1.57) ($p < 0.0001$), which would be considered to be a large effect size, and all but one outcome was positive.

Beck and Fernandez (1998) concluded that CBT was a robustly effective treatment for anger. More recent meta-analyses of anger management have reached similar conclusions by documenting the continued replication and expansion of the use of anger management. For example, Fernandez et al. (2018) searched the literature on anger management for individuals aged over 18 years from 2000 onwards.

They found 42 outcome studies, of which 25 were randomized controlled trials (RCTs). This study again documents the extensive research base for CBT for anger.

In a *tour de force*, Lee and DiGiuseppe (2018) reported a recent systematic review of meta-analyses of treatments for anger, aggression, and violence provided a useful description of the current status of research in this field. The authors conducted a systematic search of research published up to 2017. They identified 13 meta-analyses of treatment of anger and eight of aggression. CBT was the most commonly evaluated treatment for anger and aggression: Eight meta-analyses focused exclusively on CBT. Generally, effect sizes were moderate to large and generally were robust across contexts and populations. Thus, broadly similar outcomes were observed for specific populations, such as children, forensic populations, and individuals with developmental disabilities such as intellectual disabilities or autism spectrum disorders. Interestingly, effects sizes were somewhat smaller in the meta-analyses of treatment of aggression than meta-analyses of treatment of anger.

As with many meta-analyses, this paper provides a useful description of what has been done and what has been neglected. For example, the authors noted that little attention had been given to anger management classes which are primarily educational in content as opposed to individual anger management which is more likely to include skills training and individualized case formulation, even though such classes are commonly used by courts for problems such as intimate partner violence and child abuse. Second, no meta-analyses addressed domestic violence and child abuse perpetrators and only one meta-analysis addressed abusive parents and spouses. Third, there was little research on family therapy and psychodynamic therapy for aggression and violence. Fourth, the CBT packages were highly variable in content, formats, and intensity and included many different components and combinations of components. Thus, the effective components, treatment doses, and intensities of CBT for effective treatment of anger and violence remain relatively unknown. Thus, to conclude that "CBT is an effective treatment for anger, aggression, and violence" belies the many differences in CBT procedures which have been used. Finally, there was a lack of individualization of treatment based on case conceptualization and functional assessment and a lack of evidence as to whether this makes a difference to treatment outcomes.

16.1.3 Aggression Replacement Training

One of the earliest CBT approaches to aggression and anger is aggression replacement training (ART) which is a common behaviorally oriented, structured form of CBT developed in the 1970s for juveniles (Goldstein et al., 1998) which aims to improve prosocial behavior in individuals who are aggressive. It has been extensively implemented and evaluated with juveniles and extended to other populations. ART grew out of the 1970s literature on social skills training, anger management, social skills training, and training in moral reasoning applied to aggression and violence. Treatment was delivered over 10 weeks with three sessions per week and involved homework assignments to promote generalization. Over time, its use has been extended to other populations and its content expanded or developed to form other programs. The original authors have copyrighted and trademarked ART in Europe.

Although reviews of CBT for aggression and violence and some government reports have been favorable as to the effectiveness of ART, there has been no systematic review of the effectiveness of ART until the recent publication by Brännström et al. (2016). These authors searched the literature for studies that evaluated unmodified ART in adolescents or adults aged over 12 years old. They included both RCTs and non-experimental, mostly pre-post, designs. They identified 16 unique studies, four on adults and 12 on adolescents published between 1987 and 2014 which reported data on primary outcomes such as recidivism and/or other secondary outcomes. There were six RCTs and ten non-RCTs.

Brännström et al. (2016) decided not to conduct a meta-analysis because the studies were of such varying quality, many were open to various potential sources of bias and seven studies were conducted by authors who were judged to have had vested interests in ART. Although some studies found reductions in recidivism, others found no effects. Other studies may have been affected by dropout of participants who were more likely than other participants to reoffend. Outcomes on secondary variables were similarly mixed. Thus, the authors concluded that, although some studies were suggestive of positive outcomes, no conclusions could be drawn. They wrote "… we cannot say whether ART helps, whether it has no effect, or even whether it is a harmful intervention" (p. 40).

More recently, Ensafdaran et al. (2019) conducted an updated systematic review of ART in children and youth which identified ten publications reporting 11 studies. Overall, the outcomes for aggressive behavior, anger control, social skills, and moral reasoning were positive in almost all studies. One study compared ART with a token economy and found both approaches were effective, but did not differ from one another in terms of treatment effectiveness. The quality of the studies was mixed and common problems included small sample sizes and lack of experimental designs, although at least two studies were true experiments that showed positive effects of ART.

16.1.3.1 Beyond ART

ART is one of the oldest programs for aggression and violence and some of its limitations are relatively well known and so it has been the basis for expanded programs. Recently, Hornsveld et al. (2019) described an expanded ART program for the obligatory treatment of violent and sexually violent offenders.

They noted that the original ART program might be incomplete for these particularly challenging populations, who often have specific and perhaps rather different presenting problems from many outpatients. These additional problems can include more severe and repugnant forms of aggression and violence such as child rape or violent rape of women, and they may also often have additional problems, such as alcohol and drug addictions. Therefore, Hornsveld et al. added new units to the basic ART procedures, such as sports (Zwets et al., 2016), and new units to address the problems of violence and sexually violent offenders (Hornsveld et al., 2019). Evaluations of these specific new components and the revised ART package are ongoing at this time.

16.1.4 Rational-Emotive Behavior Therapy

Albert Ellis began developing the forerunner of Rational-Emotive Behavior Therapy (REBT) in the late 1950s by emphasizing that rather than eternal events, it was the way people think about those external events that resulted in misery or happiness (Ellis, 1957). Noting the ineffectiveness of his own use of traditional psychotherapy, Ellis developed a new approach, REBT, which emphasized both changing patterns of thinking and core beliefs, but also actively challenging those patterns of thinking by changing what people do through homework exercises, including desensitization and self-exposure to anger-related triggers. In REBT, therapists teach clients that responding to challenging events is a choice between adaptive or maladaptive responses. Ellis suggested that two core beliefs were related to anger. The first was the belief that other people must always treat them nicely, considerately, and fairly at all time. The second was that life should be free of hassles and challenges at all times, otherwise life will be horrible and intolerable and that people causing this harm should be punished severely (Ellis, 2003). Thus, REBT is an example of an early CBT which includes substantial cognitive and behavioral components which grew out of general dissatisfaction with the ineffectiveness of traditional psychotherapy.

REBT appears to be at least moderately effective in treating aggression and violence and related constructs such as anger, at least in some contexts, producing effect sizes comparable to other forms of CBT. For example, in a meta-analysis of REBT, David et al. (2018) reported a moderate effect size for treatment of anger based on between groups comparison (six studies, $d = 0.58$, $p < 0.01$) and a large effect size based on pre-post comparisons (two studies, $d = 0.88$, $p < 0.01$). In a systematic review forensic application of REBT, Debidin and Dryden (2011) identified 36 studies published between 1995 and 2007 of which 12 were judged to be better quality studies. They were cautiously optimistic that REBT had the potential to reduce reoffending and address some problems commonly observed in forensic settings, such as anger. Finally, Gonzalez et al. (2004) reported a meta-analysis of the application of REBT to a wide range of problems in children and adolescents and reported a large effect size for disruptive behavior (seven effect sizes, $z = 1.18$), which might have included some studies on aggression, violence, and more indirectly related constructs.

There is a large outcome literature on REBT for anger, aggression, and violence. Thus, it seems that REBT is a promising approach to treatment of aggression and violence.

16.1.5 Novaco's Anger Management Program

There are many forms of anger management, but a commonly researched version comes from Novaco's (1975). This approach shares many features with Meichenbaum's stress inoculation (Meichenbaum & Deffenbacher, 1988). The program includes five components: "(1) education about personal anger patterns, (2) self-monitoring of anger arousal and its cognitive companions, (3) the ability to alternatively construe provocation, (4) covert self-instruction to guide one's appraisal of events and to cue non-antagonistic responses, and (5) the ability to remain task oriented rather than ego oriented when provoked" (p. 1136). There are three phases of the program. The first consists of cognitive preparation, in which the client self-monitors anger to identify triggers, learns about the role of thinking in magnifying or diminishing anger, the role of perceived threat to self-worth in anger, wanting to control situations, and how antagonistic acts can escalate anger-provoking situations. In this phase the client also learns relaxation as an antidote to anger as relaxation is incompatible with the arousal that accompanies anger. (This is quite similar to Wolpe's work discussed below.) In the skills acquisition and rehearsal phase the client learns to discriminate between justified and unjustified anger and, using a template, learns to change covert self-statements in anger-provoking situations. The client also learns to consider anger-provoking situations as those which involve solving a problem, rather than a personal attack, and so focus on the task, rather than the internal stimuli. The client also learns various breathing and mental imagery techniques. Finally, in the practice and application phase, the client develops a hierarchy of anger-provoking situations and progressively exposes themselves to these situations imaginally with relaxation and subsequently in vivo. They also practice skills such as empathy and non-antagonistic interaction, which are incompatible with angry behavior, again echoing Wolpe's earlier work.

The effectiveness of Novaco's CBT program is illustrated by a RCT conducted in a typical applied clinical setting from Naeem et al. (2009). The study took place over a 2-year period in a National Health Service outpatient department. Patients entered the trial if they reported unmanageable angry feelings with or without other mental health diagnoses. Individuals with active psychosis, severe depression, addictions, or Intellectual Disabilities were excluded. There were 37 patients who completed the anger management program and there were 35 individuals in the wait list control group. Prior to treatment the groups were comparable on almost all demographics and multiple measures of anger, anxiety, and depression. Post-treatment the anger management group scored significantly lower on several measures of anger but did not differ in terms of anxiety and depression. Although follow-up was planned, insufficient numbers of patients responded to follow-up calls to present data. Although this study has significant problems with dropout prior to randomization, during treatment and the wait period for the wait list control group, and at follow-up for all participants, it demonstrated that Novaco's anger management program can readily be used in routine clinical practice for typical outpatients with anger problems. Further, for those who complete the program, they are very likely to report reduction in feelings of anger.

Novaco's approach has been adopted widely (Novaco, 2020). It has been modified for particular populations and contexts such as for use with offenders with Intellectual Disabilities (Taylor et al., 2016) among others. (See Chap. 19, below for a further discussion of applications to offenders with Intellectual Disabilities.) Thus, Novaco's program is a promising one for the treatment of aggression and violence and has a substantial evidence base with a variety of populations.

16.1.6 Summary

The various forms of CBT have a long history and extensive empirical basis in treating anger and violence so much so that CBT appears the "industry standard" treatment for aggression and violence. Further, the large quantity of evidence allows us to make conclusions with relative confidence about the effectiveness and limitation of CBT treatments of anger, aggression, and violence. Namely, CBT is an evidence-based practice for aggression and violence in that it jumped the low barrier of two or more well conducted RCTs many years ago. This large quantity of evidence also allows us to begin to make more nuanced conclusions, such as whether or not CBT is effective in a particular context, in what ways it may help and so on.

16.2 Behavior Modification

There are a wide range of behavior modification approaches to aggression, violence, and anger which date back many years. One approach comes from classical conditioning and psychopathology and the second comes from operant psychology. Classical conditioning approaches are most well known in terms of classical conditioning and respondent extinction of fear (Watson & Rayner, 1920), but it is also known that Pavlov attended psychiatric clinics on a regular basis during part of his career, leading to speculation about the role of classical conditioning more generally in the acquisition and treatment of psychopathology (Todes, 2014). The historical work that most clearly relates to anger and aggression is that of Wolpe's (1952, 1954, 1958) who described approaches based on Pavlov's work on respondent extinction to a wide range of clinical problems—not just anxiety!—including anger based on reciprocal inhibition. The second approach—operant psychology—contributed non-function-based contingency management procedures, such as token economies and classroom management procedures, such as the good behavior game among many others.

16.2.1 General Characteristics

Deriving from basic research on operant learning conducted primarily with non-humans throughout the 1930s to the 1950s. Subsequently, behaviorists extended the use of operant technology to measure and understand human behavior, such as individuals with schizophrenia (Lindsley, 1956) and some even ventured out of their Skinner box to investigate naturally occurring operant processes such as those that occur in child development (Gewirtz, 1954) and verbal behavior, including during verbal behavior during psychotherapy (Krasner, 1958). Further, in the 1950s operantly oriented behaviorists began to address socially significant problems in applied settings, such as mental hospital wards, classrooms, and residential homes resulting in a large number of then novel applications of behavior modification (Ullman & Krasner, 1965).

Behavior modification thus retained many characteristics of operant psychology such as empiricism; highly reliable direct observation of behavior, sometimes measured by devices analogous to cumulative recorders; and, an emphasis on environmental control of behavior, from antecedent stimuli, such as reinforcer deprivation and discriminative stimuli, and contingencies of reinforcement and punishment. Likewise, treatments were operationally defined, sometimes delivered through technological devices that measured behavior and delivered antecedents and consequences reliably. At this time, functional assessment and analysis tended to be done informally and perhaps not reported explicitly and systematically as in later applied behavior analysis, although hints at this framework are to be found in these publications.

Early behavior modification resulted in a broad technology of behavior change, often with specific treatments tied with specific problem behaviors, such as treatment packages for stuttering, toilet training, psychotic speech, etc. (Ullman & Krasner, 1965). Early behavior modification of aggressive behavior was highly influenced by Azrin's work (Azrin, 1970; Azrin et al., 1968) demonstrating the operant nature of aggressive behavior in both non-humans and humans. This early behavioral work addressed aggressive behavior through contingency management procedures, token economies, and group contingency procedures, such as the good behavior game among many other procedures.

16.2.2 Reciprocal Inhibition and Anger

Wolpe (1954) defined reciprocal inhibition as "the complete or partial suppression of the anxiety responses as a consequence of the simultaneous evocation of other responses physiologically antagonistic to anxiety" (p. 205). Wolpe's definition of reciprocal inhibition is not specific to anxiety, but, rather, was applicable to a range of undesirable emotions which included anger. In fact, Wolpe described four responses that could be used for reciprocal inhibition: Assertive responses, sexual responses, relaxation responses, and conditioned avoidance responses. In his description of assertive responses, Wolpe described a man who became hurt when his family criticized him and he responded with either withdrawal and sulkiness or *angry behavior*. Wolpe formulated that his anxiety inhibited expressing resentment (Cf. Estes & Skinner, 1941) and this treatment involved gradually learning to express resentment assertively which resulted in suppression of his anxiety and thus in a reduction in angry behavior. This case study became a model for more systematic evaluations of reciprocal inhibition to treat angry behavior.

Edmondson and Conger (1996) published a systematic review and meta-analysis of the anger management literature that included studies using Wolpe's approach. They identified 18 studies of which ten studies with 17 treatment groups provided sufficient information to calculate effect sizes. The overall effect sizes were medium to large with Cohen's d being −0.82, 0.60, 0.76, and 0.64 for relaxation, social skills, and cognitive-relaxation cognitive treatment, respectively. Further, the effect sizes on angry behavior were larger for social skills training ($d = 1.13$) than cognitive interventions ($d = 0.34$, 1 study only); there were no reports of effect sizes on angry behavior for cognitive-relaxation and relaxation alone. Many treatment packages now incorporate procedures that look like reciprocal inhibition procedures, such as graded exposure and social skills training, that include assertiveness training, suggesting that this approach is now widely accepted as a common component of anger management procedures.

16.2.3 Operant Treatment of Aggressive Behavior

The use of consequences to reduce aggression was one of the earliest behavioral methods to treat aggression in typical children and adolescents and children with intellectual and other disabilities both individually (Friman et al., 1986; Luce et al., 1980) and in group settings such as classrooms. Interventions included reinforcing the absence of aggression and reinforcing specific other or alternate behavior, and punishing aggressive behavior either by the presentation of aversive stimuli (positive punishment), such as physical restraint, or loss of privileges or access to pleasant activities (negative punishment).

An early example of operant treatment of aggression comes from Zeilberger et al. (1968) who trained Rory's mother to implement contingency management procedures at home. Rory was nearly 5 years old and had an average IQ. Seilberger et al. trained his mother to give clear instructions, attend to him, and give him special toys when he played appropriately and when he followed instructions. She also used time out in his bedroom, where she had ensured there were no toys or fun things. Using an ABAB experimental design, the authors demonstrated that these procedures resulted in large increases in compliance, accompanied by large reductions in yelling and aggression. These procedures are so well researched today that they are often incorporated into parent and staff training packages often alongside other more contemporary behavior management strategies to reduce aggression.

16.2.4 Token Economies

Building on earlier basic human work on schedule performance in hospital patients with schizophrenia (Lindsley, 1956), the token economy was developed to apply contingency management initially with groups of people with severe mental illness, but subsequently with a wide range of populations (Kazdin, 1982, 2012; Soares et al., 2016). Researchers have also used token economies with individuals and families and with individuals to manage one's own behavior by administering one's own token economy (Nelson et al., 1991). One of the main features of a token economy is that, unlike basic learning studies, where food can be used as a reinforcer in hungry animals, some alternative reinforcer that is robust to the effects of reinforcer satiation and that is convenient to deliver is needed. Fortunately, most humans can learn to use tokens quickly, thus circumventing the numerous problems of primary reinforcers such as relatively rapid reinforcer satiation. In addition, token economies can include a levels system whereby individuals are gradually weaned off explicit immediate contingencies mediated by others to self-managing their own behavior and more delayed and natural consequences. Thus, token economies are really seen as a broad strategy for increasing appropriate behavior initially with a very structured approach which is then followed by progressive fading out of that support which is replaced with forms of behavioral self-management. Token economies also depend on the behavior of caregivers to administer tokens and other elements of the program accurately and consistently; hence, caregiver training and supervision are also essential elements of effective token economies (Ayllon & Michael, 1959). Token economies have been applied to individuals and groups in numerous contexts, such as general and special education classrooms, prisons, and youth facilities, and to a wide range of problems from lining up for a meal to stuttering, and, indeed, aggression.

An illustration of the powerful effects token economies can have on aggression comes from LePage et al. (2003), who reported on the effects of a token economy on multiple measures of aggression over a 2-year period. The compared injury data from 1-year prior to the introduction of the token economy to the 2-year period when the token economy was in operation. The study took place in a 24-bed in-patient psychiatric unit predominantly for individuals aged 18–20 years. During baseline, 316 individuals were admitted to the unit and during the token economy 553 individuals were admitted with overall admission and discharge rates of approximately 22–26 individuals per month. Introduction of the token economy had a large impact on injuries: There was a 33% reduction in all injuries, 48% reduction in patient-patient injuries, 21% reduction in employee injuries, and 93% reduction in staff time lost due to injuries. Thus, the token economy was highly effective in reducing aggression in a challenging population, but also resulted in programmatic benefits, such as increased resident, staff and unit safety and stability, and presumed large economic benefits from avoidance of costs of staff injuries. A systematic review of seven studies published between 1993 and 2013 in in-patient psychiatric units concluded that token economies were highly effective in reducing aggression and violence in these settings (Glowacki et al., 2016).

16.2.5 Good Behavior Game

As well as token economies, there are other group contingency procedures that are designed to be practical in typical settings and that address aggression effectively. Barrish et al. (1969) developed a group contingency management procedure for use in classrooms. In the original good behavior game the teacher divided the classroom into two teams. The teacher posted behavioral rules and every time the children broke a rule, she made a mark on the chalkboard. If a team had the fewest marks or both teams had fewer than five marks, then the winning team(s) would access various positive consequences, such as wearing victory tags, having a star placed next to their name on a winner's chart, going to lunch first, etc. The study took place in a fourth grade class with 24 students and the good behavior game resulted in large improvements in classroom behavior such as reductions in calling out and out of seat behavior. (The original study did not measure aggression directly.) The procedure can be considered an example of Differential Reinforcement of Lower rates of behavior (Tingstrom et al., 2006).

The good behavior game has been widely replicated over the last 50 year (Bowman-Perrott et al., 2016; Tingstrom et al., 2006). For example, a recent meta-analysis by Bowman-Perrott et al. (2016) identified 21 small *N* experiments with 1,580 pre-K through 12th grade students. The overall effect size was large (Tau-$U = 0.87$) with substantial reductions in problem behavior which included several kinds of aggression and increases in appropriate behavior.

The good behavior game has been identified as a best practice in primary and secondary prevention of problem behavior, including aggression in young children and youth, by both researchers and several government agencies (Embry, 2002). Embry described it as "a behavioral vaccine" implying it should be administered to the entire population to prevent aggression and violence.

16.2.6 Commentary

There are many approaches to reducing aggression and violence using behavior modification. The two principal approaches are those based on Wolpe's work on reciprocal inhibition and a variety of contingency management approaches. Although Wolpe's work addressed individualization through clinical case formulation, contingency management approaches tend not to be explicit in this regard, although there is a considerable body of research supporting this approach to reducing aggression and violence.

16.3 Applied Behavior Analysis

16.3.1 General Approach

Behavioral approaches to psychopathology (Skinner, 1953; Sturmey, 1996, 2008, 2020), including aggression and violence, hypothesizes that problematic behavior comes from a kind of adaptation to society's aversive control of its members' behavior. (See Chap. 13 for a more detailed description.) When it comes to both understanding the causes of aggression and its treatment, ABA emphasizes the current environment and the immediate antecedents and consequences that influence aggression. While acknowledging the potential importance of biological, cultural, and developmental-historical causes of aggression, ABA focuses on identifying and altering the current independent variables of which aggression and other relevant behavior is a function. This is done because almost always those other variables cannot be assessed in a reliable and valid manner for a particular individual and, also, they cannot be manipulated and hence have limited utility for understanding and treating aggressive behavior. Behavior analysts also ground their interventions within the framework of behavior analytic theory and concepts. Thus, they link their interventions to behavior analytic concepts such as schedules of reinforcement, operant extinction, etc. They use small *N* experiments to assess if the treatment caused the change in behavior. Behavior analysts also describe treatments in a highly operational manner and, rather than rely on professional experience or the presence of manuals, they measure treatment delivery using observational data. Aware of the issue of generalization and maintenance of treatment effects over time, they program generalization and maintenance. Finally, they assess the meaningfulness of the behavior change by asking participants if they value the goals, methods, and outcomes of the treatment (Baer et al., 1968; Cooper et al., 2019; Wolf, 1978).

16.3.1.1 An ABA Intervention for Aggression

Let us consider this example of an ABA approach to reduce aggressive behavior that emphasizes control of aggressive behavior from the current environment. Bay-Hinitz et al. (1994) compared the effects of cooperative and competitive games on the aggressive and cooperative behavior of 70 preschool children aged 4–5 years in four classes in three preschools. They defined competitive games as those in which only some children won and cooperative games as those in which there was no winner. Aggression referred to behavior such as hitting peers and cooperative behavior included behavior such as sharing materials. As expected, competitive games resulted in increased aggression and decreased cooperation, both during the time periods when the games were played and generalization also occurred in subsequent free play periods. This example shows the use of clear definition and manipulation of a treatment (competitive vs. noncompetitive games), reliable measure of behavior that is socially important in naturalistic settings with typical change agents, and the use of small *N* experimental designs to evaluate treatment effects.

The previous example shows an effective approach to reducing aggression. It is not, however, as analytic as some ABA approaches as it did not demonstrate the role of ABA concepts such as contingencies of reinforcement. It is unclear exactly what it was about competitive games that resulted in increased aggressive behavior.

A second more explicitly analytic ABA approach to reducing aggressive behavior comes from Carr and Durand (1985) who conducted a functional analysis of disruptive and aggressive behavior in four children aged 7–13 years with severe developmental disabilities. In the first experiment, which took place before intervention, the authors conducted a functional analysis. They placed the children in four test conditions to determine the consequences maintaining aggression/disruption in order to identify the variables of which aggression was a function and in order to design the most effective function-based treatment of aggression for each child, they exposed the children to four conditions prior to intervention. The four conditions were: easy task/low attention, difficult task/low attention, easy task/high attention, and difficult task/high attention. The rates of aggression and disruption varied systematically for each child. That is some children's aggression increased only with difficult tasks and other children's aggression only increased with lack of attention.

In the second experiment, the authors evaluated treatments to teach communication responses there were or were not based on the results of the functional analyses for each child's behavior. For children whose aggression was maintained by attention, the function-based treatment was to ask "Am I doing good work?" as a way to obtain adult attention, and for children whose aggression was maintained by avoidance of difficult tasks, the function-based treatment was to ask "I don't understand" to obtain adult assistance. The authors taught the children both responses systematically to fluency. Experiment 2 demonstrated that aggression only reduced when the functionally equivalent replacement behavior, the one that matched the function of the child's aggression, was taught and increased. Thus, effective treatment did not consist of increasing *any* form of communication behavior, but *only* increasing communication that served the same function as the child's aggression. This approach has proven effective for a wide range of problem behavior in many populations (Tiger et al., 2008).

16.3.1.2 Summary

ABA's approach to reducing aggression is based on an understanding of biological evolution, cultural practices, and lifelong learning that may result in the selection of aggressive behavior and elimination or attenuation of adaptive behavior during the individual's lifespan. Unlike other behavioral approaches, ABA emphasizes pretreatment functional analysis to identify the current contingencies that select and maintain aggressive behavior and this information is then used to design idiographic, function-based interventions that match the function of each person's aggression.

ABA has identified at least three common causes of aggressive behavior each with their own implied family of treatments. These are pain-elicited aggression, operant aggression, and aggression during operant extinction. The next three sections will consider each of these models and the implications for treatment.

16.3.2 Pain-Induced Aggression

Skinner (1953) and Michael (1993) both considered the phenomenon of pain-elicited aggression from a behavior analytic approach. They suggested that pain may function as an Unconditioned Elicitor of aggressive behavior. This reflects the effects of biological evolution in that there is an evolutionary advantage and selection pressure for organisms to be aggressive to conspecifics when they experience pain. In addition, pain may function as an Unconditioned Establishing Operation. That is, pain both: (a) elicits behavior that has been reinforced in the past in the presence of pain, and (b) momentarily increases the reinforcing value of reducing pain. Finally, operant processes may be at play when pain elicits behavior such as teeth on other organism's flesh which can then be subject to its consequences to select forms of behavior that: (a) reduce pain more efficiently and (b) access other reinforcers, such as flesh to eat, conspecifics being less aggressive and moving away, and access to resources that were blocked by the presence of a conspecific.

Skinner (1953) also addressed aggression and anger when he considered the meaning of the everyday term "emotion" as exemplified by "anger." He suggested that the term "emotion" refers to behaviors that "hang together" because they occur in a common sequence and share common reinforcers and antecedents. The angry person shows a likelihood of certain kinds of behavior—striking, inflicting injury, insulting—and a low probability of other behavior—aiding, comforting, and helping. Thus, angry behavior might include, among other behavior, both raising one's arm with clenched fist and striking another person (a sequence of angry behavior) and both responses are reinforced by inflicting damage and injury to the other person or the physical environment. Such angry behavior might be shown in the presence of Unconditioned Elicitors, such as the careless person who is treading on one's toes, and also Conditioned Elicitors, such as the sight of the careless person staggering and approaching oneself immediately after having trod on one's toes. After enough pairing perhaps the second person's name or voice alone comes to elicit some part of the first person's angry behavior, such as physiological arousal or an angry glance at the second person, which results in the second person backing off, thereby negatively reinforcing the first person's angry behavior.

Curiously, although there is an extensive basic literature on pain-elicited aggression, there is relatively little treatment research that directly acknowledges this framework. Basic behavioral research has shown that pain-elicited aggression can be reduced by positive punishment procedures (Azrin, 1970), but such procedures would almost always be considered unethical in applied settings. Three obvious function-based interventions suggest themselves: (a) Eliminating the pain and the associated Conditioned Establishing Operations, (b) teaching another form of attenuating the pain, and (c) reducing pain-induced behavior through consequences such as positive punishment or extinction. In everyday life we do several things that looks like all these mechanisms. We give someone a muscle relaxant to reduce the pain of a medical procedure or put on headphones at the dentist to reduce the stimulation from the drill in part to stop the person lashing out when hurt; we listen to relaxing music and practice relaxation exercises at the dentist to reduce the pain of the procedure; and, we tell the child who tenses up and looks away when taking an injection that they are brave and give them a lollipop or restrain them tell them they are a baby if they push the nurse away and cry.

16.3.3 Reinforcement of Aggression

The notion that aggression is reinforced is a common one: It underlies everyday ideas such as acquisitive offenses related to aggression, cold, proactive aggression, the bully obtaining money and candy from the victim, and the violent spouse whose every whim is indulged to avoid violence.

There is good science to back up these everyday ideas. Nearly half of published studies on functional analysis address aggression, showing that it may be reinforced by: (a) positive reinforcers, such as attention, attention from specific people, certain forms of attention; specific tangibles, such as toys, food, drinks, etc., and access to preferred behavior, such as talking about specific topics or stereotyped behavior; and (b) negative reinforcers, such as escape from difficult academic self-help tasks or aversive environments, for example, that are too noisy (Beavers et al., 2013; Hanley et al., 2003). This literature relates most to individuals with developmental disabilities, although there are also examples involving people with dementia, acquired brain damage, etc.

These ideas have permeated many areas of practice, including many aspects of mental health (Sturmey, 2020) and forensic services. For example, Singh and Strand (2008) developed a function-based intervention for aggression in an individual with mental health needs. These authors worked with a challenging 35-year-old in-patient, Carl, who resided in a maximum security psychiatric unit because of a history of severe aggression and inappropriate sexual behavior. He has been in in-patient units for 9 years. He had multiple diagnoses including borderline intelligence, post-traumatic brain injury psychosis, and substance abuse issues. The main problem was aggression, which include property destruction and self-injury, and inappropriate sexual behavior, including public masturbation and a charge of sexual battery of his mother and sodomy against a fellow resident. (None of Carl's diagnoses had resulted in effective treatment for aggression after 35 years.)

The authors conducted the assessment and treatment from a strengths-based approach using person centered planning. Assessment of the functions of his aggression included a modified *Questions About Behavioral Function in Mental Illness* (Singh et al., 2006) the *Functional Analysis Interview* (O'Neil et al., 1997), interview with the client, assessment of topographies of aggression and chart reviews. All methods converged on the conclusion that Carl's aggression and inappropriate sexual behavior were reinforced by staff attention. Carl agreed with this formulation, but stated that he lost control of his behavior early on and could not control himself. He agreed to participate in a standardized 16-week anger management program. After 3 weeks, a specific antecedent for problem behavior was identified as staff nudging him on the shoulder to get him out of bed which put him in a bad mood when he woke up. He preferred to be woken up by staff rubbing him on his back which put him in a good mood when he awoke. This strategy alone eliminated many incidents of aggression which occurred early in the morning. Carl also liked the idea of "being cool" and learned relaxation skills related to this notion. He then began to formulate goals about his future outside of a locked in-patient facility. Carl's aggression reduced from around 10 per month to near zero levels within 4 months of intervention. His inappropriate sexual behavior was also almost completely eliminated around the same time. Follow-up at 5 months and 2 years revealed that treatment effects were maintained and he had been placed successfully in a less secure unit.

To date true functional analyses of aggression behavior in people with psychiatric disorders are rare, indeed, as is empirical demonstration of the interaction of function and treatment outcomes. Nevertheless, the functional approach to conceptualizing and treating aggression is promising.

16.3.4 Extinction-Induced Aggression

One example of understanding and treating aggression based on ABA concepts comes from the phenomenon of extinction-induced aggression. Extinction of operant behavior refers to procedures in which the consequence maintaining behavior is identified and the contingency between the response and its consequence is disrupted, either by eliminating the consequence or presenting it independently of the response on a time-based schedule. Operant extinction is accompanied by reduction in the operant behavior, increase in response variability, and increased emotional responding.

One example of emotional side effects which sometimes occurs is an increase in aggression, which has been shown in both non-humans (Azrin et al., 1966) and humans (Lerman et al., 1999). Since many behavioral interventions for aggression might include extinction, procedures have been developed to mitigate extinction-induced aggression, such as reinforcing other desirable behavior or the use of fixed-time schedules of reinforcement during extinction (Lerman et al., 1999; Vollmer et al., 1998).

16.3.5 Comment

ABA approaches to aggression and violence have made two important contributions to aggression and violence. The first is a very large body of experimental evidence—hundreds of small *N* experiments—that support the effectiveness of a function-based approach to treat aggression. Most of this literature is of course related mostly to children and adults with ID/ASD and children in public settings where observable data can be collected readily. A second important contribution of ABA is that it provides a conceptual framework that can be readily applied to conceptualization, case formulation, and treatment of aggression in other contexts such as forensic settings, etc.

16.4 Third Wave Behavior Therapy

Third wave behavior therapies are a group of psychotherapies based on differing degrees of behaviorism or broad behavioral procedures. The most widely known and most explicitly behavior analytic is ACT which is an extension of and application of Skinner's (1957) *Verbal Behavior* to understanding the role of verbal behavior in psychotherapy using Relational Frame Theory (Hayes et al., 2001). ACT analyzes the role of verbal behavior—language in everyday terms—on behavior. ACT has quite a large empirical base evaluating its use in a wide range of contexts and problems and does indeed touch upon its use to change aggression and violence. A second example of a behavior analytic therapy is Functional Analytic Psychotherapy (FAP) (Kohlenberg et al., 1993; Kohlenberg & Tsai, 1994). FAP has a much more modest empirical basis than ACT. A more recent example of a behavioral psychotherapy is mode deactivation therapy (MDT) (Apsche et al., 2002; Apsche & Ward Bailey, 2003a, b) which also has a modest empirical basis, but has been applied to reduce aggression and violence.

There are several other third wave therapies, which, although not explicitly based on behaviorism, include behavioral procedures and are perhaps distinct from traditional CBT, in that they use behavioral concepts or self-management procedures. DBT (Linehan, 1993), originally developed for Borderline Personality Disorder and related suicidal and other problematic behavior, uses behavioral concepts such as behavior chains and disruption of behavioral chains. DBT has been extended in many ways including application to anger and aggression (Frazier & Vela, 2014). Finally, mindfulness has become a very popular form of psychotherapy. Based on Buddhist philosophy and using Buddhist practices such as medication, it also uses some basic behavioral procedures such as interruption of behavior chains, self-management, and relaxation training among others. These third wave therapies have also influenced more traditional forms of CBT, both by incorporating specific third wave psychotherapy techniques into mainstream CBT, but also by providing a radical behavioral perspective on verbal processes in traditional psychotherapy, including CBT.

16.4.1 Acceptance and Commitment Therapy

ACT is based on RFT. (See Chap. 9.) From an RFT perspective, psychopathology is characterized by two problems. The first problem is cognitive fusion. Hayes (2004) described this as "This combination of features means that stimulus functions from relational frames typically dominate over other sources of behavioral regulation in humans (what we term 'cognitive fusion'), making an individual less in contact with here-and-now experience and direct contingencies" (pp. 17–18). The second is experiential avoidance and the failure of suppression. Hayes (2004) described this as "the attempt to escape or avoid private events, even when the attempt to do so causes psychological harm" (p. 16).

ACT was designed to address these two problems. ACT consists of two components: Acceptance, an approach to managing problematic thoughts and emotions, and Commitment, an action-based, relatively traditional behavior therapy approach to changing behavior. Acceptance strategies address covert behavior, such as problematic thinking and feeling. Rather than control problematic thinking and feeling, as in much CBT, acceptance encourages the person to fully experience the problematic here and now and not to suppress it in any way. There are numerous acceptance exercises, such as the "milk, milk, milk" exercise in which the client repeats a word aloud until it has no more meaning and the word is experienced as just a sound. This exercise may be done with words that are problematic for the client, such as "anxious" or "fear." Commitment involved overt behavior change using fairly traditional behavior therapy procedures such as establishing goals, achieving goals in small steps, etc. ACT's approach to overt behavior change is a little different from traditional behavior therapy or traditional CBT, however, because it focuses on goals that open up the person's experiences that have been narrowed by experiential avoidance and because it focuses on changing behavior in a way that is consonant with the person's core values (Hayes, 2004).

Several studies have evaluated the effectiveness of third wave therapies for anger, aggression, and violence. This has resulted in consideration of anger, hostility, and aggression from an ACT perspective (Berkout et al., 2019), a treatment manual (Eifert et al., 2006), and a modified form of ACT designed specifically for anger and aggression *Achieving Change Through Values-Based Behavior* (Lawrence et al., 2014). Thus, there is a slowly growing literature on ACT to treat anger, aggression, and violence.

In an early experimental study, Zarling et al. (2015) reported an RCT for interpersonal violence which compared ACT with an attention-placebo control condition. The attention control condition consisted of group therapy support and discussion which included elements likely to engender change, such as hope, modeling, sharing information, etc. but without any likely effective ACT components. One hundred individuals who reported engaging in at least two incidents of intimate partner violence in the last 6 months were randomized to ACT or placebo. Both groups received 12 weeks of treatment and follow-up data were collected at 3 and 6 months post-treatment. There were small to large reductions in psychological and physical aggression at 3 and 6 months follow-up and the ACT group also reported reduced experiential avoidance and improved emotional regulation. In addition, experiential avoidance and emotional regulation both mediated outcomes of ACT.

Other studies have also reported positive outcomes of ACT for aggression in a variety of different contexts and populations (e.g., Livheim et al., 2020). Using administrative data, Zarling et al. (2019) evaluated the rates of reoffending 12 months post-treatment in 3,474 men who were court-mandated non-randomly to one of two batterer intervention programs—either traditional combined Duluth and CBT program or *Achieving Change Through Values-Based Behavior*. The authors found that reoffending broadly defined, violent reoffending, and reoffending involving intimate partner violence were all lower in the ACT group than the traditional Duluth plus CBT therapy group, although dropout was also higher in the ACT group.

There is a steadily growing literature on ACT for the treatment of anger, aggression, and violence (Byrne et al., 2021; Yildiz, 2020). For example, Byrne and Ghráda's (2019) systematic review of third wave psychotherapies in correctional and forensic settings identified eight evaluations of ACT, one of compassion focused psychotherapy, and no adequate studies of meta-cognitive therapy and FAP. Of the eight studies of ACT, four were RCTs, two were within group, and two were between group comparisons. Five ACT studies focused on aggression, anger, and/or impulsivity. The one true RCT (Malouf et al., 2017) found no effect of ACT on measures of self-control and impulsivity. Thus, the number of well conducted studies on ACT for treatment of aggression and related concepts is slim. Thus, since Zarling et al. (2015) found significant effects of ACT on aggression for interpersonal violence and Malouf et al. (2017) for no effects on aggression-related measures, the jury is still out as to the effectiveness of ACT for aggression, but perhaps ACT is a promising approach to treating aggression and violence in some contexts, but the literature is relatively very small compared to established treatments such as CBT.

16.4.2 Functional Analytic Psychotherapy

FAP is a therapy informed by radical behaviorism which uses the client–therapist relationship and individual psychotherapy sessions as a microcosm of the clients' life (Kohlenberg & Tsai, 1994). It assumes that the behavior, including problem behavior, observed in psychotherapy mirrors that outside of therapy and that function-based interventions within the psychotherapy session can produce change outside of the therapy session (Kohlenberg & Tsai, 1994). FAP distinguishes clinically relevant behaviors that occur within sessions, rather than contrived within sessions opportunities such as may occur during social skills training. These naturally occurring clinically relevant behaviors are believed to be more powerful than imagining such occurrences or talking about such occurrences. In FAP sessions the therapist uses, identifies, or triggers problematic interactions that occur during the session and the therapist provokes the client to analyze what is going on and what avoidance has occurred. The FAP therapist attempts to use naturally occurring, uncontrived social reinforcement to increase appropriate client behavior within and outside of the therapy sessions. This may result in a shift from the client dealing with non-social problems to dealing with interpersonal problems outside of therapy, such as assertively expressing disappointment to others.

There is only a slim outcome literature on FAP (Kanter et al., 2017) which does not particularly focus on treatment of aggression and violence. More recently, however, Newring and Wheeler (2010, 2012) have extended FAP to the analysis and potential treatment and integrated risk assessment considerations with juveniles who have committed sexual violence and Callaghan's (2006) FAP manual addresses interpersonal conflict and aggression but only one experimental study has evaluated FAP for anger directly. Siroos et al. (2021) reported an experimental study in which 24 individuals with borderline personality disorder were randomized to FAP or a control group. Participants in the FAP group, but not the control group, reported statistically significant reductions in three measures of anger. Thus, the evidence to support the use of FAP for aggression so far is slim.

16.4.3 Dialectical Behavior Therapy

Another third wave therapy is DBT, a treatment originally developed for treatment of borderline personality disorder, self-harm, and suicide in women (Linehan, 1993) but which was subsequently extended to other problems including aggression and violence. DBT aims to treat emotional dysregulation and suicidal behavior by teaching mindfulness, distress tolerance, emotional regulation, and interpersonal effectiveness. DBT is delivered via weekly therapy sessions, weekly skills training, weekly therapist consultation, and 24-h phone skills coaching. Once established for treatment of Borderline Personality Disorder, research extended its use to aggression and violence which required modifications in some contexts.

In a recent systematic review and meta-analysis of the application of DBT to aggression and violence, Frazier and Vela (2014) identified 21 studies, many of which documented treatment modifications and most of which used pre-post designs. There were nine studies, eight of which were with outpatients and one with a correctional population. Seven studies involved adults with BPD. Eight of nine of these studies found DBT to be effective in reducing anger or aggression as measured with self-report instruments.

The authors noted that although there was preliminary evidence of effectiveness of DBT for anger and aggression, most of the comparison groups involved treatment as usual. There was no comparison with standardized evidence-based treatments for aggression, such as CBT, and the one study that compared DBT with another form of psychotherapy found that DBT was inferior to transference-focused psychotherapy. Thus, the outcome literature on DBT and anger and aggression is rather smaller than for broad-based CBT and has focused on individuals with BPD.

16.4.4 Mindfulness

Kabat-Zinn (1982) defined mindfulness as "paying attention in a particular way: on purpose, in the present moment, and nonjudgmentally" (p. 4). Mindfulness refers to a series of related interventions that include focusing attention on the present moment, breathing, and somatic sensations; breathing exercises, such as slowing down breathing; daily meditation; self-management strategies, such as interruption of increases in arousal, self-instruction; and increasing empathy and compassion toward others, etc. Mindfulness shares certain approaches to managing distress with both traditional behavior therapy and CBT, e.g., interruption of response chains, and use of relaxation as a preventative strategy or as a strategy to use as an alternate behavior when one becomes angered. Mindfulness also shares certain features with other third wave behavior therapy, e.g., experiencing and tolerating unpleasant thoughts and feelings, rather than trying to change or fight them. Elements of mindfulness have permeated other third wave behavior therapies and, indeed, daily life for many people and the media to the extent it could be seen as a cultural meme, commercial product, or fad.

Recently, mindfulness has become a very popular intervention in areas such as parenting (Singh et al., 2007), supporting direct care staff (Singh et al., 2009), and treatment in many mental health contexts as attested to by multiple meta-analyses of mindfulness generally and its application to specific populations, contexts, and mental health diagnoses (Bartlett et al., 2019; Dunning et al., 2019; de Abreu Costa et al., 2019; Goldberg et al., 2019; Khoury et al., 2013; Wright et al., 2009). There is also an extensive literature on its application and outcomes related to anger, aggression, and violence. An early systematic review of treatment of aggression in adolescents and adults with mindfulness procedures comes from Fix and Fix (2013) who searched the literature up to 2012. (They did not include mindfulness studies with caregivers or significant others.) They located four group studies and seven single subject studies, six of which evaluate the Soles of the Feet meditation intervention. The soles of the feet meditation consist of: (1) recognizing situations that trigger arousal or aggression; (2) finding a natural, non-aggressive posture, (3) breathing naturally, (4) experiencing rather than controlling emotions without trying to stop or control them, (5) shift attention to the soles of the feet until calm, and (6) walking away with a half-smile or other appropriate expression without aggression (Singh et al., 2003). Fix and Fix concluded that the results of the group designs were mixed. They also concluded that the results of the single subject experiments provided solid evidence of the effectiveness of mindfulness training in reducing aggression.

In a second systematic review which included studies of mindfulness and aggression in individuals aged 18 years and older, Gillions et al. (2019) identified 22 studies of which seven focused on mindfulness. (Studies they reviewed using ACT or DBT will not be considered here.) The papers delivered mindfulness over 3–20 weeks. Six studies found positive outcomes with effect sizes of 0.24–0.84, although studies were generally of low quality. (There were also four laboratory-based studies using very brief mindfulness interventions in analog situations involving aggression.)

The tremendous interest in mindfulness as a mental health intervention has resulted in an explosion of interest from researchers and practitioners alike. Since no meta-analysis of mindfulness to treat aggression or violence was identified it is hard to specifically comment on how effective mindfulness is; however, the quantity of evidence is large, but not quite so large as for typical CBT. Thus, one can be moderately confident that mindfulness interventions may be an effective treatment at least in some contexts.

16.4.5 Summary

Third wave therapies have had a large impact on psychotherapy over the last 30 years. They were developed for typical outpatient mental health issues such as depression and anxiety but over time their application has been expanded considerably to address many other problems including aggression and violence. In addition to direct application to aggression and violence applications of third wave behavior therapies to more broadly defined interpersonal problems may also be relevant to the treatment of aggression and violence in some cases.

ACT is the oldest of these third wave therapies and has the greatest quantity of research to support its use both generally and specifically for aggression and violence. Although other third wave therapies, BDT and mindfulness in particular, have a growing literature that addresses aggression and violence, the quantity of the literature is much smaller than that for ACT. Hence, of the third wave therapies that are currently available we can be most confident in the use of ACT, although its application to aggression and violence is limited at this time.

16.5 Chapter Summary

Individual psychotherapies for aggression and violence are dominated by CBT. We have the most information about outcomes for CBT and the most information in a wider variety of contexts. Although the range of outcomes for CBT—effect sizes and number of responders—is likely to be quite different in different contexts, we have more information upon which to make decisions for CBT than any other treatments. Various forms of behavioral treatment, including behavior modification and ABA, show that these procedures have the greatest support in the areas of children and adolescents, children and adults with disabilities, and smaller amounts of research for other populations with disabilities such as Alzheimer's disease, severe mental illness, and TBI. It seems relatively robust that interventions based on functional assessment and functional analysis are more likely to produce better outcomes, avoid the use of positive punishment, and more likely to involve skills teaching than behavioral approaches that do not use these pre-treatment procedures. Ideas from ABA have permeated some forms of psychotherapy including third wave behavioral procedures and some approaches to clinical case formulation (Sturmey, 2020; Sturmey et al., 2019) and forensic case formulation (Sturmey & McMurran, 2011).

The next chapter goes on to consider a wide range of other psychological interventions for aggression and violence. These include couples therapy, psychotherapy, and a variety of other psychosocial treatments.

References

de Abreu Costa, M., de Oliveira, G. S. D. A., Tatton-Ramos, T., Manfro, G. G., & Salum, G. A. (2019). Anxiety and stress-related disorders and mindfulness-based interventions: A systematic review and multilevel meta-analysis and meta-regression of multiple outcomes. *Mindfulness, 10*(6), 996–1005.

Apsche, J. A., & Ward Bailey, S. R. (2003a). Mode deactivation therapy (MDT): A theoretical case analysis (part II). *Behavior Analyst Today, 4*(4), 395.

Apsche, J. A., & Ward Bailey, S. R. (2003b). Mode deactivation therapy (MDT): A theoretical case analysis (Part III). *Behavior Analyst Today, 5*(3), 324.

Apsche, J. A., Ward, S. R., & Evile, M. M. (2002). Mode deactivation therapy (MDT): A functionally based treatment. *Behavior Analyst Today, 4*(3), 342.

Ayllon, T., & Michael, J. (1959). The psychiatric nurse as a behavioral engineer. *Journal of the Experimental Analysis of Behavior, 2*(4), 323–334.

Azrin, N. H. (1970). Punishment of elicited aggression. *Journal of the Experimental Analysis of Behavior, 14*(1), 7–10.

Azrin, N. H., Hutchinson, R. R., & Hake, D. F. (1966). Extinction-induced aggression. *Journal of the Experimental Analysis of Behavior, 9*(3), 191–204.

Azrin, N. H., Rubin, H. B., & Hutchinson, R. R. (1968). Biting attack by rats in response to aversive shock. *Journal of the Experimental Analysis of Behavior, 11*(5), 633–639.

Baer, D. M., Wolf, M. M., & Risley, T. R. (1968). Some current dimensions of applied behavior analysis. *Journal of Applied Behavior Analysis, 1*(1), 91–97.

Barrish, H. H., Saunders, M., & Wolf, M. M. (1969). Good behavior game: Effects of individual contingencies for group consequences on disruptive behavior in a classroom. *Journal of Applied Behavior Analysis, 2*(2), 119–124. https://doi.org/10.1901/jaba.1969.2-119

Bartlett, L., Martin, A., Neil, A. L., Memish, K., Otahal, P., Kilpatrick, M., & Sanderson, K. (2019). A systematic review and meta-analysis of workplace mindfulness training randomized controlled trials. *Journal of Occupational Health Psychology, 24*(1), 108.

Bay-Hinitz, A. K., Peterson, R. F., & Quilitch, H. R. (1994). Cooperative games: A way to modify aggressive and cooperative behaviors in young children. *Journal of Applied Behavior Analysis, 27*(3), 435–446.

Beavers, G. A., Iwata, B. A., & Lerman, D. C. (2013). Thirty years of research on the functional analysis of problem behavior. *Journal of Applied Behavior Analysis, 46*(1), 1–21.

Beck, R., & Fernandez, E. (1998). Cognitive-behavioral therapy in the treatment of anger: A meta-analysis. *Cognitive Therapy and Research, 22*(1), 63–74.

Berkout, O. V., Tinsley, D., & Flynn, M. K. (2019). A review of anger, hostility, and aggression from an ACT perspective. *Journal of Contextual Behavioral Science, 11*, 34–43.

Bowman-Perrott, L., Burke, M. D., Zaini, S., Zhang, N., & Vannest, K. (2016). Promoting positive behavior using the Good Behavior Game: A meta-analysis of single-case research. *Journal of Positive Behavior Interventions, 18*(3), 180–190.

Brännström, L., Kaunitz, C., Andershed, A. K., South, S., & Smedslund, G. (2016). Aggression replacement training (ART) for reducing antisocial behavior in adolescents and adults: A systematic review. *Aggression and Violent Behavior, 27*, 30–41.

Byrne, G., & Ghráda, Á. N. (2019). The application and adoption of four 'third wave' psychotherapies for mental health difficulties and aggression within correctional and forensic settings: A systematic review. *Aggression and Violent Behavior, 46*, 45–55.

Byrne, G., Ghráda, Á. N., O'Mahony, T., & Brennan, E. (2021). A systematic review of the use of acceptance and commitment therapy in supporting parents. *Psychology and Psychotherapy: Theory, Research and Practice, 94*, 378–407.

Callaghan, G. M. (2006). Functional analytic psychotherapy and supervision. *International Journal of Behavioral Consultation and Therapy, 2*(3), 416.016.1176686.

Carr, E. G., & Durand, V. M. (1985). Reducing behavior problems through functional communication training. *Journal of Applied Behavior Analysis, 18*(2), 111–126.

Cooper, J. O., Heron, T. H., & Heward, W. L. (2019). *Applied behavior analysis*. Pearson. (Global Edition).

David, D., Cotet, C., Matu, S., Mogoase, C., & Stefan, S. (2018). Fifty years of rational-emotive and cognitive-behavioral therapy: A systematic review and meta-analysis. *Journal of Clinical Psychology, 74*(3), 304–318.

Debidin, M., & Dryden, W. (2011). A systematic review of the literature on the use of rational emotive behaviour therapy in criminal justice work to reduce re-offending. *Journal of Rational-Emotive & Cognitive-Behavior Therapy, 29*(2), 120–143.

Drossel, C., & Rummel, C. (2009). Assessment and cognitive behavior therapy: Functional analysis as key process. In W. T. O'Donohue & J. E. Fisher (Eds.), *General principles and empirically supported techniques of cognitive behavior therapy* (pp. 15–41). John Wiley & Sons, Inc.

Dunning, D. L., Griffiths, K., Kuyken, W., Crane, C., Foulkes, L., Parker, J., & Dalgleish, T. (2019). Research Review: The effects of mindfulness-based interventions on cognition and mental health in children and adolescents–A meta-analysis of randomized controlled trials. *Journal of Child Psychology and Psychiatry, 60*(3), 244–258.

Edmondson, C. B., & Conger, J. C. (1996). A review of treatment efficacy for individuals with anger problems: Conceptual, assessment, and methodological issues. *Clinical Psychology Review, 16*(3), 251–275.

Eifert, G. H., Forsyth, J. P., & McKay, M. (2006). *ACT on life not on anger: The new acceptance and commitment therapy guide to problem anger*. New Harbinger Publications.

Ellis, A. (1957). Rational psychotherapy and individual psychology. *Journal of Individual Psychology, 13*(1), 38–49.

Ellis, A. (2003). Cognitive restructuring of the disputing of irrational beliefs. In W. O'Donohue, J. E. Fisher, & S. C. Hayes (Eds.), *Cognitive behavior therapy: Applying empirically supported techniques in your practice* (pp. 79–83). Wiley.

Embry, D. D. (2002). The Good Behavior Game: A best practice candidate as a universal behavioral vaccine. *Clinical Child and Family Psychology Review, 5*(4), 273–297.

Ensafdaran, F., Krahé, B., Njad, S. B., & Arshadi, N. (2019). Efficacy of different versions of Aggression Replacement Training (ART): A review. *Aggression and Violent Behavior, 47*, 230–237.

Estes, W. K., & Skinner, B. F. (1941). Some quantitative properties of anxiety. *Journal of Experimental Psychology, 29*(5), 390.

Fernandez, E., Malvaso, C., Day, A., & Guharajan, D. (2018). 21st century cognitive behavioural therapy for anger: A systematic review of research design, methodology and outcome. *Behavioural and Cognitive Psychotherapy, 46*(4), 385–404.

Fix, R. L., & Fix, S. T. (2013). The effects of mindfulness-based treatments for aggression: A critical review. *Aggression and Violent Behavior, 18*(2), 219–227. https://doi.org/10.1016/j.avb.2012.11.009

Frazier, S. N., & Vela, J. (2014). Dialectical behavior therapy for the treatment of anger and aggressive behavior: A review. *Aggression and Violent Behavior, 19*(2), 156–163.

Friman, P. C., Barnard, J. D., Altman, K., & Wolf, M. M. (1986). Parent and teacher use of DRO and DRI to reduce aggressive behavior. *Analysis and Intervention in Developmental Disabilities, 6*(4), 319–330.

Gewirtz, J. L. (1954). Three determinants of attention-seeking in young children. *Monographs of the Society for Research in Child Development, 19*(2), 1–48.

Gillions, A., Cheang, R., & Duarte, R. (2019). The effect of mindfulness practice on aggression and violence levels in adults: A systematic review. *Aggression and Violent Behavior, 48*, 104–115. https://doi.org/10.1016/j.avb.2019.08.012

Glowacki, K., Warner, G., & White, C. (2016). The use of a token economy for behaviour and symptom management in adult psychiatric inpatients: A critical review of the literature. *Journal of Psychiatric Intensive Care, 12*(2), 119–127.

Goldberg, S. B., Tucker, R. P., Greene, P. A., Davidson, R. J., Kearney, D. J., & Simpson, T. L. (2019). Mindfulness-based cognitive therapy for the treatment of current depressive symptoms: A meta-analysis. *Cognitive Behaviour Therapy, 48*(6), 445–462.

Goldstein, A. P., Glick, B., & Gibbs, J. C. (1998). *Aggression replacement training: A comprehensive intervention for aggressive youth*. Research Press.

Gonzalez, J. E., Nelson, J. R., Gutkin, T. B., Saunders, A., Galloway, A., & Shwery, C. S. (2004). Rational emotive therapy with children and adolescents: A meta-analysis. *Journal of Emotional and Behavioral Disorders, 12*(4), 222–235.

Hanley, G. P., Iwata, B. A., & McCord, B. E. (2003). Functional analysis of problem behavior: A review. *Journal of Applied Behavior Analysis, 36*, 147–185.

Hayes, S., Barnes-Holmes, D., & Roche, B. (2001). *Relational frame theory. A Post-Skinneran account of language and cognition*. Springer.

Hayes, S. C. (2004). Acceptance and commitment therapy, relational frame theory, and the third wave of behavioral and cognitive therapies. *Behavior Therapy, 35*(4), 639–665.

Hornsveld, R. H., Kraaimaat, F. W., Gijs, L. A., & Palmer, E. J. (2019). *Assessment and obligatory treatment of violent and sexually violent offenders*. Springer.

Kanter, J. W., Manbeck, K. E., Kuczynski, A. M., Maitland, D. W., Villas-Bôas, A., & Ortega, M. A. R. (2017). A comprehensive review of research on functional analytic psychotherapy. *Clinical Psychology Review, 58*, 141–156.

Kabat-Zinn, J. (1982). An outpatient program in behavioral medicine for chronic pain patients based on the practice of mindfulness meditation: Theoretical considerations and preliminary results. *General Hospital Psychiatry, 4*(1), 33–47.

Kazdin, A. E. (1982). The token economy: A decade later. *Journal of Applied Behavior Analysis, 15*(3), 431–445.

Kazdin, A. (Ed.). (2012). *The token economy: A review and evaluation*. Springer.

Khoury, B., Lecomte, T., Fortin, G., Masse, M., Therien, P., Bouchard, V., Chapleau, M.-A., Paquin, K., & Hofmann, S. G. (2013). Mindfulness-based therapy: A comprehensive meta-analysis. *Clinical Psychology Review, 33*(6), 763–771.

Kohlenberg, R. J., & Tsai, M. (1994). Functional analytic psychotherapy: A radical behavioral approach to treatment and integration. *Journal of Psychotherapy Integration, 4*(3), 175–201. https://doi.org/10.1037/h0101264

Kohlenberg, R. J., Hayes, S. C., & Tsai, M. (1993). Radical behavioral psychotherapy: Two contemporary examples. *Clinical Psychology Review, 13*(6), 579–592.

Krasner, L. (1958). Studies of the conditioning of verbal behavior. *Psychological Bulletin, 55*(3), 148–170. https://doi.org/10.1037/h0040492

Lawrence, E., Langer Zarling, A., & Orengo-Aguayo, R. (2014). *Achieving change through value-based behavior (ACTV)*. University of Iowa Foundation.

LePage, J. P., DelBen, K., Pollard, S., McGhee, M., VanHorn, L., Murphy, J., Lewis, P., Aboraya, A., & Mogge, N. (2003). Reducing assaults on an acute psychiatric unit using a token economy: A 2-year follow-up. *Behavioral Interventions: Theory & Practice in Residential & Community-Based Clinical Programs, 18*(3), 179–190.

Lee, A. H., & DiGiuseppe, R. (2018). Anger and aggression treatments: a review of meta-analyses. *Current Opinion in Psychology, 19*, 65–74.

Lerman, D. C., Iwata, B. A., & Wallace, M. D. (1999). Side effects of extinction: Prevalence of bursting and aggression during the treatment of self-injurious behavior. *Journal of Applied Behavior Analysis, 32*(1), 1–8.

Lindsley, O. R. (1956). Operant conditioning methods applied to research in chronic schizophrenia. *Psychiatric Research Reports, 5*, 118–139.

Linehan, M. M. (1993). Dialectical behavior therapy for treatment of borderline personality disorder: Implications for the treatment of substance abuse. *NIDA Research Monograph, 137*, 201–201.

Livheim, F., Tengström, A., Andersson, G., Dahl, J., Björck, C., & Rosendahl, I. (2020). A quasi-experimental, multicenter study of acceptance and commitment therapy for antisocial youth in residential care. *Journal of Contextual Behavioral Science, 16*, 119–127.

Luce, S. C., Delquadri, J., & Hall, R. V. (1980). Contingent exercise: A mild but powerful procedure for suppressing inappropriate verbal and aggressive behavior. *Journal of Applied Behavior Analysis, 13*(4), 583–594.

Malouf, E. T., Youman, K., Stuewig, J., Witt, E. A., & Tangney, J. P. (2017). A pilot RCT of a values-based mindfulness group intervention in jail inmates: Evidence for reduction in post-release risk behavior. *Mindfulness, 8*(3), 603–614.

Meichenbaum, D. H., & Deffenbacher, J. L. (1988). Stress inoculation training. *The Counseling Psychologist, 16*(1), 69–90.

Michael, J. (1993). Establishing operations. *The Behavior Analyst, 16*(2), 191–206.

Naeem, F., Clarke, I., & Kingdon, D. (2009). A randomized controlled trial to assess an anger management group programme. *Cognitive Behaviour Therapist, 2*(1), 20–31.

Nelson, J. R., Smith, D. J., Young, R. K., & Dodd, J. M. (1991). A review of self-management outcome research conducted with students who exhibit behavioral disorders. *Behavioral Disorders, 16*(3), 169–179.

Newring, K. A., & Wheeler, J. G. (2010). FAP with people convicted of sexual offenses. In J. Kanter, M. Tsai, & R. J. Kohlenberg (Eds.), *The practice of functional analytic psychotherapy* (pp. 225–246). Springer.

Newring, K. A., & Wheeler, J. G. (2012). Functional analytic psychotherapy with juveniles who have committed sexual offenses. *International Journal of Behavioral Consultation and Therapy, 7*(2–3), 102–110.

Novaco, R. W. (1975). *Anger control: The development and evaluation of an experimental treatment*. Lexington Books.

Novaco, R. W. (2020). Anger treatment with violent offenders. In J. S. Wormith, L. A. Craig, & T. E. Hogue (Eds.), *The Wiley handbook of what works in violence risk management: Theory, research and practice* (pp. 385–397). Wiley.

O'Neil, R. E., Horner, R. H., Albin, R. W., Storey, K., & Sprague, J. R. (1997). *Functional assessment and program development for problem behavior: A practical handbook* (2nd ed.). Wadsworth.

Singh, A. N., & Strand, P. S. (2008). A person-centered, strength-based treatment of aggression and sexually inappropriate behavior in mental health. *Clinical Case Studies, 7*(5), 397–408.

Singh, N. N., Wahler, R. G., Adkins, A. D., Myers, R. E., & Mindfulness Research Group. (2003). Soles of the feet: A mindfulness-based self-control intervention for aggression by an individual with mild mental retardation and mental illness. *Research in Developmental Disabilities, 24*(3), 158–169.

Singh, N. N., Matson, J. L., Lancioni, G. E., Singh, A. N., Adkins, A. D., McKeegan, G. F., & Brown, S. W. (2006). Questions About Behavioral Function in Mental Illness (QABF-MI): A behavior checklist for functional assessment of maladaptive behavior exhibited by individuals with mental illness. *Behavior Modification, 30*(6), 739–751.

Singh, N. N., Lancioni, G. E., Winton, A. S. W., Singh, J., Curtis, W. J., Wahler, R. G., et al. (2007). Mindful parenting decreases aggression and increases social behavior in children with developmental disabilities. *Behavior Modification, 31*, 749–771.

Singh, N. N., Lancioni, G. E., Winton, A. S. W., Singh, A. N. A., Adkins, A. D. A., & Singh, J. (2009). Mindful staff can reduce the use of physical restraints when providing care to individuals with intellectual disabilities. *Journal of Applied Research in Intellectual Disabilities, 22*, 194–202.

Siroos, M., Mirzaian, B., & Hasanzadeh, R. (2021). The effectiveness of functional analytical therapy on controlling, expressing and internalizing anger in patients with a borderline personality disorder. *International Journal of Health Studies, 7*(4), 12–16.

Skinner, B. F. (1953). *Science and human behavior*. McMillan.

Skinner, B. F. (1957). *Verbal behavior*. Appleton-Century-Crofts.

Soares, D. A., Harrison, J. R., Vannest, K. J., & McClelland, S. S. (2016). Effect size for token economy use in contemporary classroom settings: A meta-analysis of single-case research. *School Psychology Review, 45*(4), 379–399.

Sturmey, P. (1996). *Functional analysis in clinical psychology*. Wiley.

Sturmey, P. (2008). *Behavioral case formulation and intervention. A functional analytic approach*. Wiley-Blackwell.

Sturmey, P., & McMurran, M. (Eds.). (2011). *Forensic case formulation* (Vol. 49). John Wiley & Sons.

Sturmey, P. (Ed.). (2020). *Functional analysis in clinical treatment* (2nd ed.). Academic Press.

Sturmey, P., McMurran, M., & Daffern, M. (2019). Case formulation and treatment planning. In D. L. L. Polscheck, A. Day, & C. R. Hollin (Eds.), *The Wiley international handbook of correctional psychology* (pp. 476–487). Wiley.

Taylor, J. L., Novaco, R. W., & Brown, T. (2016). Reductions in aggression and violence following cognitive behavioural anger treatment for detained patients with intellectual disabilities. *Journal of Intellectual Disability Research, 60*(2), 126–133.

Tiger, J. H., Hanley, G. P., & Bruzek, J. (2008). Functional communication training: A review and practical guide. *Behavior Analysis in Practice, 1*(1), 16–23.

Tingstrom, D. H., Sterling-Turner, H. E., & Wilczynski, S. M. (2006). The Good Behavior Game: 1969-2002. *Behavior Modification, 30*(2), 225–253. https://doi.org/10.1177/0145445503261165

Todes, D. P. (2014). *Ivan Pavlov: A Russian life in science*. Oxford University Press.

Ullman, L., & Krasner, L. (Eds.). (1965). *Case studies in behavior modification*. Holt, Rinehart, & Winston.

Vollmer, T. R., Progar, P. R., Lalli, J. S., Van Camp, C. M., Sierp, B. J., Wright, C. S., Nastasi, J., & Eisenschink, K. J. (1998). Fixed-time schedules attenuate extinction-induced phenomena in the treatment of severe aberrant behavior. *Journal of Applied Behavior Analysis, 31*(4), 529–542.

Watson, J. B., & Rayner, R. (1920). Conditioned emotional reactions. *The American Psychologist, 3*(1), 1–14.

Wolf, M. M. (1978). Social validity: The case for subjective measurement or how applied behavior analysis is finding its heart. *Journal of Applied Behavior Analysis, 11*(2), 203–214.

Wolpe, J. (1952). Experimental neuroses as learned behaviour. *British Journal of Psychology, 43*(4), 243–268.

Wolpe, J. (1954). Reciprocal inhibition as the main basis of psychotherapeutic effects. *American Medical Association Archives of Neurology and Psychiatry, 72*, 204–226.

Wolpe, J. (1958). *Psychotherapy by reciprocal inhibition*. Stanford University Press.

Wright, S., Day, A., & Howells, K. (2009). Mindfulness and the treatment of anger problems. *Aggression and Violent Behavior, 14*(5), 396–401.

Yildiz, E. (2020). The effects of acceptance and commitment therapy in psychosis treatment: A systematic review of randomized controlled trials. *Perspectives in Psychiatric Care, 56*(1), 149–167.

Zarling, A., Lawrence, E., & Marchman, J. (2015). A randomized controlled trial of acceptance and commitment therapy for aggressive behavior. *Journal of Consulting and Clinical Psychology, 83*(1), 199–212.

Zarling, A., Bannon, S., & Berta, M. (2019). Evaluation of acceptance and commitment therapy for domestic violence offenders. *Psychology of Violence, 9*(3), 257–266. https://doi.org/10.1037/vio0000097

Zeilberger, J., Sampen, S. E., & Sloane, H. N., Jr. (1968). Modification of a child's problem behavior in the home with the mother as therapist. *Journal of Applied Behavior Analysis, 1*(1), 47–53.

Zwets, A. J., Hornsveld, R. H., Muris, P., Kanters, T., Langstraat, E., & van Marle, H. J. (2016). Psychomotor therapy as an additive intervention for violent forensic psychiatric inpatients: A pilot study. *International Journal of Forensic Mental Health, 15*(3), 222–234.

Chapter 17
Individual Therapies for Violence and Aggression. II. Other Therapies

There are a very wide range of individual psychosocial treatments for aggression and violence in addition to cognitive behavior therapy, behavior modification, and applied behavior analysis. This chapter will review a sample of these. The first group deals with the problem of aggression and violence in social groups with two or more members such as couples and families. A well-researched aspect of this approach is parent training where there are at least two well established parent training packages that may be effective in preventing or treating aggression and violence in children and adolescents. In addition there are conjoint and family therapy approaches. The other approaches discussed here include psychotherapy, counseling, individual restorative justice practices, and combat sports and martial arts.

17.1 Couples and Family Therapy

Intimate partner violence (IPV) is one of the most common forms of aggression and violence. Many practitioners and researchers have believed that rather than focusing treatment on the perpetrator, it might be better to work with partners or the entire family. Johnson (1995) has distinguished two kinds of IPV: One is more pervasive and involves violence alongside systematic control, economic coercion, isolation, and threats and the other is more situational. Johnson suggested that one treatment model may not be appropriate for both of these forms of IPV. There are several quite distinct approaches to IPV including the Duluth model, batterer intervention programs (BIPs), and conjoint therapy. Research in this area is plagued by practical and implementation problems reflecting the difficulty of working with perpetrators who are often court-mandated to therapy, the difficulties of measuring violence in the home when victims may be reluctant to report that violence for many good reasons.

P. Sturmey, *Violence and Aggression*, Advances in Preventing and Treating Violence and Aggression, https://doi.org/10.1007/978-3-031-04386-4_17

17.1.1 The Duluth Model

Pence and Paymar (1993) developed the Duluth model approach based on feminist and sociological theory and practitioner experience, rather than basing it on empirical findings. The model is based on the idea that male violence is a form of abuse to maintain male privilege learned from socially legitimized sense of male privilege, rather than a form of psychopathology. The groups are primarily educational and focus on attitude change rather than skills training or situational analysis of IPV. The Duluth model is often applied to groups of men deferred for adjudication by courts.

Empirical evaluations, which were conducted after the Duluth model was developed, have been ambivalent or negative. For example, Armenti and Babcock's (2016) meta-analysis found only two empirical studies evaluating the Duluth model. Both studies had major problems including selecting motivated participants and very high (50–70%) dropout at 21-months. Others have pointed out that this model fails to address issue of violence within lesbian and gay couples and female to male violence within heterosexual couples. These and other results have led Dutton and Corvo (2007) to conclude that the Duluth model is "A data-impervious strategy and failed paradigm" (p. 658).

17.1.2 Batterer Intervention Programs

There are a variety of BIPs developed and evaluated over the last 30 years. The content and format of BIPs vary considerably in terms of content—some are very influenced by the Duluth model, some are educational, and others are more skills-based. BIPs also vary considerably in terms of their length and use of open or closed formats (Eckhardt et al., 2013).

Earlier meta-analyses have often been stymied due to the lack of experimental studies and their poor quality. Eckhardt et al. (2013), for example, identified 30 evaluations of BIP programs and noted that although recently there have been more experimental studies, the outcomes were very mixed with many studies producing no significant outcomes. This was true for both Duluth and for CBT-type BIPs resulting in little evidence of the superiority of one approach over another. Thus, the evidence for the effectiveness of BIP programs is very limited at this time.

17.1.3 Parent Skills Training

Some parents engage in ineffective, coercive, aggressive, and violent behavior with their children and other young family members because they do not know how to raise their children any other way or because their appropriate parenting skills are weak and/or ineffective. It may be easier and in the short term more effective to resort to using these coercive strategies with their children. Indeed, some of these parents may have been raised in a similar manner themselves and learned coercive parenting practices and failed to learn assertive, non-coercive parenting from their own family caregivers. Some, indeed, may believe that coercion and positive punishment is the best way to raise children.

Parent skills training programs have a very long history. (See, for example, the description of Zeilberger et al., 1968 in the previous chapter.) Thus, there is a large and varied outcome literature that can go beyond the simple question, "Does it work?" to answer more specific and refined questions such as whether or not parent training works for parents of specific kinds of children and specific kinds of parents and whether a specific parent training package works and, if so, in what way. This developed literature has resulted in several meta-analyses of outcome studies.

An example of such a meta-analysis comes from Kaminski et al. (2008) who conducted a meta-analysis to identify effective components of parent training packages for children aged 0–7 years. Their search identified 77 studies published between 1990 and 2002. They included RCTs and pre-post group designs because RCTs tend to be associated with initial demonstration studies in university settings and pre-post group designs tend to be associated with evaluation in more typical settings. They found an overall effect size of 0.43 with a larger effect size for parent than child outcomes. Effect sizes were smaller for parent skills than parent knowledge, attitudes, and self-efficacy. Components of parent training associated with larger effect sizes included training in positive interaction skills, practicing skills with their own child, and training in time out. Training in information concerning child development may have been associated with smaller effect sizes than direct training of parent skills.

This meta-analysis demonstrated that parent training is a robustly effective approach to reducing aggression and related externalizing problem behavior in young children. It replicated the results of several similar meta-analyses (McCart et al., 2006; Postorino et al., 2017; Serketich & Dumas, 1996). Its results are also useful in designing the most effective and efficient programs of parent training for young children with aggression and related externalizing behavior problems. This literature is now sufficiently developed that it is possible to comment on the effectiveness of at least two individual treatment packages, namely the Incredible Years Parent Training Program (IYTP) and the Positive Parent Training Program (PPP).

17.1.3.1 Incredible Years Parent Training

The IYTP aims to reduce coercive parenting practices, such as yelling, and replace them with positive strategies in order to promote child development, reduce child problem behavior, and child anxiety. Parents learn basic interaction skills, like talking to their children, empathy and listening; praise and encouragement; setting clear limits such as household routines, and follow through; and basic behavior management strategies such as ignore-distract-redirect; and, only if needed, selective use of time out and other negative punishment procedures. This package has been implemented very widely in many countries.

In a recent meta-analysis of the outcome literature for IYPT, Leijten et al. (2018) focused only on RCTs. They located 14 RCTs with 1,799 participants. The authors found that IYPT was effective in achieving some of its primary aims, such as reduction in parent yelling and increased parent use of praise, but not other parental skills such as use of tangible reinforcers, child monitoring, or reducing parental laxness. They also reported that child conduct problems and Attention Deficit Hyperactivity Disorder symptoms improved, but parental mental health did not improve. There was no evidence of negative effects. The effect sizes were typically small, being of the order of 0.2–0.4.

17.1.3.2 Positive Parent Training Program

The PPP program is a five-level, triage-like parent support program that addresses parent and child needs from ages 0 to 16 years that combines social learning and public health approaches to child development. For example, level 1 includes public information and education through radio, on line, TV and on-line media, etc., whereas level 5 includes intense individual and group parent training. PPP has been extensively evaluated (Sanders et al., 2014), including with parents of children with disabilities (Ruane & Carr, 2019; Tellegen & Sanders, 2013). Here we will review the most recent and comprehensive meta-analysis of PPP.

Sanders et al. (2014) identified 116 studies conducted over 33 years with 101 studies reporting data on 16,099 families. Effect sizes were small to moderate ranging from 0.47 for child emotional and behavioral outcomes, 0.58 for parent behavior, 0.50 for observational child data with smaller effects for outcomes such as parental relationship (0.22). Significant effects were found at follow-up, but were smaller (e.g., parent observation data—0.25). Thus, like IYPT, there is a very large evidence base to support the use of PPP.

17.1.3.3 Comment

Parent training is one of the most robust and acceptable approaches to prevent and treat aggression and violence: The long standing and very large general literature and the extensive literatures on specific programs such as IYTP and PPP attest to this. A challenge is that sometimes the effect sizes are small to moderate, which is perhaps surprising, if one contrasts these outcomes with very large and consistent changes in parent behavior in the behavioral skills training literature. Thus, the issue of effectiveness shifts from the simple question "What work?" to "What can be done to make this approach work best for each parent and their children?"

17.1.4 Conjoint Therapy

As an alternative to individual therapy for perpetrators, some researchers and practitioners have developed and evaluated conjoint therapy between couples. This often addresses conflict resolution skills and developing a relationship with a shared meaning. Conjoint therapy may involve CBT, behavioral, and other effective treatment principles among other approaches.

There have been several reviews and meta-analyses of this literature (Armenti & Babcock, 2016; Carr, 2014, 2019; Eckhardt et al., 2013; Karakurt et al., 2016; Stover et al., 2009). An illustrative, recent example, comes from Karakurt et al. (2016) who conducted a meta-analysis of six studies of conjoint treatment of IPV of which three were explicitly CBT forms of conjoint therapy. The weighted mean difference between the experimental and control group was −0.84 (CI = −1.37 to −0.30). Karakurt et al. concluded that couples therapy is a viable and potentially effective treatment option.

Other reviews and empirical studies have refined this overall positive conclusion. For example, Armenti and Babcock (2016) also concluded that conjoint therapy was most likely to be effective for situational IPV where communication and relationship skills might be the most effective intervention in that circumstance. Stover et al. (2009) also concluded that it may be necessary to integrate evidence-based practices into the needs of the entire family. For example, they concluded that interventions that addressed substance abuse were associated with lower dropout rates and that manualized child trauma treatments were also potentially useful adjuncts when working with an entire family. More recently, Stover and Morgos (2013) reviewed work that could be done between fathers who commit IPV and their children. They noted the general absence of evidence-based practices in this important domain of IPV work and that more research is needed on that topic.

17.1.5 Family Therapy

If intervention for aggression and violence should address the couple, rather than the individual perpetrator, should this approach not be extended to the entire family unit (Minuchin, 1967)?

Multi-systemic systemic family therapy (MSFT) is based on an understanding of the risk and protective factors that underlie antisocial behavior. These risk factors include: (a) individual factors, such as low cognitive and social skills; (b) family factors, such as low warmth, ineffective parenting and supervision, and parental criminal behavior; (c) school factors, such as dropping out of school; and (d) community factors, such as criminal subcultures. MSFT begins by establishing a good therapeutic alliance. It then works with all family members to promote problem solving, skills, obtaining resources, and building a better family environment, better parental supervision, school participation, and blocking inter-general transmission of violence (Henggeler et al., 1996).

Henggeler et al. (1992) evaluated the effects of MSFT to address violence and delinquency. They randomized 84 youth (average age 14 years from semi-skilled working families) who were at risk for out of home placement due to criminal behavior, delinquency, and violence to with MSFT or services as usual, which included court-ordered curfews, school attendance, etc. At 59 weeks post-referral, participants in the MSFT group had large (approximately 50% or greater) reductions in arrests, days of incarceration, and self-reported delinquency compared to the control group. MSFT families reported greater cohesion and reduced aggression to peers. Better outcomes were associated both with a shorter history of delinquency, but also better peer relationships. Follow-up indicated that these group differences were maintained (Henggeler et al., 1993). The average time to re-arrest was 56 weeks and 32 weeks in the MST and control groups, respectively. Also, at 129 weeks post-referral 39% and 20% had not been re-arrested in the MST and control groups, respectively. The initial costs of MST can be offset by avoided costs such as the costs of incarceration, hospitalization, and residential treatment (Schoenwald et al., 2000).

A later meta-analysis of MST (Curtis et al., 2004) identified 11 studies with 708 participants. The overall effect size was moderate ($d = 0.55$) but was considerably larger ($d = 0.81$) with graduate therapists than community therapists ($d = 0.26$). This latter finding resulted in studies showing that improving therapy integrity and better training and supervision of therapists (Henggeler et al., 1992). Subsequent studies in Sweden did not find positive outcomes for youth (Löfholm et al., 2009; Sundell et al., 2008) and positive economic outcomes were not found (Olsson, 2010a, b). These differences might be due to the fact that the Swedish studies did not deliver sufficient quantity and intensity of treatment and they took place in a different legal context. Thus, it appears that MSFT can be a highly effective program if delivered with sufficient integrity, with adequate therapist training and supervision and, perhaps, in certain legal contexts.

The original studies of MSFT reported very promising and impressive outcomes. The failure to replicate by Swedish studies leads to confusion as to whether this is an effective program. The simplest answer is that the studies conducted in Sweden did not deliver the program with sufficient integrity and intensity and hence are not true replications. The question as to the robustness of MST can be answered by studies that deliver the treatment as designed with adequate treatment integrity.

An interesting second application of family-wide intervention to improve family functioning and prevent and reduce family violence comes from a series of studies of interventions for families in which at least one caregiver is HIV positive. The Family Strengthening Intervention program (Betancourt et al., 2011) was developed for use in low income counties. The authors interviewed children and caregivers in Rwanda in families where at least one family member was HIV-positive to identify themes, develop psychometric measures, and design a family- and strength-based intervention focused on strengthening parent-child relationships by focusing on family connectedness, communication and caregiving skills, and access to social support. This was done in collaboration with international experts and local advisors.

In a pilot study (Betancourt et al., 2020), 20 families with 39 children aged 7–17 years participated. The authors collected data pre- and post-intervention and at 6 months follow-up. They found that both child and adult participants reported high program satisfaction, increased family connectedness, better parenting, improved social support, increased child prosocial behavior, and reductions in harsh parenting.

Betancourt et al. (2017) then conducted an RCT in which they randomized 82 families with at least one HIV-positive member with 170 children to Family Strengthening Intervention or treatment as usual. Although at 3-month follow-up children reported reduced depression, there was no reduction in child conduct problems, functional impairment, family connectedness, or parenting practices. These results were rather less impressive than the pilot study, perhaps reflecting problems with treatment integrity or lack of experimental power. Thus, the Family Strengthening Intervention program is a potentially promising approach to reducing family violence in a challenging context, but which needs further research to demonstrate its effectiveness at this time.

17.2 Psychoanalytic and Psychodynamic Psychotherapy

There are a small number of reviews that have addressed the effectiveness of psychoanalytic and psychodynamic treatment for aggression (Beail, 2017; Heyman & Slep, 2007; Taubner et al., 2017a, b); hence, this approach is rarely mentioned in evidence-based practice guidelines on the treatment of aggression and violence. More recently, there have been more empirical outcome papers than in the past. For example, Fonagy (2003) conducted a cluster RCT in which they randomized nine schools to treatment as usual, regular psychiatric consultation or the Creating A Peaceful Learning Environment (CAPSLE) program, a multi-component teacher-implemented program that included a mentalization component. (Mentalization is the ability to imagine and think about the mental states of other people and understanding their intentions as well as one's own mental states and intentions.) The participants were third to fifth graders. The program took place over 2 years of treatment and 1 year of follow-up. There was some evidence of the effectiveness of CAPSLE. For example, ratings of self- and other-reported aggression decreased in the CAPSLE participants compared to the treatment as usual group at various times. There were, however, few differences between the CAPSLE and psychiatric consultation group and there were not clear benefits to the CAPSLE group in terms of increasing bystander helpfulness. Thus, it was not entirely clear if the differences were due to novelty and non-specific program characteristics or specific effects of mentalization. This study is nevertheless interesting as a theoretical counterpoint to school-wide positive behavior support interventions to reduce aggression and violence.

More recently, Taubner et al. (2017b) searched the literature between 1991 and 2015 and identified four RCTs, one non-RCT, and seven pre-post designs evaluating psychodynamic individual and group therapy, transference-focused psychotherapy, mentalization based treatment (MBT) and integrated dynamic psychotherapy and CBT. Some studies found reductions in aggression and violence and self-harm, although other studies found that psychotherapy was equivalent to CBT.

Evidence for the effectiveness of psychotherapy for aggression and violence slowly continues to accumulate. For example, McKeown and McCrory (2019) reported four case studies of mentalization to treat aggression and violence in violent women offenders using AB designs and reported some success. To date, however, there has been no systematic evaluation of the quality of these studies or meta-analysis of their outcomes; hence, given the availability of several well researched, highly effective treatments for aggression and violence, the jury remains out on the question of the effectiveness of psychoanalytic and psychodynamic therapies for aggression and violence.

17.3 Counseling

Kaplan et al. (2014) defined counseling as "a professional relationship that empowers diverse individuals, families, and groups to accomplish mental health, wellness, education, and career goals" (p. 366). Thus, counseling refers to interventions using many forms of intervention and is used in many contexts.

Counseling is widely used in schools for aggression and violence—children who have been aggressive or violent are often sent to counseling by school psychologists and other mental health professionals within the school system. Couples also sometimes elect to go to counseling and violent offenders and individuals with anger problems may be required to attend counseling as part of the legal process.

But does counseling work for aggression and violence? It is often noted that when students are sent for counseling that being sent to counseling may remove classroom demands and/or deliver desirable consequences such as a nice chat with the principal or time on the computer. Alternatively, they may receive a moral lecture from their counselor explaining to them what they already know, namely that hitting other children hurts, is disrespectful to others, and disruptive to the classroom. Likewise, couples and individuals with anger issues may be motivated to "fake good" perhaps through fear of the consequences and for various ulterior motives. In this section we will refer to counseling more narrowly to specific supportive professional relationships such as those espoused by Carl Rogers and note in passing that some of the literature uses the term to encompass specific forms of psychotherapy, such as MSFT and CBT (Erford et al., 2017).

Compared to the literature on other individual interventions such as CBT, there are surprisingly few studies of counseling to treat aggression. For example, in a meta-analysis of counseling (broadly defined) for youth with conduct disorders, almost all studies were in fact of CBT or MSFT. Not only is there a lack of evidence for the effectiveness of counseling, but there is also evidence of its ineffectiveness. For example, a RCT by Kazdin et al. (1987) evaluated the effectiveness of three conditions for antisocial child behavior. The three conditions were problem solving skills, non-directive relationship-based therapy, and an attention-placebo control group. The authors selected problem solving because they felt that the parents in these families would likely have problems in implementing standard behavioral strategies in the home. The children were in-patients in a psychiatric hospital who were aged 7–13 years. The authors took data both pre- and post-treatment and at 1-year follow-up. The authors found that problem solving was superior to both relationship-based therapy and the attention control group. The problem solving group showed greater reductions in externalizing behavior problems, aggression, behavior problems, in school and at home. They also showed increased prosocial behavior both post-treatment and at 1-year follow-up. In addition, there was a greater percentage of children in the problem solving group than the other groups who functioned in the normative range of test scores assessing behavior problems.

Further, several broad reviews of evidence-based practice do not identify counseling as an effective treatment for aggression and violence (e.g., Erford et al., 2017; Kazdin, 2015). Thus, at this time, non-directive, supportive counseling that does not incorporate some form of skills training does not appear to be an effective treatment for aggression and violence. Perhaps there are some forms of counseling that are effective, but the exact nature of the effective and ineffective counseling procedures must be clearly defined.

17.4 Restorative Justice Practices

Chapter 14, discussed restorative justice practices applied to large societal violence, such as in the Troubles in Northern Ireland and post-Apartheid in South Africa. In this section we will review the evidence for the application of a specific form of restorative justice to individuals, their victims and communities termed the restorative justice conference (RJC).

Sherman et al. (2015) pointed out that the RJC is a multi-component process involving eight elements. These components included: (1) one-on-one pre-conference screenings to discuss the process with offender and victim; (2) scheduling a conference; (3) participants sit in a circle in a room that is private and with doors that are closed; (4) the facilitator introduces participants and their emotional connection to the crime; (5) the facilitator opens the conference and asks the offender to make a description of their crime; (6) the facilitator invites everyone to describe the harms; (7) the facilitator attempts lead the group to reach consensus on repairing the harms, making a written plan signed by everyone; and (8) file the agreement (Sherman et al., pp. 2–3). It is important to note that a RJC can take place at a variety of times related to the crime and any legal actions. Thus, RJCs can take place before adjudication or many years after the crime has been committed, such as when the offender is about to be released. Over time, there have been a variety of rationales for why people should use them and how RJC might work. Hence, RJCs are a fairly varied set of procedures. Recall also that claims for the effectiveness of restorative justice included: (a) benefits for the victim, such as feeling more consulted improved empathy for the perpetrator and benefiting from perpetrator restorative practices; (b) benefits for the perpetrator, such as increased empathy for the victim, reduced reoffending, and better reintegration into society; and (c) benefits for society, such as reduced crime, a more peaceful society, and reduced crime-related costs. What is the evidence that any of these claims are true?

An example of an RCT evaluating restorative justice comes from Mills et al. (2019) who conducted an RCT to evaluate the application or restorative justice-informed approaches to domestic violence. The authors compared a standard BIP with BIP plus Circles of Peace, a restorative justice-informed approach to prevent domestic violence. Both groups participated in 12 weeks of treatment. The Circle of Peace treatment included victims, perpetrators, their friends and family and focused on repairing harm, organized dialog and a social compact which was monitored each week. The hybrid program was superior to the standard BIP in that, over a 24-month period, there was a 53% reduction in re-arrests and a 52% reduction in all crime severity, including offenses related to domestic violence. Thus, the authors concluded that BIP plus Circles of Peace was a more effective program and a viable treatment option for IPV. An important limitation of the study was that only 42% of victims attended at least one session of the Circle of Peace sessions.

There have been at least three systematic reviews and meta-analyses of individual restorative justice approaches to reducing aggression and violence. In the first, Schwalbe et al. (2012) conducted a systematic review and meta-analysis of a broad array of diversion programs for juvenile offenders. They identified 28 studies with over 19,000 youth participants published between 1980 and 2011. They found that overall diversion programs were ineffective: Recidivism was 31.4% and 36.3% for experimental and control groups, respectively (OR = 0.83, 95% CI = 0.43–1.58). These wide confidence limits indicated widely varying rates of recidivism across different studies. The only effective program was family treatment ($k = 4$, OR = 0.57, 95% CI = 0.40–0.82). Overall, restorative justice interventions were ineffective ($k = 6$, OR = 0.987, 95% CI = 0.67–1.12) except when implemented in university-based settings ($k = 3$, OR = 0.69, 95% CI = 0.51–0.95). Schwalbe et al. attributed this to two of the three studies using "highly controlled" victim–offender conferences.

In the second meta-analysis, Sherman et al. (2015) conducted a meta-analysis of RJC for reducing repeat offending. There were ten RCTs with 1,880 offenders participating. Nine of ten studies reported a decrease in recidivism over 2 years, but the overall effect size was quite small but significantly different from zero ($d = 0.155$, CI = 0.06–0.25, $p = 0.0001$). A mediator analysis found almost no effects except—surprisingly—that when evaluation was done by those independent of the treatment the effect size was modestly larger than when done by the authors. The most interesting outcomes are related to cost–benefit analyses. When the cost of crimes was weighted by the estimated costs of specific individual types of crimes, all three studies that did so found that the payback for RJC varied from 3.7 to 8.1 for each unit cost of treatment, that is the benefits outweighed the economic costs considerably.

Finally, Howell and Lipsey (2012) conducted a comprehensive meta-analysis of several hundred outcomes studies of delinquent youth. They found that restorative practices were effective resulting in an approximately 10% reduction in recidivism, which was very different from punitive-based practices such as coercive discipline and "scared strait" deterrent programs, but not quite as effective as skills building, structured counseling, and multi-component services.

The outcomes of RJCs are promising but varied. This is most likely related to treatment definition, treatment integrity issues, and gaining a better understanding of what should occur during RJC. Behavior analysis has a long history of attempting to understand non-behavioral treatments (Skinner, 1953). In this example, RJC includes many elements. Some of these elements include: (a) victim exposure to the perpetrator and talking about potentially traumatic events, potentially resembling respondent extinction; (b) development of self-management for both victims and perpetrators, as when they each set goal related to resolution and report back as to their own performances and the outcomes of the agreed plans; and (c) developing rule-governed behavior in victims and perpetrators, etc. Other aspects of RJCs which may affect outcomes could include the potential negative effects of both victims and perpetrators failing to follow through which could, under certain circumstances, mean harm to the victim and perpetrator when the perpetrator does not follow through on their promises. Additionally, issues of treatment integrity and the effects of unskilled mediators could potentially result in ineffective or iatrogenic treatment. Finally, it is interesting to note that, so far, the literature has focused on a limited range of outcomes and many of the broader claims of the potential benefits of restorative justice have not yet been extensively evaluated.

17.5 Martial Arts and Combat Sports

Combat sports, such as boxing, and martial arts, such as taekwondo, are often recommended as a treatment or preventative strategy for aggression and violence. This may be done intuitively, for example, as part of the notion of getting at-risk and delinquent adolescents off the streets where they may get into trouble and as part of community programs to promote children's and adolescent's development and prevent crime. This may also be done with the explicit or implicit notion of catharsis, that is redirecting psychic energy from violent behavior and crime into controlled, violent sports to reduce the more severe violence. Finally, sports and martial arts may also come with other social benefits, such as providing social support, pleasant activities, learning skills and self-control, and providing models of social adaptation as alternatives to models of deviant social behavior. But does it work when it comes to reducing aggression and violence?

In a recent systematic review of martial arts and combat sports, Lafuente et al. (2021) identified nine studies including three cohort studies and six RCTs. The studies were quite varied in content and population. For example, some studies evaluated traditional martial arts that include emphasis on respect, self-control, and mindfulness practices, whereas others included only skills training. Of the nine studies, only two included individuals with histories of behavioral issues or crime; participants had a wide range of previous experience with contact sports or martial arts; and participants were children, adolescents, and adults. Given these features of the literature, it is unsurprising that there was no clear and simple relationship between participating in contract sports and martial arts, leading Lefuerte et al. to conclude that "The results do not support the idea that MA&CS [martial arts and combat sports] will reduce or increase anger and aggression" (p. 1). There were hints, based on one or two studies, that traditional martial arts with greater emphasis on relaxation training, respect, self-control, and mindfulness, participants with criminal and violent histories, and studies with adult participants were more likely to report reductions in anger and aggression. These tentative results have to be tempered with the observations that sample sizes were small and the doses and intensities of sports varied widely, etc. Hence, future research might benefit from isolating specific independent variables, such as the content and intensity of the training as well as the populations studied.

17.6 Boot Camps: Just Say No!

Boot camps to treat aggression and violence in juveniles have a lot of curb appeal to many: The idea that a "short, sharp, shock" with quasi-military school training will straighten out the delinquent, who is mollycoddled by inept and unmotivated parents and professionals, and tolerant or negligent schools, is appealing to many. Many such programs often have implicit or explicit punitive overtones to them, or at least promise the parents a break from their adolescent's difficult behavior. Indeed, boot camps became popular and diffused across the USA rapidly in the 1990s in the absence of evidence to support their use at that time (Bergin, 2016). Unfortunately for the juveniles and taxpayers, reviews of boot camps for juvenile offenders find a modest amount of evidence, but the evidence is that the outcomes are bad.

A review of the evidence by Henggeler and Schoenwald (1994) concluded that, although boot camps might serve the functions of incapacitance and punishment, there is little good evidence that these programs are effective in reducing aggression and violence. Their review of the literature led them to conclude that often boot camps were used not for hardened, repeat offenders, but rather for first time offenders without a history of serious aggressive behavior. Further, they found little evidence of effectiveness and some evidence of harms. They speculated that the ineffectiveness of boot camps was due to their failure to address the core causes of delinquency. Hence, they recommended that program administrators and service funders should "Just say no" to boot camps as they are both expensive and if we are lucky only ineffective rather than positively harmful.

Subsequently, Meade and Steiner (2010) conducted a systematic review of outcomes studies for boot camps that found only slightly more positive outcomes. Of

23 outcomes studies 13 reported no effects of boot camps on recidivism. Of the four studies that reported increases in recidivism, three of those studies did not include any rehabilitation component. In contrast, in the five studies which reported a decrease in the odds of recidivism, four included a rehabilitation component. Three studies found that boot camps reduced the bed space occupied in the justice system and two of these studies found economic benefits of boot camps.

The appeal of boot camps as a solution to aggression and violence in juveniles has received minimal support from the literature. This minimal hint of effectiveness must be counterbalanced against the availability of effective alternatives and, indeed, the instances of abuse and deaths that have occurred in some boot camps (Bergin, 2016).

17.7 Other Psychosocial Interventions

There are numerous other individual psychological and psychosocial interventions for aggression and violence. The list is endless, but some common examples include play therapy, dance therapy, drama therapy, gestalt therapy, sensory therapies, motor therapies, and music therapies to name but a few. As a broad generalization these interventions have much less research to support their use and sometimes there is evidence that they are ineffective and, in a few cases, may even be harmful. Hence, systematic review of play therapy (Ray et al., 2015), dance therapy (Strassel et al., 2011), drama therapy, music therapy, art therapy (Baker et al., 2018) report few studies and even fewer that address aggression. Systematic reviews of gestalt therapy and aggression are hard to find. Thus, many of these fringe therapies simply lack evidence to support their use in reducing aggression and violence and so at this time should be avoided.

17.8 Summary

This chapter and the previous chapter have reviewed psychosocial interventions for aggression and violence categorized by type of therapy. As might be anticipated, reflecting the severity and ubiquitous nature of the problem, there is an enormous literature. As a broad generalization, individual CBT and different forms of behavioral interventions dominate the literature and have the greatest support from empirical studies, expert panels, and governmental agencies in many contexts. There is considerably less evidence supporting fringe therapies such as art therapy and evidence that popular treatments such as boot camps are largely ineffective and sometimes dangerous or even lethal. Although such broad generalizations are useful in deciding which treatments to use, it is also important to consider what works in specific populations and settings. The next chapter addresses some of the more important populations and settings.

References

Armenti, N. A., & Babcock, J. C. (2016). Conjoint treatment for intimate partner violence: A systematic review and implications. *Couple and Family Psychology: Research and Practice, 5*(2), 109–123. https://doi.org/10.1037/cfp0000060

Baker, F. A., Metcalf, O., Varker, T., & O'Donnell, M. (2018). A systematic review of the efficacy of creative arts therapies in the treatment of adults with PTSD. *Psychological Trauma Theory Research Practice and Policy, 10*(6), 643–651.

Beail, N. (2017). Anger in people who have intellectual and developmental disabilities. In P. Sturmey (Ed.), *The Wiley handbook of violence and aggression* (pp. 1–12). Wiley.

Bergin, T. (2016). *The evidence enigma: Correctional boot camps and other failures in evidence-based policymaking*. Routledge.

Betancourt, T. S., Meyers-Ohki, S. E., Stevenson, A., Ingabire, C., Kanyanganzi, F., Munyana, M., Mushashi, C., Teta, S., Fayida, I., Cyamatare, F. R., Stulac, S., & Beardslee, W. R. (2011). Using mixed-methods research to adapt and evaluate a family strengthening intervention in Rwanda. *African Journal of Traumatic Stress, 2*(1), 32.

Betancourt, T. S., Ng, L. C., Kirk, C. M., Brennan, R. T., Beardslee, W. R., Stulac, S., Mushashi, C., Nduwimana, E., Mukunzi, S., Nyirandagijimana, B., Kalisa, G., Rwabukwisi, C. F., & Sezibera, V. (2017). Family-based promotion of mental health in children affected by HIV: A pilot randomized controlled trial. *Journal of Child Psychology and Psychiatry, 58*(8), 922–930.

Betancourt, T. S., Franchett, E., Kirk, C. M., Brennan, R. T., Rawlings, L., Wilson, B., Yousafzai, A., Wilder, R., Mukunzi, S., Mukandanga, J., Ukundineza, C., Godfrey, K., & Sezibera, V. (2020). Integrating social protection and early childhood development: Open trial of a family home-visiting intervention, Sugira Muryango. *Early Child Development and Care, 190*(2), 219–235.

Carr, A. (2014). The evidence base for couple therapy, family therapy and systemic interventions for adult-focused problems. *Journal of Family Therapy, 36*(2), 158–194.

Carr, A. (2019). Couple therapy, family therapy and systemic interventions for adult-focused problems: The current evidence base. *Journal of Family Therapy, 41*(4), 492–536.

Curtis, N. M., Ronan, K. R., & Borduin, C. M. (2004). Multisystemic treatment: A meta-analysis of outcome studies. *Journal of Family Psychology, 18*(3), 411–419.

Dutton, D. G., & Corvo, K. (2007). The Duluth model: A data-impervious paradigm and a failed strategy. *Aggression and Violent Behavior, 12*(6), 658–667.

Eckhardt, C. I., Murphy, C. M., Whitaker, D. J., Sprunger, J., Dykstra, R., & Woodard, K. (2013). The effectiveness of intervention programs for perpetrators and victims of intimate partner violence. *Partner Abuse, 4*(2), 196–231.

Erford, B. T., Bardhoshi, G., Ross, M., Gunther, C., & Duncan, K. (2017). Meta-analysis of counseling outcomes for youth with Conduct Disorders. *Journal of Counseling & Development, 95*(1), 35–44.

Fonagy, P. (2003). Towards a developmental understanding of violence. *The British Journal of Psychiatry, 183*(3), 190–192.

Garrido, V., & Morales, L. A. (2007). Serious (violent or chronic) juvenile offenders: A systematic review of treatment effectiveness in secure corrections. *Campbell Systematic Reviews, 3*(1), 1–46.

Henggeler, S. W., & Schoenwald, S. K. (1994). Boot camps for juvenile offenders: Just say no. *Journal of Child and Family Studies, 3*(3), 243–248.

Henggeler, S. W., Melton, G. B., & Smith, L. A. (1992). Family preservation using multisystemic therapy: An effective alternative to incarcerating serious juvenile offenders. *Journal of Consulting and Clinical Psychology, 60*(6), 953.

Henggeler, S. W., Melton, G. B., Smith, L. A., Schoenwald, S. K., & Hanley, J. H. (1993). Family preservation using multisystemic treatment: Long-term follow-up to a clinical trial with serious juvenile offenders. *Journal of Child and Family Studies, 2*(4), 283–293.

Henggeler, S. W., Cunningham, P. B., Pickrel, S. G., Schoenwald, S. K., & Brondino, M. J. (1996). Multisystemic therapy: An effective violence prevention approach for serious juvenile offenders. *Journal of Adolescence, 19*(1), 47–61.

Heyman, R. E., & Slep, A. M. S. (2007). Therapeutic treatment approaches to violent behavior. In D. J. Flannery, A. T. Vazsonyi, & I. D. Waldman (Eds.), *The Cambridge handbook of violent behavior and aggression* (pp. 602–617). Cambridge University Press. https://doi.org/10.1017/CBO9780511816840.032

Howell, J. C., & Lipsey, M. W. (2012). Research-based guidelines for juvenile justice programs. *Justice Research and Policy, 14*(1), 17–34.

Johnson, M. P. (1995). Patriarchal terrorism and common couple violence: Two forms of violence against women. *Journal of Marriage and Family Therapy, 57*, 282–294.

Kaminski, J. W., Valle, L. A., Filene, J. H., & Boyle, C. L. (2008). A meta-analytic review of components associated with parent training program effectiveness. *Journal of Abnormal Child Psychology, 36*(4), 567–589.

Kaplan, D. M., Tarvydas, V. M., & Gladding, S. T. (2014). 20/20: A vision for the future of counseling: The new consensus definition of counseling. *Journal of Counseling & Development, 92*(3), 366–372.

Karakurt, G., Whiting, K., Van Esch, C., Bolen, S. D., & Calabrese, J. R. (2016). Couples therapy for intimate partner violence: A systematic review and meta-analysis. *Journal of Marital and Family Therapy, 42*(4), 567–583.

Kazdin, A. E. (2015). Psychosocial treatments for conduct disorder in children and adolescents. In P. E. Nathan & J. M. Gorman (Eds.), *A guide to treatments that work* (pp. 141–173). Oxford University Press.

Kazdin, A. E., Esveldt-Dawson, K., French, N. H., & Unis, A. S. (1987). Problem-solving skills training and relationship therapy in the treatment of antisocial child behavior. *Journal of Consulting and Clinical Psychology, 55*(1), 76–85. https://doi.org/10.1037/0022-006X.55.1.76

Lafuente, J., Zubiaur, M., & Gutiérrez-García, C. (2021). Effects of martial arts and combat sports training on anger and aggression: A systematic review. *Aggression and Violent Behavior, 58*, 101611.

Leijten, P., Gardner, F., Landau, S., Harris, V., Mann, J., Hutchings, J., Beecham, J., Bonin, E. M., & Scott, S. (2018). Research Review: Harnessing the power of individual participant data in a meta-analysis of the benefits and harms of the Incredible Years parenting program. *Journal of Child Psychology and Psychiatry, 59*(2), 99–109.

Lipsey, M. W., & Wilson, D. P. B. (1998). Effective intervention for serious juvenile offenders. In R. Loeber & D. P. Farrington (Eds.), *Serious and violent juvenile offenders. Risk factors and successful intervention* (pp. 313–345). Sage.

Löfholm, C. A., Olsson, T., Sundell, K., & Hansson, K. (2009). Multisystemic therapy with conduct-disordered young people: Stability of treatment outcomes two years after intake. *Evidence & Policy: A Journal of Research, Debate and Practice, 5*(4), 373–397.

McCart, M. R., Priester, P. E., Davies, W. H., & Azen, R. (2006). Differential effectiveness of behavioral parent-training and cognitive-behavioral therapy for antisocial youth: A meta-analysis. *Journal of Abnormal Child Psychology, 34*(4), 525–541.

McGuire, J., Bilby, C. A., Hatcher, R. M., Hollin, C. R., Hounsome, J., & Palmer, E. J. (2008). Evaluation of structured cognitive–behavioural treatment programmes in reducing criminal recidivism. *Journal of Experimental Criminology, 4*(1), 21–40.

McKeown, A., & McCrory, S. (2019). Life Minus Violence-Enhanced (LMV-E) with women in custody: Single-case experimental design. *Journal of Criminological Research, Policy and Practice, 5*(4), 295–306. https://doi.org/10.1108/JCRPP-08-2019-0055

Meade, B., & Steiner, B. (2010). The total effects of boot camps that house juveniles: A systematic review of the evidence. *Journal of Criminal Justice, 38*(5), 841–853.

Minuchin, S. (1967). *Families of the slums: An exploration of their structure and treatment*. Basic Books.

Mills, L. G., Barocas, B., Butters, R. P., & Ariel, B. (2019). A randomized controlled trial of restorative justice-informed treatment for domestic violence crimes. *Nature Human Behaviour, 3*(12), 1284–1294.

Olsson, T. M. (2010a). Intervening in youth problem behavior in Sweden: A pragmatic cost analysis of MST from a randomized trial with conduct disordered youth. *International Journal of Social Welfare, 19*(2), 194–205.

Olsson, T. M. (2010b). MST with conduct disordered youth in Sweden: Costs and benefits after 2 years. *Research on Social Work Practice, 20*(6), 561–571.

Pence, E., & Paymar, M. (1993). *Education groups for men who batter: The Duluth model.* Springer.

Postorino, V., Sharp, W. G., McCracken, C. E., Bearss, K., Burrell, T. L., Evans, A. N., & Scahill, L. (2017). A systematic review and meta-analysis of parent training for disruptive behavior in children with autism spectrum disorder. *Clinical Child and Family Psychology Review, 20*(4), 391–402.

Ray, D. C., Armstrong, S. A., Balkin, R. S., & Jayne, K. M. (2015). Child-centered play therapy in the schools: Review and meta-analysis. *Psychology in the Schools, 52*(2), 107–123.

Ruane, A., & Carr, A. (2019). Systematic review and meta-analysis of Stepping Stones Triple P for parents of children with disabilities. *Family Process, 58*(1), 232–246.

Sanders, M. R., Kirby, J. N., Tellegen, C. L., & Day, J. J. (2014). The Triple P-Positive Parenting Program: A systematic review and meta-analysis of a multi-level system of parenting support. *Clinical Psychology Review, 34*(4), 337–357.

Schoenwald, S. K., Henggeler, S. W., Brondino, M. J., & Rowland, M. D. (2000). Multisystemic therapy: Monitoring treatment fidelity. *Family Process, 39*(1), 83–103.

Schwalbe, C. S., Gearing, R. E., MacKenzie, M. J., Brewer, K. B., & Ibrahim, R. (2012). A meta-analysis of experimental studies of diversion programs for juvenile offenders. *Clinical Psychology Review, 32*(1), 26–33.

Serketich, W. J., & Dumas, J. E. (1996). The effectiveness of behavioral parent training to modify antisocial behavior in children: A meta-analysis. *Behavior Therapy, 27*(2), 171–186.

Sherman, L. W., Strang, H., Mayo-Wilson, E., Woods, D. J., & Ariel, B. (2015). Are restorative justice conferences effective in reducing repeat offending? Findings from a Campbell systematic review. *Journal of Quantitative Criminology, 31*(1), 1–24.

Skinner, B. F. (1953). *Science and human behavior*. Simon & Schuster.

Stover, C. S., & Morgos, D. (2013). Fatherhood and intimate partner violence: Bringing the parenting role into intervention strategies. *Professional Psychology: Research and Practice, 44*(4), 247–256. https://doi.org/10.1037/a0031837

Stover, C. S., Meadows, A. L., & Kaufman, J. (2009). Interventions for intimate partner violence: Review and implications for evidence-based practice. *Professional Psychology: Research and Practice, 40*(3), 223–233.

Strassel, J. K., Cherkin, D. C., Steuten, L., Sherman, K. J., & Vrijhoef, H. J. (2011). A systematic review of the evidence for the effectiveness of dance therapy. *Alternative Therapies in Health & Medicine, 17*(3), 50–59.

Sundell, K., Hansson, K., Löfholm, C. A., Olsson, T., Gustle, L. H., & Kadesjö, C. (2008). The transportability of multisystemic therapy to Sweden: Short-term results from a randomized trial of conduct-disordered youths. *Journal of Family Psychology, 22*(4), 550–560.

Taubner, S., Rabung, S., Bateman, A., & Fonagy, P. (2017a). Psychoanalytic concepts of violence and aggression. In P. Sturmey (Ed.), *The Wiley handbook of violence and aggression* (pp. 1–14). Wiley.

Taubner, S., Fonagy, P., Bateman, A., & Rabung, S. (2017b). Psychodynamic treatment of violence and aggression: Empirical evidence and new approaches. In P. Sturmey (Ed.), *The Wiley handbook of violence and aggression* (pp. 1–11). Wiley.

Tellegen, C. L., & Sanders, M. R. (2013). Stepping Stones Triple P-Positive Parenting Program for children with disability: A systematic review and meta-analysis. *Research in Developmental Disabilities, 34*(5), 1556–1571.

Zeilberger, J., Sampen, S. E., & Sloane, H. N., Jr. (1968). Modification of a child's problem behaviors in the home with the mother as therapist. *Journal of Applied Behavior Analysis, 1*(1), 47–53.

Chapter 18
Treatment of Violence and Aggression in Children, Adolescents, Families, and Schools

When considering the psychotherapy outcome literature there are always at least two ways of slicing the pie depending upon the question to be answered: One way is to ask which therapies work, as in the preceding two chapters, and another is to look at what works for specific populations and contexts. This and the next chapter address the second question by examining the outcome literature for aggression and violence first for children, and adolescents and their families and caregivers and schools, and then for various adult populations, including the victims of aggression and violence.

Prevention and treatment of aggression and violence in children and adolescents are appealing in many ways. Many children and adolescents may seem relatively less blameworthy than adults. In addition, from the point of view of preventing lifelong suffering and the potential for providing the possibility of greatest economic benefits to society, there may be potential for maximum lifelong benefits to the individual, potential victims, and society generally.

This outcome literature is very extensive: Over 10 years ago, McGuire (2008) listed 70 meta-analyses published between 1985 and mid-2007 alone, and so there is no attempt to be comprehensive in this chapter. Rather, this chapter will selectively look at treatment of aggression and violence in children and adolescents, such as delinquents; violence within the family, including violence against children and seniors within the family.

18.1 Intervention in The Cradle of Violence

The stereotype of a "typical family" is one in which two loving heterosexual and monogamous parents care and protect their children who in turn love their parents back and return their love and affection in many other ways. Such a stereotype was probably never true: As noted elsewhere, families are a varied feast with many

P. Sturmey, *Violence and Aggression*, Advances in Preventing and Treating Violence and Aggression, https://doi.org/10.1007/978-3-031-04386-4_18

different configurations. Families have widely differing degrees of aggression and violence between family members, leading some to name the family as "The cradle of violence" (Steinmetz & Straus, 1973). This section will consider several forms of violence within the family including abuse of infants, children and youth, and elder abuse. The reader is referred to Chap. 17, for review of outcome research on the Duluth model, batterer intervention programs, parent skills training, and conjoint therapy.

18.1.1 Family Violence Against Infants and Young Children

Aggression and violence toward children by parents and other family members are particularly disturbing as we hope that family caregivers love and nurture their children; yet, sadly, about half of all children worldwide experience aggression and violence and even filicide (Hillis et al., 2016) often in the context of their own families. Human infanticide is a well-recognized part of human culture and history and is reflected in the ambiguous legal status of infants as humans and ambiguity over the culpability of family members and others who kill infants. It is also reflected in cultural practices to protect the newborn, such as safe haven laws that permit leaving abandoned babies at hospital, baby hatches, where mothers may leave their babies at convents and similar places, and free delivery of unwanted babies.

In a systematic review, Tanaka et al. (2017) reported that neonaticide by the mother occurs at a rate of between 0.12 (Paris 1991–2008) and 27.2 (Dar es Salam, 2005) per 100,000 live births. It is associated with maternal characteristics such as young age, being single, and a first birth. These data are largely based on state records and so probably under-report the phenomenon. According to Putkonen et al. (2016), the most common motive for maternal neonaticide was to get rid of the child and was not associated with maternal psychosis or suicidality. Humans infants and younger children, and children with disabilities or exceptionality are especially likely to be killed (Bayat, 2015) and the perpetrators are often seen as less culpable and receive more lenient sentences in mock jury trials (Bottoms et al., 2011; MacKay & Covell, 2013).

"I could just shake you!" is a common expression of frustration, but shaking infants is highly dangerous and even lethal and may result in brain damage and associated developmental problems sometimes referred to as "shaken baby syndrome." It is characterized by subdural hematoma, retinal hemorrhages, and encephalopathy in the absence of overt signs of head trauma. Shaking babies may result in brain damage, intellectual and physical disability, negative social outcomes, and death. For example, Antonietti et al. (2019) followed up 80 babies following probable shaken baby syndrome recognized during initial hospitalization as infants. Three infants died and at 26-month follow-up a third had neurological problems, such as cerebral palsy, multiple disabilities, and severe intellectual disabilities. A quarter were in foster care for an average of about 3 years. Infants are at risk for many severe outcomes due to violence from their family members. In addition to human suffering, the lifelong medical and other costs of violence to infants are very large (Miller et al., 2018). What can be done to prevent and treat this form of aggression and violence?

There have been several approaches to prevention and intervention for infants at risk for violence. Many studies have noted that infant crying is a powerful aversive stimulus to caregivers and termination of crying is a powerful negative reinforcer for a variety of caregiver behaviors such as picking up and comforting the infant (Bruzek et al., 2009; Thompson et al., 2011). Infant crying is also a common antecedent for infant abuse (Barr et al., 2006; Talvik et al., 2008) which should be addressed in prevention programs. Prevention is challenging because the problem of abusive head trauma is relatively infrequent and thus, prevention programs must access very large numbers of caregivers to identify at-risk infants for intervention. Thus, scaling up prevention over long time periods is important due to the diffuse nature of the problem.

One study that addressed this comes from Lopez-Bushnell et al. (2017) who evaluated a three component, province-wide program in British Columbia over an 8-year period. They attempted to deliver the program to all caretakers of all 354,777 newborns between 2009 and 2018. There were three program components. First, was a booklet and DVD describing and normalizing infant crying, describing the dangers of shaking babies, and basic strategies to respond to crying, such as recruiting helpers, which was offered to both mothers, fathers and other family members. The second component was a follow-up phone call or home visit to discuss talking points related to handling crying. Finally, there was an annual community-wide education. Following intervention there was a 35% reduction in hospital admissions for babies with abusive head trauma over an 8-year implementation period. In a subsequent study, caregivers were more knowledgeable about crying and more likely to report being more likely to walk away from inconsolable crying (Fujiwara et al., 2012).

There are many studies using broadly similar approaches. Geeraert et al. (2004) reported a meta-analysis of 40 studies which aimed to prevent abuse and neglect in infants and children aged 0–3 years published between 1975 and 2002. They calculated 587 effect sizes with a mean $d = 0.29$ ($p < 0.001$). Outcomes were quite varied. Effect sizes were about the same for both reduction in abuse ($d = 0.26$) and risk factors ($d = 0.29$). The authors concluded that programs were effective and the results were similar to previous meta-analysis in this area, but the magnitude of the effects was modest.

These studies suggest that violence against infants and young children, including infanticide, is an enduring part of human society that is both disturbing, and personally and financially costly. There are a variety of intervention approaches, some of which can be delivered on a large scale and appear to be modestly effective.

18.2 School-Based Interventions

Schools are not merely places of education. Rather, they are the social agents of change in numerous ways delivering not only education, but a host of other social interventions such as nutritional programs, parent support and training, and mental health interventions for children and their families. As such they are one of the main social agents in controlling aggression and teaching and promoting other, more acceptable behavior and values such as conflict resolution, social skills and peaceful interactions (Cf. Skinner, 1953).

Schools are challenged by aggression and violence in a number of ways. Early in education, nurseries, kindergartens, and preschools, along with the family, can be the vehicle to move children through the developmental stages when they are most likely to be aggressive and to give them a repertoire of skills to deal effectively with situations that may precipitate aggression and violence in the future. During adolescence, when bullying and gang violence may increase, schools become an important social vehicle to prevent and reduce these problems. In the past and in many schools today, these problems have been dealt with through threats and delivery of positive punishment, removal of the offending child or young adult from education, and placement in specialized programs and the justice system. Present educational ideology preaches rejection of these approaches in favor of preventive and positive strategies that address not only the individual student, but also the school system.

18.2.1 General Prevention in Schools

Unsurprisingly, there is a very large literature on school-based interventions to reduce aggression (Beck & Fernandez, 1998; Bennett & Gibbons, 2000; McCart et al., 2006; Sukhodolsky et al., 2004; Wilson et al., 2003). More recently, for example, Barnes et al. (2014) systematically reviewed school-based individual prevention and treatment for children and adolescents delivered either universally across entire schools or in small group formats. Universally delivered strategies use the teaching expertise of school staff who are familiar with presenting pedagogical material to classrooms of students in a manner that may not stigmatize high risk students and thus may reach the entire schools relatively quickly. Universally delivered strategies, however, may have challenges in terms of treatment integrity across a relatively large number of change agents and a long period of time, may not appear relevant to some students and families, may not provide relevant treatment individualization, and may use verbal instruction and explanation at the expense of modeling and feedback. Small group instruction involves pulling at-risk students out of the classroom and may target relevant students, gives teachers relief from managing their problem behavior in a larger classroom, perhaps providing some greater degree of individualization and concentration of resources for those students who are most likely to benefit from the intervention.

In their meta-analysis, Barnes et al. (2014) identified 25 experiments or quasi-experiments with 30,309 students, mostly in kindergarten through grade 5. They evaluated CBT for aggression in school-aged children and adolescents using CBT principles, behavior therapy, and cognitive mediation through self-talk using universal and small group delivery of interventions in studies published between 1992 and 2012. The overall unweighted mean Cohen's d was −0.14 with a wide range of outcomes: Thus, the range of unweighted Cohen's d was −0.3.23 to −0.62 and three quarters of the studies showed decreases in aggression, but one quarter showed *increases*. Interestingly, demographic and personal factors had no impact on outcome: The only mediator was treatment type, namely that universal outcomes were associated with an effect size of $d = -0.18$, whereas small group interventions had an effect size of $d = 0.00$. The authors concluded that universal interventions were effective, but that the effect size was notably smaller than previous meta-analyses. They attributed this smaller effect size to moving from demonstration projects often conducted in association with universities, researchers, and relatively rich resources to larger scale business-as-usual studies in settings with fewer resources and multiple priorities. As in other outcome literatures, the issue of translating early research demonstration studies to routine services is a challenge, suggesting the importance of future research identifying factors that can be manipulated to enhance outcomes, such as treatment dose, treatment integrity, and maximizing use of the most effective components of an effective treatment package.

Schools have some obvious potential in preventing aggression and violence: They can access large numbers of children, can do so early on in child development, and have many experienced and potentially effective change agents. Prevention and treatment of aggression and violence in the schools are truly challenging because of the scale of the problem and the magnitude of effort that is required across many sites and staff.

18.2.2 Schools and Gangs

A second important school-based intervention for violence is interventions related to gang membership. Efforts to do so can target gang membership at several points including preventing joining gangs, getting members to leave gangs, and suppression of current gang activity. Although there are many popular and commercial products to reduce gang membership, claims for effectiveness must be backed up with evidence. Sometimes palatable and politically endorsed and popular programs either have no evidence to support their effectiveness or have been positively demonstrated not to work.

An example of the gradual evolution of a school-based, gang prevention program is the Gang Resistance and Education Training (GREAT) program. Initially, this program was designed by Phoenix-based police based around the Drug Abuse Resistance Education (DARE) program. Although use of the initial GREAT model spread rapidly, subsequent evaluations demonstrated that it was ineffective. This may have been because it was based on the ineffective DARE model, used education and information, rather than skills training, was administered to all students, rather than focused on at-risk children, did not have a clear theoretical model, and was not empirically grounded. Subsequently, identification of alternate program models resulted in a substantially revised GREAT program (Esbensen et al., 2011).

Esbensen et al. (2011) conducted an RCT of the GREAT program in which they randomized 95 classrooms from 31 schools from multiple cities to GREAT or no treatment. There were 3,820 student participants who were mostly 11 years old. The goals of the revised GREAT program were to help youth avoid gang membership, reduce violence and criminal activity, and help youth develop a positive attitude toward law enforcement. The revised GREAT program consisted of education, skills training using role play, and a project to improve the school or local environment. Impressively, data included annual data collection over a 4-year period. Students in the GREAT program were less likely to report being in gangs. There was a 39% reduction in the odds ratio at post-treatment and a 24% reduction at 4-year follow-up.

This study had some important limitations. There were many self-report outcomes measures and multiple tests of significance of which a few were significant but the effect sizes were very small. Further, the effects were not replicable at all sites. Some have argued that this approach is "a breakthrough" (Howell, 2018), but caution might be warranted concerning the use of self-report data and the important issue of treatment integrity across multiple sites and many people delivering the treatment.

18.3 Delinquent Adolescents

Delinquent adolescents represent a particular challenge for the effective treatment of aggression and violence. There are many studies in this area including specific programs that were developed early on for delinquent adolescents, such as Achievement Place. In addition, concern over effective treatment and protecting the public has also led to the development of treatments provided in collaboration with the courts and justice system and interventions in collaboration with the police.

As might be expected with such a common and widespread social problem, there have been many studies on the effective treatment of many aspects of adolescent delinquents. This is reflected in the fact that there have been several systematic reviews and meta-analyses of interventions for delinquent adolescents. Table 18.1 provides a summary of some of these reviews and meta-analyses. There are some notable generalizations that can be made based on these meta-analyses. First, there is a very large evidence-base to guide treatment choice so that we can be confident about conclusions. Second, all meta-analyses and reviews in the table concluded that several treatments were effective in reducing recidivism, other than punitive-based approaches such as boot camps and fear-based "scaring strait" approaches. (See Chap. 17.) The only punitive-based approach that reduced recidivism was

Table 18.1 A table of some illustrative meta-analyses of treatments of juvenile offenders. Howell and Lipsey (2012) are presented as a representative publication of a series of meta-analyses by Lipsey and colleagues. *BPT* behavioral parent training, *CBT* cognitive behavior therapy, *TAU* treatment as usual

Meta-analysis	Number of studies and participants	Aims and types of treatments and programs	Effect sizes	Conclusions
McCart et al. (2006)	$k = 71$ studies of which 30 were for BPT and 31 were for CBT published between 1973 and 2004. No overall n reported	Aimed to compare the relative effectiveness of BPT and CBT	Overall ES post-treatment was 0.40 (CI 0.37, 0.47) and at follow-up was 0.22 (CI 0.11, 0.32); for BPT ES was 0.47 (CI 0.34, 0.61) and for CBT was 0.35 (0.25, 0.47). There was a slight improvement in parental adjustment with BPT (ES = 0.33)	Both BPT and CBT can be effective in treating aggressive behavior in youth and perhaps BPT is also effective in reducing parental distress. CBT may have been less effective than BPT for younger children
Garrido and Morales (2007)	$k = 17$ documents (journal articles, book chapters, reports, etc.) with 30 effect sizes; $n = 7509$ (3685 in treatment and 3824 in comparison groups) published between 1970 and 1996	Aimed to compare the effects of treatment on both frequency and severity of reoffending. Punitive programs were harmful. Differences between institutional and community programs were unclear, but institutional programs might have negative outcomes not measured by this meta-analysis.	log OR = 1.235 ($r = 0.064$, i.e., a 6.4% reduction in general recidivism in experimental over control groups) ($p = 0.006$) and log OR = 1.074 for serious recidivism; both effect sizes were significantly different from 0	Interventions for both offending and serious offending do work compared to TAU. Cognitive, CBT, and multi-focus programs were effective. Effect sizes were very similar to Lipsey and Wilson (1998). Effect sizes were relatively small because most comparisons were with TAU
McGuire (2008)	Narrative review of treatment of aggression including applications to youth	Aimed to synthesize results of approximately 70 meta-analyses, including those addressing youth violence	Effective interventions included interpersonal skills training, structured counseling, behavioral interventions, and achievement place	At least four interventions can reduce recidivism in youth

(continued)

Table 18.1 (continued)

Meta-analysis	Number of studies and participants	Aims and types of treatments and programs	Effect sizes	Conclusions
Howell and Lipsey (2012)	$k = 548$ independent study samples from 361 reports with juveniles aged 12–21 with at least one delinquency outcome with either experimental or quasi-experimental designs published 1958–2002	Aimed to summarize comprehensively the outcomes literature	Reported clear differences in effect sizes in reducing recidivism depending upon program type. Negative effect sizes (i.e., increases in recidivism) were associated with punitive programs focusing on disciplinary programs such as boot camps and fear-based programs, such as “scaring strait”; positive effects were found for intense surveillance in the absence of treatment such as intense probation. The largest positive effects were observed for restorative, skill-building, counseling, and multimodal approaches with the latter three being associated with an approximately 10–15% reduction in recidivism. Behavioral and CBT programs were associated with the largest outcomes with approximately 20–25% reduction in recidivism, higher quality more extensive programs (up to a point of diminishing returns) had larger effect sizes	Behavioral and CBT programs were most effective, but other positive programs including restorative, counseling, and multimodal treatments were also quite effective. Punitive programs were harmful. Differences between institutional and community programs were unclear, but institutional programs might have negative outcomes not measured by this meta-analysis.

intense surveillance, such as highly supervised probation (Howell & Lipsey, 2012), but it is unclear if this approach results in positive youth gains. Thus, these meta-analyses demonstrate that there are treatments for juvenile delinquents that work.

The treatments that have the most extensive data to support their use are CBT and behavioral interventions such as behavioral parent training (BPT), behavioral contracting, social skills training, improving educational performance, etc. Other approaches such as restorative justice approaches, structured counseling, and multi-modal approaches have broadly comparable effect sizes. There are many qualifications to these observations. First, most of the effect sizes are small to moderate. Although such small to moderate effect sizes are worthwhile having probably in social terms, and most likely in economic terms, there is considerable room for improvement, especially at the level of implementation for each individual. Further, the outcomes of these meta-analyses are often quite varied, indicating that other variables are important in determining treatment outcomes, although what these mediating variables beyond treatment type, quality, and integrity are is sometimes unclear, as meta-analyses are hobbled in analyzing outcome mediators by the data that are available, rather than the data that might be important or meaningful. Nevertheless, there are considerable grounds for modest optimism in treatment of aggression and violence among delinquent youth.

18.3.1 Prevention of Youth Violence

Given the extensive nature of youth violence (Fowler et al., 2018; Lyons et al., 2016), it is unsurprising it has received extensive attention from researchers and policy makers. This includes efforts at prevention before aggression and violence had occurred.

An overview of the panorama of prevention programs comes from two reports from the Center for Disease Control and Prevention (David-Ferdon & Simon, 2014; David-Ferdon et al., 2015) which identified multiple effective programs for prevention programs. These authors organized prevention programs into seven approaches, each with their own descriptions and specific programs. The first approach was Universal School-based Youth Violence Prevention, which consisted of providing school children and staff information on violence, and ways to change thinking and behaving including non-violent skills, such as conflict resolution. Examples included anger management programs and the Good Behavior Game. The second approach was parenting skills and family relationship approaches which included supporting and training parents and other family members in communication, problem solving, monitoring, and behavior management skills. Examples included programs such as Incredible Years and Triple P programs. (See Chap. 17.) Third was intensive family-focused approaches, which was providing therapeutic services for high risk youth and their families. Examples included skills training, academic support, behavioral support, and multisystemic family therapy for at-risk youth. The fourth approach was policy, environmental, and structural approaches which involved community-wide interventions to enhance safety, affect violence and risk factors. Examples included collaborations between private and public entities and changing the physical environment through physical redesign. Fifth, street outreach and community mobilization approaches include connecting trained staff to at-risk youth to mediate conflicts and change beliefs about the acceptability of violence. Examples of this approach included community interventions to disrupt the use of guns and retaliatory violence. The sixth approach was early childhood home visitation which was providing education and training to professionals and family members of disadvantaged children to foster strong families. Examples included programs that partner in home support and families and the Triple P program. The final approach was early childhood education, which included high quality early education for low income children. There are many examples of such early intervention approaches which might lay a strong developmental foundation to prevent aggression and violence in youth.

18.3.2 Achievement Place

One approach to delinquency with an extensive history and research has been Achievement Place. Early research on token economies rapidly extended their application from back wards of psychiatric hospitals in the late 1950s to residential and other programs for delinquents (e.g., Tyler & Brown, 1968). One approach which incorporates a token economy as a key component is Achievement Place, which is a behavioral program developed in the 1960s. It has a long history of implementation, development, and evaluation and is interesting, in part, because the program has survived so long and has an extensive history of replication.

In the original study, Phillips (1968) reported the evaluation of the use of a residential home for three adolescent boys who had been referred for residential treatment by the court following minor offenses such as theft, fighting, and disruption. The home had two house parents who were the author of this article and his wife who administered the program jointly. There was a set of schedules and chores which were considered to be similar to a typical family. The house parents administered a points system, which included points for appropriate behavior, such as remaining on schedule, completing chores, completing homework, and social behaviors, such as good manners. Points were later exchanged for backup reinforcers including allowances, bikes, games, and snacks. Residents could earn points for these things and could lose points for aggression, cursing, and being late. Phillips reported five experiments showing that Achievement Place was effective in reducing aggressive statements, reducing poor grammar, and increasing doing chores, punctuality, and homework.

Since the original study there has been extensive research on Achievement Place. For example, Fixsen and Blase (2019) published a 50-year history of the development and dissemination of Achievement Place. The first 10 years after Phillips' (1968) study was characterized by development and refinement and extension of the original procedures, such as improving interactions and changing residents' behavior at school, and defining clearly the essential features of the program including a clear program description, its functions, the operational definitions of these functions and assessment of procedural integrity primarily through measurement of the performance of the home parents. This resulted in a comprehensive program handbook defining the procedures and infrastructure of Achievement Place.

By 1974 there were 64 replication homes, but with varying outcomes. Thirty-three homes were closed, had abandoned the model, or stopped training their staff in the relevant procedures. Closure was much more likely in sites that were distant from the original site. Whereas 58% of local homes in Kansas remained open, only 23% of out-of-state homes remained so. These experiences resulted in greater emphasis on procedures to ensure program integrity through ongoing support and certification of each home and the development of an association to support the program. A key change was moving toward developing sites with multiple homes which were physically near each other, rather than geographically scattered programs. This modification to Achievement Place resulted in key personnel in place and more readily available to support a cluster of homes and a system of external validation of program performance. These organizational changes resulted in greater rates of site certification. In the mid-1970s, independent evaluation comparing Achievement Place homes with other matched homes indicated that the Achievement Place homes were 7% cheaper per diem, the costs per resident were approximately 20% cheaper than other homes, that satisfaction and school outcomes were higher than other homes, but that there were no differences in rates of delinquency at 1 and 3 years after discharge. Following the 1980s there continued to be extensive numbers of replication sites.

The authors have differed as to their conclusions as to the effects of Achievement Place on recidivism. Kirigin et al. (1982) presented data on comparisons of rates of alleged offenses and percentage of youth engaged in criminal activity in Achievement Place homes with comparison homes and found lower rates of criminal behavior during the treatment program but not subsequently. Kingsley (2006) reanalyzed the follow-up data and found odds ratios of 0.49 and 0.41 for boys and girls, respectively. Further, the quantity of data on recidivism is much smaller than for other outcomes in Achievement Place. Thus, although Achievement place appears to have robust outcomes early on when the adolescents are resident in the homes the longer-term outcomes are less clear at this time.

18.3.3 Adolescent Sex Offenders

There are numerous existing specialized treatment programs for adolescent sex offenders reflecting societies concern over the offenses that these individuals have already committed and the potential future offenses that might be prevented. It is unclear as to whether adolescent sex offenders should be considered a unique and specific group, part of the broader group of adolescent offenders. These models have implications for whether programs should be generic or specialized, but the view that adolescent sexual offenders are a unique group requiring specialized programs has dominated the literature.

Thus, Kettrey and Lipsey (2018) conducted a systematic review and meta-analysis of treatment programs for juvenile sexual offenders. They searched the literature using an existing comprehensive database and online search strategies. Studies have to include adolescent sex offenders, be published since 1950 and be either a RCT or involve baseline matching on variables relevant to the risk of sexual offending. They identified only eight studies that met these criteria of which only one was a RCT. Programs generally contained features of CBT including relapse prevention, cognitive restructuring, changing dysfunctional thinking, although one study included multisystemic therapy. Studies reported primarily recidivism data.

The overall effect size for sexual offending was not different from zero, but the effect size for general offending was 0.58 (95% CI 0.42, 0.81). These results indicated that there was no evidence supporting the effectiveness of programs for sexual reoffending, but there was evidence that these programs resulted in a reduction in general reoffending. The authors highlighted the most prominent feature of this literature, which was the lack of experimental studies. Although there have been at least one non-experimental studies published since this meta-analysis (Jenkins et al., 2020), there does indeed seem to be a dearth of experimental or quasi-experimental studies evaluating these programs.

18.3.4 Summary

Aggression and violence perpetrated by children and adolescents is a common and socially significant problem, yet one that is quite responsive to intervention. The biggest appeal of work in this area is not only its immediate benefits to all concerned, but also the possibility of lifelong benefits to the individuals, their family and society.

18.4 Elder Abuse Within the Family

There are many studies documenting the prevalence and risk factors of elder abuse by family members. A systematic review of abuse of seniors by Cooper et al. (2008) estimated that approximately 6% of seniors experienced abuse within their relationships. Around 5–20% reported physical abuse by a family member. Finally, approximately 5–50% of family members reported engaging in abuse of a senior family member. These wide ranges of figures reflected the time periods of studies and methods used. Individual victim risk factors for abuse included degree of cognitive disability, dependence (Yan & Tang, 2004), behavior problems, psychiatric disorders, poor physical health, and frailty. Abuser risk factors included caregiver burden; their own mental health issues; relationship issues, such as poor family relationships; and environmental factors, such as low social support (Johannesen & LoGiudice, 2013). Storey (2020) also conducted a review of risk factors and extended the previous studies by identifying perpetrator characteristic, such as chronic illness, mental health, and substance abuse issues, financial dependency on the senior and related anger, stress and life events, negative attitudes and unrealistic expectations of the elder family member and poor relationships and social isolation.

An example of an intervention study that took place in family settings comes from Drossel et al. (2011), who evaluated the effects of Dialectical Behavior Therapy on prevention of elder abuse. The authors held weekly manualized Dialectical Behavior Therapy sessions in groups. The participants were 16 full-time caregivers aged 38–87 years old. The study used a pre-post design and used self-report measures of distress, well-being, and fatigue. Six caregivers also took part in booster sessions. Two participants were reported for elder neglect during study. Caregivers reported improvements in their own social functioning, emotional well-being, depression, and personal energy. They also reported greater use of problem solving. There are other psychosocial interventions for caregivers' stress management, such as mindfulness, for which there is some evidence that they affect outcomes, such as caregiver depression, which might decrease elder abuse (Fearing et al., 2017).

Although there have been many programs and a couple of systematic reviews and meta-analyses to address elder abuse in nursing homes and other medical and residential facilities, elder abuse within the family has received much less attention. For example, Ploeg et al.'s (2009) systematic review of interventions for elder abuse within the family identified only a couple of relevant studies. More recently, Fearing et al. (2017) conducted a systematic review of treatment programs for elder abuse in community settings of research published between 2009 and 2015. The main finding again was the lack of studies. They located only nine quantitative, but not necessarily experimental studies. These studies included three evaluations of psychological interventions for family caregivers of individuals with dementia, two studies of multidisciplinary team interventions, two studies of forensic center and conservatorship interventions, and two studies concerning elder abuse intervention programs for caregivers. Fearing et al. also graded the quality of the evidence in each study. The only intervention that had strong evidence from one relatively well designed study was psychological interventions for carers of family members with dementia. These interventions included eight individual sessions delivered by trained therapists to family carers that covered caregiver coping, anxiety, and depression which resulted in improved caregiver affect and non-significant trends toward less anxiety and self-reported abusive behavior.

Thus, evidence is accumulating toward evidence-based practices for preventing and intervening in elder abuse by family members. Unfortunately, it is an important but sadly a relatively neglected area of research and practice in prevention of aggression and violence (Marshall et al., 2020).

18.5 Chapter Summary

Despite our optimistic images of the family as the cradle of nurturance, it is too easy to see it as the cradle of violence, given the propensity of so many family members to be violent to one another, especially to the most vulnerable members of the family one is supposed to love and who are supposed to love you back. Fortunately, intervention for violent family members or entire families that engage in violent behavior is possible and desirable in order to benefit all concerned and to avoid or attenuate the harm done to family members and there are many evidence-based options available to do so.

Schools are one of society's major vehicles to achieve many outcomes including prevention and treatment of aggression and violence. There are many prevention strategies which represent good teaching and classroom management strategies with many benefits, but schools can also be involved in tertiary interventions such as working with delinquents and adolescents in gangs. Adolescents with histories of delinquency or who are at risk for delinquency and adolescents with histories of sexual violence may benefit from specialized treatment programs, but only if those programs are well evaluated and well run. Finally, we must acknowledge the existence of elder abuse by other family members. To date there is relatively little research to guide practice in this area. In the next chapter we continue our survey of prevention and treatment of aggression and violence in adult populations.

References

Antonietti, J., Resseguier, N., Dubus, J. C., Scavarda, D., Girard, N., Chabrol, B., & Bosdure, E. (2019). The medical and social outcome in 2016 of infants who were victims of shaken baby syndrome between 2005 and 2013. *Archives de Pédiatrie, 26*(1), 21–29.

Barnes, T. N., Smith, S. W., & Miller, M. D. (2014). School-based cognitive-behavioral interventions in the treatment of aggression in the United States: A meta-analysis. *Aggression and Violent Behavior, 19*(4), 311–321.

Barr, R. G., Trent, R. B., & Cross, J. (2006). Age-related incidence curve of hospitalized Shaken Baby Syndrome cases: Convergent evidence for crying as a trigger to shaking. *Child Abuse & Neglect, 30*(1), 7–16.

Bayat, M. (2015). The stories of 'snake children': Killing and abuse of children with developmental disabilities in West Africa. *Journal of Intellectual Disability Research, 59*(1), 1–10.

Beck, R., & Fernandez, E. (1998). Cognitive-behavioral therapy in the treatment of anger: A meta-analysis. *Cognitive Therapy and Research, 22*(1), 63–74.

Bennett, D. S., & Gibbons, T. A. (2000). Efficacy of child cognitive-behavioral interventions for antisocial behavior: A meta-analysis. *Child and Family Behavior Therapy, 22*(1), 1–15.

Bottoms, B. L., Kalder, A. K., Stevenson, M. C., Oudekerk, B. A., Wiley, T. R., & Perona, A. (2011). Gender differences in jurors' perceptions of infanticide involving disabled and non-disabled infant victims. *Child Abuse & Neglect, 35*(2), 127–141.

Bruzek, J. L., Thompson, R. H., & Peters, L. C. (2009). Resurgence of infant caregiving responses. *Journal of the Experimental Analysis of Behavior, 92*(3), 327–343. https://doi.org/10.1901/jeab.2009-92-327

Cooper, C., Selwood, A., & Livingston, G. (2008). The prevalence of elder abuse and neglect: A systematic review. *Age and Ageing, 37*(2), 151–160.

David-Ferdon, C., & Simon, T. R. (2014). *Preventing youth violence: Opportunities for action*. Centers for Disease Control and Prevention. https://stacks.cdc.gov/view/cdc/23501, downloaded 6/8/2022.

David-Ferdon, C., Simon, T. R., Spivak, H., Gorman-Smith, D., Savannah, S. B., Listenbee, R. L., & Iskander, J. (2015). CDC grand rounds: Preventing youth violence. *MMWR. Morbidity and Mortality Weekly Report, 64*(7), 171.

Drossel, C., Fisher, J. E., & Mercer, V. (2011). A DBT skills training group for family caregivers of persons with dementia. *Behavior Therapy, 42*(1), 109–119.

Esbensen, F. A., Peterson, D., Taylor, T. J., Freng, A., Osgood, D. W., Carson, D. C., & Matsuda, K. N. (2011). Evaluation and evolution of the Gang Resistance Education and Training (GREAT) program. *Journal of School Violence, 10*(1), 53–70.

Fearing, G., et al. (2017). A systematic review on community-based interventions for elder abuse and neglect. *Journal of Elder Abuse & Neglect, 29*(2–3), 102–133. https://doi.org/10.1080/08946566.2017.1308286

Fixsen, D. L., & Blase, K. A. (2019). The teaching-family model: The first 50 years. *Perspectives on Behavior Science, 42*(2), 189–211.

Fowler, K. A., Jack, S. P., Lyons, B. H., Betz, C. J., & Petrosky, E. (2018). Surveillance for violent deaths—National violent death reporting system, 18 states, 2014. *MMWR Surveillance Summaries, 67*(2), 1.

Fujiwara, T., Yamada, F., Okuyama, M., Kamimaki, I., Shikoro, N., & Barr, R. G. (2012). Effectiveness of educational materials designed to change knowledge and behavior about crying and shaken baby syndrome: A replication of a randomized controlled trial in Japan. *Child Abuse & Neglect, 36*(9), 613–620.

Garrido, V., & Morales, L. A. (2007). Serious (violent or chronic) juvenile offenders: A systematic review of treatment effectiveness in secure corrections. *Campbell Systematic Reviews, 3*(1), 1–46.

Geeraert, L., Van den Noortgate, W., Grietens, H., & Onghena, P. (2004). The effects of early prevention programs for families with young children at risk for physical child abuse and neglect: A meta-analysis. *Child Maltreatment, 9*(3), 277–291.

Hillis, S., Mercy, J., Amobi, A., & Kress, H. (2016). Global prevalence of past-year violence against children: A systematic review and minimum estimates. *Pediatrics, 137*(3), e20154079.

Howell, J. C. (2018). What works with gangs: A breakthrough. *Criminology & Public Policy, 17*(4), 991–999.

Howell, J. C., & Lipsey, M. W. (2012). Research-based guidelines for juvenile justice programs. *Justice Research and Policy, 14*(1), 17–34.

Jenkins, C. S., Grimm, J. R., Shier, E. K., van Dooren, S., Ciesar, E. R., & Reid-Quiñones, K. (2020). Preliminary findings of problematic sexual behavior-cognitive-behavioral therapy for adolescents in an outpatient treatment setting. *Child Abuse & Neglect, 105*, 104428. https://doi.org/10.1016/j.chiabu.2020.104428

Johannesen, M., & LoGiudice, D. (2013). Elder abuse: A systematic review of risk factors in community-dwelling elders. *Age and Ageing, 42*(3), 292–298.

Kettrey, H. H., & Lipsey, M. W. (2018). The effects of specialized treatment on the recidivism of juvenile sex offenders: A systematic review and meta-analysis. *Journal of Experimental Criminology, 14*(3), 361–387.

Kingsley, D. E. (2006). The teaching-family model and post-treatment recidivism: A critical review of the conventional wisdom. *International Journal of Behavioral Consultation and Therapy, 2*(4), 481–497. https://doi.org/10.1037/h0101002

Kirigin, K. A., Braukmann, C. J., Atwater, J. D., & Wolf, M. M. (1982). An evaluation of teaching-family (Achievement Place) group homes for juvenile offenders. *Journal of Applied Behavior Analysis, 15*(1), 1–16.

Lopez-Bushnell, K., Torrez, D., Robertson, J. V., Torrez, C., & Strickler, L. (2017). Shaken Baby Syndrome: A hospital-based education and prevention program in the intermediate care and the newborn intensive care nurseries. *Journal of Neonatal Nursing, 23*(3), 142–150.

Lipsey, M. W., & Wilson, D. B. (1998). Effective intervention for serious juvenile offenders: a synthesis of research. In R. Loeber, & D. P. Farrington (Eds.), *Serious & violent juvenile offenders: Risk factors and successful interventions* (pp. 313–345). Sage.

Lyons, B. H., Fowler, K. A., Jack, S. P., Betz, C. J., & Blair, J. M. (2016). Surveillance for violent deaths—National violent death reporting system, 17 states, 2013. *Morbidity and Mortality Weekly Report. Surveillance Summaries, 65*(10), 1–42.

MacKay, M. E., & Covell, K. (2013). What about the rights of the infant with disabilities? Responses to infanticide as function of infant health status. *Canadian Journal of Disability Studies, 2*(2), 35–57.

Marshall, K., Herbst, J., Girod, C., & Annor, F. (2020). Do interventions to prevent or stop abuse and neglect among older adults work? A systematic review of reviews. *Journal of Elder Abuse & Neglect, 32*(5), 409–433.

McCart, M. R., Priester, P. E., Davies, W. H., et al. (2006). Differential effectiveness of behavioral parent-training and cognitive-behavioral therapy for antisocial youth: A meta-analysis. *Journal of Abnormal Child Psychology, 34*, 525–541. https://doi.org/10.1007/s10802-006-9031-1

McGuire, J. (2008). A review of effective interventions for reducing aggression and violence. *Philosophical Transactions of the Royal Society, B: Biological Sciences, 363*(1503), 2577–2597.

Miller, T. R., Steinbeigle, R., Lawrence, B. A., Peterson, C., Florence, C., Barr, M., & Barr, R. G. (2018). Lifetime cost of abusive head trauma at ages 0–4, USA. *Prevention Science, 19*(6), 695–704.

Phillips, E. L. (1968). Achievement Place: Token reinforcement procedures in a home-style rehabilitation settings for predelinquent boys. *Journal of Applied Behavior Analysis, 1*(3), 213–223.

Ploeg, J., Fear, J., Hutchison, B., MacMillan, H., & Bolan, G. (2009). A systematic review of interventions for elder abuse. *Journal of Elder Abuse & Neglect, 21*(3), 187–210.

Putkonen, H., Amon, S., Weizmann-Henelius, G., Pankakoski, M., Eronen, M., Almiron, M. P., & Klier, C. M. (2016). Classifying filicide. *International Journal of Forensic Mental Health, 15*(2), 198–210.

Skinner, B. F. (1953). *Science and human behavior*. Simon & Schuster.

Steinmetz, S. K., & Straus, M. A. (1973). The family as cradle of violence. *Society, 10*, 50–56. https://doi.org/10.1007/BF02695281

Storey, J. E. (2020). Risk factors for elder abuse and neglect: A review of the literature. *Aggression and Violent Behavior, 50*, 101339.

Sukhodolsky, D. G., Kassinove, H., & Gorman, B. S. (2004). Cognitive-behavioral therapy for anger in children and adolescents: A meta-analysis. *Aggression and Violent Behavior, 9*(3), 247–269.

Talvik, I., Alexander, R. C., & Talvik, T. (2008). Shaken baby syndrome and a baby's cry. *Acta Paediatrica, 97*(6), 782–785.

Tanaka, C. T., Berger, W., Valença, A. M., Coutinho, E. S., Jean-Louis, G., Fontenelle, L. F., & Mendlowicz, M. V. (2017). The worldwide incidence of neonaticide: A systematic review. *Archives of Women's Mental Health, 20*(2), 249–256.

Thompson, R. H., Bruzek, J. L., & Cotnoir-Bichelman, N. M. (2011). The role of negative reinforcement in infant caregiving: An experimental simulation. *Journal of Applied Behavior Analysis, 44*(2), 295–304.

Tyler, V. O., Jr., & Brown, G. D. (1968). Token reinforcement of academic performance with institutionalized delinquent boys. *Journal of Educational Psychology, 59*(3), 164–168. https://doi.org/10.1037/h0025874

Wilson, S. J., Lipsey, M. W., & Derzon, J. H. (2003). The effects of school-based intervention programs on aggressive behavior: A meta-analysis. *Journal of Consulting and Clinical Psychology, 71*(1), 136.

Yan, E. C. W., & Tang, C. S. K. (2004). Elder abuse by caregivers: A study of prevalence and risk factors in Hong Kong Chinese families. *Journal of Family Violence, 19*(5), 269–277.

Chapter 19
Treatment of Violence and Aggression in Adults

Aggression and violence in adults is a common, socially and personally important and expensive problem. Consequently, there are numerous treatment studies, predominantly of psychopharmacology, cognitive behavior therapy, and behavioral approaches. These approaches have been applied to a wide range of populations including violent adult offenders, individuals with psychiatric disorders, individuals with developmental disorders including intellectual disabilities and autism, individuals with traumatic brain injury (TBI), and a variety of residential settings. In addition, there is a literature on treatment of victims of aggression and violence and their sequelae, including infants, children, adolescents, and adults, some of whom show post-traumatic stress disorder.

19.1 Violent Adult Offenders

When individuals commit violent and/or sexual crimes what should society do? Incarceration of criminals is often only a temporary solution as criminals will be let out of prison after some time. Should they have received treatment in prison and after prison to protect the public from future harms? If so, what kind of treatment should it be? Do such treatments work and if so do they work sufficiently well to protect the public from enough future harms for the treatment to be said to be effective—not merely better than no treatment—but sufficiently effective for the satisfaction of society? This section looks at these questions for a variety of violent adult offenders.

As might be expected, there are a large number of studies of treatment of aggression and violence in adults, which is reflected in numerous meta-analyses. A recent meta-analysis of this literature serves to summarize studies in this area. Gannon et al. (2019) conducted a comprehensive meta-analysis of 70 studies with 55,604 participants in evaluations of three types of treatments: Sexual offenses, domestic violence,

P. Sturmey, *Violence and Aggression*, Advances in Preventing and Treating Violence and Aggression, https://doi.org/10.1007/978-3-031-04386-4_19

Table 19.1 A recidivism rates for programs for sexual offenses, domestic violence, and general violence programs from Gannon et al. (2019)

Program type (number of studies)	Change in experimental (%)	Change in control (%)	Absolute difference (%)	Relative difference (%)	Follow-up (months)
Sexual violence ($k = 44$)	9.5%	14.1%	4.6%	32.6%	76
Domestic violence ($k = 14$)	15.5%	24.2%	8.7%	36.0%	62
General violence ($k = 4$)	29.0%	38.3%	9.3%	24.3%	25

and general violence. This meta-analysis focused on recidivism data in papers published up to 2018. In addition, they also looked at the effects of having involvement of professional therapists on outcomes. The overall effects were quite promising. The offense-specific recidivism rate for all programs was 13.4% for treated individuals and 19.4% for non-treated individuals over a 66-month follow-up period which represented an absolute decrease of 6.0% and a relative decrease of 30.9%.

Table 19.1 summarizes the outcomes for each of the three types of programs. This table shows that the literature is quite large for sexual and domestic violence, but not for general violence. That all comparisons show positive effects of treatment and although the absolute differences were small and the recidivism rate in the experimental group remains socially significant, the programs achieve a reduction in recidivism of about a quarter to a third over period of 5–6 years for sexual and domestic violence and a reduction in recidivism of about a quarter over about 2 years for general violence. In addition, when comparing programs with input from licensed psychologists as therapists compared to others, almost all comparisons found that treatment with professional input was more effective than treatment by non-professionals. Other meta-analyses have reached similarly basically optimistic conclusions concerning the effectiveness of psychological treatments for violent offenders (Jolliffe & Farrington, 2007; Papalia et al., 2019).

These meta-analyses give reasons for optimism for the treatment of aggression and violence in adults which is enhanced by professional input. The reduction of 6% in reoffending for a relative reduction of around a third is large and worth having socially and economically. At the same time, the recidivism rate of 13.4% at 66 months follow-up is still sizable and indicates that there is much room for improvement in the treatment of violent adult offenders.

19.2 Psychiatric Disorders

Some individuals with psychiatric disorders show violence and aggressive behavior, and related problem behavior is often a trigger for admission to restrictive and expensive in-patient treatment. In addition, aggression and violence in this population may lead to complex legal issues related to criminal intent and may result in admission of a relatively small number of individuals into secure forensic units. This section will consider common psychosocial and pharmacological approaches to intervention.

19.2.1 Psychosocial Interventions

Many early studies of the token economy conducted with individuals with psychiatric disorders addressed aggression and violence. For example, LePage et al. (2003) reported that the introduction of a token economy resulted in a 33% reduction in injuries over a 2-year period ($p < 0.05$) including reductions in patient-to-patient injuries, staff injuries and staff time lost to injuries from aggression over a 2-year period. Further, a systematic review and meta-analysis of token economies for adults with mental illness published between 1999 and 2013 by Glowacki et al. (2016) identified seven studies including three randomized controlled trials (RCTs). Although the quantity and quality of the data were limited compared to other meta-analyses, the authors were still able to conclude that "Using a token economy may be a means of reducing violence and aggression on inpatient units" (p. 126).

A recent systematic review of non-pharmacological treatment of aggression and violence in individuals with mental health issues comes from Rampling et al. (2016). These authors located 23 studies most of which were described as pilot studies and used quasi-experimental designs. Interventions evaluated included cognitive behavior therapy (CBT) for psychosis, CBT for AntiSocial Personality Disorder, the reasoning and rehabilitation cognitive skills program, Dialectical Behavior Therapy, enhanced thinking skills, anger/aggression focused therapy, schema therapy, and animal assisted therapy. The RCTs were scattered across these different types of therapy and the quality of study was quite varied. Nevertheless the authors were able to conclude that CBT and reasoning and rehabilitation therapy might result in short to medium term reductions in aggression in persons with mental illness.

More recently, Quinn and Kolla's (2017) systematic review of treatment of violence in schizophrenia included a review of psychosocial treatments. They identified two evaluations of CBT, only one of which measured violence directly. Haddock et al. (2007) compared Novaco's CBT with social activity therapy for the treatment of violence in in-patients and outpatients with psychosis in a single RCT. Social activity therapy consisted of finding interesting pleasurable activities and having staff help the participants carry them out. Data included incidents of aggression and violence collected by masked assessors and the authors collected 6- and 12-month follow-up. Novaco's CBT program was effective in reducing the number of incidents of aggression and violence and number of participants engaging in aggression and violence during treatment and at follow-up. There were no changes in staff ratings of anger and aggression, although there were reductions in staff ratings of risk in the CBT group. There were also reductions in delusions, but not other measures of psychopathology only in the CBT group. The authors concluded that the program was effective and warranted further research. Thus, there are few well designed outcome studies of psychosocial interventions for violence in individuals with mental illness.

19.2.2 Pharmacological Interventions

Violence is relatively common in individuals with psychosis, especially upon first admission to an in-patient unit (Large & Nielssen, 2011). Thus, psychotropic medication is used very commonly for individuals with mental health needs and especially among individuals with severe mental illness, such as schizophrenia. Aggression and violence rather than schizophrenia per se is often a reason for admission to in-patients units and they are also an important focus of treatment. Hence, the question of whether psychotropic medication in general, antipsychotic medication, and different classes of antipsychotic medications in particular are effective treatments for aggression and violence in individuals with mental illness had been addressed in several systematic reviews (Citrome, 2012; Quinn & Kolla, 2017; Topiwala & Fazel, 2011) and case series (Fond et al., 2016).

A recent meta-analysis by van Schalkwyk et al. (2018) of aggression in adults found 21 randomized, placebo-controlled trials with 2,610 participants. The main agents evaluated were risperidone and haloperidol ($k = 8$), quetiapine ($k = 5$), aripiprazole ($k = 4$) and tiapride, perphenazine and milenperone (all k's = 1). The most common diagnoses were dementia ($k = 9$), bipolar mood disorder ($k = 7$), psychotic disorders ($k = 4$), and intellectual disability ($k = 3$). There was a small but statistically significant overall effect size (SMD = 0.29, CI 0.22–0.36, $z = 8.5$, $p < 0.001$). The outcomes were not affected by type of agent or diagnosis. That is a diagnosis of a psychotic disorder was not particularly predictive of treatment response to antipsychotics. There was also a very slight but statistically significant trend that higher medication doses had slightly larger effect sizes ($p = 0.038$). Van Schalkwyk et al. (2018) noted that these effect sizes were rather smaller than those reported in other meta-analyses of psychosocial treatments of broadly comparable problems, although this meta-analysis did not make comparisons directly. Thus, the authors were cautious about their use because of the personal costs of negative side effects of these agents. Further, although there are many RCTs on this important topic, the quality of the RCTS is variable with only half of such trials in schizophrenia meeting minimal reporting and design standards (Aran et al., 2019).

A common problem in psychiatry is the immediate management of severe aggression and violence often seen during admission to in-patient units and with a small number of in-patients who are challenging to manage due to long-standing aggression. One pharmacological approach is "as needed" (prn) medication, also known as "rapid tranquilization" or "chemical restraint." The practice remains controversial as it is often done without consent, may inhibit all behavior, rather than aggression alone, may be associated with negative side effects and is sometimes used when alternative, safer, and more humane and dignified alternative should have been considered (Muir-Cochrane et al., 2020) (see previous section in earlier chapters on this topic).

19.2.3 Summary

Although aggression and violence are common in some people with mental health diagnoses, the outcome literature is surprisingly small and of mixed quality, although there is more research on psychopharmacology than psychosocial interventions. The effect sizes for psychopharmacology, although statistically significant, are also small, but come with the well-known short- and long-term risks of these medications. Thus, there is much to be done in developing psychosocial and pharmacological interventions for aggression and violence in individuals with mental illness.

19.3 Intellectual Disabilities and Autism Spectrum Disorders

Reflecting the high risk of aggression and violence in this population, there is an extensive literature on individual treatment of aggression and violence in individuals with ID and ASD. Most of this literature is based on ABA, but there is also over 20 years of work on modifying CBT, including anger management, and third wave behavioral therapies and mindfulness to work with individuals with mild and moderate ID, including those who are aggressive and/or have anger issues. In addition there is an extensive literature on treatment of offenders with ID and or ASD (Griffiths et al., 2018) (see below). There are a wide range of therapies that are commonly used, such as Sensory Integrative therapy (SIT), some of which have been evaluated and shown to be ineffective, less effective than ABA or harmful. This section will not review them in any detail.

19.3.1 Applied Behavior Analysis

There is an extensive literature on ABA treatment of aggression (Beavers et al., 2013; Hanley et al., 2003). For example, Beavers et al. (2013) found that 47.5% of ABA articles published between 2001 and 2012 addressed aggression and 81.6% of these articles included participants with developmental disabilities and 37.3% of them included participants with ASD. There have been several recent reviews and meta-analyses of evidence-based practice and aggression for individuals with ID/ASD which have reviewed the effectiveness of ABA (Brosnan & Healy, 2011; Didden et al., 2016; Heyvaert et al., 2012). For example, Heyvaert et al. (2012)

conducted a comprehensive systematic review and meta-analysis of treatment of a variety of problem behavior in persons with intellectual disabilities which included separate reporting of results for treatment of aggression. The authors identified 137 previously published meta-analyses and reviews and hand searched relevant variables to identify 285 studies which included 598 participants. The overall effect size for all problem behavior was very large ($z = -13.16$, $p < 0.0001$). Following separate analyses of potential moderator effects a final regression model revealed that the effect size was somewhat smaller for treatment of aggression than for treatment of other problem behaviors, such as self-injury, and that effect sizes were more effective with individuals with ASD, older participants, and antecedent interventions. The results are useful both in pointing out the need for further research toward more effective treatment of aggression and also in highlighting that antecedent interventions might be more effective than other treatment approaches.

Most of these reviews and meta-analyses have concurred that ABA treatment for aggression can be effective. Further, they often concur that interventions based on functional assessment and analysis produce larger effect sizes than those that are not, in part by avoiding contraindicated interventions and in part by focusing on positive preventative treatment strategies. (See below for a discussion of anger management and mindfulness for offenders with ID.)

19.3.2 Psychopharmacology

Psychopharmacology, including polypharmacy, is used extensively with individuals with ID for a wide range of problem behavior, including aggression and violence, and for mental health issues. Reviews of the prevalence of psychotropic medication report that approximately 5–15% of individuals with ID show aggressive behavior (Cooper et al., 2009; Emerson et al., 2001). Around a third or more take psychotropic medication (Tsiouris et al., 2013) which is often associated with aggressive and other externalizing problem behavior (Tsakanikos et al., 2007) and in some studies of clinic-referred populations approximately three quarters experience polypharmacy (Lunsky & Modi, 2018). Despite guidelines to reduce and eliminate inappropriate use of psychopharmacology for over a decade, especially antipsychotics, this guidance is largely not followed (Henderson et al., 2020). Evaluations of psychotropic medications often revolve around important practice questions, such as whether or not the medications are well tolerated and safe, and questions of effectiveness, but very few studies have addressed questions of economics.

There are many reviews and some meta-analyses of the effectiveness of psychotropic medication in individuals with developmental disabilities which have addressed aggression (McQuire et al., 2015; Molina-Ruiz et al., 2017; Sawyer et al., 2014; Willner, 2015). There is conduction because these reviews divide the literature up in different ways. Thus, some reviews focus on children, adults, or all age groups; some focus on only intellectual disability, only ASD, or both; some review all psychotropic medications and others focus on specific agents or classes of agents; finally, other reviews do not focus specifically on developmental disabilities but include studies on these populations as part of a review of broader populations. In addition to evaluations of the effectiveness of psychotropic medication, there is also a literature on the effects of dose reduction and discontinuation (Sheehan & Hassiotis, 2017).

A review of psychopharmacology specifically for aggression in individuals with ASD by Fitzpatrick et al. (2016) identified eight psychotropic medications that had been evaluated including the antipsychotics haloperidol, risperidone, aripiprazole, olanzapine, lurasidone; the antiepileptic, valproic acid; and others agents including *N*-acetylcysteine and naltrexone. Haloperidol has been evaluated in at least two RCTs. Although haloperidol is associated with reduced aggression compared to placebo, it has fallen out of favor because (a) of negative side effects such as sedation and associated suppression of adaptive behavior, dystonia, restlessness, heart irregularities (Hassaballa & Balk, 2003) and rarely a life-threatening condition called neuroleptic malignant syndrome and (b) the availability of second generation antipsychotic medications which some consider to be safer as shown by an increased risk of mortality with haloperidol compared to other agents (Lao et al., 2020).

19.3.2.1 Risperidone, Autism, and Aggression

Risperidone is the most studied agent in the treatment of ASD generally and with aggression specifically, which resulted in it being approved by the FDA for treatment of agitation in individuals with ASD. Recall that Fitzpatrick et al. (2016) identified three RCTs plus the Research Units of Pediatric Psychopharmacology (RUPP) network studies of risperidone. These studies have reported that risperidone is superior to placebo in reduction of broadly defined irritability or externalizing behavior problems, which encompasses aggression. Risperidone is also associated with sedation, weight gain, increased appetite, and enuresis.

The most comprehensive evaluation of risperidone with individuals with ASD comes from the RUPP network who conducted an initial series of evaluations of risperidone alone and then compared medications alone with combined medication and parent training. In the first group of studies, Shea et al. (2004) conducted an 8-week RCT comparing risperidone with placebo in 101 participants with ASD aged 5–17 years. They found risperidone was superior to placebo in treatment of irritability. These effects were largely maintained at follow-up and a substantial proportion of non-responders enrolled in an extended open-label trial responded (RUPP, 2005). In a placebo discontinuation trial, participants who were taken off risperidone relapsed and were more likely to show aggression on placebo than risperidone. In the second group of studies participants in the combined group did better than the medication group (Aman et al., 2009) but these differences washed out at 1-year follow-up. Both groups remained superior to baseline at 1 year and parents in the combined group continued to pursue parent training whereas parents in the medication group largely did not (Arnold et al., 2012). A recent meta-analysis (Mano-Sousa et al., 2021) has confirmed these results on the effects of risperidone on broadly defined irritability including aggression, but has tempered these findings due to the effects of weight gain and very few adult participants.

The evaluations of risperidone for irritability in individuals with ASD have been highly influential in practice and resulted in the FDA approved risperidone for treatment of irritability in individuals with ASD (CenterWatch, 2009) resulting in widespread use of risperidone for aggression, often in the absence of other effective interventions, although the literature on adults with ASD is very limited. Others have been more cautious about the use of risperidone for aggression. For example, it may not be possible to conduct double-blind trials of risperidone because the initial side effects (sedation and weight gain) may break the double blind, the long-term effects of weight gain could be serious and place children at risk for medical problems associated with early onset chronic obesity and other long-term side effects are unknown (Singh et al., 2005).

19.3.3 Comment

The psychopharmacology literature is difficult to interpret comprehensively, because information on psychopharmacological treatment of aggression is sometimes hidden by studies that report on externalizing behavior generally or general reviews of psychopharmacology which do not focus specifically on aggression. In addition, there are few studies comparing psychotropic medications with other treatments, such as ABA and CBT, or evaluating the interaction of treatments. Yoo et al. (2003) reported a small *N* experiment in which positive reinforcement attenuated the performance-inhibiting effects of risperidone. Cox and Virués-Ortega (2016) reviewed several such studies.

Psychotropic medications are widely used for aggression in individuals with ASD and many practitioners and family members continue to use them. At this time, however, the research evidence to support their effectiveness remains inconclusive due to the poor quality of studies and the availability of safer alternative interventions.

19.3.4 Other Psychosocial Interventions

There are now a wide array of traditional psychotherapies and other therapies, including sensory and other therapies which have been used with individuals with ID and ASD. Generally, the quantity of research on these approaches is much smaller and often the quality of the research is varied or poor, in part reflecting the fact that research into these approaches is at the early stages of development.

Professional opinion is divided over some treatments, such as Sensory Integration Therapy (SIT), where some have concluded that SIT has been evaluated and found to be ineffective (Davis et al., 2013) or inferior to standard treatments, although others have concluded that there have been no fair evaluations of SIT because the quantity and quality of SIT delivered in research do not correspond to best practices. In addition, treatment of ASD is especially vulnerable to treatment fads which may be merely wasteful and ineffective or in some cases positively harmful (Foxx & Mulick, 2015; Zane et al., 2008).

19.3.5 Offenders with ID

There is a longstanding interest in offenders with ID. This dates back to both the eugenic view that individuals with ID are a criminal menace to be controlled. More recently there has been a steady stream of books (Craig et al., 2010; Griffiths et al., 2018; Lindsay, 2009; Lindsay et al., 2004, 2019; Lofthouse et al., 2018) and reviews on this topic (Browne & Smith, 2018; Jones & Chaplin, 2020; Morrissey et al., 2017a, b). Offenders with ID are commonly found in contemporary prisons, probation services, and elsewhere in the justice system. For example, Hayes et al. (2007) found that in a sample of 140 prisoners, 7.1% had IQs below 70 and 23.6% had IQs between 70 and 79 and 10.1% has quotients on measures of adaptive behavior below 70 and 33.3% had quotients on adaptive behavior between 70 and 79, indicating much higher prevalences of ID in prisons than would be expected, given an overall prevalence of around 1% in the general population. Broadly similar data have been found in probation services (Mason & Murphy, 2002). The higher than expected prevalence of ID in prison and probation services has stimulated interest in treatment of problems related to offending. There are two broad areas of interest: Treatments of anger and aggression related to offending and rehabilitation programs for offenders with ID.

19.3.5.1 Anger Management

Several authors have modified standard anger management treatment packages for use with offenders with ID (Taylor & Novaco, 2005). For example, Taylor et al. (2005) randomized 40 detained men with mild-borderline ID to either 18 sessions of individualized manualized modified anger management or wait list control. Both immediately post-treatment and at 4-month follow-up, self-reports of anger were lower in the treatment than the control group. Staff reports of anger trended in the same direction but were not statistically significant.

Subsequently, there have been at least six systematic reviews and meta-analyses of anger management studies which have included studies with offenders (Ali et al., 2015; Borsay, 2013; Browne & Smith, 2018; Hamelin et al., 2013; Nicoll et al., 2013; Vereenooghe & Langdon, 2013). In the most recent systematic review, conducted using PRISMA guidelines, Browne and Smith (2018) identified 16 studies of interventions for anger and aggression. Two were RCTS, two were non-RCTs, and 12 were classified as case series. The studies involved 274 participants in treatment and 52 in control groups most of whom were labeled as having mild-borderline ID. Most dependent variables were self-reports of anger and aggression. The treatments included CBT, Dialectic Behavior Therapy, mindfulness and behavioral skills training. Two British research groups conducted much of the research. Browne and Smith concluded that, although there was evidence of short-term reductions in anger and aggression in most studies and for all treatment approaches, limitations on study design, such as lack of experimental design, treatment integrity and follow-up limited confidence in the conclusion that treatments were effective.

19.3.5.2 Offender Rehabilitation Programs

The second area of activity has been rehabilitation programs for offenders with ID. There have been several such population-specific programs which often have involved tailoring standard CBT programs for offenders generally for offenders with ID (Hickman et al., 2017; Waugh et al., 2014).

One example is the Social Problem Solving and Offence Related Thinking (SPORT) program (Lindsay et al., 2011) which included a package of CBT strategies modified for individuals with ID including cognitive restructuring, role play, and problem solving. In a pre-post evaluation with ten offenders with ID, Lindsay et al. (2011) found that participants became more favorably disposed to problem solving and less impulsive after participating in the SPORT program.

A second broadly similar program is the Equipping Youth to Help one Another Program (EQUIP), a program developed for adolescents and adults (Gibbs et al., 1995) that has its roots in both Aggression Replacement Training (Goldstein et al., 1998), developmental psychology models of moral development, and CBT. There are three kinds of sessions in EQUIP: Anger management and cognitive errors, social skills training, and social decision-making training. EQUIP has been modified for use with offenders with mild intellectual disabilities (Langdon et al., 2013) by simplifying language and homework tasks. Langdon et al. delivered EQUIP to seven offenders, three with ID and four with Asperger syndrome, who lived in a locked forensic program. The program was delivered over 12 weeks with four, 1-h sessions per week delivered by three facilitators. Using a pre-post design, Langdon et al. found that after EQUIP training, participants showed improved scores on psychometric measures of moral reasoning and problem solving related to overcoming obstacles while minimizing harm, and distorted cognitions, but there was no overall improvement in broad problem solving or anger. Tearle and Holt (2018) and Tearle et al. (2020) have recently demonstrated that this modified EQUIP program can also be implemented with individuals living in community settings with broadly similar outcomes.

A third program is virtual immersion therapy, an approach to teaching social skills to offenders with ID using video simulated social situations. This allows the therapist to coach the individual through these social situations using immediate feedback. A pilot study with 12 participants found some preliminary evidence of improved emotional stimuli, and problem solving in some, but not all participants. Unfortunately, changes were not maintained at follow-up.

These rehabilitation programs for offenders with ID are promising because they are based on principles that are accepted by many and because they have some data suggesting they result in some positive change. Evaluation of these programs is at an early stage. So far, the number of studies of such programs is quite small and we lack data on aggression and reoffending and on long-term follow-up.

19.3.6 Offenders with Autism Spectrum Disorders

Whereas there is an extensive literature on offenders with ID, the parallel literature on offenders with ASD is slim and generally recent in comparison. Sturmey (2019) reviewed this literature and found case studies, such as the Sandy Hook massacre, which was committed by an individual with ASD who, absent effective education and health services for most of his life and with clear indications of severe violence, shot and killed 20 first-graders and six staff.

A systematic review of the prevalence of ASD in the criminal justice system (King & Murphy, 2014) found widely varying prevalences from 3% to 27%, but uniformly higher than the expected approximately 1% prevalence of ASD in the general population, suggesting that individuals with ASD are somewhat more at risk for conviction of offenses leading to incarceration. Others have published similar case series giving prominence to a wide range of comorbid psychopathology (Allely, 2020; Newman & Ghaziuddin, 2008) and speculating that certain features of autism, such as preoccupation with objects, obsessional features, and social insensitivity, might predispose some individuals with ASD to certain kinds of crimes, such as stealing, stalking, and sexual offending. Woodbury-Smith et al. (2010) compared 23 offenders with ASD with 23 matched non-offenders with ASD on a series of measures of recognizing emotions, theory of mind, etc. The only difference between the groups they found was that offenders with ASD were somewhat less likely than the control group to recognize fear in others. This suggested offenders with ASD may be less sensitive to social punishment than other individuals with ASD and that this might predispose them to sexual offenses.

Although the general outcome literature on offenders and on offenders with ID might provide useful guidelines when selecting treatments for aggression in offenders with ASD and ID, there is a small and limited literature on treatment of offenders with ASD. For example, in a systematic review of CBT for offenders with ASD, Melvin et al. (2017) found 13 empirical papers with 30 of 75 participants being individuals with ASD. The literature consisted of case studies and case series, but no RCTs. A PRISMA systematic review of zoophilia and ASD (Allely, 2020) located only three studies and a systematic review of treatment of public masturbation also found only a small number of studies (Mann & Travers, 2019).

Reviews of psychopharmacology for this population are few. Those that exist have only found limited non-experimental research in this area (Coskun et al., 2007; Coskun & Mukaddes, 2008; Hellmans et al., 2007).

There is also a small and growing literature on ABA and offenders with developmental disabilities including ASD including treatment of inappropriate public sexual behavior which may result in prosecution in some cases (Davis et al., 2016; Mann & Travers, 2019), although many of these studies lack experimental rigour (Mann & Travers, 2019). Recently, there has been progress in residential treatment of teenage offenders with ASD using multi-component ABA methods such as skills training, token economy, and function-based treatment for aggressive behavior (Pritchard et al., 2012, 2018). This appears a promising approach, but requires independent replications and long-term follow-up.

19.3.6.1 Comment

Prevention and rehabilitation of offenders with ID and/or ASD is a socially very important question that presents many research design, practical and ethical challenges. The last 30 years have seen a growing and now substantial literature on offenders with ID, especially in the area of anger management, and to a lesser extent, rehabilitation programs, although now several models for how to approach rehabilitation now exist. There is little evidence for the effectiveness of psychopharmacology and minimal evidence for extending ABA to this population at this time. Unlike the literature on offenders with ID, the literature on offenders with ASD is smaller, relatively new and underdeveloped. In both examples, the challenges in conducting research on this topic mean that the research is necessarily imperfect although it still provides the best guidelines to practice we have, absent evidence for alternate approaches or better evidence.

19.3.7 Summary

The effectiveness of psychological treatments of aggression and violence in individuals with ID and ASD is quite well established: Some behavioral methods, especially those based on ABA and using functional analysis have an extensive evidence-base; in addition, the literature on anger management and mindfulness has grown and evidence is accumulating of their effectiveness. The literature on offenders with ID and/or ASD has also grown considerably, but there are fewer well-controlled studies in this specific population.

19.4 Traumatic Brain Injury

TBI includes conditions where brain damage is acquired after birth, rather than from genetic syndromes, etc., and typically includes physical injuries from trauma, road accidents, near drowning, etc., and strokes. Many people, especially younger individuals, often recover well from TBI, although more individuals show some aggression and violence immediately after an injury which is often related to confusion and disorientation. A minority continues to show significant aggression and violence. About one-third to a half of children show some broadly defined behavioral and emotional problems in the first year after TBI (Li & Liu, 2013) as do around one a quarter of adults (Ylvisaker et al., 2007), although prevalence rates vary from around 10% to nearly 100% of cases (Byrne & Coetzer, 2016). Severity and persistence of aggression may relate to the extent and locations of the person's brain damage.

19.4.1 Psychosocial Interventions

At least three recent systematic reviews and meta-analyses have addressed evidence-based practice in the area of psychosocial interventions for aggression following TBI. Almost all interventions were behavioral and there is a substantial quantity of research in this area. For example, Ylvisaker et al. (2007) conducted a systematic review of behavioral interventions for children and adults following TBI. They located 65 studies with 172 participants and divided interventions into contingency management, positive behavior interventions, and combined treatments. Most participants were aged 6–39 years and many studies were conducted several years after the initial traumatic brain injury. The most common target behavior was various forms of verbal and physical aggression. Outcomes were uniformly positive for all three types of behavioral interventions in terms of reduction of target behaviors, and, where reported, generalization, increases in skills, and follow-up data. Although these authors were very positive about these outcomes, they did caution that the uniform positive results might reflect submission and publication bias and so may not generalize well to practice.

In a second systematic review of treatment of TBI, Cattelani et al. (2010) included many studies which treated aggression. They concluded that comprehensive-holistic rehabilitation programs were most effective because of the relatively large number of RCTS. They also concluded that ABA and cognitive behavior therapy were effective treatment options, but were cautious since they did not find RCTs for these approaches. Although many studies addressed a wide range of externalizing behavior problems, this study did not separate out data on aggression and violence.

Recall also that Heinicke and Carr (2014) published a meta-analysis of applied behavior analytic approaches to skills training and analysis and treatment of aggression in individuals with TBI. They found good evidence of the effectiveness of this approach based on 112 small *N* experiments on ABA approaches to TBI published between 1984 and 2010, which included a substantial number of studies of treatment of aggression. Sixty-four percent of participants were adults, 70% were male, and the median time post-injury was 4.6 years. Aggression was present in 19% of participants and related target behaviors include inappropriate vocalizations (19%), disruptive behavior (12%), and restraint (2%). Of these studies, 81% used only behavioral interventions. The most common skills training interventions included reinforcement, antecedents interventions, and self-management and the most commonly used behavioral reductive strategy was differential reinforcement. Unlike other behavioral literatures, the majority (70%) did not implement functional assessments or analyses and those that did typically used only one single descriptive assessment.

The authors applied American Psychological Association standards of evidence-based practice for skills training. They concluded that reinforcement was an efficacious procedure (Percentage of Non-overlapping Data points [PND] = 90%, 13 studies, 34 participants) and that antecedent interventions were also an efficacious procedure (PND = 96%, 12 studies, 25 participants). Additionally, behavioral self-management met criteria for being probably efficacious (PND = 92.1%, 8 studies, 20 participants). Other behavioral and non-behavioral interventions lacked sufficient quantity of evidence to meet criteria for being efficacious or probably efficacious. With respect to behavior reductive strategies differential reinforcement (PND = 90%), antecedent intervention (PND = 71%), and punishment (PND = 89%) and extinction (PND = 87%) were probably efficacious. Other behavioral and non-behavioral interventions had insufficient evidence to establish their effectiveness for behavior reduction.

Finally, in 2016, Byrne and Coetzer reported a systematic review and meta-analysis of psychological interventions specifically for aggression in individuals with TBI. They identified four group designs, of which three were pre-post designs and one was an RCT, and seven were small *N* experiments. The treatments in the group designs included various formats of CBT and group anger management and the treatments in the small *N* experiments included functional analysis, management of antecedents and consequences and some forms of self-management. The omnibus effect size for the group designs was moderate ($d = -0.48$, four studies) and was still significant at follow-up ($d = -0.30$, two studies). All but one small *N* experiment reported moderate to large effect sizes. Only one study reported follow-up data which had a small effect size. The authors concluded that psychological treatments were an established evidence-based practice for TBI.

19.4.2 Psychopharmacology

There have been at least two systematic reviews and meta-analyses of psychopharmacology for aggression and agitation in individuals following TBI (Fleminger et al., 2003; Hicks et al., 2019). Here we will consider only the most recent one.

Hicks et al. (2019) used PRISMA guidelines to conduct a systematic review and meta-analysis. They searched five databases up to December 2018 and retained ten studies of which five were RCTs and five were case series. There were mixed, but positive findings for amantadine, an anti-Parkinsonism medication, for irritability (two studies) and aggression (two studies). There were large individual differences within these studies. That is, some individuals responded well and others did not. One study found some effects of methylphenidate on anger, but the evidence for propranolol was weak. Other agents that have been evaluated but less systematically included carbamazepine, valproic acid, quetiapine, and sertraline. There were few negative side effects reported. Based on three studies all conducted by Hicks and others, Hicks et al. recommended amantadine for treatment of irritability and aggression.

19.4.3 Summary

These systematic reviews and meta-analyses of treatment of TBI found a quite large literature on behavioral treatment of aggression consisting of tens of studies and over a hundred participants. Many of these studies reported treatment efficacy of aggression and violence. The most common effective psychological treatment was ABA and there was some evidence for the efficacy of cognitive behavior therapy. There was little evidence for the effectiveness of cognitive rehabilitation due to the lack of studies.

The reviews of psychopharmacology found no convincing evidence of effectiveness. Many of these studies were limited in a number of ways, but perhaps the most clinically significant limitation was the focus on reducing aggression and violence rather than improving the person's overall functioning and lifestyle.

19.5 Residential Settings

There are many congregate residential settings for youth, and individuals with various disabilities including people with ID/ASD, individuals with TBI, dementias and other disabilities. Staff in these settings are sometimes aggressive or violent to residents. Indeed, oversight of abuse and neglect, including aggression and violence from staff to residents, in residential facilities and potential cover-ups from individuals, management, unions, and local police, and inept and ineffectual oversight of such facilities are long-term problems dating back many years. These problems, however, continue today in contemporary residential settings including community settings. These populations are especially vulnerable to staff aggression and violence because they may not be able to identify and report staff violence readily, may not be believed when they do so, or face no response, a severely delayed response, or enormous bureaucratic hurdles when they do report abuse. They may also be subject to direct or indirect retaliation from the staff who abused them or from their colleagues.

19.5.1 Youth Residential Services

Residential services for youth cater for a diverse group of youngsters including those referred through family and criminal courts, child welfare, department of education, and mental health services. Often, individuals within one service may have multiple needs, such as criminal activity, addictions, mental health, and borderline/mild ID. These services may be run by state, not-for-profit and for-profit agencies (Egozy & Cox, 2017). Aggression that families, schools, and outpatient services cannot handle is often a reason for admission.

Recent data on the current status of US youth residential facilities is hard to find, but in 2010 Sedlak reported the results of interviews with 7,073 youth. Sedlak found that 29% of youth in residential settings reported being beaten up or threatened while in a youth facility. These incidents were mostly from other youth they knew but 24% of these incidents were from staff. Approximately 10% of youth reported being sexually victimized by staff including 10% of males and 7% of females. These figures varied tremendously across facilities and states and over time. There are currently no centralized and standardized Federal data on restraints, shackles, pepper spray, and solitary confinement in youth facilities in the USA, so obtaining data on these practices is not possible. Thus, violence is probably quite common in some youth facilities and staff violence is sometimes a significant component of that problem.

From time to time, scandals of staff abuse at youth facilities hit the headlines and sometimes continue for decades. For example, violence at the Rikers Island facility in New York City has become an ongoing scandal for many years. Problems included brawls, gang slashings, riots, and guard violence to inmates (Buser, 2015), including staff beating an inmate for allegedly resisting while shackled to a gurney (Schwirtz & Winerip, 2014). In addition, as reported by the former Chief Medical Officer, health care was so bad and sometimes negligent at Rikers that people died unnecessarily resulting in expensive and protracted lawsuits in which Rikers is often found guilty (Human Rights Defense Center, 2017).

Conditions such as those at Rikers Island and other youth facilities have prompted some to call for total reform or abandonment of youth facilities for alternate treatment models. Krisberg (2017), for example, documented how some states successfully abandoned filthy, dangerous, and violent youth facilities and instituted total institutional reform. Krisberg reported that in Massachusetts, for example, these efforts often involved closure of all old facilities and setting up of newer, often smaller facilities, with more humane, rehabilitative, and educational practices. Such attempts are often met with strong resistance from unions and some state politicians at first who had financial and other incentives to maintain the status quo and by people living in neighborhoods where new facilities might be located. Other states have attempted to replicate these efforts with varying degrees of success and commitment.

19.5.2 Residential Services for Seniors

As noted earlier, seniors may be subject to violence from family members, but seniors may also be subject to violence from staff in residential services. Yon et al. (2017) reported a systematic review and meta-analysis of the prevalence of elder abuse in various community settings. The authors identified 52 studies from 28 different countries. They found that the overall prevalence of all types of elder abuse was 15.7% and the prevalence of physical abuse was 2.6% and the overall prevalence of sexual abuse was 0.9%. In a subsequent similar review of residential settings, Yon et al. (2019) identified nine relevant studies and shockingly reported that approximately 64% of staff reported elder abuse (broadly defined) and 14% of seniors experienced physical and 2% of seniors experienced sexual abuse in residential settings. Yon et al. (2019) identified risk factors including individual risk factors, such as being female, aged over 74 years, higher dependency, poorer health, and needing assistance in activities of daily living; and organizational factors, such as staff shortages, time pressure, and low ratio of nurses to residents. Sexual violence towards elders, including within the family is less frequent, perhaps being around 2% in most studies, although it may be more common in certain situations (Bows, 2018).

There have been several approaches to reducing violence toward seniors in residential settings. Burnes et al. (2017) painted a broad picture of services that can be used to prevent aggression against seniors in residential settings. These can include primary prevention in the form of staff and family education, screening to identify at-risk individuals, supporting caregivers, programs to help seniors manage their own money and helplines for seniors and caregivers alike. When aggression has already occurred services may use mandatory reporting, adult protective services, and community support and multidisciplinary teams to prevent further abuse.

Many interventions focus on skills training and stress management in direct care staff. For example, Drossel et al. (2011) evaluated the effects of Dialectic Behavior Therapy on prevention of elder abuse. The authors held manualized Dialectic Behavior Therapy sessions which were held in weekly groups. The participants were 16 full-time caregivers aged 38–87 years old. The study used a pre-post design and used self-report measures of distress, well-being, fatigue, etc. Six caregivers also took part in booster sessions. Two participants were reported for elder neglect during study. Caregivers reported improvements in their own social functioning, emotional well-being, depression, energy and reported greater use of problem solving. There are other psychosocial interventions for caregivers' stress management, such as mindfulness, for which there is some evidence that they affect outcomes, such as caregiver depression, which might decrease elder abuse (Fearing et al., 2017).

The evidence-based for reducing abuse of seniors in residential services is relatively slim. Much has focused on supporting carestaff in various ways through education and skills training. Clearly, more research is needed in this area.

19.5.3 Residential Services for People with ID/ASD

Violence and abuse have long been recognized as a facet of the lives of many individuals with intellectual and other developmental disabilities (Sobsey, 1994). Residential staff violence to individuals with ID/ASD was often part of institutional scandals in the USA and UK and elsewhere since the 1950s; indeed part of the motivation to implement deinstitutionalization was to prevent such abuse and provide better oversight in more publicly accessible and visible settings. Yet, despite deinstitutionalization and the hope that staff abuse of residents was a phenomenon of the past, there have been repeated examples of staff violence to residents with ID/ASD in community settings in many countries, which are sometimes associated with coercive behavior management practices, including misuse of restraints and seclusion (Murphy, 2020; Sturmey, 2015).

A systematic review of research on victimization of people with a broad range of disabilities, Hughes et al. (2011) found that both women and men with disabilities had high rates of abuse and more recent papers have produced similar high prevalence estimates (Krnjacki et al., 2016). Results specifically on physical abuse have also found high but very varied rates of violence, but these reviews did not break down the data specifically for residential services and violence from staff to residents (Horner-Johnson & Drum, 2006).

One can distinguish three approaches to abuse prevention: Skills training with individuals, staff training, and organizational interventions. Several studies have evaluated teaching individuals with disabilities to recognize, refuse approaches, and report incidents of abuse, including staff violence. For example, Lund's (2011) review of abuse prevention programs identified that public abuse prevention programs such as community-based domestic violence programs do serve a small number of individuals with ID. In addition, Lund identified a small number of papers on behavioral, cognitive, and psychoeducational approaches to abuse prevention, but these papers were quite limited in number and quality. A typical example comes from Lund and Hammond (2014) who implemented the Stop Abuse For Everyone (SAFE) curriculum, which consisted of four modules: Financial abuse, neglect, physical and verbal abuse, and sexual abuse. Although the authors reported that they could implement the curriculum in an acceptable way they did not report any outcome data. To date there are few well conducted studies with good outcome and follow-up data on this topic.

It might be useful to consider two kinds of situations in which residential staff violence occurs to individuals with ID/ASD: Individual relatively isolated acts of violence toward individuals with ID/ASD ("rotten apples") and unit- and system-wide abuse ("rotten barrels"). Relatively isolated incidents of staff abuse may occur in a well run service when specific provocative situations occur and the staff member does not know alternative strategies to use or does not believe them to be effective or right. Such incidents can be understood through individual clinical case formulation. Here, individual motivations (e.g., striking a person with a disability when angry and facing aversive client behavior, such as spitting on a staff member), and system factors, such as lack of training in interpersonal skills, stress management, poor shift patterns, and aversive supervisory practices that lead to physical and emotional exhaustion in staff. Such incidents could perhaps be dealt with by interventions for individual staff members, such as reassignment, stress management, or other skills training as well as system-wide interventions to improve training and oversight.

In addition to individual incidents of aggression to residents are system-wide aggression and violence in residential services. A recent example was the Whorlton Hall scandal in the UK (Murphy, 2020). Whorlton Hall was a residential service for around 24 individuals with ID and ASD, although only around 12 places were used at any one time. The building was an old and rambling historical home located in a geographically isolated area owned by different for profit businesses over time that contracted with cities around the country to provide assessment and treatment services for individuals with ID/ASD and severe problem behavior.

On May 22, 2019 the TV series *Panorama* ran an expose with undercover video footage of verbal and physical abuse of residents at this service. Yet, paradoxically, the service had been inspected repeatedly and recently by the national Care Quality Commission (CQC), the English national agency responsible for oversight of such services. The CQC had in fact largely given the service very positive rating during the previous 5 years of inspections.

Following the media expose the government commissioned an independent report. Murphy's (2020) report identified numerous problems both with the home itself and with CQC's oversight. The home had very high staff turnover, which included high turnover of the senior service managers who apparently often left once they realized that it was not possible to be effective. The home routinely used agency staff who did not know the resident's or their treatment plans and often had weak or no effective staff training. There was little effective oversight of behavior management practices and restraint use. A team of professionals was present only 1 day/week and, hence, not only was professional input of insufficient quantity, but the professionals probably had little detailed knowledge of the residents or day to day running of the service. The management team was very problematic: It referred to itself as "the cunt club." The managers socialized together and gave each other paid overtime. When staff complained internally or whistleblowers reported abuse externally, the management often reassigned them to undesirable shifts. In contrast, staff who were found guilty of abuse were temporarily reassigned to other services, but often were returned to work with the residents they had abused.

The CQC conducted numerous annual and other inspections, sometimes in response to whistleblower complaints. The CQC did not identify any abuse over a 5-year period. The inspection process was flawed. Inspectors were accompanied by facility staff some of whom were problematic staff. Staff who might complain were scheduled to be absent during inspections. Inspections were too short and mostly announced and sometimes conducted by people who lacked the skills necessary to conduct rigorous inspections, did not have easy access to the service's basic performance data, such as critical incidents and staff turnover. Publications of reports were often very delayed often rendering them of little value to improve the service.

Placements at Whorlton hall were funded by local cities who presumably did not provide adequate residential placements for these individuals. The cities that funded individual resident placements were also largely absent from the service, perhaps because many of them were very distant from the home or perhaps because "out of sight" meant "out of mind" for these cities who had problems providing services for these challenging individuals. Finally, family members were also unable to observe services as many individuals were placed at large distances away from their families due to the lack of local placements. Some felt that the company owners placed profits over people and the "light touch" regulation was an inappropriate strategy to ensure resident safety. This residential scandal appears eerily similar to the earlier Winterbourne residential scandal in the UK (Tozer & Plomin, 2013) and other similar residential scandals in the UK and US (Sturmey, 2015).

From the perspective of this book, it is interesting to ask how such an abusive culture evolved. Murphy's (2020) report documented the progressive increase in problems over a 5-year period, suggesting that staff and manager behavior gradually changed. The CQC did not respond to this decline in service quality, perhaps because they did not have ready access to data presented in a meaningful manner and because onsite inspections were too few and too cursory. Factors, such as geographical isolation resulting in difficulties in recruiting and retaining staff, excessive reliance on agency staff, a physically unsuitable building that might limit surveillance were always there. So other factors, perhaps relating to changing in management practices as different corporations taking over management of the services, or changes in CQC oversight mechanisms during this time might be a fruitful place to look for the reasons for the evolution of a culture of violence in residential care. Perhaps the absence of prompt intervention early during the deterioration of the service quality from the CQC could have prevented this deterioration.

19.5.4 Summary

The high rates of staff aggression and violence to various individuals in residential settings are a continuing social problem that deinstitutionalization and inadequate oversight of community services has failed to address effectively. Some factors that are common across these different contexts include the risk factors of resident behavior that is aversive and/or effortful to caregivers, such as residents' abusive behavior, non-compliance, and aggression to caregivers, and situations that are aversive to caregivers, such as dealing with residents who move slowly, do not cooperate, or dealing with unpleasant smells, etc. An obvious yet under-emphasized aspect of many of the situations associated with staff violence to residents as that staff violence often leads to the termination or attenuation of aversive resident behavior.

In many situations, both the caregivers themselves and their supervisors lack basic interaction, behavior management, self-management, and stress management skills. Further, supervisors and administrators often lack basic supervision and training skills. Hence, may either give up, resort to ineffective and coercive staff training and supervision practices. The residential services themselves often fail to have effective staff training programs in place. In other more ominous situations, residential services may have a culture of violence in which staff, supervisors, and managers passively condone or even encourage staff violence to residents.

This literature has not yet sufficiently addressed broader issues identified in the risk factor literature, such as lack of staff and supervision, inadequate program design, and inadequate and ineffective oversight. When dealing with staff violence, many residential services focus on informational and educational strategies, such as training staff when hired and annually. This training often includes information as to what constitutes abuse and how to report it. Many residential services also use reactive strategies. For example, following an allegation of staff violence, they may suspend staff or limit their duties to minimize the opportunities for further resident harm and conduct an investigation. Following an investigation they may reinstate a staff member if the allegation is unsubstantiated or take negative personnel action, such as reassignment to alternate duties, demotion, termination, and reporting the staff to a state or professional register. Unfortunately, there is little evidence that this approach is effective in preventing or reducing staff violence toward the various vulnerable populations and at this time there is little external pressure from accreditation and funding agencies to adopt those practices that might improve this situation.

19.6 Treatment of Victims

Victims of aggression and violence, including those who observe violence, may suffer a variety of physical and mental harms. There are many different populations and contexts in which this may occur and there is a very broad and diverse literature on their treatment. Thus, it is difficult to make any broad unqualified statement concerning the impact of aggression and violence on victims.

In reviewing this literature with an eye to treatment implications, it is useful to consider those factors that can and cannot be changed. For example, demographics factors, including gender, age, and ethnicity, and history, such as previous incidents of aggression and violence, number and severity of incidents, and duration of exposure to aggression and violence, may be important factors in the development and understanding of a client's response to aggression and violence, but they are all factors that cannot be changed during intervention. Conversely, factors such as current avoidance, exposure to perpetrators, social support, treatment and management of perpetrator behavior, current stress management skills, and alcohol and substance use, etc. are all variables that could potentially be changed during treatment and hence are more important for developing treatments.

A second important consideration is that there are large individual differences in response to exposure to aggression and violence. The range of aggression and violence that people are exposed to is very large in terms of intensity (mild to nearly fatal), frequency (individual incident to long-term continuous exposure), and whether the exposure is direct or indirect, as when people are injured, or witness or hear about violence to others. In addition, exposure to violence might also provide opportunities to learn resilience and with it effectively under certain circumstances. Thus, one might expect that outcomes from exposure to aggression and violence might be quite varied, even in response to severe violence. Evidence for that conclusion comes from Bright and Bowland (2008) who evaluated women who had experienced childhood abuse and neglect from siblings, parents, and other family members and intimate partner violence IPV from one or more partners. These women were a subsample of participants from a larger intervention study. Using a standard measure of Post-traumatic Stress Disorder (PTSD) symptoms (Foa, 1995), they found that seven had mild, nine had moderate, 14 moderate to severe, and three had severe symptoms of PTSD. Interestingly, when contrasting the histories of women with mild and severe current PTSD, there were no obvious differences in history and severity of PTSD. This suggested that there were factors other history were important in determining current functioning.

19.6.1 Infants and Young Children as Victims

There are a number of studies that have looked at the effects of aggression and violence in infants and young children that are useful in both describing what has to be understood, but also in identifying potential targets for individual intervention. Some studies have reported on the consequences of aggression and violence in infants and children but that suggest a complex web of causation. Exposure of infants and young children to violence is not an isolated event, but takes place in the context of numerous other relevant variables. A review by Mueller and Tronick (2019) identified multiple factors, including poor ante-natal and perinatal care, chronic stress, impaired stress regulation, atypical and risky parenting practices, etc. that may affect the infant or young child's development in many ways. Absent the infant or young child's ability to regulate their own behavior and emotions well, infants are dependent upon their caretakers to do so. If those caretakers do not do so, this places them at risk for atypical development. These factors are not unidirectional but may interact one with another affecting both developmental outcomes and risk of future violence. For example, a child who has observed family violence may be withdrawn, irritable, and cry more and hence be less responsive to caregiver interaction, thereby resulting in reduced opportunities for both the infant to learn and for the caregiver to learn more effective and rewarding parenting skills. Thus, the outcomes of violence against infants and young children are quite varied (Fong et al., 2019). Some may experience mild and transitory aggression and some may be relatively mildly affected. At the other end of the spectrum, others may have severe lifelong physical, developmental, and emotional consequences, requiring medical intervention, special education, and rehabilitation, depending upon their needs.

Several studies have addressed the needs of young children who have been exposed to relatively moderate aggression and violence. An example of treating young children who were exposed to marital violence comes from a study by Lieberman et al. (2005) who conducted a RCT of child-parent psychotherapy compared to case management. The authors randomized 75 mothers of children aged 3–5 years who had witnessed or heard marital violence. Families were ethnically mixed (i.e., 91% non-White). Children were mostly referred to the study by courts, domestic violence agencies, physicians, etc. To participate in the study the perpetrator could not live at home, the child could not have intellectual disability or autism, and the mothers could not be perpetrators or have major mental health or addiction problems. Child-parent psychotherapy consisted of weekly joint mother-child sessions in which the child played with developmentally appropriate toys designed to elicit "trauma play" and social interaction, reduce problem behavior, foster developmentally appropriate play, and develop a joint narrative that resolves the trauma. Mothers also received individual sessions as needed. The control group received case management services to enable initial help to seek referral to clinics for individual psychotherapy plus monthly phone calls. Data were collected pre- and post-treatment and showed improved outcomes in both children, in terms of reduced behavior problems and trauma symptoms, and mothers in terms of psychiatric symptoms and PTSD. An important positive feature of this study was that it addresses the needs of both young children and their mothers and hence had the potential to change the system of mother–child interaction broadly. Thus, this intervention resulted in improvement for both children who had witnessed violence and for their mothers.

19.6.2 Adolescents and Intimate Partner Violence

There are many examples of adolescents being exposed to aggression and violence. One example is violence during adolescent dating is common and many young women experience IPV when they date or when they cohabit. Systematic reviews of adolescent dating violence have found that between a quarter to three quarters of adolescents reported violence in dating relationships, including physical sexual and psychological violence (Jennings et al., 2017; Rubio-Garay et al., 2017; Taquette & Monteiro, 2019). The consequences of adolescent violence during dating relationships include depression, anxiety, health problems, such as sexually transmitted infection and HIV, and increased risk for adult IPV (Taquette & Monteiro, 2019).

Interventions in this area have been quite varied. For example, Jennings et al. (2017) reported a systematic review of treatment of dating violence in young people aged 15–30 years. They identified 42 treatment studies which included 34 RCTs.

Treatments included skills training to promote healthy relationships, video models, role play, discussion, watching a play, discussions, etc. The authors classified the study effectiveness as "yes", "mixed," or "no." Mixed outcomes referred to studies that reported immediate but no long-term effects, effects on some but not all outcomes or only in certain subgroups. Overall, studies were quite effective with 42% being classified as effective, 56% as mixed, and only one study (3%) reporting no intervention effects. This meta-analysis did not disaggregate the outcomes by treatment type or report effect sizes, thus, we cannot comment on the relative effectiveness of any specific approach to treatment based on this study.

In a more recent meta-analysis restricted to RCTs, De Koker et al. (2014) also found evidence that a variety of programs that were based on such diverse approaches as feminist theory, social learning theory, education, newsletters, and classroom-based activities, often based on combinations of gender power inequality, and some skills-based learning could reduce both perpetration of IPV and reduce victimization during IPV among adolescents. These results suggest that violence in relationships for adolescents and young adults is quite amenable to treatment by a variety of different approaches.

19.6.3 Adult Victims

Many adults experience aggression and violence both directly and indirectly through witnessing aggression and violence. Approximately 23% of women and 19% of men are victims of IPV (Desmarais et al., 2012a, b) and approximately 28% of women and 22% of men perpetuate violence against partners.

Personal outcomes are quite varied: Some are made stronger and some people's lives are devastated by their experience of aggression and violence. When treating adults with negative outcomes of aggression and violence, therapists and service must consider a number of important factors. The first factor is balancing client safety with treatment efficacy and efficiency. Some clients may still face real ongoing threats as when the perpetrator of violence is still in the person's family, workplace, social circle, or neighborhood. Examples include returning children with potentially violent families, returning partners who were victims of IPV to their relationship, etc. A second is that intervention should give the person strategies to deal as effectively and quickly as possible with the negative outcomes of aggression and violence. This might include stress management skills, relaxation training, increasing social support, and reducing avoidance and self-exposure to feared situations where it is safe to do so. Finally, a common goal is constructing a new, positive and valued life following violence which could include goals such as increasing social activities, leaving violent relationships, getting a new job, and undertaking and completing legal actions, etc. (e.g., Cook et al., 2011).

This section will focus on treatment of adults who have experienced violence. Of course, it is not possible to cover all populations. Instead we will focus on adult victims of IPV and adults who experience PTSD following violence. These were chosen because they are two of the more common groups of adult victims of violence. There are numerous books on this important topic including books addressing specific populations, workbooks for victims to use on their own or with support, and advice for attorneys when working with victims of violence.

19.6.3.1 Victims of Intimate Partner Violence

Victims of IPV face many challenges. Not only do they have to manage the personal outcomes of violence for themselves, but they often have to consider the safety of other vulnerable family members, such as children or older family members, who also might be victims or might become victims of violence. They may have to consider whether to stay in their current relationship and current home or break up and move on to live elsewhere. They may have to face economic challenges, such as poverty, or a partner who controls or denies access to money and other resources. They may also lack the assertiveness, and other skills and motivation to move on to a new life. Finally, they may also face legal, physical, mental health professionals who are sometimes insensitive to their problems, and a bureaucratic system that may be slow to respond to their needs. Thus, treatment of victims of violence deals with a full range of outcomes and may deal with a broad spectrum of treatment goals.

In Eckhardt et al.'s (2013) meta-analysis of interventions for victims of IPV, they reported outcomes for brief and extended treatments of IPV victims separately. With respect to brief interventions, they found 16 studies of brief interventions which all involved female participants and typically involved brief counseling for less than 3 h that often took place in medical settings such as perinatal care settings and emergency rooms. Typically, the content involved advice concerning safety and information and referral to available services. Nine of 16 studies were RCTS, but outcomes were weak and mixed, which perhaps is unsurprising, given the very time limited nature of the interventions.

Eckhardt et al. (2013) also found 15 studies evaluating more extended treatment of victims of IPV. Treatments included more extended advocacy for victims, variations on CBT, programs that addressed PTSD, sometimes through exposure therapy, and sometimes through post-shelter goals. These programs were more distinctly effective than brief interventions, focusing on a broad array of outcomes, such as depression, revictimization, etc. Although the extended programs were more effective than comparison conditions, reoffending remained "alarmingly high" (p. 226). Thus, with respect to treatment of victims of IPV there remains much to be done.

19.6.3.2 Violence and PTSD

Some individual victims of violence, especially individuals who have experienced more severe, and longstanding violence, may suffer PTSD, including military veterans and civilians during wartime, victims of persecution during civilian times, terrorism, child abuse, political detainment and terrorism, physical assault and domestic abuse, genocide, and organized violence and rape victims amongst others. Until the late 1980s there was limited outcome research on violence-related PTSD, in part because of the failure to identify PTSD and to label the problems in terms of some generic anxiety or mood disorder.

Following early theoretical work by Edna Foa and her colleagues on cognitive behavioral conceptualizations of PTSD (Foa et al., 1989), Foa extended exposure and response treatment, previously evaluated with phobias and Obsessive-Compulsive Disorder and to treatment of PTSD following rape. For example, Foa et al. (1991) randomized 45 female rape victims who met DSM-III criteria for PTSD to four treatments. The first was prolonged exposure (PE). This involved gathering information, developing a treatment rationale, treatment planning, and seven sessions of reliving the rape experience in imagination and describing it aloud. Sessions were recorded and patients instructed to listen to the recordings at least once per day and to self-expose in vivo to feared situations that both patient and therapist believed to be safe. The second treatment was stress inoculation training (SIT). This consisted of breathing, education, muscle relaxation, thought stopping, covert modeling, with no instructions to self-expose. The third treatment was supportive counseling. This involved information gathering, general problem solving, unconditional support for the patient, daily diaries of attempts at problem solving, redirection of discussion of trauma to solving current problems. There were no instructions for anxiety management or self-exposure. Finally, participants in the Wait list control were contacted periodically to ensure safety and were then randomized to either PE or SIT. On multiple outcome measures both at post-treatment and 3.5 months follow-up, there were large differences between the groups. Although all groups improved from pre- to post-treatment, at post-treatment individuals in the SIT group reported fewer symptoms than the supportive counseling and wait list groups. At follow-up there was no difference between the three treatment groups. There were no differences from post-treatment to follow-up, although there was a non-significant trend ($p < 0.07$) for the PE group to continue to improve. The authors concluded that PE was the most effective treatment due to the continued improvement between the end of treatment and follow-up.

These early studies were provocative and stimulated a large literature on psychological treatment of violence-related PTSD. Thus, Lewis et al. (2020) in an update of an earlier Cochrane review reported a systematic review and meta-analysis of manualized psychological treatment of PTSD which identified 114 RCTs with 8171 participants in studies mostly of individuals traumatized by interpersonal violence including military combat ($k = 27$), sexual assault/rape ($k = 11$), war and persecution ($k = 8$), childhood sexual abuse ($k = 7$), etc. Twenty-nine different types of psychological therapies were evaluated and the most commonly used treatments were trauma-focused CBT and PE. The authors concluded that there was increased and robust evidence supporting the effectiveness of trauma-focused CBT and EMDR. The strongest evidence was for PE and two forms of CBT.

Evidence-based treatment of PTSD in victims of violence, including traumatized veterans and rape victims is controversial in part because many palatable, conventional treatments, such as counseling and emotional and personal support, may be largely ineffective whereas distressing treatments, such as exposure therapy, may be effective. Indeed, many therapists may initially oppose the use of exposure therapy, although efforts to disseminate these evidence-based practices have been implemented, for example, for veterans with PTSD (Karlin et al., 2010).

19.6.4 Summary

Aggression and violence in adults cause enormous human suffering as well as economic costs. It is important that it is acknowledged that the outcomes are highly varied, reflecting the type and extent of the aggression and violence, people's response to it, including the availability of intervention services and personal characteristics. There are now many effective treatment options available which may be both personally effective and economically viable, but, as in other areas, dissemination of effective practices is currently very patchy.

19.7 Chapter Summary

This chapter gives us cautious optimism that in a wide range of contexts individual treatment is potentially helpful in reducing or eliminating aggression and violence in specific populations that are often considered quite challenging due to the chronic nature of their disabilities and aggression and violence. Yet, at least for some individuals, we can be optimistic about the possibility of making meaningful differences in the lives of these individuals. In addition, there are a range of effective interventions for victims of violence that are often effective for a wide range of populations and different degrees of severity of violence.

References

Ali, A., Hall, I., Blickwedel, J., & Hassiotis, A. (2015). Behavioural and cognitive behavioural interventions for outwardly directed aggressive behaviour in people with intellectual disabilities. *Cochrane Database of Systematic Reviews, 4*, CD003406. https://doi.org/10.1002/14651858.CD003406.pub4

Allely, C. S. (2020). Autism spectrum disorder, bestiality and zoophilia: A systematic PRISMA review. *Journal of Intellectual Disabilities and Offending Behaviour, 11*(2), 75–91.

Aman, M. G., McDougle, C. J., Scahill, L., & et al., & Research Units on Pediatric Psychopharmacology Autism Network. (2009). Medication and parent training in children with Pervasive Developmental Disorders and serious behavior problems: Results from a randomized clinical trial. *Journal of the American Academy of Child & Adolescent Psychiatry, 48*, 1143–1154.

Aran, G., Hicks, C., Demand, A., Johnson, A. L., Beaman, J., Bailey, Y., et al. (2019). Treating schizophrenia: The quality of evidence behind treatment recommendations and how it can improve. *BMJ Evidence-Based Medicine, 25*(4), 138–142.

Arnold, L. E., Aman, M. G., Li, X., Butter, E., Humphries, K., Scahill, L., et al. (2012). Research Units of Pediatric Psychopharmacology (RUPP) Autism Network randomized clinical trial of parent training and medication: One-year follow-up. *Journal of the American Academy of Child & Adolescent Psychiatry, 51*(11), 1173–1184.

Beavers, G. A., Iwata, B. A., & Lerman, D. C. (2013). Thirty years of research on the functional analysis of problem behavior. *Journal of Applied Behavior Analysis, 46*(1), 1–21.

Borsay, C. (2013). Anger management interventions for adults with learning disabilities living in the community: A review of recent (2000–2010) evidence. *British Journal of Learning Disabilities, 41*, 8–44. https://doi.org/10.1111/j.1468-3156.2011.00720.x

Bows, H. (2018). Sexual violence against older people: A review of the empirical literature. *Trauma, Violence & Abuse, 19*(5), 567–583.

Bright, C. L., & Bowland, S. E. (2008). Assessing interpersonal trauma in older adult women. *Journal of Loss & Trauma, 13*, 373–393.

Brosnan, J., & Healy, O. (2011). A review of behavioral interventions for the treatment of aggression in individuals with developmental disabilities. *Research in Developmental Disabilities, 32*(2), 437–446.

Browne, C., & Smith, I. C. (2018). Psychological interventions for anger and aggression in people with intellectual disabilities in forensic services. *Aggression and Violent Behavior, 39*, 1–14.

Burnes, D., Zhao, R., & Pillemer, K. (2017). Violence and aggression against older adults: What works? What works best? In P. Sturmey (Ed.), *The Wiley handbook of violence and aggression* (pp. 1–12). Wiley.

Buser, M. E. (2015). *Lockdown on Rikers: Shocking stories of abuse and injustice at New York's notorious jail*. St. Martin's Press.

Byrne, C., & Coetzer, R. (2016). The effectiveness of psychological interventions for aggressive behavior following acquired brain injury: A meta-analysis and systematic review. *NeuroRehabilitation, 39*(2), 205–221.

Cattelani, R., Zettin, M., & Zoccolotti, P. (2010). Rehabilitation treatments for adults with behavioral and psychosocial disorders following acquired brain injury: A systematic review. *Neuropsychology Review, 20*(1), 52–85.

CenterWatch. (2009). *Drug information: FDA-approved drugs*. US Food and Drug Administration. Retrieved from http://www.centerwatch.com/drug-information/fda-approved-drugs/year/2009

Citrome, L. (2012). A systematic review of meta-analyses of the efficacy of oral atypical antipsychotics for the treatment of adult patients with schizophrenia. *Expert Opinion on Pharmacotherapy, 13*(11), 1545–1573.

Cook, J. M., Dinnen, S., & O'Donnell, C. (2011). Older women survivors of physical and sexual violence: A systematic review of the quantitative literature. *Journal of Women's Health, 20*(7), 1075–1081.

Cooper, S. A., Smiley, E., Jackson, A., Finlayson, J., Allan, L., Mantry, D., & Morrison, J. (2009). Adults with intellectual disabilities: Prevalence, incidence and remission of aggressive behaviour and related factors. *Journal of Intellectual Disabilities Research, 53*(3), 217–232. https://doi.org/10.1111/j.1365-2788.2008.01127.x

Coskun, M., & Mukaddes, N. M. (2008). Mirtazapine treatment in a subject with autistic disorder and fetishism. *Journal of Child and Adolescent Psychopharmacology, 18*(2), 206–209.

Coskun, O., Ücler, S., Cavdar, L., & Inan, L. E. (2007). Effect of valproic acid on withdrawal therapy in patients with overuse of chronic daily headache medications. *Journal of Clinical Neuroscience, 14*(4), 334–339.

Cox, A. D., & Virués-Ortega, J. (2016). A review of how psychotropic medication can affect the motivation of challenging behavior. *International Journal of Developmental Disabilities, 62*(3), 192–199.

Craig, L. A., Lindsay, W. R., & Browne, K. D. (Eds.). (2010). *Assessment and treatment of sexual offenders with intellectual disabilities: A handbook*. John Wiley & Sons.

Davis, T. N., Dacus, S., Strickland, E., Copeland, D., Chan, J. M., Blenden, K., et al. (2013). The effects of a weighted vest on aggressive and self-injurious behavior in a child with autism. *Developmental Neurorehabilitation, 16*(3), 210–215.

Davis, T. N., Machalicek, W., Scalzo, R., Kobylecky, A., Campbell, V., Pinkelman, S., et al. (2016). A review and treatment selection model for individuals with developmental disabilities who engage in inappropriate sexual behavior. *Behavior Analysis in Practice, 9*(4), 389–402.

De Koker, P., Mathews, C., Zuch, M., Bastien, S., & Mason-Jones, A. J. (2014). A systematic review of interventions for preventing adolescent intimate partner violence. *Journal of Adolescent Health, 54*(1), 3–13.

Desmarais, S. L., Reeves, K. A., Nicholls, T. L., Telford, R. P., & Fiebert, M. S. (2012a). Prevalence of physical violence in intimate relationships, Part 1: Rates of male and female victimization. *Partner Abuse, 3*(2), 140–169.

Desmarais, S. L., Reeves, K. A., Nicholls, T. L., Telford, R. P., & Fiebert, M. S. (2012b). Prevalence of physical violence in intimate relationships, Part 2: Rates of male and female perpetration. *Partner Abuse, 3*(2), 170–198.

Didden, R., Lindsay, W. R., Lang, R., Sigafoos, J., Deb, S., Wiersma, J., et al. (2016). Aggressive behavior. In N. N. Singh (Ed.), *Handbook of evidence-based practices in intellectual and developmental disabilities* (pp. 727–750). Springer.

Drossel, C., Fisher, J. E., & Mercer, V. (2011). A DBT skills training group for family caregivers of persons with dementia. *Behavior Therapy, 42*(1), 109–119.

Eckhardt, C. I., Murphy, C. M., Whitaker, D. J., Sprunger, J., Dykstra, R., & Woodard, K. (2013). The effectiveness of intervention programs for perpetrators and victims of intimate partner violence. *Partner Abuse, 4*(2), 196–231.

Egozy, O., & Cox, A. (2017). Youth facilities and violence: An overview. In P. Sturmey (Ed.), *The Wiley handbook of violence and aggression* (pp. 1–11). Wiley.

Emerson, E., Kiernan, C., Alborz, A., Reeves, D., Mason, H., Swarbrick, R., et al. (2001). The prevalence of challenging behaviors: A total population study. *Research in Developmental Disabilities, 22*(1), 77–93. https://doi.org/10.1016/S0891-4222(00)00061-5

Fearing, G., et al. (2017). A systematic review on community-based interventions for elder abuse and neglect. *Journal of Elder Abuse & Neglect, 29*(2–3), 102–133. https://doi.org/10.1080/08946566.2017.1308286

Fitzpatrick, S. E., Srivorakiat, L., Wink, L. K., Pedapati, E. V., & Erickson, C. A. (2016). Aggression in autism spectrum disorder: Presentation and treatment options. *Neuropsychiatric Disease and Treatment, 12*, 1525–1538.

Fleminger, S., Greenwood, R. R., & Oliver, D. L. (2003). Pharmacological management for agitation and aggression in people with acquired brain injury. *Cochrane Database of Systematic Reviews*, (4), CD003299. https://doi.org/10.1002/14651858.CD003299.pub2

Foa, E. B. (1995). *The posttraumatic diagnostic scale (PDS) manual* (pp. 1–5). National Computer Systems.

Foa, E. B., Steketee, G., & Rothbaum, B. O. (1989). Behavioral/cognitive conceptualizations of post-traumatic stress disorder. *Behavior Therapy, 20*(2), 155–176.

Foa, E. B., Rothbaum, B., Riggs, D., & Murdock, T. (1991). Treatment of posttraumatic stress disorder in rape victims: A comparison between cognitive-behavioral procedures and counseling. *Journal of Consulting and Clinical Psychology, 59*, 715–723.

Fond, G., Boyer, L., Favez, M., Brunel, L., Aouizerate, B., Berna, F., et al. (2016). Medication and aggressiveness in real-world schizophrenia. Results from the FACE-SZ dataset. *Psychopharmacology, 233*(4), 571–578.

Fong, V. C., Hawes, D., & Allen, J. L. (2019). A systematic review of risk and protective factors for externalizing problems in children exposed to intimate partner violence. *Trauma, Violence & Abuse, 20*(2), 149–167.

Foxx, R. M., & Mulick, J. A. (Eds.). (2015). *Controversial therapies for autism and intellectual disabilities: Fad, fashion, and science in professional practice* (2nd ed.). Routledge.

Gannon, T. A., Olver, M. E., Mallion, J. S., & James, M. (2019). Does specialized psychological treatment for offending reduce recidivism? A meta-analysis examining staff and program variables as predictors of treatment effectiveness. *Clinical Psychology Review, 73*, 101752.

Gibbs, J. C., Potter, G. B., & Goldstein, A. P. (1995). *The EQUIP program: Teaching youth to think and act responsibly through a peer-helping approach.* Research Press.

Glowacki, K., Warner, G., & White, C. (2016). The use of a token economy for behaviour and symptom management in adult psychiatric inpatients: A critical review of the literature. *Journal of Psychiatric Intensive Care, 12*(2), 119–127.

Goldstein, A. P., Glick, B., & Gibbs, J. C. (1998). *Aggression replacement training: A comprehensive intervention for aggressive youth.* (Revised). Research Press.

Griffiths, D. M., Thomson, K., Ioannou, S., Hoath, J., & Wilson, R. J. (2018). *Sex Offending Behavior of Persons with an Intellectual Disability: A Multi-component Applied Behavior Analytic Approach.* NADD Press.

Haddock, G., Barrowclough, C., Shaw, J. J., Dunn, G., Novaco, R. W., & Tarrier, N. (2007). Cognitive-behavioural therapy v. social activity therapy for people with psychosis and a history of violence: Randomised controlled trial. *British Journal of Psychiatry, 194*(2), 152–157.

Hamelin, J., Travis, R., & Sturmey, P. (2013). Anger management and intellectual disabilities: A systematic review. *Journal of Mental Health Research in Intellectual Disabilities, 6*(1), 60–70.

Hanley, G. P., Iwata, B. A., & McCord, B. E. (2003). Functional analysis of problem behavior: A review. *Journal of Applied Behavior Analysis, 36*(2), 147–185.

Hassaballa, H. A., & Balk, R. A. (2003). Torsade de pointes associated with the administration of intravenous haloperidol: A review of the literature and practical guidelines for use. *Expert Opinion on Drug Safety, 2*(6), 543–547.

Hayes, S., Shackell, P., Mottram, P., & Lancaster, R. (2007). The prevalence of intellectual disability in a major UK prison. *British Journal of Learning Disabilities, 35*(3), 162–167.

Heinicke, M. R., & Carr, J. E. (2014). Applied behavior analysis in acquired brain injury rehabilitation: A meta-analysis of single-case design intervention research. *Behavioral Interventions, 29*(2), 77–105.

Hellmans, H., Colson, K., Verbraken, C., et al. (2007). Sexual behavior in higher functioning male adolescents and young adults with autism spectrum disorders. *Journal of Autism and Developmental Disorders, 37*, 260–269.

Henderson, A., Mcskimming, P., Kinnear, D., McCowan, C., McIntosh, A., Allan, L., & Cooper, S. A. (2020). Changes over a decade in psychotropic prescribing for people with intellectual disabilities: Prospective cohort study. *BMJ Open, 10*(9), e036862.

Heyvaert, M., Maes, B., Van den Noortgate, W., Kuppens, S., & Onghena, P. (2012). A multilevel meta-analysis of single-case and small-n research on interventions for reducing challenging behavior in persons with intellectual disabilities. *Research in Developmental Disabilities, 33*, 766–780. https://doi.org/10.1016/j.ridd.2011.10.010

Hickman, G., Thrift, S., Dhaliwal, R., & Taylor, C. (2017). Sixteen years of the Brooklands Thinking Skills Offender Programme. *Journal of Intellectual Disabilities and Offending Behaviour, 8*(3), 132–143. https://doi.org/10.1108/JIDOB-03-2017-0002

Hicks, A. J., Clay, F. J., Hopwood, M., James, A. C., Jayaram, M., Perry, L. A., et al. (2019). The efficacy and harms of pharmacological interventions for aggression after traumatic brain injury–systematic review. *Frontiers in Neurology, 10*, 1169. https://doi.org/10.3389/fneur.2019.01169

Horner-Johnson, W., & Drum, C. E. (2006). Prevalence of maltreatment of people with intellectual disabilities: A review of recently published research. *Mental Retardation and Developmental Disabilities Research Reviews, 12*(1), 57–69.

Hughes, R. B., Lund, E. M., Gabrielli, J., Powers, L. E., & Curry, M. A. (2011). Prevalence of interpersonal violence against community-living adults with disabilities: A literature review. *Rehabilitation Psychology, 56*(4), 302.

Human Rights Defense Center. (2017, January 10). New York: Nearly $10 million in settlements over denial of medical care at Rikers island. *Prison Legal News*, 54.

Retrieved August 30, 2021, from https://www.prisonlegalnews.org/news/2017/jan/10/new-york-nearly-10-million-settlements-over-denial-medical-care-rikers-island/

Jennings, W. G., Okeem, C., Piquero, A. R., Sellers, C. S., Theobald, D., & Farrington, D. P. (2017). Dating and intimate partner violence among young persons ages 15–30: Evidence from a systematic review. *Aggression and Violent Behavior, 33*, 107–125.

Jolliffe, D., & Farrington, D. P. (2007). A systematic review of the national and international evidence on the effectiveness of interventions with violent offenders. *Ministry of Justice Series, 16*, 7. Retrieved August 28, 2021, from https://www.crim.cam.ac.uk/sites/www.crim.cam.ac.uk/files/violmoj.pdf

Jones, E., & Chaplin, E. (2020). A systematic review of the effectiveness of psychological approaches in the treatment of sex offenders with intellectual disabilities. *Journal of Applied Research in Intellectual Disabilities, 33*(1), 79–100.

Karlin, B. E., Ruzek, J. I., Chard, K. M., Eftekhari, A., Monson, C. M., Hembree, E. A., ... & Foa, E. B. (2010). Dissemination of evidence-based psychological treatments for posttraumatic stress disorder in the Veterans Health Administration. *Journal of Traumatic Stress, 23*(6), 663–673.

King, C., & Murphy, G. H. (2014). A systematic review of people with autism spectrum disorder and the criminal justice system. *Journal of Autism and Developmental Disorders, 44*(11), 2717–2733.

Krisberg, B. A. (2017). *Juvenile justice and delinquency*. SAGE Publications.

Krnjacki, L., Emerson, E., Llewellyn, G., & Kavanagh, A. M. (2016). Prevalence and risk of violence against people with and without disabilities: Findings from an Australian population-based study. *Australian and New Zealand Journal of Public Health, 40*(1), 16–21.

Langdon, P. E., Murphy, G. H., Clare, I. C., Palmer, E. J., & Rees, J. (2013). An evaluation of the EQUIP treatment programme with men who have intellectual or other developmental disabilities. *Journal of Applied Research in Intellectual Disabilities, 26*(2), 167–180.

Large, M. M., & Nielssen, O. (2011). Violence in first-episode psychosis: A systematic review and meta-analysis. *Schizophrenia Research, 125*(2–3), 209–220.

Lao, K. S., Wong, A., Wong, I. C., Besag, F., Chang, W. C., Lee, E. H., ... & Chan, E. W. (2020). Mortality risk associated with haloperidol use compared with other antipsychotics: An 11-year population-based propensity-score-matched cohort study. *CNS Drugs, 34*(2), 197–206.

LePage, J. P., DelBen, K., Pollard, S., McGhee, M., VanHorn, L., Murphy, J., et al. (2003). Reducing assaults on an acute psychiatric unit using a token economy: A 2-year follow-up. *Behavioral Interventions: Theory & Practice in Residential & Community-Based Clinical Programs, 18*(3), 179–190.

Lewis, C., Roberts, N. P., Andrew, M., Starling, E., & Bisson, J. I. (2020). Psychological therapies for post-traumatic stress disorder in adults: Systematic review and meta-analysis. *European Journal of Psychotraumatology, 11*(1), 1729633.

Li, L., & Liu, J. (2013). The effect of pediatric traumatic brain injury on behavioral outcomes: A systematic review. *Developmental Medicine and Child Neurology, 55*(1), 37–45.

Lieberman, A. F., Van Horn, P., & Ippen, C. G. (2005). Toward evidence-based treatment: Child-parent psychotherapy with preschoolers exposed to marital violence. *Journal of the American Academy of Child & Adolescent Psychiatry, 44*(12), 1241–1248.

Lindsay, W. R. (2009). *The treatment of sex offenders with developmental disabilities: A practice workbook*. John Wiley & Sons.

Lindsay, W. R., Taylor, J. L., & Sturmey, P. (Eds.). (2004). *Offenders with developmental disabilities*. John Wiley & Sons.

Lindsay, W. R., Hamilton, C., Moulton, S., Scott, S., Doyle, M., & McMurran, M. (2011). Assessment and treatment of social problem solving in offenders with intellectual disability. *Psychology, Crime & Law, 17*(2), 181–197.

Lindsay, W. R., Craig, L. A., & Griffiths, D. (Eds.). (2019). *The Wiley Handbook on what works for offenders with intellectual and developmental disabilities: An evidence-based approach to theory, assessment, and treatment*. John Wiley & Sons.

Lofthouse, R., Totsika, V., Hastings, R. P., & Lindsay, W. R. (2018). *The Wiley handbook on offenders with intellectual and developmental disabilities: Research, training, and practice*. Wiley.

Lund, E. M. (2011). Community-based services and interventions for adults with disabilities who have experienced interpersonal violence: A review of the literature. *Trauma, Violence & Abuse, 12*(4), 171–182.

Lund, E. M., & Hammond, M. (2014). Single-session intervention for abuse awareness among people with developmental disabilities. *Sexuality and Disability, 32*(1), 99–105.

Lunsky, Y., & Modi, M. (2018). Predictors of psychotropic polypharmacy among outpatients with psychiatric disorders and intellectual disability. *Psychiatric Services, 69*(2), 242–246.

Mann, L. E., & Travers, J. C. (2019). A systematic review of interventions to address inappropriate masturbation for individuals with Autism Spectrum Disorder or other developmental disabilities. *Review Journal of Autism and Developmental Disorders, 7*, 205–218. https://doi.org/10.1007/s40489-019-00192-9

Mano-Sousa, B. J., Pedrosa, A. M., Alves, B. C., Galduróz, J. C. F., Belo, V. S., Chaves, V. E., & Duarte-Almeida, J. M. (2021). Effects of risperidone in autistic children and young adults: A systematic review and meta-analysis. *Current Neuropharmacology, 19*(4), 538–552.

Mason, J., & Murphy, G. (2002). Intellectual disability amongst people on probation: Prevalence and outcome. *Journal of Intellectual Disability Research, 46*(3), 230–238.

McQuire, C., Hassiotis, A., Harrison, B., & Pilling, S. (2015). Pharmacological interventions for challenging behaviour in children with intellectual disabilities: a systematic review and meta-analysis. *BMC psychiatry, 15*(1), 1–13.

Melvin, C. L., Langdon, P. E., & Murphy, G. H. (2017). Treatment effectiveness for offenders with autism spectrum conditions: A systematic review. *Psychology, Crime & Law, 23*(8), 748–776.

Molina-Ruiz, R. M., Martín-Carballeda, J., Asensio-Moreno, I., & Montañés-Rada, F. (2017). A guide to psychopharmacological treatment of patients with intellectual disability in psychiatry. *The International Journal of Psychiatry in Medicine, 52*(2), 176–189.

Morrissey, C., Geach, N., Alexander, R. T., Chester, V., Devapriam, J., Duggan, C., et al. (2017a). Researching outcomes from forensic services for people with intellectual or developmental disabilities: A systematic review, evidence synthesis and expert and patient/carer consultation. *Health Services and Delivery Research, 5*(3). https://doi.org/10.3310/hsdr05030

Morrissey, C., Langdon, P. E., Geach, N., Chester, V., Ferriter, M., Lindsay, W. R., et al. (2017b). A systematic review and synthesis of outcome domains for use within forensic services for people with intellectual disabilities. *British Journal of Psychiatry, Open, 3*(1), 41–56.

Mueller, I., & Tronick, E. (2019). Early life exposure to violence: Developmental consequences on brain and behavior. *Frontiers in Behavioral Neuroscience, 13*, 156. https://doi.org/10.3389/fnbeh.2019.00156

Muir-Cochrane, E., Oster, C., & Grimmer, K. (2020). International research into 22 years of use of chemical restraint: An evidence overview. *Journal of Evaluation in Clinical Practice, 26*(3), 927–956.

Murphy, G. H. (2020). *CQC inspections and regulation of Whorlton Hall 2015-2019: An independent review*. CQC. Retrieved May 16, 2020, from https://www.cqc.org.uk/sites/default/files/20020218_glynis-murphy-review.pdf

Newman, S. S., & Ghaziuddin, M. (2008). Violent crime in Asperger syndrome: The role of psychiatric comorbidity. *Journal of Autism and Developmental Disorders, 38*(10), 1848–1852.

Nicoll, M., Beail, N., & Saxon, D. (2013). Cognitive behavioural treatment for anger in adults with intellectual disabilities: A systematic review and meta-analysis. *Journal of Applied Research in Intellectual Disabilities, 26*(1), 47–62.

Papalia, N., Spivak, B., Daffern, M., & Ogloff, J. R. (2019). A meta-analytic review of the efficacy of psychological treatments for violent offenders in correctional and forensic mental health settings. *Clinical Psychology: Science and Practice, 26*(2), e12282.

Pritchard, D., Graham, N., Ikin, A., Penney, H., Kovacs, L., Mercer, D., et al. (2012). Managing sexually harmful behaviour in a residential special school. *British Journal of Learning Disabilities, 40*(4), 302–309.

Pritchard, D., Penney, H., & Mace, F. C. (2018). The ACHIEVE! program: A point and level system for reducing severe problem behavior. *Behavioral Interventions, 33*(1), 41–55.

Quinn, J., & Kolla, N. J. (2017). From clozapine to cognitive remediation: A review of biological and psychosocial treatments for violence in schizophrenia. *The Canadian Journal of Psychiatry, 62*(2), 94–101.

Rampling, J., Furtado, V., Winsper, C., Marwaha, S., Lucca, G., Livanou, M., & Singh, S. P. (2016). Non-pharmacological interventions for reducing aggression and violence in serious mental illness: A systematic review and narrative synthesis. *European Psychiatry, 34*, 17–28.

Research Units on Pediatric Psychopharmacology Autism Network. (2005). Risperidone treatment of autistic disorder: Longer-term benefits and blinded discontinuation after 6 months. *American Journal of Psychiatry, 162*(7), 1361–1369.

Rubio-Garay, F., López-González, M. A., Carrasco, M. Á., & Amor, P. J. (2017). The prevalence of dating violence: A systematic review. *Papeles del Psicólogo, 38*(2), 135–147.

Sawyer, A., Lake, J. K., Lunsky, Y., Liu, S. K., & Desarkar, P. (2014). Psychopharmacological treatment of challenging behaviours in adults with autism and intellectual disabilities: A systematic review. *Research in Autism Spectrum Disorders, 8*(7), 803–813.

Schwirtz, M., & Winerip, M. (2014). No charges for Rikers officers in 2012 beatings of 2 inmates. *New York Times*. Retrieved August 30, 2021, from https://www.nytimes.com/2014/08/06/nyregion/no-charges-over-inmates-beaten-at-rikers-island.html

Shea, S., Turgay, A., Carroll, A., Schulz, M., Orlik, H., Smith, I., & Dunbar, F. (2004). Risperidone in the treatment of disruptive behavioral symptoms in children with autistic and other pervasive developmental disorders. *Pediatrics, 114*(5), e634–e641.

Sheehan, R., & Hassiotis, A. (2017). Reduction or discontinuation of antipsychotics for challenging behaviour in adults with intellectual disability: A systematic review. *The Lancet Psychiatry, 4*(3), 238–256.

Singh, A. N., Matson, J. L., Cooper, C. L., Dixon, D., & Sturmey, P. (2005). The use of risperidone among individuals with mental retardation: Clinically supported or not? *Research in Developmental Disabilities, 26*(3), 203–218.

Sobsey, R. (1994). *Violence and abuse in the lives of people with disabilities: The end of silent acceptance?* Brookes.

Sturmey, P. (2015). *Reducing restraint and restrictive behavior management practices*. Springer.

Sturmey, P. (2019). Treatment outcomes for people with Autistic Spectrum Disorder in forensic settings. In W. R. Lindsay, L. A. Craig, & D. Griffiths (Eds.), *The Wiley handbook on what works for offenders with intellectual and developmental disabilities: An evidence-based approach to theory, assessment, and treatment* (pp. 373–387). John Wiley & Sons.

Taquette, S. R., & Monteiro, D. L. M. (2019). Causes and consequences of adolescent dating violence: A systematic review. *Journal of Injury and Violence Research, 11*(2), 137.

Taylor, J. L., & Novaco, R. W. (2005). *Anger treatment for people with developmental disabilities: A theory, evidence and manual based approach.* John Wiley & Sons.

Taylor, J. L., Novaco, R. W., Gillmer, B. T., Robertson, A., & Thorne, I. (2005). Individual cognitive-behavioural anger treatment for people with mild-borderline intellectual disabilities and histories of aggression: A controlled trial. *British Journal of Clinical Psychology, 44*(3), 367–382.

Tearle, S., & Holt, R. R. (2018). A community EQUIP treatment group. *Journal of Intellectual Disabilities and Offending Behaviour, 9*, 69. https://doi.org/10.1108/jidob-03-2018-0004

Tearle, S., Sam, S., & Holt, R. R. (2020). Collaborative case report: Participatory action research into using EQUIP to support community discharge. *Journal of Intellectual Disabilities and Offending Behaviour, 11*(1), 23–34. https://doi.org/10.1108/JIDOB-05-2019-0010

Topiwala, A., & Fazel, S. (2011). The pharmacological management of violence in schizophrenia: A structured review. *Expert Review of Neurotherapeutics, 11*(1), 2.

Tozer, C., & Plomin, J. (2013). The abuse of vulnerable adults at Winterbourne View Hospital: The lessons to be learned. *The Journal of Adult Protection, 15*, 182–190. https://doi.org/10.1108/JAP-05-2013-0020

Tsakanikos, E., Costello, H., Holt, G., Sturmey, P., & Bouras, N. (2007). Behaviour management problems as predictors of psychotropic medication and use of psychiatric services in adults with autism. *Journal of Autism and Developmental Disorders, 37*(6), 1080–1085.

Tsiouris, J. A., Kim, S., Brown, W. T., Pettinger, J., & Cohen, I. L. (2013). Prevalence of psychotropic drug use in adults with Intellectual Disability: Positive and negative findings from a large scale study. *Journal of Autism and Developmental Disorders, 43*, 719–731. https://doi.org/10.1007/s10803-012-1617-6

van Schalkwyk, G. I., Beyer, C., Johnson, J., Deal, M., & Bloch, M. H. (2018). Antipsychotics for aggression in adults: A meta-analysis. *Progress in Neuro-Psychopharmacology and Biological Psychiatry, 81*, 452–458.

Vereenooghe, L., & Langdon, P. E. (2013). Psychological therapies for people with intellectual disabilities: A systematic review and meta-analysis. *Research in Developmental Disabilities, 34*(11), 4085–4102. https://doi.org/10.1016/j.ridd.2013.08.030

Waugh, A., Gudjonsson, G. H., Rees-Jones, A., & Young, S. (2014). A feasibility study of the Reasoning and Rehabilitation Mental Health Programme (R&R2MHP) in male offenders with intellectual disability. *Criminal Behaviour and Mental Health, 24*, 222.

Willner, P. (2015). The neurobiology of aggression: Implications for the pharmacotherapy of aggressive challenging behaviour by people with intellectual disabilities. *Journal of Intellectual Disability Research, 59*(1), 82–92.

Woodbury-Smith, M., Clare, I., Holland, A. J., Watson, P. C., Bambrick, M., Kearns, A., & Staufenberg, E. (2010). Circumscribed interests and 'offenders' with autism spectrum disorders: A case-control study. *The Journal of Forensic Psychiatry & Psychology, 21*(3), 366–377.

Ylvisaker, M., Turkstra, L., Coehlo, C., Yorkston, K., Kennedy, M., Sohlberg, M. M., & Avery, J. (2007). Behavioural interventions for children and adults with behaviour disorders after TBI: A systematic review of the evidence. *Brain Injury, 21*(8), 769–805.

Yon, Y., Mikton, C. R., Gassoumis, Z. D., & Wilber, K. H. (2017). Elder abuse prevalence in community settings: A systematic review and meta-analysis. *The Lancet Global Health, 5*(2), e147–e156.

Yon, Y., Ramiro-Gonzalez, M., Mikton, C. R., Huber, M., & Sethi, D. (2019). The prevalence of elder abuse in institutional settings: A systematic review and meta-analysis. *European Journal of Public Health, 29*(1), 58–67.

Yoo, J. H., Williams, D. C., Napolitano, D. A., Peyton, R. T., Baer, D. M., & Schroeder, S. R. (2003). Rate-decreasing effects of the atypical neuroleptic risperidone attenuated by conditions of reinforcement in a woman with mental retardation. *Journal of Applied Behavior Analysis, 36*(2), 245–248.

Zane, T., Davis, C., & Rosswurm, M. (2008). The cost of fad treatments in autism. *Journal of Early and Intensive Behavior Intervention, 5*(2), 44–51. https://doi.org/10.1037/h0100418

Chapter 20
Future Directions in Research and Practice in Violence and Aggression: Reasons for Cautious Optimism

Studying aggression and violence could readily give rise to pessimism. It involves frequent exposure to situations and material that are very disturbing and depressing as one focuses on the worst of human behavior. Human aggression and violence is very common and so many of us are perpetrators or have the potential to be perpetrators of aggression and violence. The personal and economic cost and loss of human potential is enormous and continues. These are all disturbing and depressing facts.

It is frustrating that so many solutions are available. Yet, attending to and using objective evidence effectively is difficult and done insufficiently frequently. The media dishes up verbal and physical fighting, violence, murder, and bloodletting as entertainment for us all. Entertainment and daily snuff news sell soap based on violent entertainment. Many people perversely enjoy talking about the worst of human behavior. Dramatic detail, domestic violence, upticks in gun shootings over vacations, weekends, and during COVID, and terrorist and domestic mass killings all dominate the daily news at this time. So, there are plenty of vivid and sometimes gory details to distract us from the facts. When reading about or experiencing the worst of violence, biological determinism and its associated pessimism are very tempting.

Rationally, there is much evidence to contradict this pessimism about human aggression and violence. The broad sweep of history and recent history indicates that in the process of civilization humanity has become much less violent. This is true despite geographically and temporally local upticks in violence. This is true despite the monstrous wars which, over the last century have killed millions of people, have also decreased substantially since World War II, with many notable exceptions. It is hard to remember that most of the world has lived in relative peace for the last 75 years.

This chapter will review what can be done to prevent and treat aggression and violence using the selectionist framework that this book has used. Namely, it will review what can be done to prevent and treat violence and aggression from biological, cultural, and individual learning perspectives.

P. Sturmey, *Violence and Aggression*, Advances in Preventing and Treating Violence and Aggression, https://doi.org/10.1007/978-3-031-04386-4_20

20.1 Biological Evolution and Selection of Human Aggression

The logic of this volume requires that we should consider the possibility of influencing human aggression and violence through biological evolution in a deliberate and planned manner. The evidence that genetics influence aggression in many non-human animal species, especially in cases where humans intervene, to explicitly breed aggression or non-aggression is relatively convincing and robust across species. If that observation is paired with the finding that there is at least some evidence for genetic influences on human aggression, it raises this possibility. This could include breeding both more and less aggressive humans. Has anyone tried this?

In 1869, the British eugenicist Francis Galton believed that humanity could and *should* influence the genetic makeup of future generations (Forrest, 1974). The subsequent eugenic movement was active in the UK, the USA, Europe, Commonwealth countries, and elsewhere from the late nineteenth century to the 1970s. Discussion of eugenics focuses mostly on negative eugenics, i.e., prevention and control of marriage and childbirth, sterilization, and murder of individuals, such as individuals with intellectual disabilities, psychiatric disorders, people who were Jewish, Slavic, and Romany, and individuals who were chronic criminals and misfits, who deemed to be genetically inferior or undesirable. Eugenics, however, also advocated the so-called positive eugenics, that is increasing fertility of the so-called genetically desirable individuals through payment and other material benefits for having larger numbers of children in certain families. In most contexts this referred to individuals of North European ancestry and in Nazi Germany referred to individuals with the so-called Aryan or Scandinavian physical characteristics and no family history of the so-called undesirable traits and genetic conditions. In some parts of Central and Latin America a form of positive eugenics was present in the early twentieth century extolling the virtues of indigenous and mixed-race groups alongside denigration of Latinos of African and Chinese descent (Schell, 2010).

Although some positive Eugenic programs were implemented in the USA, such as encouraging larger, White families and negative eugenics such as sterilization of the so-called undesirable women (Gallagher, 1999), the most explicit attempt to use positive genetics, which was implemented alongside negative genetics, comes from Nazi Germany. One example of positive eugenics was payment and other material benefits for approved Aryan families with no family history of mental health or disability issues to have at least four children. In addition, the *Lebensborn* ("fount of life") program involved encouragement of genetically approved Aryan women to have children with officers in the Schutzstaffel (SS), Hitler's paramilitary organization. These SS officers had already their family histories examined and been approved of as being genetically desirable. There was a network of *lebensborn* programs throughout Germany and Nazi occupied Europe (Ericsson & Simonsen, 2005). In countries such as Norway approximately 8000 children were born as a result of this program. These children were placed in foster homes and in foster families and their histories hidden from them. After World War II many of the women were vilified, sometimes subject to public violence, denied citizenship, and passports for many years by the Norwegian government. Many of these children, who were viewed as potentially dangerous to the state, since they might form a future pro-Nazi group in Norway. Some moved to Sweden, Germany, and elsewhere and most were institutionalized and subject to extensive neglect and sometimes abuse (Olsen, 2005). In addition, Nazi Germany abducted approximately 200,000 Polish children who were believed to be of Aryan decent and these too were raised in foster homes or with German families and their personal histories largely obliterated (Brüne, 2007).

The Nazi positive eugenic programs appear to focus largely on physical features, family history of genetic conditions, and apparent ethnic history. The use, if any, of selection of behavioral traits related to aggression in the positive eugenic programs, if present, does not appear to have been explicit (Brüne, 2007; Ericsson & Simonsen, 2005; Olsen, 2005). Nazi negative eugenics perhaps addressed elimination of undesirable aggression through serilization and murder of the so-called socially undesirables.

There is little evidence that it is possible to change human aggression and violence through deliberate and planned manipulation of the human gene pool, although genetic changes may have taken place without explicit planning over a long period of time, such as smaller body and brain size (Wrangham, 2019). When we look at human selection of morphological and behavioral traits in domestic animals we see that the number of animals that have been domesticated is very small indeed. Two of the characteristics of animals that have been domesticated are relatively short intergenerational time and readily identifiable characteristics to select (Diamond, 1999). Any attempt to select high or low aggressive traits in humans is likely to be stymied by both of these challenges. Whether future research can identify a single gene or a few genes for either aggression or prosocial behavior that could be the basis for selection of human traits remains to be seen, but again, such possibilities still face the two practical challenges identified earlier. Thus, it seems that influencing human aggression and violence through planned genetic interventions and biological evolution is an unlikely and repugnant prospect.

20.2 Cultural Evolution and Design

It seems that cultural practices may reduce aggression and violence while also maintaining the survival of a culture. If the work of Eisner and Pinker and other more specific examples of cultural interventions are correct, then we have substantial reasons to be very optimistic. Human aggression and violence are susceptible to changes in cultural practices over the long term since they have reduced considerably and specific interventions to change cultural practices including both social and physical environments.

As we saw in Chaps. 6 and 7, on cultural evolution and elsewhere in this volume, Skinner (1953) argued that cultures intervene to change behavior all the time, but often do so in a haphazard manner or even in ways that are ineffective or harmful in the long term, but beneficial to some of the players in such endeavors at least in the short term. Unfortunately, when public figures are explicit in their actions, they open themselves to pushback from others. When politicians have advocated for contingency management practices to control such varied social problems as traffic congestion and pollution through congestion charges, to reduce drug addiction using monetary payments for clean urine samples and reduction in obesity using payment for exercise or weight loss, their good intentions are often met with countercontrol from other public figures and the media. When the government proposes congestion charges, members of the public and others with vested interested will claim that such procedures will not work, are an unfair tax on the poor, and that everyone from the police, nurses, ambulance drivers, and people who live in the affected areas deserve exemptions. They will also claim that people in adjacent neighborhoods will suffer when traffic is pushed from one neighborhood to another. When politicians propose contingency management for drug addiction or obesity, other politicians can grandstand in front of the media and point out that no one paid them to refrain from drug or lose weight and instead propose other more acceptable programs based on personal experience, intuition, or vested interests. Often such programs are unevaluated, ineffective, harmful, or just a vehicle to divert public money to friends and family. Thus, explicit attempts to design cultures are often punished and are undoubtedly very effortful. Yet, when politicians appoint friends and family members with little or no experience in such matters and they adopt intuitively appealing programs, little happens, at least early on, although years later people may express regret at the millions of dollars wasted on ineffective, wasteful, and harmful programs. Thus, we should take Skinner's (1953) message seriously that there is no virtue in leaving cultural design to chance. Even when planned cultural design fails even in the short term—as it often must—if it is conducted in the manner Skinner suggested—as part of continuous experimentation and evaluation—then the process will be self-correcting over time.

If we examine planned behavioral utopias based on *Walden Two* (Skinner, 1969), we find there are not many of them, most do not last very long with the possible exceptions of Comunidad Los Horcones and Twin Oaks and there are no clear and convincing data on reductions in aggression or increases in prosocial behavior yet although there may still be things that can be learned from these examples. The common elements of culture-wide contingency management to support prosocial behavior, clear management procedures, and experimentation in cultural design are worth considering. Culture-wide contingency management already exists: Stores and services use loyalty schemes, productivity bonuses, and credit cards practice it with rebates to change customers' behavior to reallocate increasing spending from one product to another and to increase spending. Various forms of sin taxes are also widely used. Some versions of contingency management are acceptable and some are not when they are judged to be too coercive and demanding or they simply do not provide a sufficient magnitude of easily available reinforcement. Thus, culture-wide contingency management for other non-aggressive forms of behavior is a possibility which is often adopted today in some circumstances.

A weakness of this literature is the lack of explicit mechanisms for cultural replication. Informal examples exist with daughter colonies of organizations such as Twin Oaks, and cultural practices such as those of the Hutterites. More formal and clearly documented examples exist as in replications of Achievement Place (Kirigin et al., 2020) which addresses aggression and violence directly for pre-delinquents and delinquents, and the Princeton Child Development Institute (McClannahan & Krantz, 1997) which does so indirectly through skills training and other environmental design procedures for individuals with autism. Other behavior analytic approaches to reduce aggression and violence through cultural design of existing cultures (e.g., Mattaini, 2010) are at an early stage of development: While they use behavior analytic concepts, they currently lack data to support their use to reduce aggression. As both Skinner (1962) and Chance (2007) acknowledged, cultural design is challenging, not only in terms of the effort needed to implement such changes, but also in terms of the difficulties in anticipating the future.

The evidence on current interventions is messy and incomplete, but many societies have many programs designed to prevent aggression and violence, some of which are likely to have significant impacts on a large scale, even if some are ineffective, even harmful, too expensive, unsustainable, etc. Persistence in cultural interventions may pay off under some circumstances. Examples, such as the theory-driven work of Mark Mattaini to reduce violence in South Chicago gangs, and Medellin, Colombia; school-based interventions, such as teaching prosocial behavior to preschoolers, and school-wide positive behavioral interventions; and evidence-based approaches to soccer violence and police violence. These are all examples of societal interventions, which, though lacking complete and convincing data at this time, are all promising approaches to cultural design.

20.2.1 Families: Cradles of Violence or Nests of Peace?

If families are the cradle of violence, the primary location and social institution where aggression and violence occur and is learned, then for most of humanity families are the location where prevention and treatment of aggression and violence should have its focus. Early intervention has the potential to have the largest impact both because interventions could impact the largest number of children, the largest number of at-risk children, and have the potential to impact aggression and violence across the lifespan. In addition, other important areas of concern include aggression and violence among adult family members, including adult aggression to children, intimate partner violence, and, aggression and violence to older adults within the family.

20.2.1.1 Raising Pro-Social, Non-violent Children

Family caretakers can assist their children in acquiring skills such as communication, prosocial skills, tolerance of frustration, problem solving, assertiveness, responding to aggression from other people, and behavioral self-management which are probably paramount in preventing aggression and violence, unlikely to be harmful in any significant way, often of modest cost, and can be done quite readily. In addition, caregivers should also address the acquisition of secondary reinforcers related to the acquisition of prosocial behavior and aggression. This could include teaching children to enjoy other people being happy, not enjoying seeing other people suffering, and also devaluing the reinforcing value of items or activities that are commonly associated with aggression and violence, such as access to toys and high status items, etc. Thus, it is not only important for young children to learn to share valued items. They should also learn to be happy to see others enjoying the items they share and happy to give up preferred items, not only because they have learned to wait patiently, but also because they are not so attached to those items as much as they might be, but also because they have learned that sharing its reinforcing activity.

The large-scale nature of intervention in families is highly challenging due to the very large number of caretakers who need support. Support can include primary prevention and secondary and tertiary treatment. Thus, triaging interventions and economic analysis to maximizing efficiency of family-based interventions are essential. Primary prevention of family aggression and violence to children involves interventions aimed at the entire population of caretakers and potential caretakers. It can include society-wide direct and indirect education, and training of community members, through methods such as modeling concerning appropriate parenting both in terms of what child skills should be addressed and how to teach and support these skills. Many public health programs attempt to address primary prevention of aggression and violence through changes in attitudes and beliefs, such as persuading caregivers that coercive parental disciplinary practices are wrong, and that there are better effective non-coercive options for parents. Some media programs such as *Supernanny* offer models of some positive approaches to caregiver management of child problem behavior, such as clarifying children's desirable and undesirable behavior, having an organized household, clarifying responsibilities for all caregivers, scheduling child activities, and positive reinforcement; sometimes time out is also modeled in certain circumstances. Such educational programs attempt to balance instruction for caregivers with the need to entertain the audience. Some countries use national media outlets to manage the behavior of community members in various ways. Just as *Sesame Street* has tried to teach prosocial behavior to children through modeling and vicarious reinforcement, state media can also provide models for appropriate parenting as part of a national strategy to prevent violence and aggression toward children. Secondary intervention to treat and prevent existing aggression in children can include screening for existing problems, identifying and changing risk factors, and training caregivers. The content of caregiver training is again a topic with substantial evidence and is well researched. Broadly speaking, behavioral and cognitive behavior therapy (CBT) approaches have substantial amounts of evidence to support their use in changing parent and child behavior in many contexts. They often address specific parent skills, such as clarifying and prioritizing appropriate and inappropriate child behavior, giving clear instructions, correct use of positive reinforcement, and reduction of coercive caregiver practices. Additionally, basing interventions on functional assessments and analyses is more likely to produce desirable changes in both aggression and prosocial behavior than interventions that are not based on functional assessments and analyses. Finally, tertiary prevention of child aggression and violence within the family can address relapse prevention and minimizing child harm, child disability, and costs.

There is quite good agreement among researchers as to what skills parents should teach their children to prevent and treat aggression and violence and how to support parents to do this. The majority of researchers agree that caregivers should foster prosocial child behavior with other family members, patience, and behavioral self-management, such as learning to wait, etc. There is also good agreement that parent training can be effective in achieving these child outcomes if the right methods are used to train parents. Thus, there is widespread agreement that didactic approaches, whether written or verbal, are least likely to produce changes in either caretaker and child behavior, but modeling and feedback, if well designed, are more likely to produce changes in caretaker and parent and child behavior. Further, these findings are relatively robust across different kinds of families and different cultures and thus these approaches may be readily adapted to under-researched, novel, and emerging family contexts. Families are changing rapidly both in terms of who the primary caretakers are and social roles for women and men, parents, siblings, grandparents and other family members, friends and paid staff who are primary caretakers. For example, the development of paternal leave and fathers who are the primary caregivers in some countries and some jobs suggests that some societies are changing in the roles for men and women with respect to raising children.

Prevention of intergenerational transmission of family violence is also important. This can be addressed by identifying new parents with histories of family violence and offering services to address the effects of their own experience of family violence on their own adaptation and to support them in their relationship with their own children by supporting a positive emotional relationship with their children and where needed giving them the parenting and other skills they need to be effective, non-violent parents and to prevent the intergenerational transmission of aggression and violence to the next generation. In a recent critical synthesis of 77 studies, Isobel et al. (2019) identified two main focuses of research on intergenerational transmissions of aggression and violence. These were resolution and treatment of parental trauma and support and teaching of parent–child interaction skills and attachment.

An early start to prevent child aggression and violence is likely to be beneficial in multiple ways both in directly dealing with situations and people who provoke aggression and violence, but also in protecting young children from the development of aggression and violence early on. Early intervention has the potential to have the greatest impact on the individual children since such interventions may delay or completely prevent the negative consequences of child aggression and violence. Additionally, an early start also has the potential for the greatest economic benefits.

20.2.1.2 Reducing Violence Between Caretakers and Children

The second point of entry to reduce aggression and violence within the family is efforts to reduce aggression between caregivers and their children by reducing parental aggression and violence directly and by reducing the risk factors that can be modified. Barth (2009), for example, identified parental alcohol and substance abuse, mental illness, domestic violence, child problem behavior, and family poverty as five important risk factors that could be modified through parental training to reduce caregiver aggression and violence to their children. As noted in previous chapters, parental training is a robust strategy that is often likely to help and evidence-based parent training programs have been evaluated, identified, and disseminated.

As Barth (2009) identified, however, there are multiple risk factors for caregiver violence toward children. Thus, each family might need an individually tailored package of interventions. One family, if the primary problem is lack of basic parenting skills, might only need training and support in basic positive parenting skills and organization of the family home. Another family may experience multiple problems and parental training alone may be insufficient and fail to address the multitude of family problems. Such families may need help with addiction, poverty, job training, money management, and intervention for mental health issues in addition to skills training. In such complex but not uncommon cases, clinicians may need support and training to make adequate formulations of the family's problems, prioritize interventions, and identify keystone points of entry to prevent caregiver aggression and violence toward their children. In some cases, intervention may be insufficiently effective and in order to protect the children from caregiver violence, the children may have to be placed in alternate arrangements such as foster care and residential treatment, which too sometimes comes with risks of aggression and violence toward children.

20.2.1.3 Violence Between Adult Family Members

The most common form of aggression and violence between adult family members is Intimate Partner Violence (IPV), which is predominantly violence from male partners toward female partners but also includes violence toward senior family members and violence toward parents from some adolescent children.

The outcome literature for interventions for IPV is mixed and clouded by ideologically based claims. Gender-role based interventions have little evidence with mixed outcomes (Miller et al., 2013; Karakurt et al., 2019), whereas various forms of CBT and related interventions are effective and outcomes are enhanced if the intervention addresses comorbid issues such as substance abuse and trauma (Karakurt et al., 2019). Thus, in order to meet our ethical obligations to victims of family aggression and violence, which include using interventions that are effective, we should prefer CBT-type interventions that address both aggression and violence, and alcohol and drug abuse, and trauma.

20.2.1.4 Summary

If the family is indeed the cradle of violence, it can also become the nest of peace. It should be the main focus of prevention and treatment of aggression and violence. This is because so much socially significant aggression and violence take place there and hence offer the possibility of the largest impact of prevention and intervention, and because the family offers the possibility of early intervention with children to prevent and treat aggression early, thereby offering the possibility of long-term benefits to all. Finally, there is much aggression toward women within the family and hence as a social justice issue, the family is an appropriate location to prevent violence toward women.

One useful outcome of research in family aggression and violence is that many of the issues, risk factors, and interventions for raising non-violent children, IPV, and violence toward senior family members overlap considerably. Thus, interventions that address building better interpersonal relationships within families, especially in areas related to disagreement, problem solving, conflict resolution, and rebuilding trust and relationships after aggression and violence, are all strategies that are likely to ameliorate the effects of aggression and violence within family. In addition, interventions that address alcohol and drug abuse, limiting access to firearms, poverty, education, etc. may have broad benefits on several important kinds of family aggression and violence.

Family members who are victims of aggression and violence and/or who observe aggression and violence may experience negative outcomes for extensive periods of time. These may include ongoing violence and threats of violence including during separation (Johnston & Campbell, 1993); lack of trust and deceit within the family; family breakup, such as moving to shelters, and children being placed in care; and divorce with potential acrimonious disputes over child custody and child access and perhaps reconstitution of the family in some form. Some of these individual family members who are at risk for a broad range of maladjustment and psychopathology may need various forms of informal, non-professional, as we as formal, professional support and treatment at various points throughout their life. A good example might be when the children of violent families begin intimate relationships and raise children themselves.

20.2.2 Better Education Can Buffer Against Aggression and Violence

From kindergarten and preschool to college, there are many opportunities for interventions related to social and emotional education relating to violence and aggression. Early education is often the first time that young children encounter new peers in group settings with unfamiliar adults on a regular basis with little opportunity to avoid interactions with novel and unfamiliar others. Hence, early education is an opportunity and a challenge to prevent aggression and violence. Some examples of strategies likely to help include specific social, communication, and problem solving skills that all children can benefit from and that are incorporated into tier one Positive Behavioral Supports. These strategies and others using Evidence-Based Practices (EBPs) can be incorporated into culturally acceptable practices in early education, since it is widely recognized that early education should focus on both educational and social goals for children. Further, since a minority of children who are aggressive between the ages of 2 and 5 years continue to be aggressive, early intervention can focus on identification and intervention for these at-risk children. Aggression in the first few grades of education is statistically unusual, and, again, offers another opportunity for intervention relatively early in life. During adolescence, the two main opportunities for prevention and treatment of aggression and violence are predeliquency and delinquency and IPV in teenage couples who were dating. In both of these examples, there are effective evidence-based programs and, although the effect sizes are small, the economic paybacks for school-based interventions for pre-delinquency and delinquency are very large. Thus, although there is much room left to improve programs in these areas, we already have a good basis for effective interventions at this time.

20.2.2.1 Violence on University Campuses

Sexual violence on university campuses is relatively common both within the context of IPV during dating and in public contexts such as frat house culture. This problem is well known. Some 40 years ago, Makepeace (1981) surveyed 202 college students. Around 10% of these students reported experience threats, pushing, and slapping and 1–4% reported being punched, assaulted with a weapon, choked, and other more serious forms of violence. Approximately 91% of victims were female and alcohol was involved in 35–46% of incidents, very few of which (13%) were reported to authorities. Subsequently, a systematic review by Stoner and Cramer (2019) reported prevalence rates between approximately 5% and 58%, again with substantial under-reporting to authorities due to shame, not wanting others to know, concern about not being believed, and belief that it was not a crime.

Despite being well known for many years, there is relatively little evidence for EBPS for sexual violence on university campuses. There is some evidence that certain practices might be effective in preventing aggression and violence on campus. For example, there are multiple randomized controlled trials of bystander training, some of which are of relatively high quality. Several recent meta-analyses all agree that bystander training can be effective in changing bystander prosocial beliefs and likelihood of intervening (Evans et al., 2019; Jouriles et al., 2018; Kettrey & Marx, 2019a; McMahon, 2015), but not necessarily in reducing sexual assault (Kettrey & Marx, 2019b). Outcomes diminished over time and longer duration courses were associated with larger effects (Mujal et al., 2021) suggesting that interventions should be repeated rather than being once-off events.

Campus response to violence including sexual violence has become an important issue with multiple ongoing scandals regarding widespread sexual assault and failure of universities to respond effectively, adequately, and in a timely manner, with evidence of cover ups by some universities. Unfortunately, there is little known concerning the effectiveness of administrative policies and actions at this time (McMahon, 2015).

20.2.2.2 Failure to Adopt Evidence-Based Practices in Education

An important part of prevention in educational settings is better training of teachers specifically and other educational professionals generally in EBPs that result in better acquisition of all educational skills, including social and emotional skills relating to prevention of aggression and violence. This should include the use of EBPs methods including greater use of modeling and feedback included in teacher training and careful selection of mentors who are skilled and can be appropriate models for trainee teachers. For example, university teacher training programs should consider requiring all teachers and other educational professionals to have training in behavioral principles at a Registered Behavior Technician or equivalent and continuing support in behavior management to implement the basics of universal hygiene classroom management practices from a competent fellow professional.

A second issue related to adequate training of teachers and other educational professionals is that too often universities are not held accountable for their delivery of professional training, including professional training, related to educational professionals. Good quality practical experience is often of very limited quantity and often handed over to practicum sites with little effective oversight. Often such practica are without any requirement for the novice professional to acquire any particular skill. Further, such practica are usually evaluated in terms of the number of hours physically present at a site and perhaps global ratings of impressions of performance and student satisfaction, but not the acquisition of some professional skill set to a required level of competence.

Trainee professional problems are often defined in terms of sins of commission, such as yelling at or threatening a student or arguing with a colleague or parent at a meeting. Few sins of omission such as teaching badly and ineffectively will result in a trainee not meeting requirements for professional licensure, although insufficient practicum hours, missing supervision meetings, and failure to hand in required paperwork on time may slow down graduation and access to a professional license. Post-licensure, educational professionals of all types have minimal formal performance requirements. Those that exist again tend to require certain numbers of hours and certain content areas, such as ethics, novel technology, or specific issues of the day, but unfortunately they do not require acquisition of competence or observation of performance, and, like university-based education, mechanisms for maintaining professional competence post-training are only held accountable in a weak fashion.

20.2.3 *Controlling Access to Alcohol, Other Drugs, and Weapons*

One of the most robust associations with aggression and violence is alcohol consumption and other drug use. Both individuals who drink a lot and locations where alcohol is consumed are associated with aggression and violence. There is evidence that limiting licensing hours for liquor stores, bars, and clubs results in reduced aggression and violence. Implementing such policies is quite difficult since such activities are so popular and there is so much economic interest in alcohol and related consumption that when such policies are implemented there is pushback against them. Also, it is also easy to demonize individuals who are aggressive or violent and drunk or high and to exonerate the majority of drinkers and drug users who are not as flagrantly aggressive or violent or simply not caught. It is also difficult to identify equally fun and cheap activities that can easily substitute for alcohol consumption and maintain everyone's economic interest in the problem of excessive alcohol consumption. Perhaps designing environments to promote drinking in safer locations, etc. in combination with closing down unsafe locations in an effective manner might be options to consider.

A further aspect of reduction of aggression and violence at home is control of alcohol consumption in the family home environment, both generally and also specifically in at-risk families. Several models for this approach exist (Easton et al., 2007, 2018; Kraanen et al., 2013). For example, Fine et al. (2021) developed a treatment package to reduce violence toward women in the family home. This study was a double-blind, randomized controlled trial conducted in Zambia with families with moderate to severe IPV and alcohol use by the male partner that was described as hazardous. The treatment package included a broad array of CBT strategies including relaxation, behavioral activation, and increasing access to positive reinforcement, changing thinking patterns, in vivo exposure to triggers, CBT strategies to modify alcohol use, and safety plans. This package, delivered by non-professional staff, resulted in reductions in both IPV ($d = 0.49$) and male alcohol consumption ($d = 0.43$). Further, these effects were moderated by female alcohol and other substance abuse. That is, reduction in male alcohol consumption and IPV was larger if the female substance and alcohol abuse also decreased. Thus, family-based interventions that address both alcohol, other drugs and IPV may be effective in reducing IPV.

Access to weapons is a strong and preventable predictor of aggression, violence, and homicide. This is especially true when in combined with drinking alcohol and consuming drugs, and/or in high risk family settings. Preventing access to weapons, including prevention of access to weapons in individuals with histories of violence, alcohol and drug misuse, and mental health issues related to aggression and violence should be seriously considered and may sometimes be effective in reducing aggression and violence (Carpenter et al., 2021). Both gun buyback programs, effective firearms education, gun locks, safe storage (Violano et al., 2018) and laws to limit access to guns can be effective in reducing gun-related violence.

The acceptability of such strategies varies tremendously from one country or culture to another and within each country or culture. In the USA the gun lobby and the gun manufacturing industry appeal to second amendment rights to bear arms, and legal opposition to research into guns and violence remains powerful forces in American politics. These guns rights advocates call for increased access and use of guns, such as open gun carry policies, arming classroom teachers as a solution to school-based violence, guns in the home ostensibly to protect homeowners from violent intruders, and access to powerful weapons such as military-style firearms and staches of enormous quantities of bullets. Such an increase in ready access to firearms has the potential to *increase* rather than decrease violence and homicide. Other countries have been wary about increased gun use and have controlled it carefully and effectively, perhaps sometimes resulting in reduced violence and homicide.

20.2.4 Effective, Non-coercive Police Service and Justice Systems

One of the most visible and important agents of control of social behavior is the police and justice system. Historically, control of public aggression and violence through centralization of power to courts and nations, and eventually through the emergence of city, national, and international policing agencies may have been key in establishing safe public spaces as evidenced by historical reductions in public violence. However, the status of specific evidence-based policing and policing practices are complex and incomplete. For example, McGuire et al.'s (2021) report of a series of systematic reviews found that research had at least partial answers to some but not all important questions. (See Chap. 15 for more details of these reviews.) Briefly, two practices were identified with the best evidence available. There were: (a) adoption of current evidence-based practices such as regulation of the night-time economy through control of access to alcohol, licensure of social venues, and data-based policing and (b) victim support and counseling. Other areas of policing had mixed evidence that was sometimes largely positive, evidence that certain practices may be harmful, or in other cases there was insufficient evidence to make any conclusions.

Recently, there has been much concern regarding police violence in the USA and elsewhere, especially regarding police homicide of Black men, excessive use of restraint and restraint-related deaths while citizens are in police custody, and subsequent cover-ups by police departments and local and state politicians. Subsequently, police violence against citizens has become a current hot topic in the media that has garnered considerable attention for several years. Some reduction in police violence can be achieved through body-worn cameras and monitoring and training of police behavior, but the behavior of individual officers must be constrained so they do not turn off or obscure the cameras, hide or destroy the evidence, or fail to report and completely investigate such incidents. Related movements to defund the police refer to many different policies and currently lack clear evidence to support their use although funding of community agencies to promote prosocial behavior, community participation, and cohesion exists.

Courts too can evaluate their practices and adopt evidence practices and there is a large corpus of evidence on disposal and sentencing practices that can affect numerous socially important outcomes, such as recidivism. This is especially important as, not only are there interventions that are useless and ineffective, but there are also interventions that some may find appealing such as "scared straight " programs, which involve predelinquents and delinquents to visit prisons and interact with prisoners in an effort to deter future criminality which are actually harmful and worse than doing nothing (Petrosino et al., 2013).

20.2.5 Control of Media

If instruction and modeling are variables that influence the acquisition and maintenance of both prosocial behavior and aggression and violence, can we use various forms of public campaigns and educational efforts to change people's behavior? Can we consider where to deliberately use media to promote desirable, non-violent behavior? The control of media is a sensitive topic, but we already control media in terms of standards for violence, sexual behavior, smoking, and drug use. Are there better ways to control the media to promote prosocial behavior and reduce violence?

Potential examples include promoting non-violence in children's caregivers including family members and educational staff, prosocial behavior during dating and intimate relationships, and at work. This could be done by embedding such models in public media, such as soap operas and music videos, and media designed explicitly for educational and training purposes. In addition, it is possible to control and limit access to models of violent behavior through parental, school, industry, and other control of child access to internet sites. Unfortunately, access to violent imagery and models is rampant. An example comes from the *Squid Game*, a currently worldwide popular show in 2021, in which 201 people are mowed down by gunfire at the end of the first episode. We can all look forward to another 250 or so deaths by the end of the series. Similarly, violent pornography of all kinds is freely available at almost no cost through our phones and computer. Perhaps more insidiously is the pervasive use of arguing, threatening, yelling, pushing, and shoving in the general media and reality TV, YouTube memes such as "Karens," etc., and news presented and consumed as entertainment which present models of less intense forms of aggression as ways to solve life's problems but thereby model precursors to violent behavior.

Control of access to violent media is unlikely to be a solution that can be implemented in any effective way. Perhaps the best option using media is to promote prosocial behavior through outlets such as the BBC's *Cbeebies*, NPR's *Sesame Street*, and similar outlets for children and families in other countries and through the use of modeling in educational materials perhaps as part of national campaigns to prevent aggression and violence.

20.2.6 Summary

Unlike biological interventions, much can be done to continue efforts to redesign society to be a less violent one. The most promising interventions—because of the large quantity of evidence and the possibility of long-term benefits to individuals, their families and society more broadly—are interventions with families with young children. These interventions should be multi-component intervention packages that address both parenting skills and multiple risk factors such as alcohol and substance abuse, poverty, parental stress, and other malleable risk factors. To be most effective and economical, these interventions should be triaged into varying levels of intensity for primary, secondary, and tertiary prevention. The second most important area for psychosocial interventions are evidence-based interventions in the educational system which can focus on both aggressive and at-risk children in the early grades, pre-delinquency and delinquency, and IPV programs in later high-school grades. Finally, higher education too should adopt bystander training and develop more effective administrative procedures to prevent on campus violence.

In addition to interventions in families and educational settings, society should adopt evidence-based practices to prevent and treat aggression and violence in adults. These interventions are predominantly cognitive behavior therapy (CBT) and behavioral therapies, both for non-offenders and offenders (see below).

20.3 Selection of Individual People's Behavior During the Lifespan

20.3.1 Learning During the Lifespan

Although genetics and culture play important roles in determining aggression and violence behavior and their opposites, learning during the lifespan—and especially operant learning—takes what each individual organism inherits from its genetics and culture and refines behavior in an infinite number of ways and can sometimes reverse existing aggression and violence. Thus, in the area of lifelong learning, much could be done to prevent aggression and violence. As we have seen earlier, much of this can and should be done by all family caretakers and educational staff where the potential for the greatest payoff for investment exists.

20.3.2 Individual Therapies

One important way in which individual behavior is selected and refined is through various forms of therapy. This is particularly important for more severe forms of aggression and violence where referral to mental health services and different forms and intensities of mental health services can be an important influence on the path torrid or away from aggression and violence. Given the importance of aggression and violence and related mental health it is unsurprising that there is so much research on this topic that can be used to guide services as to which therapies to select and how to implement and evaluate them. Chapters 16 and 17 offer more detailed summaries of this approach to prevent and treat aggression and violence.

Individual interventions are showing promise in many contexts, including with some forms of aggression and violence which formerly were viewed pessimistically as the most treatment resistant. Evidence for the effectiveness of CBT, including various forms of anger management, applied behavior analysis, behavior modification all show considerable promise in this regard. An important caveat to this overgeneralization is that the effect sizes and percentage of people responding to such treatments vary considerably from context to context. For example, one might expect rather larger effect sizes for anger and aggression and violence earlier in life and of more recent onset than for long-standing, treatment-resistant severe aggression and violence in more challenging populations such as incarcerated offenders with multiple concurrent problems.

In contrast to CBT and various forms of behavioral interventions, outcomes for other forms of psychosocial interventions are mixed, sometimes weak, sometimes based on insufficient evidence, and sometimes certain forms of therapies have been evaluated and positively shown to be ineffective or even harmful. For example, various approaches to parent training that typically include elements of cognitive and behavior therapy are often quite effective. In contrast, popular treatments such as the Duluth model are evaluated and probably ineffective, while effective alternatives are available, and treatments such as psychoanalytic-based catharsis are probably positively harmful. Society in general and service providers have ethical obligations related to honesty, commitment to be benevolent to their clients, and not wasting limited public funds and the fees that individual clients pay, to always select the treatments most likely to help, and least likely not to harm.

Delivery of evidence-based practice in routine services remains highly challenging in many contexts, yet we now know it is possible. Over time society may seize the available personal and economic advantages of such approaches. This can be done through better control and regulation of professional training which could hold providers of professional training—both traditional higher education and schools of professional practice—accountable for the effectiveness of the professional training they provide. In addition, regulation, oversight, and control of funding of professional practice and organizations providing treatment, such as mental health services, forensic services, police and justice systems, prisons and probation services, might all benefit individuals and society which all have the potential to result in further reduction in aggression and violence.

20.4 Individual Biological Treatments

Although used very widely, there is little convincing high quality evidence to support prevention and treatment of aggression and violence through individual biological treatment. The predominant individual biological treatment is psychotropic medication and here the outcomes are at best mixed. Often the outcomes focus on reduction in ratings of aggression without simultaneously measuring prosocial behavior and often do not address social significance of any change. The evidence for other individual biological treatments, such as diets, neurosurgery, castration, etc. is even more limited and in some cases pose severe ethical dilemmas regarding consent and coercion.

20.5 Support and Treatment of Victims and Building Resilience

In the drama of aggression and violence it is easy to forget the impact on victims, their friends and their families, and the importance of rebuilding lives following aggression and violence. As noted earlier in this volume, the needs of victims are highly variable and hence support should be tailored individually.

Effective treatment of victims of aggression and violence is controversial and some ideologically palatable interventions and practices have not been shown to be effective when effective alternatives exist. Thus, counseling and debriefing after aggression and violence and interventions based on the Duluth model all lack support. Restorative justice practices, when done correctly, and CBT that includes building up positive experiences and skills training, and exposure therapy are all promising interventions to support victims of aggression and violence.

Also important is building resilience throughout the lifespan. Some exposure to some forms of aggression and violence is inevitable for everyone, even if society and caregivers were to make strenuous efforts to reduce aggression and violence exposure to aggression and violence in some form is inevitable. Hence, dealing with stressors generally and the effects of aggression and violence in particular can be part of supporting victims of aggression and violence.

20.6 The Challenge of Implementing What Works

The field of evidence-based practices has correctly identified many therapies and other practices that seem to work at least to some extent and that are economically viable, sometimes in specific contexts and populations. Yet, despite the evidence of effectiveness and economic viability, society is often reluctant to take up these practices and is often equally happy to continue practices that are unproven, useless, or even harmful. In response to this problem, researchers and some parts of society have taken up the challenge of studying the reluctance and opposition to effective practices and to take what has been learned to disseminate effective practices over time. Let us now consider one example of dissemination of something that seems to work and then consider some of the challenges in disseminating evidence-based practices.

20.6.1 An Example of Effective Dissemination: The National Academic Centers of Excellence for Youth Violence Prevention

Despite the challenge of implementing EBPs relating to prevention of aggression and violence, several groups have been effective in doing so. The US Federal government funds collaborations between academic centers and public services and in 2011 a special issue of the *American Journal of Community Psychology* reviewed the then current programs (Backer & Guerra, 2011) including reports from individual programs under the rubric of mobilizing communities. As in many other fields EBPs are known to academic researchers and some practitioners, but the field of practice rarely takes them up, is challenged in maintaining what works over time, and implements other unevaluated or ineffective practices. One approach that Backer and Guerra (2011) use as a framework for translation of research to practice is *Community Mobilization*. The first step is to identify EBPs, which can refer to evidence-based *programs*, which are already developed, defined, and manualized sometimes proprietary programs or to evidence-based *practices*, which refer to specific treatment principles or procedures, such as CBT, parent training, etc. One challenge of dissemination is that some practitioners and administrators are unaware of what EBPs are. To address this, the US federal government funded multiple centers for the dissemination of EBPs in practice.

Effective violence prevention programs for youth have been developed and identified. These are programs that are science-based, comprehensive, and involve school, families, and community. They use four strategies: parent-family based, home visiting, cognitive, and mentoring programs. Community mobilization may begin with relatively informal coalitions built around a research question which may evolve into coalitions of multiple participants, such as community agencies, researchers, government agencies, and universities. This is a complex and multi-stage endeavor including "(1) Understanding community context (e.g., assessing community assets and needs); (2) Collaborative planning (e.g., developing a vision, mission, objectives, strategies and action plans); (3) Developing leadership and enhancing participation (e.g., building relationships, recruiting participants); (4) Community action and intervention (e.g., designing interventions, advocacy); (5)

Evaluating community initiatives (e.g., program evaluation, documentation of community and systems change); and (6) Promoting and sustaining the initiative (e.g., social marketing, obtaining grants)" (p. 33).

One example of two projects with contrasting outcomes comes from Parker et al. (2011), who told the tale of two cities: One that readily implemented EBPs for aggression and violence and one that wasted 5 years of resources and never got there thereby continuing to expose its citizens to the costs of preventable aggression and violence. The authors used a stage framework similar to the one described in the previous paragraph. In City A, the mayor initiated a project in response to a problem he perceived as important—youth violence. An academic partner was invited to and regular meetings were held with schools, country and city agencies, etc. The mayor had hoped for a proposal within 90 days. A strategic plan was indeed developed which included a youth and family center and the team moved on to implementation quickly. Parker et al. suggested that the project was a success because the city was ready to proceed from the beginning, and, with the support of the mayor, could move on to implementation and implementation immediately. All stakeholders were signed up immediately and they participated consistently. The relatively quick establishment of the youth and family center meant that they could move quickly on to consolidation and professionalization of the plan.

Attempts to implement EBPs at city B were a longer and a sadder story of canceled meetings, people not showing up, leaders sending substitutes to meetings, leadership turnover at many participating organizations, failure to collect data, vacillation, budget cuts, failure to respond to needs surveys, competition between rival organizations and existing models, denial of the problems, lack of participation from local business, sending out incorrect forms with subsequent low response rates, etc. At the end of 5 years, the team wrote their final report of "what ifs" with little or no actual accomplishments. Subsequently, the project was not refunded. Parker et al. identified the lack of leadership, leadership, and membership turnover creating instability and shifting focus as insurmountable objects to the implementation of EBPs.

Engaging in translational work to prevent aggression and violence presents challenges to all concerned. It seems unlikely that programs developed in the context of research, many of which have been evaluated in only one or two well conducted studies in a limited range of contexts, will need modification to be applied successfully in particular new contexts. This may mean compromises on everyone's part. The research may have to live with their precious manualized program being modified to the new context while retaining its essential effective components and treatment integrity and the practitioners may have to re-organize their services and retrain their current practitioners to adopt new practices and give up old ones.

20.6.2 Challenges to Implementation of Evidence-Based Practices

As noted above, there are many examples of clearly established effective EBPs related to prevention and treatment of aggression and violence as well as some programs that are robustly harmful that should be avoided. As in other domains of EBP, implementation and adoption of EBPs into government policies has a very long history and over time, driven in part by the model of evidence-based medicine, has become progressively more widespread, and has permeated many fields of academic study and social policy. Implementation of EBPs is not straightforward for many reasons, leading some authors to describe the situation with implementation of EBPs as one of "broken experimentation and sham evidence-based practice" (Underhill, 2020, p. 150). What are some of the issues in using and translating EBPs into policies and practices?

Recently, Underhill (2020) reviewed some of the issues regarding governmental use of evidence and adoption of evidence-based policy making (EBPM). In an ideal world, Underhill argued, governments would indeed adopt EBPs in order to meet their stated policy goals in order to be maximally effective in achieving those goals and to avoid waste. Models for such an evidence-based approach come from sports where "moneyball" has become a byword for data-based decisions such as dumping overvalued players and recruiting under-valued players, recruiting players who have key skills missing in the team, etc., and evaluating all decisions using data-based approaches. Governments have also used "moneyball" approaches as a model for government policies. Governments have used science to identify and review existing evidence relating to the question at hand, synthesize and interpret the evidence and evaluate the implementation of EBMP and then adjust the implementation in the light of the data. (Note this is very similar to the procedures described in Skinner, 1953, 1962 and Los Horcones, 1992.)

Implementation of EBPM is, of course, not straightforward for many different kinds of reasons. There are limits on rationality and science. Further, governments may wish to grant differing degrees of authority to technocrats and scientists, especially if the pronouncements of experts are unwelcome and contradict politicians' claims. Witness, for example, the frequent sidelining and undermining of science and scientists during the COVID epidemic by any government. (Britain's Lord Chancellor Michael Grove in 2017 pronounced that "I think the people in this country have had enough of experts from organizations with acronyms saying that they know what is best and getting it consistently wrong.") Governments may also fail to use the existing evidence honestly through selective uptake, and deliberate sidelining of the existing evidence, distorting the evidence and avoiding EBPM. This is sometimes done by continuing experimentation on EBPs as a sleight of hand to avoid actual policy implementation. This may sometimes be motivated by numerous nefarious, dishonest, and inappropriate reasons resulting in continued funding of current programs. Finally, they may also use budget ratcheting as when an evaluation of one program that shows a procedure is ineffective or harmful is used to completely defund all present and future research in that particular policy area. For example, an ineffective specific program to prevent IPV among low income families might be given as an excuse to cut or defund all such programs. Underhill argued that there are many costs in failing to implement EBPM with honesty and integrity. These can include harms such as wanted public money and resources, misleading the public about what works, eroding the legitimacy of scientific approaches to applied problems, and harming participants in applied research who, at the least had their time wasted but also may have been needlessly exposed to harms since the research was adopted.

Despite these limitations, Underhill (2020) concluded that adoption of EBPM is "worth the candle" and its integrity can be protected in several ways. First, Underhill argued that EBPM should only be used in certain circumstances. That is if evidence is unlikely to make a difference in policy implementation, for example, when policy decisions are driven primarily by values rather than evidence, then an evidence-based approach is wasteful, but when policy implementation is genuinely driven and open to influence of evidence, then more effort should be put into EBPM in these circumstances. Underhill went on to argue that if EBP is done well with high quality research and with maximal protection against biases through mechanisms such as trial preregistration, inviting replication by potentially hostile and critical researchers, and, spelling out prior to the research how policy will change in the light of the research findings before the research is conducted. In this way EBPM may be effective in honestly achieving the desired policy outcomes.

There are some examples of EBPM regarding aggression and violence with examples of effective implementation and evaluation of science-informed policies that can be models of how to do it well (see above). More commonly and unfortunately, government policy on aggression and violence is done by muddling through a miasma of conflicting demands and priorities on politicians' time and limited resources. Nevertheless there is good information on what works in many circumstances and models of how to disseminate such practices effectively, if agencies and governments wish to do so.

20.7 Summary

Aggression and violence are commonplace, incredibly harmful to all, and expensive. But there are four main reasons for cautious optimism. First, human aggression and violence have reduced considerably both over the long historic time and recently. Second, we know where to intervene to maximize outcomes, namely families and schools. Third, there are a wide array of EBPs that have known effectiveness and there is research identifying interventions to avoid because they are ineffective or harmful. Fourth, there are some examples of national policies and dissemination of EBPs that can be used as the basis for large-scale implementation.

The benefits of doing this to individuals and society are already and will be enormous.

References

Backer, T. E., & Guerra, N. G. (2011). Mobilizing communities to implement evidence-based practices in youth violence prevention: The state of the art. *American Journal of Community Psychology, 48*(1–2), 31–42.

Barth, R. P. (2009). Preventing child abuse and neglect with parent training: Evidence and opportunities. *The Future of Children, 19*(2), 95. https://doi.org/10.1353/foc.0.0031

Brüne, M. (2007). On human self-domestication, psychiatry, and eugenics. *Philosophy, Ethics, and Humanities in Medicine, 2*(1), 1–9.

Carpenter, S., Borrup, K., & Campbell, B. T. (2021). Gun buyback programs in the United States. In M. Crandall, S. Bonne, J. Bronson, & W. Kessel (Eds.), *Why we are losing the war on gun violence in the United States* (pp. 173–186). Springer.

Chance, P. (2007). The ultimate challenge: Prove B. F. Skinner wrong. *The Behavior Analyst, 30*(2), 153–160. https://doi.org/10.1007/BF03392152

Diamond, J. M. (1999). *Guns, germs, and steel.* Penguin.

Easton, C. J., Mandel, D. L., Hunkele, K. A., Nich, C., Rounsaville, B. J., & Carroll, K. M. (2007). A cognitive behavioral therapy for alcohol-dependent domestic violence offenders: An integrated substance abuse-domestic violence treatment approach (SADV). *The American Journal on Addictions, 16*, 24–31. https://doi.org/10.1080/10550490601077809

Easton, C. J., Crane, C. A., & Mandel, D. (2018). A randomized controlled trial assessing the efficacy of cognitive behavioral therapy for substance-dependent domestic violence offenders: An integrated substance abuse-domestic violence treatment approach (SADV). *Journal of Marital and Family Therapy, 44*, 483–498. https://doi.org/10.1111/jmft.12260

Ericsson, K., & Simonsen, E. (Eds.). (2005). *Children of World War II: The hidden enemy legacy.* Bloomsbury Publishing.

Evans, J. L., Burroughs, M. E., & Knowlden, A. P. (2019). Examining the efficacy of bystander sexual violence interventions for first-year college students: A systematic review. *Aggression and Violent Behavior, 48*, 72–82.

Fine, S. L., Kane, J. C., Murray, S. M., Skavenski, S., Paul, R., & Murray, L. K. (2021). Moderator effects in a randomized controlled trial of the Common Elements Treatment Approach (CETA) for intimate partner violence and hazardous alcohol use in Zambia. *Drug and Alcohol Dependence, 228*, 108995.

Forrest, D. W. (1974). *Francis Galton: The life and work of a Victorian genius*. Taplinger.

Gallagher, N. L. (1999). *Breeding better Vermonters: The eugenics project in the green mountain state*. University Press of New England.

Isobel, S., Goodyear, M., Furness, T., & Foster, K. (2019). Preventing intergenerational trauma transmission: A critical interpretive synthesis. *Journal of Clinical Nursing, 28*(7–8), 1100–1113.

Johnston, J. R., & Campbell, L. E. (1993). Parent-child relationships in domestic violence families disputing custody. *Family Court Review, 31*(3), 282–298.

Jouriles, E. N., Krauss, A., Vu, N. L., Banyard, V. L., & McDonald, R. (2018). Bystander programs addressing sexual violence on college campuses: A systematic review and meta-analysis of program outcomes and delivery methods. *Journal of American College Health, 66*(6), 457–466.

Karakurt, G., Koç, E., Çetinsaya, E. E., Ayluçtarhan, Z., & Bolen, S. (2019). Meta-analysis and systematic review for the treatment of perpetrators of intimate partner violence. *Neuroscience & Biobehavioral Reviews, 105*, 220–230.

Kettrey, H. H., & Marx, R. A. (2019a). The effects of bystander programs on the prevention of sexual assault across the college years: A systematic review and meta-analysis. *Journal of Youth and Adolescence, 48*(2), 212–227.

Kettrey, H. H., & Marx, R. A. (2019b). Does the gendered approach of bystander programs matter in the prevention of sexual assault among adolescents and college students? A systematic review and meta-analysis. *Archives of Sexual Behavior, 48*, 2037–2053. https://doi.org/10.1007/s10508-019-01503-1

Kirigin, K. A., Phillips, E. L., Timbers, G. D., Fixsen, D. L., & Wolf, M. M. (2020). Achievement place: The modification of academic behavior problems of delinquent youths in a group home setting. In B. C. Etzel, J. M. LeBlanc, & D. M. Baer (Eds.), *New developments in behavioral research: Theory, method and application* (pp. 473–487). Routledge.

Kraanen, F. L., Vedel, E., Scholing, A., & Emmelkamp, P. M. G. (2013). The comparative effectiveness of Integrated treatment for Substance abuse and Partner violence (IStoP) and substance abuse treatment alone: A randomized controlled trial. *BMC Psychiatry, 13*, 189. https://doi.org/10.1186/1471-244X-13-189

Los Horcones, C. (1992). Natural reinforcement: A way to improve education. *Journal of Applied Behavior Analysis, 25*(1), 71–75.

Makepeace, J. M. (1981). Courtship violence among college students. *Family Relations, 30*, 97–102.

Mattaini, M. A. (2010). Cultural analysis and social change in Medellín. *Behavior and Social Issues, 19*(1), 1–5.

McClannahan, L. E., & Krantz, P. J. (1997). Princeton Child Development Institute. *Behavioral Science, 7*, 65–68. https://doi.org/10.5210/bsi.v7i1.301

McGuire, J., Evans, E., & Kane, E. (2021). *Evidence-based policing and community crime prevention*. Springer.

McMahon, S. (2015). Call for research on bystander intervention to prevent sexual violence: The role of campus environments. *American Journal of Community Psychology, 55*, 472–489. https://doi.org/10.1007/s10464-015-9724-0

Miller, M., Drake, E., & Miller, M. (2013). *What works to reduce recidivism by domestic violence offenders?* Washington State Institute for Public Policy.

Mujal, G. N., Taylor, M. E., Fry, J. L., Gochez-Kerr, T. H., & Weaver, N. L. (2021). A systematic review of bystander interventions for the prevention of sexual violence. *Trauma, Violence & Abuse, 22*(2), 381–396.

Olsen, K. (2005). Under the care of Lebensborn: Norwegian war children and their mothers. In K. Ericsson & E. Simonsen (Eds.), *Children of World War II: The hidden enemy legacy* (pp. 15–34). Bloomsbury Publishing.

Parker, R. N., Alcaraz, R., & Payne, P. R. (2011). Community readiness for change and youth violence prevention: A tale of two cities. *American Journal of Community Psychology, 48*(1–2), 97–105.

Petrosino, A., Turpin-Petrosino, C., Hollis-Peel, M. E., & Lavenberg, J. G. (2013). Scared straight and other juvenile awareness programs for preventing juvenile delinquency: A systematic review. *Campbell Systematic Reviews, 9*(1), 1–55.

Schell, P. A. (2010). Eugenics policy and practice in Cuba, Puerto Rico, and Mexico. In A. Bashford & P. Levine (Eds.), *The Oxford handbook of the history of eugenics* (pp. 477–492). Oxford University Press.

Skinner, B. F. (1953). *Science and human behavior*. Simon and Schuster.

Skinner, B. F. (1962). The design of cultures. In H. Hoagland & R. W. Burhoe (Eds.), *Evolution and Man's progress* (pp. 124–136). Columbia University Press.

Skinner, B. F. (1969). *Walden two*. Hackett Publishing.

Stoner, J. E., & Cramer, R. J. (2019). Sexual violence victimization among college females: A systematic review of rates, barriers, and facilitators of health service utilization on campus. *Trauma, Violence & Abuse, 20*(4), 520–533. https://doi.org/10.1177/1524838017721245

Underhill, K. (2020). Broken experimentation, Sham evidence-based policy. *Yale Law & Policy Review, 38*(3), 150–227. Retrieved from https://digitalcommons.law.yale.edu/ylpr/vol38/iss1/3

Violano, P., Bonne, S., Duncan, T., Pappas, P., Christmas, A. B., Dennis, A., Goldberg, S., Greene, W., Hirsh, M., Shillinglaw, W., Robinson, B., & Crandall, M. (2018). Prevention of firearm injuries with gun safety devices and safe storage: An Eastern Association for the Surgery of Trauma Systematic Review. *Journal of Trauma and Acute Care Surgery, 84*(6), 1003–1011.

Wrangham, R. W. (2019). Hypotheses for the evolution of reduced reactive aggression in the context of human self-domestication. *Frontiers in Psychology, 10*, 1914.

Index

P. Sturmey, *Violence and Aggression*, Advances in Preventing and Treating Violence and Aggression , https://doi.org/10.1007/978-3-031-04386-4

B

Q

R

S

www.ingramcontent.com/pod-product-compliance
Lightning Source LLC
Chambersburg PA
CBHW072313070526
44736CB00003B/84
9783031043857